Universitext

J. Bliedtner W. Hansen

Potential Theory

An Analytic and Probabilistic Approach to Balayage

Springer-Verlag
Berlin Heidelberg New York Tokyo

Jürgen Bliedtner
Fachbereich Mathematik, Universität Frankfurt
Robert-Mayer-Strasse 6–10, D-6000 Frankfurt, FRG

Wolfhard Hansen
Fakultät für Mathematik, Universität Bielefeld
Universitätsstrasse, D-4800 Bielefeld, FRG

Mathematics Subject Classification (1980): 31–XX, 60Jxx, 35Jxx, 35Kxx

ISBN-13: 978-3-540-16396-1 e-ISBN-13: 978-3-642-71131-2
DOI: 10.1007/978-3-642-71131-2

Library of Congress Cataloging-in-Publication Data. Bliedtner, Jürgen, 1941– Potential theory. (Universitext) Bibliography: p. Includes indexes. 1. Potential, Theory of. I. Hansen, Wolfhard, 1940–. II. Title. QA404.7.B57 1986 515.7 86-3772

2141/3140-543210

To Brigitte and Luise

Contents

Introduction

During the last thirty years potential theory has undergone a rapid development, much of which can still only be found in the original papers. This book deals with one part of this development, and has two aims. The first is to give a comprehensive account of the close connection between analytic and probabilistic potential theory with the notion of a balayage space appearing as a natural link. The second aim is to demonstrate the fundamental importance of this concept by using it to give a straight presentation of balayage theory which in turn is then applied to the Dirichlet problem. We have considered it to be beyond the scope of this book to treat further topics such as duality, ideal boundary and integral representation, energy and Dirichlet forms.

The subject matter of this book originates in the relation between classical potential theory and the theory of Brownian motion. Both theories are linked with the Laplace operator. However, the deep connection between these two theories was first revealed in the papers of S. KAKUTANI [1], [2], [3], M. KAC [1] and J.L. DOOB [2] during the period 1944-54: This can be expressed by the fact that the harmonic measures which occur in the solution of the Dirichlet problem are hitting distributions for Brownian motion or, equivalently, that the positive hyperharmonic functions for the Laplace equation are the excessive functions of the Brownian semigroup. This equivalence allows potential theoretic results and notions (such as balayage, fine topology, polar set, thinness and regular point) to be given a probabilistic interpretation.

This equivalence also led J.L. DOOB [3],[4],[5] to treat the Dirichlet problem for the heat equation using a combination of analytic and probabilistic methods. Based on these results and earlier attempts by G. TAUTZ [1], [2], M. BRELOT [9]-[12] and H. BAUER [3],[5] developed the potential theory of harmonic spaces. Starting from basic properties of harmonic functions (sheaf property, local solvability of the Dirichlet problem, convergence properties) an extensive theory was built up which covers a wide class of linear second order elliptic and parabolic partial differential equations. In particular, a common treatment of the Dirichlet problem for arbitrary open sets was given. This development culminated in the monograph of C. CONSTANTINESCU - A. CORNEA [4]. Meanwhile, on the probabilistic side,

chapter V we obtain further examples by studying subspaces, subordination by convolution semigroups, products, and images. In particular, our list of standard examples is completed with Riesz potentials and the heat equation.

The second part of the book proper starts with chapter VI. Balayage of functions and balayage of measures are of central interest here. All proofs are given within the framework of balayage spaces, i.e. the proofs are analytic. However, all important notions are also interpreted probabilistically; we hope this will widen the appeal of the theory, and at the same time lead to a deeper understanding of many of the statements. Of course, it will be noted that even many analytic proofs are (and perhaps have to be) guided by probabilistic intuition. The interested reader is encouraged to find genuine probabilistic proofs of some of the results in order to see whether analytic subtleties can be avoided by considering suitable sets of paths and applying the strong Markov property. As an application of balayage theory different types of Dirichlet problems are studied in chapter VII. First the method of Perron-Wiener-Brelot is adapted to balayage spaces - where of course functions on the boundary of an open set have to be replaced by functions on its complement. The rich structure of the cone of continuous real potentials allows us to develop a Choquet type theory for cones of continuous superharmonic functions. This leads to a solution of the weak Dirichlet problem and yields important approximation theorems. The final chapter establishes that nice linear second order elliptic or parabolic differential operators generate harmonic spaces and that consequently their potential theory is covered by the present general theory.

Nearly all of the sections contain exercises. Their principal aim is to increase the familiarity of the reader with the material. We have added bibliographical notes including some historical remarks which are far from being complete. We apologize for all omissions and inaccuracies. For the convenience of the reader the last page of the book is a guide to our standard examples.

The material for this book has evolved out of courses for graduate students at the universities of Frankfurt and Bielefeld. The reader is assumed to be familiar with basic facts from functional analysis (e.g. Hahn-Banach theorem), measure theory (e.g. Radon measures on locally compact spaces), and probability theory (e.g. conditional expectations). However, no knowledge of potential theory is presupposed. Nevertheless, we expect that the book has something to offer to the expert, whether it be through simplified proofs or even because of the results themselves.

It is a pleasure to thank Dorothea Burghardt, Christa Draeger, and Hannelore Sternberg for their superb job in typing the final manuscript.

fundamental papers of G.A. HUNT [1],[2],[3] during 1957-58 marked the beginning of a potential theory for Markov processes. A compact presentation of this theory can be found in the book of R.M. BLUMENTHAL - R.K. GETOOR [1].

Papers of P.A. MEYER [1] in 1963 and N. BOBOC - C. CONSTANTINESCU - A. CORNEA [1] in 1967 showed that, by analogy with the relation between classical potential theory and Brownian motion, every harmonic space admits corresponding Markov processes. But it was only in 1978 that J. BLIEDTNER - W. HANSEN [5] could characterize the class of Markov processes which are associated with harmonic spaces (see also J.C. TAYLOR [5]).

The introduction of harmonic spaces was motivated by the properties of solutions of certain differential equations. This is reflected in the fact that the corresponding Markov processes always have continuous paths, i.e. they are diffusions. However, whether the paths are continuous or not plays no important rôle in the potential theory of Markov processes. Moreover, even within the theory of harmonic spaces itself it turned out that potentials are more important than harmonic functions and that the really crucial properties of the cone of continuous potentials hold in a far more general setting.

This observation led the authors to combine investigations of G. MOKOBODZKI - D. SIBONY [1]-[4] with approaches used in the book of C. CONSTANTINESCU - A. CORNEA [4] and to introduce the notion of a balayage space. Not only does this concept elucidate the connections between analytic and probabilistic potential theory, it also allows a clear and direct presentation of balayage theory. This theory originates in H. POINCARÉ's method of balayage (balayer (french) = to sweep) to study the equilibrium problem now solved by M. BRELOT's reduction technique. All this can be carried out in the general framework of a balayage space without losing known results for harmonic spaces and without more complicated proofs. In particular, different types of Dirichlet problems can be treated in an elegant manner. A decisive additional advantage of this approach is that Riesz potentials and Markov chains become further standard examples and their potential theory can thus be covered.

In chapter 0 we give a concise presentation of the fact that the excessive functions for the Brownian semigroup are the same as the positive hyperharmonic functions for the Laplace operator. This discussion serves to motivate the general treatment of excessive functions for sub-Markov semigroups (chapter II) and of positive hyperharmonic functions for a family of harmonic kernels (chapter III). The necessary prerequisites from functional analysis can be found in chapter I. The first part of the book closes with a short introduction to Markov processes (chapter IV). The setup here is essentially geared to the main result, which shows that the various descriptions of potential theory using either balayage spaces, families of harmonic kernels, sub-Markov semigroups, or Markov processes are all equivalent. For better insight three standard examples are treated along the way, namely classical potential theory, translation in $\mathbb{R}$, and discrete potential theory. In

Basic Notations

$\mathbb{N}$	set of natural integers
$\mathbb{N}_o$	$\mathbb{N} \cup \{0\}$
$\mathbb{Z}$	set of integers
$\mathbb{Q}$	set of rational numbers
$\mathbb{R}$	set of real numbers
$\mathbb{R}_+, \mathbb{R}_+^*$	set of real numbers ≥ 0 , > 0
$\overline{\mathbb{R}}$	extended real line
$<\cdot,\cdot>, \Vert\cdot\Vert$	scalar product, norm on $\mathbb{R}^n$
λ^n	Lebesgue measure on $\mathbb{R}^n$
$\lim\limits_{s\downarrow t} \; (\lim\limits_{s\uparrow t})$	$\lim\limits_{s\to t, s>t} \; (\lim\limits_{s\to t, s<t})$
X	locally compact space with countable base
$\bar{A}, \mathring{A}, A^*$	closure, interior, boundary of A
$\mathfrak{P}(X)$	set of subsets of X
$\mathcal{U}$	base of relatively compact open subsets of X
$\mathcal{B}(X)$	set of all Borel numerical functions on X (subsets of X)
l.s.c. (u.s.c.)	lower (upper) semicontinuous
$\mathcal{C}(X)$	space of continuous real functions on X
$\mathcal{C}_o(X)$	space of continuous real functions on X vanishing at infinity
$\mathcal{K}(X)$	space of continuous real functions on X with compact support
A_b, A^+, A_r	set of all bounded, positive, real functions in A
$\Vert f \Vert$	uniform norm of $f \in \mathcal{B}_b(X)$
$\{f \geq g\}$	$\{x \in X : f(x) \geq g(x)\}$
$S(f)$	support of f, i.e. $\overline{\{f \neq 0\}}$
1_A	characteristic function of a set A
$f_{\mid A}$	restriction of f on A or $1_A f$
$f \in A/\mathcal{B}$	f is A - $\mathcal{B}$ - measurable
$\mathcal{M}_+(X)$	set of all (positive Radon) measures on X
$\mathcal{M}_+^1(X)$	set of all probability measures on X
supp (μ)	support of a measure μ
$\mu_{\mid A}$	restriction of μ on A or $1_A\mu$
ε_x	unit mass (Dirac measure) at x

0. Classical Potential Theory

In this chapter we shall see that classical potential theory and Brownian motion are mathematically equivalent. After a short description of basic properties of harmonic and hyperharmonic functions in section 1, we shall introduce the Brownian semigroup in section 2 and prove some important properties. In section 3 the identity of the set of excessive functions with respect to Brownian semigroup and the set of positive hyperharmonic functions on $\mathbb{R}^n$ will be shown, which means that in a technical sense these two theories are identical. The importance of this result will become clear later on, especially in chapter VI when the probabilistic model of Brownian semigroup, namely Brownian motion, will be available.

1. Harmonic and Hyperharmonic Functions

Classical potential theory is based on the *Laplace equation* $\Delta h = 0$ on $\mathbb{R}^n$, $n \geq 1$, where

$$\Delta := \sum_{i=1}^{n} \frac{\partial^2}{\partial x_i^2}$$

is the *Laplace operator*. Let U be an open subset of $\mathbb{R}^n$ and $h \in C^2(U)$, i.e. let h be a real function on U which is twice continuously differentiable. If $\Delta h = 0$ then h is called *harmonic on* U. We note that the set $H(U)$ of all harmonic functions on U is a linear subspace of $C(U)$.

Suppose now that U is non-empty and relatively compact, and let $f \in C(U^*)$. Then the classical *Dirichlet problem* consists in finding a harmonic continuation of f on U, i.e. a function $h \in H(U)$ such that $\lim_{x\to z} h(x) = f(z)$ for every $z \in U^*$. If this Dirichlet problem is solvable for every $f \in C(U^*)$ then U is called *regular*.

The following boundary minimum principle shows that given a boundary function f there exists at most one solution of the corresponding Dirichlet problem.

1.1. PROPOSITION. Let $h \in C^2(U)$ such that $\Delta h \leq 0$ on U and $\liminf_{x \to z} h(x) \geq 0$ for every $z \in U^*$. Then $h \geq 0$.

Proof. 1. Suppose first that even $\Delta h < 0$ on U and $\inf h(U) < 0$. Then there exists a point $a \in U$ such that

$$h(a) = \inf h(U) .$$

For every $1 \leq i \leq n$ the function

$$t \mapsto h(a_1, \ldots, a_{i-1}, t, a_{i+1}, \ldots, a_n)$$

is defined and twice continuously differentiable in a neighborhood of a_i and has a minimum at a_i. Hence

$$\frac{\partial^2 h}{\partial x_i^2}(a) \geq 0$$

contradicting $\Delta h < 0$. Thus $h \geq 0$ if $\Delta h < 0$ on U.

2. If $\Delta h \leq 0$ and $\inf h(U) < 0$, we may choose $r > 0$ and $\varepsilon > 0$ such that $\| x \| \leq r$ for every $x \in U$ and $\inf h(U) + \varepsilon r^2 < 0$. Defining $\tilde{h}$ by

$$\tilde{h}(x) = h(x) + \varepsilon (r^2 - \| x \|^2) \qquad (x \in U)$$

we then have $h \in C^2(U)$,

$$\Delta\tilde{h} = \Delta h - 2\varepsilon n < 0$$

and $\tilde{h} \geq h$, hence for every $z \in U^*$

$$\liminf_{x \to z} \tilde{h}(x) \geq \liminf_{x \to z} h(x) \geq 0 .$$

Thus on one hand $\tilde{h} \geq 0$ by (1) whereas on the other hand $\inf \tilde{h}(U) < 0$ by the choice of ε. This contradiction shows that $h \geq 0$. ∎

In the following we shall see that every open ball in $\mathbb{R}^n$ is regular. So let $a \in \mathbb{R}^n$, $r > 0$ and

$$B_r(a) = \{x \in \mathbb{R}^n : \|x-a\| < r\}, \quad S_r(a) = \{x \in \mathbb{R}^n : \|x-a\| = r\} .$$

Let $\sigma_{a,r}$ denote the surface measure on $S_r(a)$ normed to 1. For every $x \in B_r(a)$, we define the *Poisson kernel* $P_x : S_r(a) \to \mathbb{R}$ by

$$P_x(y) := P_x^{a,r}(y) := r^{n-2} \frac{r^2 - \|x-a\|^2}{\|x-y\|^n}$$

and for every lower bounded Borel measurable function $f : S_r(a) \to \overline{\mathbb{R}}$ we define the *Poisson integral* $Hf : B_r(a) \to \overline{\mathbb{R}}$ by

$$Hf(x) := H_{a,r}f(x) := \int f\, P_x \, d\sigma_{a,r} .$$

We note that $Hf \geq 0$ if $f \geq 0$.

<u>1.2. PROPOSITION.</u> Let $f : S_r(a) \to \overline{\mathbb{R}}$ be lower bounded and Borel measurable. If f is $\sigma_{a,r}$-integrable then Hf is harmonic on $B_r(a)$. If $f \in C(S_r(a))$ then $\lim_{x \to z} Hf(x) = f(z)$ for every $z \in S_r(a)$. In particular, $B_r(a)$ is regular.

Proof. It is easily verified that for every $y \in S_r(a)$ the function $x \mapsto P_x(y)$ is harmonic on $B_r(a)$. Furthermore, the functions $(x,y) \mapsto \frac{\partial P_x(y)}{\partial x_i}$ and $(x,y) \mapsto \frac{\partial^2 P_x(y)}{\partial x_i \partial x_j}$ are continuous on $B_r(a) \times S_r(a)$. Hence we may change the order of integration and differentiation and obtain $Hf \in C^2(B_r(a))$,

$$\Delta Hf(x) = \int f(y)\, \Delta P_x(y) \sigma_{a,r}(dy) = 0$$

for every $x \in B_r(a)$. Thus Hf is harmonic on $B_r(a)$.

We next claim that $H1 = 1$. Let $0 < r' < r$ and $x', x'' \in S_{r'}(a)$. Then there exists a rotation S of $\mathbb{R}^n$ such that $S(a) = a$ and $x'' = Sx'$. Since $\sigma_{a,r}$ is invariant with respect to rotations we obtain

$$H1(x'') = H1(Sx') = r^{n-2} \int \frac{r^2 - \|Sx' - a\|^2}{\|Sx' - y\|^n} \sigma_{a,r}(dy)$$

$$= r^{n-2} \int \frac{r^2 - \|x' - a\|^2}{\|Sx' - Sy\|^n} \sigma_{a,r}(dy) = H1(x') .$$

Hence the function $H1$ has a constant value $\alpha_{r'}$ on $S_{r'}(a)$. Considering the harmonic function $H1 - \alpha_{r'}$ we conclude by (1.1) that $H1 - \alpha_{r'} = 0$ on $B_{r'}(a)$. In particular, $\alpha_{r'} = H1(a) = r^{n-2} \int \frac{r^2}{r^n} d\sigma_{a,r} = 1$.

Suppose now that $f \in C(S_r(a))$. Let $z \in S_r(a)$ and $\varepsilon > 0$. Then there exist $f', f'' \in C(S_r(a))$ and $\delta > 0$ such that $f' = 0$ on $B_{2\delta}(z) \cap S_r(a)$, $|f''| \leq \varepsilon$ and

$$f = f(z) + f' + f'' .$$

Since $H1 = 1$ we have

$$Hf = f(z) + Hf' + Hf''$$

where $|Hf''| \leq H|f''| \leq H\varepsilon = \varepsilon$. Furthermore, for every $x \in B_r(a) \cap B_\delta(z)$ and every $y \in S_r(a) \setminus B_{2\delta}(z)$,

$$P_x(y) = r^{n-2} \frac{r^2 - \|x-a\|^2}{\|x-y\|^n} \leq r^{n-2} \frac{r^2 - \|x-a\|^2}{\delta^n}$$

Hence for every $x \in B_r(a) \cap B_\delta(z)$,

$$|Hf'(x)| = \Big| \int_{S_r(a)\setminus B_{2\delta}(z)} f' P_x \, d\sigma_{a,r} \Big| \leq \|f'\| \frac{r^{n-2}}{\delta^n} (r^2 - \|x-a\|^2) .$$

This shows that

$$\lim_{x \to z} Hf'(x) = 0$$

and consequently $\limsup_{x \to z} |Hf(x) - f(z)| \leq \varepsilon$. Since $\varepsilon > 0$ is arbitrary, we conclude that

$$\lim_{x \to z} Hf(x) = f(z) .$$ ┘

1.3. COROLLARY (Mean value property). Let U be an open subset of $\mathbb{R}^n$ and $h \in C(U)$. Then h is harmonic on U if and only if for every $a \in U$ and $r > 0$ with $\overline{B}_r(a) \subset U$ and every $x \in B_r(a)$,

$$h(x) = \int h \, P_x \, d\sigma_{a,r} .$$

Proof. If h is harmonic on U then $h|_{B_r(a)} = H_{a,r}(h|_{S_r(a)})$ for every $a \in U$ and $r > 0$ with $\overline{B}_r(a) \subset U$ by (1.1) and (1.2). If conversely this mean value property holds then by (1.2) h is harmonic on every ball $B_r(a)$ with $\overline{B}_r(a) \subset U$ and hence harmonic on U. ┘

Let U be an open subset of $\mathbb{R}^n$. A lower semi-continuous (l.s.c.) function $v : U \to]-\infty, +\infty]$ is called *hyperharmonic* if

$$\int v \, P_x \, d\sigma_{a,r} \leq v(x)$$

for every $x \in B_r(a)$ and $a \in U$, $r > 0$ such that $\overline{B}_r(a) \subset U$. By $^*H(U)$ we shall denote the set of all hyperharmonic functions on U. By (1.3), we have

$$^*H(U) \cap (-\,^*H(U)) = H(U) .$$

It is immediate that $^*H(U)$ is a min-stable convex cone. Furthermore, if (v_n) is an increasing sequence in $^*H(U)$ then $\sup v_n \in {}^*H(U)$.

1.4. PROPOSITION. Let $v \in C^2(U)$. Then v is hyperharmonic if and only if $\Delta v \leq 0$.

Proof. Suppose that $\Delta v \leq 0$ and let $a \in U$ and $r > 0$ such that $\overline{B}_r(a) \subset U$. Define $f : B_r(a) \to \mathbb{R}$ by

$$f(x) = v(x) - H(v|_{S_r(a)}) \qquad (x \in B_r(a)) .$$

Then $f \in C^2(B_r(a))$, $\Delta f = \Delta v \leq 0$ on $B_r(a)$ and $\lim_{x \to z} f(x) = v(z) - v(z) = 0$ for every $z \in S_r(a)$. Hence $f \geq 0$ by (1.1), i.e. for every $x \in B_r(a)$

$$\int v\, P_x \, d\sigma_{a,r} \leq v(x) .$$

Thus $v \in {}^*H(U)$.

Suppose now conversely that v is hyperharmonic on U and let V be the set of all $x \in U$ such that $\Delta v(x) > 0$. Then V is an open subset of U. Applying the preceding considerations to the restriction of $-v$ on V we conclude that $-v$ is hyperharmonic on V. Hence v is harmonic on V, i.e. $\Delta v = 0$ on V. Thus $V = \emptyset$, $\Delta v \leq 0$ on U. ∎

1.5. EXAMPLE. We define a function $N : \mathbb{R}^n \times \mathbb{R}^n \to \overline{\mathbb{R}}$ by

$$N(x,y) = \begin{cases} \infty & , x = y \\ \| x-y \|^{2-n} & , x \neq y \end{cases}$$

if $n \geq 3$,

$$N(x,y) = \begin{cases} \infty & , x = y \\ -\log \| x-y \| , & x \neq y \end{cases}$$

if $n = 2$, and

$$N(x,y) = -|x-y|$$

if $n = 1$. If $n \geq 3$ ($n = 2$ resp.) the function N is called *Newtonian kernel* (*logarithmic kernel* resp.). For every $x \in \mathbb{R}^n$ the function $N^x : y \mapsto N(x,y)$ is hyperharmonic on $\mathbb{R}^n$, harmonic on $\complement\{x\}$.

Indeed, suppose first that $n \geq 2$. A straightforward computation shows that N^x is harmonic on $\complement\{x\}$. Furthermore, $\lim_{y\to x} N^x(y) = \infty$. Hence N^x is continuous. Let $a \in \mathbb{R}^n$, $r > 0$ and $h = H^{a,r}(N^x|_{S_r(a)})$. We have to show that $h \leq N^x$ on $B_r(a)$. If $x \notin \overline{B}_r(a)$ then evidently $h = N^x$ on $B_r(a)$ since N^x is harmonic on $\complement\{x\}$. Consider now the case $x \in B_r(a)$. Then N^x is continuous and real valued on $S_r(a)$, hence h is harmonic on $B_r(a)$. In particular, there exists $r' > 0$ such that $\overline{B}_{r'}(x) \subset B_r(a)$ and $N^x \geq h$ on $\overline{B}_{r'}(x)$. Considering the function $N^x - h$ on the open set $B_r(a) \setminus \overline{B}_{r'}(x)$ and applying (1.1) we conclude that $N^x \geq h$. Suppose finally that $x \in S_r(a)$. Then, for every $n \in \mathbb{N}$, let

$$h_n = H(\inf(N^x,n)|_{S_r(a)}) .$$

By (1.1), $N^x - h_n \geq 0$ for every n and hence

$$h = \sup h_n \leq N^x .$$

Suppose finally that $n = 1$. Let $a \in \mathbb{R}$ and $r > 0$. Then $B_r(a) =]a-r,a+r[$, $S_r(a) = \{a-r, a+r\}$, and the definition of the Poisson integral reduces to

$$H_{a,r}f(x) = \frac{(a+r)-x}{2r} f(a-r) + \frac{x-(a-r)}{2r} f(a+r) .$$

Hence a function $f : U \to]-\infty,+\infty]$ is hyperharmonic on an open subset U of $\mathbb{R}$ if and only if it is concave on every interval contained in U. In particular, every function $y \mapsto -|x-y|$ is hyperharmonic on $\mathbb{R}$, harmonic on $\complement\{x\}$.

1.6. COROLLARY. If $n \geq 3$ then ${}^*\mathcal{H}^+(\mathbb{R}^n)$ is linearly separating.

Proof. If $x,y \in \mathbb{R}^n$, $x \neq y$, and $\lambda \in \mathbb{R}_+$ then $N^x(x) = \infty \neq \lambda N^x(y)$. ┘

1.7. EXERCISES. 1. For all $x,y \in B_{r/2}(a)$ and $f \in \mathcal{B}^+(S_r(a))$

$$H_{a,r}f(x) \leq 3^{n+1} H_{a,r}f(y) .$$

2. Let U be a domain in $\mathbb{R}^n$ and let (h_n) be an increasing sequence in $\mathcal{H}(U)$ such that $\sup h_n(x_0) < \infty$ for some point $x_0 \in U$. Then $\sup h_n \in \mathcal{H}(U)$.

3. Let U be a domain in $\mathbb{R}^n$, let K be a compact subset of U.

Then there exists a constant $c > 0$ such that $h(x) \leq c\, h(y)$ for every $h \in H^+(U)$ and any pair of points $x, y \in K$.

2. Brownian Semigroup

For every $t > 0$ we define a real function g_t on $\mathbb{R}^n$ $(n \geq 1)$ by

$$g_t(x) := \left(\frac{1}{2\pi t}\right)^{n/2} e^{-\frac{\|x\|^2}{2t}}$$

and a kernel P_t on $\mathbb{R}^n$ by

$$P_t f = g_t * f,$$

i.e.

$$P_t f(x) = \int g_t(x-y)\, f(y)\, \lambda^n(dy) \qquad (x \in \mathbb{R}^n)$$

for every $f \in \mathcal{B}^+(\mathbb{R}^n)$. We note that for every $x \in \mathbb{R}^n$

$$P_t 1(x) = \left(\frac{1}{2\pi t}\right)^{n/2} \int e^{-\frac{\|x-y\|^2}{2t}} \lambda^n(dy)$$

$$= \prod_{i=1}^{n} \left(\frac{1}{\sqrt{2\pi t}} \int_{-\infty}^{+\infty} e^{-\frac{(x_i - y_i)^2}{2t}} dy_i\right) = 1 ,$$

i.e. $P_t 1 = 1$. Furthermore, we claim that

$$P_s P_t = P_{s+t}$$

for every $s, t > 0$. Since $g_s * (g_t * f) = (g_s * g_t) * f$ we have to show that

$$g_s * g_t = g_{s+t} .$$

It is immediately seen that it suffices to consider the case $n = 1$ which we shall do now. Then for every $x \in \mathbb{R}$

$$g_s * g_t(x) = \frac{1}{\sqrt{2\pi s}} \cdot \frac{1}{\sqrt{2\pi t}} \int_{-\infty}^{+\infty} e^{-\frac{(x-y)^2}{2s}}\, e^{-\frac{y^2}{2t}}\, dy$$

$$= \frac{1}{\sqrt{2\pi s}} \cdot \frac{1}{\sqrt{2\pi t}}\, e^{-\frac{x^2}{2s}} \int_{-\infty}^{+\infty} e^{-\frac{s+t}{2st} y^2 + \frac{x}{s} y}\, dy$$

where

$$\int_{-\infty}^{+\infty} e^{-\frac{s+t}{2st}y^2 + \frac{x}{s}y}\, dy = e^{\frac{tx^2}{2s(s+t)}} \int_{-\infty}^{+\infty} e^{-\frac{s+t}{2st}\left(y - \frac{t}{s+t}x\right)^2} dy$$

$$= e^{\frac{tx^2}{2s(s+t)}} \sqrt{2\pi \frac{st}{s+t}} \; .$$

Since

$$-\frac{x^2}{2s} + \frac{tx^2}{2s(s+t)} = -\frac{x^2}{2s}\left(1 - \frac{t}{s+t}\right) = -\frac{x^2}{2(s+t)}$$

we hence obtain

$$g_s * g_t(x) = \frac{1}{\sqrt{2\pi(s+t)}} e^{-\frac{x^2}{2(s+t)}} = g_{s+t}(x) \; .$$

Because of this property the family $(P_t)_{t>0}$ is called a *semigroup*, it is the *Brownian* semigroup on $\mathbb{R}^n$.

2.1. LEMMA. For every $t > 0$, $x \in \mathbb{R}^n$ and $1 \leq i \leq n$

$$\frac{\partial g_t}{\partial x_i}(x) = -\frac{x_i}{t} g_t(x)$$

$$\frac{\partial^2 g_t}{\partial x_i^2}(x) = \left(\frac{x_i^2}{t^2} - \frac{1}{t}\right) g_t(x) \; ,$$

$$\frac{\partial g_t}{\partial t}(x) = \frac{1}{2}\left(\frac{\|x\|^2}{t^2} - \frac{n}{t}\right) g_t(x) \; .$$

In particular ,

$$\frac{1}{2}\Delta g_t = \frac{\partial}{\partial t} g_t \; .$$

Proof. Straightforward. ┘

2.2. PROPOSITION. For every $f \in B_b(\mathbb{R}^n)$ and every $t > 0$, $P_t f \in C^2(\mathbb{R}^n)$ and

$$\frac{1}{2}\Delta P_t f = \frac{\partial}{\partial t} P_t f \; .$$

Proof. Since $\frac{1}{2}\Delta g_t = \frac{\partial}{\partial t} g_t$ by (2.1) it suffices to justify the change of the order of integration and partial differentiation.

Let $r > 1$ and define $g \in B^+(\mathbb{R}^n)$ by

$$g(y) = \begin{cases} (\frac{r}{2\pi})^{n/2} & , y \in B_{2r}(0) \\ (\frac{r}{2\pi})^{n/2} \; e^{-\frac{\|y\|^2}{8r}} & , y \in \complement B_{2r}(0) . \end{cases}$$

Let $x \in B_r(0)$ and $\frac{1}{r} < t < r$. If $y \in \complement B_{2r}(0)$ then $\|y\| \leq 2 \|x-y\|$ and hence

$$- \frac{\|x-y\|^2}{2t} \leq - \frac{\|y\|^2}{8r} .$$

So for every $y \in \mathbb{R}^n$,

$$g_t(x-y) \leq g(y) .$$

Furthermore, $\|y-x\| \leq r + \|y\|$ and therefore

$$\left| \frac{\partial g_t}{\partial x_i} (x-y) \right| \leq r \; (r + \|y\|) \, g(y)$$

$$\left| \frac{\partial^2 g_t}{\partial x_i^2} (x-y) \right| \leq r^2 \; (r + \|y\|)^2 \; g(y) .$$

Since $y \mapsto (r + \|y\|)^2 \; g(y)$ is integrable the statement follows. ┘

<u>2.3. PROPOSITION.</u> Let $f \in \mathcal{B}_b(\mathbb{R}^n)$. Then for every $z \in \mathbb{R}^n$

$$\liminf_{x \to z} f(x) \leq \liminf_{(x,t) \to (z,0)} P_t f(x) \leq \limsup_{(x,t) \to (z,0)} P_t f(x) \leq \limsup_{x \to z} f(x) .$$

If f is uniformly continuous then $\lim_{t \downarrow 0} \| P_t f - f \| = 0$.

Proof. Using the transformation $y \mapsto \frac{y-x}{\sqrt{t}}$ we obtain that for every $t > 0$ and $x \in \mathbb{R}^n$

$$P_t f(x) = \int g_1(y) \; f(\sqrt{t} \cdot y + x) \; \lambda^n(dy) .$$

Hence by Fatou's lemma

$$\liminf_{x \to z} f(x) = \int g_1(y) \liminf_{(x,t) \to (z,0)} f(\sqrt{t} \cdot y + x) \; \lambda^n(dy)$$

$$\leq \liminf_{(x,t) \to (z,0)} \int g_1(y) \; f(\sqrt{t} \cdot y + x) \; \lambda^n(dy) = \liminf_{(x,t) \to (z,0)} P_t f(x) .$$

Similarly for lim sup.

Suppose now that f is uniformly continuous and let $\varepsilon > 0$. Then there exists a compact subset K of $\mathbb{R}^n$ such that $\int_{\complement K} g_1 \, d\lambda < \varepsilon$ and $t_0 > 0$ such that $|f(\sqrt{t}\cdot y+x) - f(x)| < \varepsilon$ for every $x \in \mathbb{R}^n$ and $0 < t < t_0$, $y \in K$. Thus for every $x \in \mathbb{R}^n$ and $0 < t < t_0$

$$|P_t f(x) - f(x)| = \left| \int g_1(y) \, (f(\sqrt{t}\cdot y+x) - f(x)) \, \lambda^n(dy) \right|$$

$$\leq 2 \, \| f \| \int_{\complement K} g_1 \, d\lambda^n + \varepsilon \int_K g_1 \, d\lambda^n \leq (2 \, \| f \| + 1) \, \varepsilon .$$ ┘

2.4. PROPOSITION. Let $f \in \mathcal{B}_b(\mathbb{R}^n)$ and let K be a compact subset of $\mathbb{R}^n$ such that $f = 0$ on a neighborhood U of K. Then $\frac{1}{t} P_t f$ converges to zero uniformly on K as t tends to zero.

Proof. There exists $\varepsilon > 0$ such that $B_\varepsilon(x) \subset U$ for every $x \in K$. Then for every $x \in K$

$$|P_t f(x)| \leq \| f \| \, P_t(x, \complement B_\varepsilon(x)) = \| f \| \, P_t(0, \complement B_\varepsilon(0)) ,$$

where

$$P_t(0, \complement B_\varepsilon(0)) = \int_{\{\sqrt{t}\| y \| \geq \varepsilon\}} g_1(y) \, \lambda^n(dy) .$$

If $\sqrt{t} \, \| y \| \geq \varepsilon$ then $\frac{1}{t} \leq \frac{1}{\varepsilon^2} \| y \|^2$. Since the function $y \mapsto \| y \|^2 g_1(y)$ is integrable on $\mathbb{R}^n$ an application of Lebesgue's convergence theorem yields that $\lim_{t \to 0} \frac{1}{t} P_t(0, \complement B_\varepsilon(0)) = 0$ finishing the proof. ┘

2.5. EXERCISE. Let $f \in \mathcal{B}_b(\mathbb{R}^n)$, let K be a compact subset of $\mathbb{R}^n$ such that $f = 0$ on a neighborhood of K, and let $m \in \mathbb{N}$. Then $\frac{1}{t^m} P_t f$ converges to zero uniformly on K as t tends to zero.

3. Excessive Functions

A function $f \in \mathcal{B}^+(\mathbb{R}^n)$ is called *excessive* (with respect to the Brownian semigroup $(P_t)_{t>0}$) if $\sup_{t>0} P_t f = f$. Let E be the set of all excessive functions. By S we shall denote the set of all *supermedian* functions on $\mathbb{R}^n$, i.e. the set of all $f \in \mathcal{B}^+(\mathbb{R}^n)$ such that $P_t f \leq f$ for every $t > 0$. Obviously, S is min-stable, i.e. $\inf(f,g) \in S$ for all $f, g \in S$.

3.1. LEMMA. For every $f \in S$ the mapping $t \mapsto P_t f$ is decreasing.

Proof. For every $0 < t < s$,

$$P_s f = P_{t+(s-t)} f = P_t P_{s-t} f \leq P_t f .$$

3.2. THEOREM. $*H^+(\mathbb{R}^n) = E = \{\sup f_n : (f_n) \subset E \cap C_0(\mathbb{R}^n) \text{ increasing}\}$.

Proof. 1. Let $t > 0$. Since g_t is invariant under rotations with center 0, there exists a measurable function $\varphi_t : \mathbb{R}_+ \to \mathbb{R}_+$ such that for every $f \in B^+(\mathbb{R}^n)$

$$P_t f(0) = \int f g_t \, d\lambda^n = \int_0^\infty \left(\int f \, d\sigma_{0,r}\right) \varphi_t(r) \, dr .$$

We note that $\int_0^\infty \varphi_t(r) \, dr = 1$ since $P_t 1 = 1$.

Given $f \in B^+(\mathbb{R}^n)$ and $x \in \mathbb{R}^n$ we define $f_x \in B^+(\mathbb{R}^n)$ by $f_x(y) = f(x+y)$. Then

$$P_t f(x) = P_t f_x(0) = \int_0^\infty \left(\int f_x \, d\sigma_{0,r}\right) \varphi_t(r) dr = \int_0^\infty \left(\int f \, d\sigma_{x,r}\right) \varphi_t(r) \, dr .$$

Consider now $u \in *H^+(\mathbb{R}^n)$. Then for every $x \in \mathbb{R}^n$

$$P_t u(x) = \int_0^\infty \left(\int u \, d\sigma_{x,r}\right) \varphi_t(r) \, dr \leq u(x) \int_0^\infty \varphi_t(r) \, dr = u(x) .$$

Hence $u \in S$.

We finally choose an increasing sequence (f_n) in $K^+(X)$ such that $\sup f_n = u$. By (2.3) and (3.1), we obtain for every $n \in \mathbb{N}$,

$$f_n = \lim_{t \to 0} P_t f_n \leq \lim_{t \to 0} P_t u = \sup_{t > 0} P_t u$$

and hence

$$u = \sup_n f_n \leq \sup_{t>0} P_t u \leq u ,$$

i.e. u is excessive.

2. Let $f \in E$. Then for every $n \in \mathbb{N}$

$$f_n := \inf(f, n, nN^0) \in S \cap B_b^+(\mathbb{R}^n)$$

and hence by (2.2) and (3.1)

$$\frac{1}{2} \Delta P_t f_n = \frac{\partial}{\partial t} P_t f_n \leq 0 .$$

Using (1.4) we therefore obtain that every $P_t f_n$, $t > 0$, $n \in \mathbb{N}$, is hyperharmonic. For every $n \in \mathbb{N}$ let

$$p_n = P_{\frac{1}{n}} f_n \ .$$

Then (p_n) is an increasing sequence in ${}^*\mathcal{H}^+(\mathbb{R}^n) \cap C(\mathbb{R}^n)$ such that $f = \sup p_n$. Hence $f \in {}^*\mathcal{H}^+(\mathbb{R}^n)$. Moreover $p_n \in C_0(\mathbb{R}^n)$ for every $n \in \mathbb{N}$ since

$$p_n \leq n P_{\frac{1}{n}} N^0 \leq n N^0 .$$

┘

3.3. EXERCISE. If $n \leq 2$ then every positive hyperharmonic function on $\mathbb{R}^n$ is constant. (Hint: Given $f \in {}^*\mathcal{H}^+(\mathbb{R}^n) \cap C^2(\mathbb{R}^n)$, $x_0 \in \mathbb{R}^n$, and $a < f(x_0)$, consider functions $f - a - \varepsilon N^{x_0}$ and use (1.1) in order to show that $f \geq a$.)

I. General Preliminaries

In this chapter we develop some of the prerequisites in functional analysis for later chapters of this book. We derive a few basic results from Choquet theory of function cones (section 1) and a related general minimum principle (section 2) whereas section 3 gives an up-to-date version of Choquet's capacitability theorem. In section 4 we present Bernstein's theorem concerning the identity of completely monotone functions and Laplace transforms of measures on $\mathbb{R}_+$. Finally, we prove Stampacchias's general projection theorem for Hilbert spaces which will be needed in chapter VIII.

1. Function Cones

One of the main tools in this book will be the representation of positive linear functionals on certain cones of real continuous functions on a locally compact space X with a countable base by measures on X. We start with some general definitions.

Given $F \subset \mathcal{B}(X)$ we define

$$S(F) = \{\sup f_n : (f_n) \subset F \text{ increasing}\}$$

and

$$W(F) = \{\inf(f_1,\ldots,f_n) : n \in \mathbb{N},\ f_i \in F\} .$$

F is called *σ-stable* (min-*stable* resp.) provided $S(F) = F$ ($W(F) = F$ resp.). We say that F is *linearly separating* if for every pair of points $x \neq y$ of X and every $\lambda \in \mathbb{R}_+$ there exists $f \in F$ such that $f(x) \neq \lambda f(y)$.

A real function f on X is said to be *dominated* by a real function $g \geq 0$

arly separating and that there exists a strictly positive function $f_0 \in F \cap C(X)$. Then every function $f \in F$ which is positive on $Ch_F X$ is positive on X.

Proof. 1. Assume that $1 \in F$ and let $\mathcal{A}$ be the set of all non-empty compact subsets A of X such that for any $x \in A$ and any $\mu \in M_x(F)$,

$$\mu(X \setminus A) = 0 .$$

For example, $X \in \mathcal{A}$. The set $\mathcal{A}$ is inductively ordered by the converse inclusion relation. Hence, by Zorn's lemma, for every set $A \in \mathcal{A}$ there exists a minimal set $A' \in \mathcal{A}$ such that $A' \subset A$.

Now let $f \in F$ and assume that $\alpha = \inf f(X) < 0$. Let

$$A = \{f = \alpha\} .$$

Then $A \in \mathcal{A}$ and $\mu(X) = 1$ for every $x \in A$ and $\mu \in M_x(F)$. Indeed, A is a non-empty compact subset of X and if $x \in A$ and $\mu \in M_x(F)$ then

$$\alpha \leq \int \alpha \, d\mu \leq \int f \, d\mu \leq \alpha ,$$

hence $\mu(X) = 1$ and $\int (f-\alpha) d\mu = 0$, i.e. $\mu(X \setminus A) = 0$. Let $A' \in \mathcal{A}$ be minimal such that $A' \subset A$. Suppose that A' contains more than one point. Since F separates the points of X there exists $g \in F$ such that $g_{|A'}$ is non-constant. Let $\beta = \inf g(A')$ and

$$A'' = \{x \in A' : g(x) = \beta\} .$$

Then A'' is a non-empty compact set. If $x \in A''$ and $\mu \in M_x(F)$ then

$$\beta = \int_{A'} \beta \, d\mu \leq \int_{A'} g \, d\mu = \int g \, d\mu \leq g(x) = \beta ,$$

hence $\int_{A'} (g-\beta) d\mu = 0$ and therefore

$$\mu(X \setminus A'') = \mu(X \setminus A') + \mu(A' \setminus A'') = 0.$$

Thus $A'' \in \mathcal{A}$ contradicting the minimality of A'. So A' reduces to a singleton $\{x\}$. Obviously $M_x(F) = \{\varepsilon_x\}$, i.e. $x \in Ch_F X$, and $f(x) = \alpha < 0$.

2. Consider now the general case. Let $f_0 \in F \cap C(X)$, $f_0 > 0$, and

$$F_0 = \{\frac{f}{f_0} : f \in F\} .$$

Then F_0 satisfies the assumptions of (1), and for every $x \in X$

on X (notation: $f \in o(g)$) if for every $\varepsilon > o$ there exists a compact subset K of X such that

$$|f(x)| \leq \varepsilon g(x) \quad \text{for every} \quad x \in \complement K .$$

If G is a set of real positive functions on X we denote by $o(G)$ the set $\bigcup_{g \in G} o(g)$.

A convex cone $S \subset C(X)$ is called a *function cone* if S satisfies the following conditions:

(F_1) There exists an $s_o \in S$ such that $s_o > o$ on X ;

(F_2) S^+ is linearly separating;

(F_3) S is *adapted*, i.e. $S \subset o(S^+)$.

If moreover S is a vector space then S is called a *function space.*

Note that (F_3) is trivially satisfied if X is compact. In general, condition (F_3) will allow us to carry out essential parts of proofs solely on compact sets.

1.1. REMARKS.

1. If S is a function cone on X then S^+ and $W(S)$ are function cones on X . The vector space

$$C_S(X) := \{f \in C(X) : \exists\, s \in S^+ \text{ with } |f| \leq s\}$$

of all S-*bounded* real continuous functions on X is a function space containing $K(X)$. More generally, every convex cone T such that $S^+ \subset T \subset C_S(X)$ is a function cone. Furthermore, $S_{|F}$ is a function cone on F for every closed subset F of X .

2. (F_1) and (F_3) state that for every $s \in S$ there exists a strictly positive $t \in S$ such that $s \in o(t)$, i.e. $\frac{s}{t} \in C_o(X)$.

3. Every function cone $P \subset C^+(X)$ evidently satisfies the following conditions:

(F_1') For every $x \in X$ there exists $p \in P$ such that $p(x) > o$.

(F_3') For every $p \in P$, $\varepsilon > o$ and $x \in X$ there exist $q \in P$ and $K \subset X$ compact such that $q(x) < \varepsilon$ and $p \leq q$ on $\complement K$.

Conversely, let $P \subset C^+(X)$ be a min-stable convex cone satisfying (F_1'), (F_2),

(F_3'). Then

$$P_\sigma := \{ \sum_{n=1}^{\infty} p_n \in C(X) : (p_n) \subset P \}$$

is a function cone on X.

Indeed, using (F_1') and the fact that X has a countable base we may choose a sequence (p_n) in P such that $X = \bigcup_{n=1}^{\infty} \{p_n > 0\}$. Then $p := \sum_{n=1}^{\infty} \alpha_n p_n \in P_\sigma$ for a suitable sequence $(\alpha_n) \subset \mathbb{R}_+^*$, and $p > 0$, Hence P_σ satisfies (F_1). Moreover P_σ satisfies (F_2) since $P \subset P_\sigma$. Since P is min-stable, condition (F_3') implies at once the following condition:

(F_3'') For every $p \in P$, $\varepsilon > 0$ and $K \subset X$ compact there exists $q \in P$ and $K' \subset X$ compact such that $q < \varepsilon$ on K and $p \leq q$ on $\complement K'$.

Indeed, by (F_3') for every $x \in K$ there exist $q_x \in P$ and $K_x \subset X$ compact such that $q_x(x) < \varepsilon$ and $p \leq q_x$ on $\complement K_x$. Since K is compact there exist finitely many points $x_1, \ldots, x_n \in K$ such that $q := \inf(q_1, \ldots, q_n)$ satisfies $q < \varepsilon$ on K. Moreover, $K' = K_{x_1} \cup \ldots \cup K_{x_n}$ is compact and $p \leq q$ on $\complement K'$.

P_σ also satisfies (F_3''). Indeed, if $p = \sum_{n=1}^{\infty} p_n \in P_\sigma$, $\varepsilon > 0$ and $K \subset X$ compact, there is an $m \in \mathbb{N}_0$ such that $\sum_{n=m+1}^{\infty} p_n < \frac{\varepsilon}{2}$ on K. Choosing $q \in P$ and $K' \subset X$ compact such that $q < \frac{\varepsilon}{2}$ on K and $\sum_{n=1}^{m} p_n \leq q$ on $\complement K'$ we have $q' = q + \sum_{n=m+1}^{\infty} p_n \in P_\sigma$, $q' < \varepsilon$ on K and $p \leq q'$ on $\complement K'$.

We finally prove that P_σ satisfies (F_3). Let $p \in P_\sigma$ and choose an *exhaustion* (K_n) of X, i.e. let (K_n) be a sequence of compact subsets of X such that $\bigcup_{n=1}^{\infty} K_n = X$ and $K_n \subset \overset{\circ}{K}_{n+1}$ for every $n \in \mathbb{N}$. Then for every $n \in \mathbb{N}$ there exist $q_n \in P_\sigma$ and $k_n \in \mathbb{N}$ such that $q_n < \frac{1}{2^n}$ on K_n and $p \leq q_n$ on $\complement K_{k_n}$. We obviously may assume that the sequence (k_n) is increasing. Then $q := \sum_{n=1}^{\infty} q_n \in P_\sigma$ such that $p \in o(q)$. Indeed, let $\varepsilon > 0$ and $m \in \mathbb{N}$ such that $\frac{1}{m} \leq \varepsilon$. For every $x \in \complement K_{k_m}$ we obtain

$$q(x) \geq \sum_{n=1}^{m} q_n(x) \geq mp(x) \geq \frac{1}{\varepsilon} p(x) .$$

4. Using (3) and $C^+(X) = (C^+(X))_\sigma$ it follows that $C(X)$ is a function space. If $f \in C_0(X)$ then $g = \sum_{n=1}^{\infty} \inf(\frac{1}{2^n}, |f|) \in C_0^+(X)$ and $f \in o(g)$. Thus $C_0(X)$ is a function space.

5. Let $P \subset C_0^+(X)$ be a min-stable convex cone satisfying (F_1'), (F_2) and such that $\inf(p,1) \in P$ for every $p \in P$. Then P_σ is a function cone by (3).

For the remainder of this section let S be a function cone on X. The following two approximation properties and the resulting representation of positive linear functionals on S by measures illustrate the importance of the notion of a function cone.

1.2. PROPOSITION. Let $s_0 \in S$, $s_0 > 0$. For every $f \in K^+(X)$ and every $\varepsilon > 0$ there exist $s, t \in W(S)^+$ such that

$$0 \leq s - t \leq f \leq s - t + \varepsilon s_0 .$$

Proof. Let $t_0 \in S$, $t_0 > 0$ such that $s_0 \in o(t_0)$ and $t_0 \leq s_0$ on $S(f)$. If

$$T = \{\frac{s}{t_0} : s \in W(S)^+ , \ s \leq \alpha s_0 \ \text{for some} \ \alpha > 0\}$$

then T is a min-stable, linearly separating convex cone in $C_0^+(X)$. By a version of the approximation theorem of Stone-Weierstrass, $T - T$ is dense in $C_0(X)$ with respect to the topology of uniform convergence.

Let $f \in K^+(X)$ and $\varepsilon > 0$. Since $\frac{f}{t_0} \in K^+(X)$ there exist $s, s' \in W(S)^+$ such that

$$\left|\frac{f}{t_0} - \frac{s-s'}{t_0}\right| \leq \frac{\varepsilon}{2} ,$$

i.e. $s-s' - \frac{\varepsilon}{2} t_0 \leq f \leq s-s' + \frac{\varepsilon}{2} t_0$. If $t := \inf(s, s' + \frac{\varepsilon}{2} t_0)$ then $t \in W(S)^+$ and

$$0 \leq s - t \leq f \leq s - t + \varepsilon t_0 .$$

Since $t_0 \leq s_0$ on $S(f)$ we finally obtain that $f \leq s - t + \varepsilon s_0$. ┘

1.3. COROLLARY. Let $f \in C_S^+(X)$ and $t_0 \in S^+$ such that $t_0 > 0$ and $f \in o(t_0)$. Then for every $\varepsilon > 0$ there exist $s, t \in W(S)^+$ such that

$$0 \leq s - t \leq f \leq s - t + \varepsilon t_0 .$$

Proof. Given $\varepsilon > 0$ we have $\varphi := (f - \frac{\varepsilon}{2} t_0)^+ \in K^+(X)$, hence there exist $s, t \in W(S)^+$ such that $0 \leq s - t \leq \varphi \leq s - t + \frac{\varepsilon}{2} t_0$ and therefore

$0 \le s - t \le f \le s - t + \varepsilon t_0$. ┘

<u>1.4. PROPOSITION</u> (G. CHOQUET). Let $T : W(S) \to \mathbb{R}$ be an additive, positively homogeneous, increasing functional. Then there exists a unique measure μ on X such that $T(s) = \int s \, d\mu$ for every $s \in W(S)$.

Proof. T can be extended to a positive linear functional on $W(S) - W(S)$ which will be denoted by T, too. By (1.2) for every $f \in K^+(X)$

$$L(f) := \sup \{T(t) : t \in W(S)^+ - W(S)^+ , \ t \le f\}$$
$$= \inf \{T(t) : t \in W(S)^+ - W(S)^+ , \ t \ge f\} .$$

Hence L defines a positively homogeneous, additive functional on $K^+(X)$, i.e. a measure μ on X.

Let $s \in W(S)^+$. Then

$$\int s \, d\mu = \sup \{\int f \, d\mu : f \in K^+(X), \ f \le s\}$$
$$= \sup \{L(f) : f \in K^+(X), \ f \le s\} \le T(s) .$$

For the converse inequality choose $t \in W(S)^+$ such that $\frac{s}{t} \in C_0^+(X)$.
If $\varepsilon > 0$, then $f := (s - \varepsilon t)^+ \in K^+(X)$ such that $s - \varepsilon t \le f \le s$, hence

$$T(s) - \varepsilon T(t) = T(s - \varepsilon t) \le L(f) = \int f \, d\mu \le \int s \, d\mu .$$

Therefore $T(s) = \int s \, d\mu$.

If $s \in W(S)$ choose $t \in W(S)^+$ such that $s + t \ge 0$. Then

$$T(s) + T(t) = T(s+t) = \int (s+t) d\mu = \int s \, d\mu + \int t \, d\mu ,$$

thus $T(s) = \int s \, d\mu$.

By (1.2), the measure μ is uniquely determined. ┘

The richness of a function cone $P \subset C^+(X)$ is also reflected by the following result which enables us to verify equalities involving functions $q \in P$ by checking them for a particular $p \in P$.

<u>1.5. PROPOSITION.</u> Let $P \subset C^+(X)$ be a min-stable function cone and $p_0 \in P$, $p_0 > 0$. Then there exists a *strict* $p \in P_\sigma$ such that $p \le p_0$, i.e. two measures $\mu, \nu \in M_+(X)$ coincide if

(1) $\int p\, d\mu = \int p\, d\nu < \infty$,

(2) $\int q\, d\mu \leq \int q\, d\nu$ for every $q \in P$.

Proof. There exists a sequence (f_n) in $K^+(X)$ separating $M_+(X)$. We may suppose that $f_n \leq p_0$ for every $n \in \mathbb{N}$. By (1.1) there exist sequences (p_m), (q_m) in P such that for all $k,n \in \mathbb{N}$

$$|f_n - (p_{n_k} - q_{n_k})| \leq \frac{1}{k} p_0$$

for some $n_k \in \mathbb{N}$. Since P is min-stable we may suppose that $p_m \leq p_0$, $q_m \leq p_0$ for every $m \in \mathbb{N}$. Then

$$p := \frac{1}{3}\left(p_0 + \sum_{m=1}^{\infty} \frac{1}{2^m}(p_m + q_m)\right) \in P_\sigma$$

and $p \leq p_0$. Let $\mu,\nu \in M_+(X)$ satisfying (1) and (2). Then $\mu(p_0) = \nu(p_0) < \infty$ and $\mu(p_m) = \nu(p_m)$, $\mu(q_m) = \nu(q_m)$ for every $m \in \mathbb{N}$. Hence $\mu(f_n) = \nu(f_n)$ for every $n \in \mathbb{N}$, i.e. $\mu = \nu$. ┘

1.6. EXAMPLE. Let P be the set of all increasing functions $p \in C^+(\mathbb{R})$ satisfying $\lim_{x\to-\infty} p(x) = 0$. Then $P = P_\sigma$ is a function cone by (1.1.3). A function $p \in P$ is strict if and only if p is strictly increasing. Indeed, if $p \in P$ and $x,y \in \mathbb{R}$ such that $x < y$ and $p(x) = p(y)$ then $\varepsilon_x \neq \varepsilon_y$, $\varepsilon_x(p) = \varepsilon_y(p) < \infty$, and $\varepsilon_x(q) \leq \varepsilon_y(q)$ for every $q \in P$, i.e. p is not strict. Conversely, let $p \in P$ be strictly increasing and let $\mu,\nu \in M_+(X)$ such that $\mu(p) = \nu(p) < \infty$ and $\mu(q) \leq \nu(q)$ for every $q \in P$, Fix $a \in \mathbb{R}$, $\varepsilon > 0$, and define

$$q = \inf(p,p(a)) - \inf(p,p(a-\varepsilon)) .$$

Then $q,p-q \in P$, hence $\mu(q) = \nu(q) < \infty$. Since

$$1_{[a,\infty[} \leq \frac{q}{p(a)-p(a-\varepsilon)} \leq 1_{]a-\varepsilon,\infty[}$$

we obtain that

$$\mu([a,\infty[) \leq \frac{\mu(q)}{p(a)-p(a-\varepsilon)} \leq \frac{\nu(q)}{p(a)-p(a-\varepsilon)} \leq \nu(]a-\varepsilon,\infty[)$$

and similarly $\nu([a,\infty[) \leq \mu(]a-\varepsilon,\infty[)$. Letting ε tend to zero we conclude that $\mu([a,\infty[) = \nu([a,\infty[< \infty$ and thus $\mu = \nu$.

The following lemmas will be quite useful in the following chapters.

1.7. LEMMA. Let F be a family of l.s.c. numerical functions on X. Then there exists a countable family $F_0 \subset F$ such that $\sup F_0 = \sup F$.

Proof. For every $s \in \mathbb{Q}$ there exists a countable family $G_s \subset F$ such that

$$\bigcup_{f \in G_s} \{f > s\} = \bigcup_{f \in F} \{f > s\},$$

i.e. $\{\sup G_s > s\} = \{\sup F > s\}$. Let $F_0 = \bigcup_{s \in \mathbb{Q}} G_s$. Then $F_0 \subset F$ is countable, and if $x \in X$ and $a < \sup F(x)$ then there exists $s \in \mathbb{Q}$ such that $a < s < \sup F(x)$ and hence $\sup F_0(x) \geq \sup G_s(x) > s > a$. Thus $\sup F_0 = \sup F$. ┘

1.8. LEMMA (G. CHOQUET). Let F be a family of numerical functions on X. Then there exists a countable family $F_0 \subset F$ such that $\widehat{\inf F_0} = \widehat{\inf F}$.*)

Proof. Let V be a countable base of X. For every $V \in V$, there exists a countable family $F_V \subset F$ such that

$$\inf \{f(y) : y \in V, f \in F\} = \inf \{f(y) : y \in V, f \in F_V\}.$$

Then $F_0 := \bigcup_{V \in V} F_V$ satisfies the requirement. It suffices to show that

$$\widehat{\inf F_0}(x) \leq \widehat{\inf F}(x)$$

holds for every $x \in X$ such that $\widehat{\inf F_0}(x) > -\infty$. So fix $x \in X$ and $a \in \mathbb{R}$ such that $a < \widehat{\inf F_0}(x)$. There exists $V \in V$ such that $x \in V$ and $a < \inf F_0(y)$ for every $y \in V$, consequently

$$a \leq \{\inf f(y) : y \in V, f \in F_V\} = \inf \{f(y) : y \in V, f \in F\},$$

which yields $a \leq \inf F$ on V. Thus $a \leq \widehat{\inf F}(x)$. ┘

1.9. LEMMA. Suppose that X is compact and let F be a convex set of lower bounded, l.s.c. numerical functions on X such that for every measure $\mu \neq 0$ on X there is an $f \in F$ with $\mu(f) > 0$. Then there exists an $f \in F$ such that $f > 0$ on X.

---

*)For every $f : X \to \overline{\mathbb{R}}$, we denote by $\hat{f}$ the lower semi-continuous regularization of f, i.e. $\hat{f}(x) := \liminf_{y \to x} f(y)$, $x \in X$.

Proof. Define $G = \{g \in C(X) : \exists \lambda \in \mathbb{R}_+^*, f \in F \text{ with } g \leq \lambda f\}$ and $G_0 = \{g \in C(X) : g > 0\}$. Then G and G_0 are convex cones. Furthermore G_0 is open in $C(X)$ endowed with the topology of uniform convergence.

Assume, $G \cap G_0 = \emptyset$. Then by the separation theorem for convex sets there exists a linear functional $T \neq 0$ on $C(X)$ and $\alpha \in \mathbb{R}$ such that

$$\sup_{g \in G} T(g) \leq \alpha \leq \inf_{g \in G_0} T(g) .$$

Since G and G_0 are cones, we have $\alpha = 0$, so T is a measure μ on X. By assumption, there is an $f \in F$ such that $\mu(f) > 0$. Since for every $g \in G$ we have $\mu(g) \leq 0$ we obtain the contradiction $\mu(f) \leq 0$. Therefore, $G \cap G_0 \neq \emptyset$, and the assertion follows. ∎

1.10. LEMMA. Suppose that X is compact, let $m \in \mathbb{N}$, and for every $1 \leq i \leq m$ let $(f_{in})_{n \geq 1}$ be a uniformly lower bounded sequence of l.s.c. numerical functions on X such that $\liminf_{n \to \infty} f_{in} > 0$. Then there exist $n \in \mathbb{N}$ and $\alpha_1, \ldots, \alpha_n \in [0,1]$ such that $\sum_{j=1}^{n} \alpha_j = 1$ and $\sum_{j=1}^{n} \alpha_j f_{ij} > 0$ for every $1 \leq i \leq m$.

Proof. For every $n \in \mathbb{N}$ let

$$f_n = \inf_{1 \leq i \leq m} f_{in} .$$

Then (f_n) is a uniformly lower bounded sequence of l.s.c. numerical functions on X such that $\liminf_{n \to \infty} f_n > 0$. Let $\mu \in M_+(X)$, $\mu \neq 0$. Then

$$0 < \int \liminf_{n \to \infty} f_n \, d\mu \leq \liminf_{n \to \infty} \int f_n \, d\mu .$$

Hence there exists $n \in \mathbb{N}$ such that $\mu(f_n) > 0$. By (1.9) we obtain $n \in \mathbb{N}$ and $\alpha_1, \ldots, \alpha_n \in [0,1]$ such that $\sum_{j=1}^{n} \alpha_j = 1$ and $\sum_{j=1}^{n} \alpha_j f_j > 0$. Since $f_{ij} \geq f_j$ for all i,j, the statement now follows immediately. ∎

1.11. COROLLARY. Suppose that X is compact and let F be a convex subset of $C(X)$. Let t be an upper bounded, u.s.c. numerical function on X and let s be a lower bounded, l.s.c. numerical function on X such that $t < s$. Finally, let (f_n) be a uniformly bounded sequence in F which converges pointwise to a function f satisfying $t < f < s$. Then there exists a function $g \in F$

such that $t < g < s$.

Proof. For every $n \in \mathbb{N}$ let

$$f_{1n} = f_n - t , \quad f_{2n} = s - f_n .$$

Then (f_{1n}) and (f_{2n}) are uniformly lower bounded sequences of l.s.c. numerical functions on X such that $\lim_{n\to\infty} f_{1n} = f - t > 0$ and $\lim_{n\to\infty} f_{2n} = s - f > 0$. By (1.10), there exist $n \in \mathbb{N}$ and $\alpha_1,\dots,\alpha_n \in [0,1]$ such that $\sum_{j=1}^{n} \alpha_j = 1$ and $\sum_{j=1}^{n} \alpha_j f_{ij} > 0$ for $i = 1,2$. Then $g := \sum_{j=1}^{n} \alpha_j f_j \in F$ and obviously $g - t > 0$ and $s - g > 0$. ┘

1.12. EXERCISES. 1. If X is compact and F is a cone in $C^+(X)$ then every strictly positive function $u \in S(S(F))$ is contained in $S(F)$.

2. Let $X = [0,1[$ and let F be the linear space of all $f \in C(X)$ such that $\lim_{x\to 1} f(x)$ exists (in $\mathbb{R}$). Then F is not a function space and there exists a positive linear functional on F which cannot be represented by a measure on X.

2. Choquet Boundary

Minimum principles play a fundamental rôle in potential theory. We shall present a general version of Bauer's minimum principle for function cones which will imply among others the well known boundary minimum principle for hyperharmonic functions (see section (III.4)).

Let X be a locally compact space with countable base and let F be a convex cone of l.s.c. numerical functions $> -\infty$ on X . For any $x \in X$ let $M_x(F)$ denote the set of all measures $\mu \in M_+(X)$ such that $-\infty < \mu(f) \leq f(x)$ for every $f \in F$. Then always $\varepsilon_x \in M_x(F)$. The set

$$Ch_F X = \{x \in X : M_x(F) = \{\varepsilon_x\}\}$$

is called the *Choquet boundary* of X with respect to F .

2.1 PROPOSITION (H. BAUER). Suppose that X is compact, that F is line-

$$M_x(F_0) = \{\frac{f_0\mu}{f_0(x)} : \mu \in M_x(F)\} ,$$

hence $Ch_{F_0} X = Ch_F X$.

Let $f \in F$ such that $f \geq 0$ on $Ch_F X$. Then $\frac{f}{f_0} \in F_0$ with $\frac{f}{f_0} \geq 0$ on $Ch_F X$, thus $\frac{f}{f_0} \geq 0$ on X by the first case and so $f \geq 0$ on X. ┘

For the remainder of this section assume that F contains a function cone $P \subset C^+(X)$ such that every function $f \in F$ is lower P-bounded, i.e. $f \geq -p$ for some $p \in P$.

<u>2.2. THEOREM.</u> If $f \in F$ such that $f \geq 0$ on $Ch_F X$ then $f \geq 0$ on X.

Proof. Let $f \in F$, $f \geq 0$ on $Ch_F X$. By assumption on F there exist $p, q \in P$ such that $-p \leq f$, $q > 0$, and $p \in o(q)$. Let $\varepsilon > 0$ and

$$g = f + \varepsilon q .$$

Then $g \in F$ and

$$K := \{g \leq 0\}$$

is a compact subset of $X \setminus Ch_F X$. Let $x \in K$. Then there exists a measure $\mu \in M_x(F)$ such that $\mu \neq \varepsilon_x$. Let ν be the restriction of μ on K. Then for every $u \in P$

$$\int u \, d\nu \leq \int u \, d\mu \leq u(x) .$$

Moreover

$$\int g \, d\nu < \int g \, d\nu + \int_{\complement K} g \, d\mu = \int g \, d\mu \leq g(x) ,$$

where $\mu(\complement K) > 0$ implies that $\int_{\complement K} g \, d\mu > 0$, hence $\int g \, d\nu < g(x)$, $\nu \neq \varepsilon_x$. If however $\mu(\complement K) = 0$ then $\nu = \mu \neq \varepsilon_x$. Choosing

$$G = (P + \mathbb{R}_+ g)_{|K}$$

we therefore have $Ch_G K = \emptyset$. By (2.1) we conclude that $g_{|K} = 0$, i.e. $g \geq 0$. Thus $f \geq 0$. ┘

We now meet the important concept of reducing functions for the first time. For every $\varphi \in C_P(X)$, let

$$^{F}R_{\varphi} = \inf\{f \in F : f \geq \varphi\} .$$

The following propositions will already show how useful this definition is. Especially, we shall obtain a manageable characterization of the Choquet boundary.

2.3. PROPOSITION. For every $\varphi \in C_P(X)$ and every $x \in X$,

$$^{F}R_{\varphi}(x) = \max\{\int \varphi \, d\mu : \mu \in M_x(F)\} .$$

Proof. Let $\varphi \in C_P(X)$ and $x \in X$. Then for every $\mu \in M_x(F)$ and every $f \in F$ satisfying $f \geq \varphi$

$$\int \varphi \, d\mu \leq \int f \, d\mu \leq f(x) ,$$

hence

$$\int \varphi \, d\mu \leq {}^{F}R_{\varphi}(x) .$$

On the other hand the function $\psi \mapsto {}^{F}R_{\psi}(x)$ is a sublinear functional on $C_P(X)$. Hence by Hahn-Banach theorem there exists a linear functional T on $C_P(X)$ such that $T(\psi) \leq {}^{F}R_{\psi}(x)$ for every $\psi \in C_P(X)$ and $T(\varphi) = {}^{F}R_{\varphi}(x)$. If $\psi \leq 0$ then obviously $T(\psi) \leq {}^{F}R_{\psi}(x) \leq 0$. Hence T is positive. Applying (I.1.4) to the function space $C_P(X)$ we obtain a measure $\mu \in M_+(X)$ such that $T(\psi) = \int \psi \, d\mu$ for every $\psi \in C_P(X)$.

If $f \in F$ then for every $\psi \in C_P(X)$ satisfying $\psi \leq f$,

$$\int \psi \, d\mu = T(\psi) \leq {}^{F}R_{\psi}(x) \leq f(x) ,$$

hence

$$\int f \, d\mu \leq f(x) ,$$

i.e. $\mu \in M_x(F)$. Moreover,

$$\int \varphi \, d\mu = T(\varphi) = {}^{F}R_{\varphi}(x) .$$

2.4. COROLLARY. For every $\varphi \in C_P(X)$, ${}^{F}R_{\varphi} = \varphi$ on $Ch_F(X)$. If p is a strict element of P then $Ch_F(X) = \{{}^{F}R_{-p} = -p\}$.

Proof. The first statement follows immediately from (2.3) and the definition of $Ch_F X$. So let p be a strict element of P and $x \in X$ such that ${}^{F}R_{-p}(x) = -p(x)$. Let $\mu \in M_x(S)$. Then $\mu \in M_x(S)$ and

$$\int(-p)d\mu \leq {}^{F}R_{-p}(x) = -p(x) ,$$

hence $\int p\,d\mu = p(x)$. Therefore $\mu = \varepsilon_x$, $x \in Ch_F(X)$. ┘

2.5. PROPOSITION. Let $\varphi, \psi \in C_P(X)$ such that $\varphi \leq \psi \leq {}^F R_\varphi$. Then ${}^F R_\varphi = {}^{F-\mathbb{R}_+\psi} R_\varphi$. In particular, $Ch_{F-\mathbb{R}_+\psi} X \subset \{{}^F R_\varphi = \varphi\}$.

Proof. Obviously ${}^F R_\varphi \geq {}^{F-\mathbb{R}_+\psi} R_\varphi$. Let $f \in F$ and $\alpha \in \mathbb{R}_+$ such that

$$g := f - \alpha\psi \geq \varphi .$$

Then $f \geq \alpha\psi + \varphi \geq (\alpha+1)\varphi$, hence $\frac{f}{\alpha+1} \geq {}^F R_\varphi$, $\frac{f}{\alpha+1} \geq \psi$. Therefore $\frac{\alpha}{\alpha+1} f \geq \alpha\,\psi = f-g$, $g \geq \frac{f}{\alpha+1}$. Thus $g \geq {}^F R_\varphi$. The second statement now follows from (2.4). ┘

For the remainder of this section suppose that F is contained in $C_P(X)$ and let $\hat{F}$ denote the set of all l.s.c., lower P-bounded functions f on X such that $\mu(f) \leq f(x)$ for all $x \in X$ and $\mu \in M_x(F)$. Obviously, $\hat{F}$ is min-stable and $F \subset \hat{F}$. The elements of $\hat{F}$ are sometimes called *F-concave*.

2.6. LEMMA. Let $f \in \hat{F}$ and $\varphi \in C_P(X)$ such that $\varphi \leq f$. Then there exists a function $g \in \hat{F} \cap C_P(X)$ such that $\varphi \leq g \leq f$.

Proof. Let $p_0 \in P$, $p_0 > 0$, and let (K_n) be an exhaustion of X . Then there exist an increasing sequence (φ_n) in $C_P(X)$ and a decreasing sequence (g_n) in $\hat{F} \cap C_P(X)$ such that

$$\varphi \leq \varphi_n \leq \inf\,(f, g_n)$$

for every $n \geq 1$ and

$$g_n \leq \varphi_n + 2^{-n} p_0 \quad \text{on } K_n$$

for every $n \geq 2$. Indeed, choose $\varphi_1 = \varphi$ and $g_1 \in P$ such that $\varphi \leq g_1$. Suppose that $n \in \mathbb{N}$ and $\varphi_n \in C_P(X)$ and $g_n \in \hat{F} \cap C_P(X)$ are already chosen such that $\varphi_n \leq \inf\,(f, g_n)$. By (2.3),

$${}^F R_{\varphi_n}(x) = \sup_{\mu \in M_x(F)} \int \varphi_n\,d\mu \leq \sup_{\mu \in M_x(F)} \int f\,d\mu \leq f(x) .$$

Using a finite covering of K_{n+1} we obtain a function $g_{n+1} \in \hat{F} \cap C_P(X)$ such that $\varphi_n \leq g_{n+1} \leq g_n$ on X and $g_{n+1} < f + 2^{-(n+1)} p_0$ on K_{n+1} . We may choose

$\varphi_{n+1} \in C_p(X)$ such that $\varphi_n \leq \varphi_{n+1} \leq \inf(f, g_{n+1})$ on X and $\varphi_{n+1} = \sup(\varphi_n, g_{n+1} - 2^{-(n+1)} p_o)$ on K_{n+1}. This shows the existence of the sequences (φ_n) and (g_n).

For every $n \in \mathbb{N}$,

$$g_n - g_{n+1} \leq g_n - \varphi_n \leq 2^{-n} p_o \quad \text{on} \quad K_n .$$

Hence the decreasing sequence (g_n) converges locally uniformly to a function $g \in \hat{F} \cap C_p(X)$. Obviously, $\varphi \leq g$. Moreover, $g_n \leq \varphi_n + 2^{-n} p_o \leq f + 2^{-n} p_o$ on K_n for every $n \in \mathbb{N}$, hence $g \leq f$. ┘

2.7. PROPOSITION. For every $f \in \hat{F}$ there exists an increasing sequence (g_n) in $\hat{F} \cap C_p(X)$ such that $f = \sup g_n$.

Proof. There exists a sequence (φ_n) in $C_p(X)$ such that $\sup \varphi_n = f$. By (2.6), there exists $g_1 \in \hat{F} \cap C_p(X)$ such that $\varphi_1 \leq g_1 \leq f$. If $n \in \mathbb{N}$ and g_n is already chosen we use (2.6)again to obtain $g_{n+1} \in \hat{F} \cap C_p(X)$ such that $\sup(g_n, \varphi_{n+1}) \leq g_{n+1} \leq f$. ┘

2.8. EXERCISES. 1. Let U be a relatively compact open subset of $\mathbb{R}^n$ and let $H(U)$ be the set of all continuous real functions on $\overline{U}$ which are harmonic on U.

a) $H(U)$ is a function space such that $Ch_{H(U)} \overline{U} \subset U^*$.

b) If U is regular then $Ch_{H(U)} \overline{U} = U^*$.

c) If $n \geq 2$ and $U = B_r(a) \setminus \{a\}$, then $H(U) = H(B_r(a))$ and $Ch_{H(U)} \overline{U} = S_r(a) = U^* \setminus \{a\}$. (Hint: Given $h \in H(U)$, consider $\pm H_{a,r}(h|_{S_r(a)}) + \varepsilon(N^a - H_{a,r}(N^a|_{S_r(a)}))$ and apply (0.1.1).)

2. Let K be a non-empty compact convex subset of $\mathbb{R}^n$ and let $A(K)$ denote the set of all affine real functions on K. Then $A(K)$ is a function space and $Ch_{A(K)} K$ is the set K_e of all extreme points of K.

3. Let F be a convex cone of l.s.c. numerical functions on X containing a function cone $P \subset C^+(X)$ such that every function in F is lower P-bounded. Then $\overline{Ch_F X}$ is the smallest closed subset A of X such that any function $f \in F$

satisfying $f \geq 0$ on A is positive, i.e. $\overline{Ch_F X}$ is the *Šhilov boundary* of X with respect to F.

3. Analytic Sets and Capacitances

As always let X be a locally compact space with countable base. Many statements involving measures on X can be obtained using the well known fact that measures $\mu \in M(X)$ are regular, i.e. that

$$\mu(A) = \sup\{\mu(K) : K \text{ compact} \subset A\}$$

for every $A \in \mathcal{B}(X)$. In classical potential theory the Newtonian capacity furnishes an important example of a function $c : \mathfrak{P}(\mathbb{R}^n) \to \overline{\mathbb{R}}_+$ which is not additive, but nevertheless has certain properties in common with measures, namely it is monotone and continuous for increasing sequences of subsets as well as for decreasing sequences of compact subsets (see section (V.4) for the definition and further details). The general theory of capacities developed by G. CHOQUET [1] shows that these properties already imply that $c(A) = \sup\{c(K) : K \text{ compact} \subset A\}$ for every $A \in \mathcal{B}(\mathbb{R}^n)$.

So a function $c : \mathfrak{P}(X) \to \overline{\mathbb{R}}_+$ is called a (Choquet) *capacity* (on X) if the following two properties hold:

1. $\sup c(A_n) = c(\bigcup_{n=1}^{\infty} A_n)$ for every increasing sequence (A_n) of subsets of X.
2. $\inf c(K_n) = c(\bigcap_{n=1}^{\infty} K_n)$ for every decreasing sequence (K_n) of compact subsets of X.

A subset A of X is called c-*capacitable* if

$$c(A) = \sup\{c(K) : K \text{ compact} \subset A\}.$$

It is called *universally capacitable* if it is c-capacitable for every capacity c on X.

<u>3.1. EXAMPLE.</u> For every measure μ on X the outer measure μ^* is a capacity on X. A subset A of X is μ^*-capacitable if and only if it is

μ-measurable.

As in the original paper of Choquet the proof for the fundamental statement that every Borel subset of X is universally capacitable will be given in two steps. We shall first show that every analytic subset of X is universally capacitable and then prove that every Borel subset of X is analytic.

In order to define analytic sets we introduce the following notation. For every locally compact space Z let $K_{\sigma\delta}(Z)$ denote the set of all countable intersections of countable unions of compact subsets of Z, i.e.

$$K_{\sigma\delta}(Z) = \{ \bigcap_{i=1}^{\infty} \bigcup_{j=1}^{\infty} K_{ij} : K_{ij} \text{ compact} \subset Z \} .$$

A subset A of X is called *analytic* if there exists a locally compact space Y with countable base and a set $B \in K_{\sigma\delta}(X \times Y)$ such that A is the image of B under the projection π_X of $X \times Y$ on X. Since every locally compact space Y with countable base is a subspace of a compact space with countable base we may even assume that Y is compact. Let $A(X)$ denote the set of all analytic subsets of X.

In section (VI.6) we shall need that Borel sets are capacitable in a sense which is even more general, they are hypercapacitable. The definition of a hypercapacitable set can be given very suggestively in terms of a topological game between two players where player I chooses capacitances on X and player II chooses subsets of X. A subset C of $\mathfrak{P}(X)$ is called a *capacitance* (on X) provided the following two properties hold:

1. If $A \in C$ and $A \subset A' \subset X$ then $A' \in C$.
2. If (A_n) is an increasing sequence in $\mathfrak{P}(X)$ such that $\bigcup_{n=1}^{\infty} A_n \in C$ then $A_n \in C$ for some $n \in \mathbb{N}$.

<u>3.2. EXAMPLES.</u> Obviously $\mathfrak{P}(X)$, $\{A \in \mathfrak{P}(X) : A \neq \emptyset\}$, $\{A \in \mathfrak{P}(X) : A$ uncountable$\}$ are capacitances on X. If c is a capacity on X and $\alpha \in \mathbb{R}$ then $\{A \in \mathfrak{P}(X) : c(A) > \alpha\}$ is a capacitance on X.

If C is a capacitance on X and $A \in C$ then there exists a relatively compact subset A_1 of A such that $A_1 \in C$. Indeed, it suffices to choose a sequence (K_n) of compact subsets of X which is increasing to X. Then $(A \cap K_n)$ is increasing to A and hence $A \cap K_n \in C$ for some $n \in \mathbb{N}$.

Given a subset A of X, the game is now the following. Player I chooses a capacitance C_1 such that $A \in C_1$ and player II chooses a relatively compact subset A_1 of A such that $A_1 \in C_1$. Then player I chooses a capacitance C_2 such that $A_1 \in C_2$ and player II chooses a subset A_2 of A_1 such that $A_2 \in C_2$. And so on. We shall say that player II wins the game if the compact set $\bigcap_{n=1}^{\infty} \bar{A}_n$ is contained in A. The set A is called *hypercapacitable* if player II can win against any possible choice of capacitances by player I. Of course, we expect the following result.

<u>3.3. PROPOSITION.</u> Every hypercapacitable subset of X is universally capacitable.

Proof. Let A be a hypercapacitable subset of X, let c be a capacity on X and $\alpha < c(A)$. Since player I may choose

$$C_n = \{B \in \mathfrak{P}(X) : c(B) > \alpha\}$$

there exists a decreasing sequence (A_n) of relatively compact subsets of A such that $c(A_n) > \alpha$ for every $n \in \mathbb{N}$ and $K := \bigcap_{n=1}^{\infty} \bar{A}_n \subset A$. Then K is a compact subset of A and

$$c(K) = \inf c(\bar{A}_n) \geq \inf c(A_n) \geq \alpha .$$

Thus A is c-capacitable. ┘

<u>3.4. THEOREM.</u> Every analytic subset of X is hypercapacitable.

Proof. Let $A \in \mathcal{A}(X)$. Then there exists a compact space Y with countable base and compact subsets K_{ij} of $X \times Y$, $i, j \in \mathbb{N}$, such that $\pi_X(B) = A$ where

$$B = \bigcap_{i=1}^{\infty} \bigcup_{j=1}^{\infty} K_{ij} .$$

We may assume that for every $i \in \mathbb{N}$ the sequence $(K_{ij})_{j \in \mathbb{N}}$ is increasing.

Now let C_1 be a capacitance on X such that $A \in C_1$. Since the sequence

$(B \cap K_{1j})_{j\in\mathbb{N}}$ is increasing to B and hence the sequence $(\pi_X(B\cap K_{1j}))_{j\in\mathbb{N}}$ is increasing to A there exists $m_1 \in \mathbb{N}$ such that

$$A_1 := \pi_X(B\cap K_{1m_1}) \in C_1 .$$

Obviously A_1 is a relatively compact subset of A. Let C_2 be a capacitance on X such that $A_1 \in C_2$. Since the sequence $(B\cap K_{1m_1} \cap K_{2j})_{j\in\mathbb{N}}$ is increasing to $(B\cap K_{1m_1})$ there exists $m_2 \in \mathbb{N}$ such that

$$A_2 := \pi_X(B\cap K_{1m_1} \cap K_{2m_2}) \in C_2 .$$

And so on. Thus player II may always choose the sequence (A_n) in such a way that

$$A_n = \pi_X(B\cap K_{1m_1} \cap \ldots \cap K_{nm_n})$$

for some sequence (m_n) in $\mathbb{N}$. Let

$$K_n = \bigcap_{i=1}^{n} K_{im_i} .$$

Since (K_n) is a decreasing sequence of compact subsets of $X\times Y$ we have $\bigcap_{n=1}^{\infty} \pi_X(K_n) = \pi_X(\bigcap_{n=1}^{\infty} K_n)$ and hence

$$\bigcap_{n=1}^{\infty} \overline{A}_n \subset \bigcap_{n=1}^{\infty} \pi_X(K_n) = \pi_X(\bigcap_{n=1}^{\infty} K_n) \subset \pi_X(B) = A .$$

Therefore A is hypercapacitable. ┘

3.5. COROLLARY (G. CHOQUET). Every analytic subset of X is universally capacitable.

3.6. REMARK. It can be shown that the analytic subsets of X are exactly the hypercapacitable subsets of X (C. DELLACHERIE [4]).

We shall now prove that $A(X)$ is stable under countable union and countable intersection, and this implies immediately the desired inclusion $B(X) \subset A(X)$.

3.7. PROPOSITION. Let (A_n) be a sequence in $A(X)$. Then $\bigcap_{n=1}^{\infty} A_n, \bigcup_{n=1}^{\infty} A_n \in A(X)$.

Proof. There exist compact spaces Y_n with countable base and $B_n \in K_{\sigma\delta}(X\times Y_n)$ such that $\pi_X(B_n) = A_n$.
Then $Y := \prod_{n=1}^{\infty} Y_n$ is a compact space with countable base. For every $n \in \mathbb{N}$ let φ_n

denote the projection of $X \times Y$ on $X \times Y_n$. Obviously for every compact subset K of $X \times Y_n$ the inverse image $\varphi_n^{-1}(K)$ is a compact subset of $X \times Y$ and hence

$$B'_n := \varphi_n^{-1}(B_n) \in K_{\sigma\delta}(X \times Y).$$

Therefore

$$B' := \bigcap_{n=1}^{\infty} B'_n \in K_{\sigma\delta}(X \times Y).$$

We claim that

$$\pi_X(B') = \bigcap_{n=1}^{\infty} A_n$$

and hence $\bigcap_{n=1}^{\infty} A_n \in A(X)$. Indeed, obviously $\pi_X(B') \subset \pi_X(B'_n) = A_n$ for every $n \in \mathbb{N}$. Conversely, let $x \in \bigcap_{n=1}^{\infty} A_n$. Then for every $n \in \mathbb{N}$ there exists $y_n \in Y_n$ such that $(x,y_n) \in B_n$. Choosing $y = (y_n)_{n \in \mathbb{N}}$ we have $(x,y) \in Y$ and $\varphi_n(x,y) = (x,y_n) \in B_n$ for every $n \in \mathbb{N}$, i.e. $(x,y) \in B'$. Thus $x \in \pi_X(B')$.

Moreover it is easily checked that

$$B'' := \bigcup_{n=1}^{\infty} B'_n \times \{n\} \in K_{\sigma\delta}(X \times Y \times \mathbb{N}).$$

Thus $\bigcup_{n=1}^{\infty} A_n = \pi_X(B'') \in A(X)$. ┘

3.8. COROLLARY. Every Borel subset of X is analytic.

Proof. Let

$$\mathcal{D} = \{A \in A(X) : \complement A \in A(X)\}.$$

Obviously $A \in \mathcal{D}$ if and only if $\complement A \in \mathcal{D}$. If (A_n) is a sequence in $\mathcal{D}$ then by (3.7)

$$\bigcup_{n=1}^{\infty} A_n \in A(X), \quad \complement \bigcup_{n=1}^{\infty} A_n = \bigcap_{n=1}^{\infty} \complement A_n \in A(X),$$

and hence $\bigcup_{n=1}^{\infty} A_n \in \mathcal{D}$. If U is an open subset of X then U and $\complement U$ are countable unions of compact subsets of X. If K is a compact subset of X then $K \times \{1\} \in K_{\sigma\delta}(X \times \{1\})$, hence $K = \pi_X(K \times \{1\}) \in A(X)$. Therefore $\mathcal{D}$ contains all open subsets of X. Thus $\mathcal{B}(X) \subset \mathcal{D} \subset A(X)$. ┘

This is all we really need on analytic sets. However, it may be interesting to note that in fact $\mathcal{B}(X) = \mathcal{D}$ (see (3.12.2)). Moreover, it can be shown that there exist Borel subsets of $\mathbb{R}^2$ such that the images under the projections on $\mathbb{R}$

are not Borel subsets of $\mathbb{R}$. However, projections of Borel sets will always yield analytic sets. In fact, images of analytic sets even under arbitrary Borel measurable mappings are always analytic as the next proposition shows. This leads to the possibility of characterizing analytic sets in many different ways and may be considered as a second motivation for the introduction of analytic sets.

3.9. LEMMA. Let X_1, X_2 be locally compact spaces with countable base and $A_1 \in A(X_1)$, $A_2 \in A(X_2)$. Then $A_1 \times A_2 \in A(X_1 \times X_2)$.

Proof. Let (K_n) be an exhaustion of X_1 . There exists a compact space Y with countable base and a set $B_2 \in K_{\sigma\delta}(X_2 \times Y)$ such that $\pi_{X_2}(B_2) = A_2$. Let $n \in \mathbb{N}$. Then $K_n \times B_2 \in K_{\sigma\delta}(K_1 \times X_2 \times Y)$ and hence

$$K_n \times A_2 = \pi_{X_1 \times X_2}(K_n \times B_2) \in A(X_1 \times X_2) .$$

Hence $X_1 \times A_2 = \bigcup_{n=1}^{\infty} (K_n \times A_2) \in A(X_1 \times X_2)$. Similarly $A_1 \times X_2 \in A(X_1 \times X_2)$ and therefore

$$A_1 \times A_2 = (X_1 \times A_2) \cap (A_1 \times X_2) \in A(X_1 \times X_2) . \qquad \lrcorner$$

3.10. PROPOSITION. Let $A \in A(X)$ and let φ be a Borel measurable mapping of X into another locally compact space Y with countable base. Then $\varphi(A) \in A(Y)$.

Proof. Since the mapping

$$\psi : (x,y) \mapsto (\varphi(x),y)$$

of $X \times Y$ into $Y \times Y$ is Borel measurable and the diagonal D of $Y \times Y$ is closed, we know that

$$\Gamma := \{(x,\varphi(x)) : x \in X\} = \psi^{-1}(D) \in B(X \times Y) .$$

Hence $\Gamma \in A(X \times Y)$ by (3.8). Moreover $A \times Y \in A(X \times Y)$ by (3.9) and therefore

$$B := (A \times Y) \cap \Gamma \in A(X \times Y) .$$

Obviously $\pi_Y(B) = \varphi(A)$. There exists a compact space Z with countable base and a set $C \in K_{\sigma\delta}(X \times Y \times Z)$ such that $\pi_{X \times Y}(C) = B$. Thus $\varphi(A) = \pi_Y(C) \in A(Y)$. $\lrcorner$

3.11. COROLLARY. For every subset A of X the following properties are equivalent:

(1) A is analytic.

(2) There exists a locally compact space Y with countable base, a set $B \in \mathcal{B}(Y)$ and a continuous mapping $\varphi : Y \to X$ such that $\varphi(B) = A$.

(3) There exists a locally compact space Y with countable base, a set $B \in \mathcal{B}(Y)$ and a Borel measurable mapping $\varphi : Y \to X$ such that $\varphi(B) = A$.

3.12. EXERCISES. 1. Let c be a cpacity on X. Then for every $A \in \mathcal{A}(X)$ there exists $B \in \mathcal{B}(X)$ such that $A \subset B$ and $c(A) = c(B)$. (Hint: Show that the function $A \mapsto \inf\{c(B) : B \in \mathcal{B}(X), A \subset B\}$ is a capacity on X.)

2. Let A_1 and A_2 be disjoint analytic subsets of X. Then there exist disjoint Borel subsets B_1 and B_2 of X such that $A_1 \subset B_1$ and $A_2 \subset B_2$. In particular, a subset A of X is a Borel set if and only if the sets A and $\complement A$ are analytic. (Hint: Show that the subsets C of $X \times X$ such that $B_1 \cap B_2 \neq \emptyset$ for all $B_1, B_2 \in \mathcal{B}(X)$ satisfying $C \subset B_1 \times B_2$ form a capacitance on $X \times X$. Let π_1 (π_2 resp.) be the projection $(x_1,x_2) \mapsto x_1$ ($(x_1,x_2) \mapsto x_2$ resp.) from $X \times X$ on X and note that $\pi_1(\overline{C}) \cap \pi_2(\overline{C}) \neq \emptyset$ for every relatively compact $C \in \mathcal{C}$.)

3. Let c be a capacity on X which is subadditive, i.e. which satisfies $c(A \cup B) \leq c(A) \cup c(B)$ for all subsets A,B of X. Let (A_n) be a sequence of subsets of X and define $B_n = \bigcup_{m=n}^{\infty} A_m$, $n \in \mathbb{N}$. Then $\sum_{n=1}^{\infty} c(A_n)\gamma^n \leq \sum_{n=1}^{\infty} c(B_n)\gamma^n \leq (1 - \frac{1}{\gamma})^{-1} \sum_{n=1}^{\infty} c(A_n)\gamma^n$ for every $\gamma > 1$.

4. Let $\mathcal{K}$ be the class of all compact subsets of X and let c be a positive real function on $\mathcal{K}$ having the following three properties:

(a) If $K,L \in \mathcal{K}$ and $K \subset L$, then $c(K) \leq c(L)$.

(b) Given $K \in \mathcal{K}$ and $\varepsilon > 0$, there exists an open neighborhood U of K such that $c(L) < c(K) + \varepsilon$ for every $L \in \mathcal{K}$ contained in U.

(c) $c(K \cup L) + c(K \cap L) \leq c(K) + c(L)$ for all $K,L \in \mathcal{K}$.

For every subset A of X let

$$c_*(A) = \sup\{c(K) : K \text{ compact} \subset A\},$$

$$c^*(A) = \inf\{c^*(U) : U \text{ open} \supset A\}.$$

Then $c_*(K) = c^*(K) = c(K)$ for every $K \in \mathcal{K}$ and $c_*(A) = c^*(A)$ for every $A \in \mathcal{A}(X)$.

4. Laplace Transforms

For a measure μ on $\mathbb{R}_+$ such that the function $t \mapsto e^{-xt}$ is integrable with respect to μ for all $x > 0$, the function

$$L\mu : x \mapsto \int_0^\infty e^{-xt} \, d\mu(t)$$

is called the *Laplace transform* of μ. Let uns note that $L\mu$ is obviously a decreasing function on $\mathbb{R}_+$, $\lim_{x\to 0} L\mu(x) = \mu(\mathbb{R}_+)$, $\lim_{x\to\infty} L\mu(x) = \mu(\{0\})$.

4.1. PROPOSITION (Uniqueness of Laplace transforms). Let μ,ν be measures on $\mathbb{R}_+$ and $x_0 \in \mathbb{R}_+$ such that $\int_0^\infty e^{-xt} \, d\mu(t) = \int_0^\infty e^{-xt} \, d\nu(t) < \infty$ for every $x \geq x_0$. Then $\mu = \nu$.

Proof. Let $\mu_0 = e^{-x_0 t} \mu$, $\nu_0 = e^{-x_0 t} \nu$. Then clearly $L\mu_0 = L\nu_0$ and $\mu_0(\mathbb{R}_+) = \nu_0(\mathbb{R}_+) < \infty$. Every function $t \mapsto e^{-xt}$, $x \geq 0$, has a unique continuous extension to $[0,\infty]$ (by 1 if $x = 0$, by 0 if $x > 0$). Let A denote the set of all linear combinations of these continuous extensions. Considering μ_0 and ν_0 as measures on $[0,\infty]$ we then have $\mu_0(f) = \nu_0(f)$ for every $f \in A$. Apparently A is a subalgebra of $C([0,\infty])$ containing the constants and separating the points of the compact space $[0,\infty]$. Therefore, we know by the Stone-Weierstrass approximation theorem that A is dense in $C([0,\infty])$ with respect to uniform convergence. Hence $\mu_0(f) = \nu_0(f)$ for every $f \in C([0,\infty])$, i.e. $\mu_0 = \nu_0$ and $\mu = \nu$. ┘

A C^∞-function $f :]0,\infty[\to \mathbb{R}$ is called *completely monotone* if

$$(-1)^n f^{(n)} \geq 0$$

for all $n \geq 0$.

4.2. THEOREM (S. BERNSTEIN). A function $f :]0,\infty[\to \mathbb{R}$ is completely monotone if and only if there exists a measure μ on $\mathbb{R}_+$ such that $f = L\mu$.

Changing the order of integration and derivation it is easily verified that every Laplace transform $L\mu$ is completely monotone. The converse can be deduced from the general representation theorem by Choquet and we refer the reader to G. CHOQUET [3] for the details.

4.3. PROPOSITION. Let μ,ν be measures on $\mathbb{R}_+$ such that the function $t \mapsto e^{-xt}$ is integrable with respect to μ and ν for all $x > 0$. Then the convolution $\mu * \nu$ exists and

$$L(\mu * \nu) = L\mu \cdot L\nu .$$

Proof. For every $x > 0$,

$$\int_0^\infty \int_0^\infty e^{-x(s+t)} d\mu(s) d\nu(t) = \int_0^\infty e^{-xt} \left(\int_0^\infty e^{-xs} d\mu(s)\right) d\nu(t) = L\mu(x) \int_0^\infty e^{-xt} d\nu(t) = L\mu(x) L\nu(x) < \infty .$$

If $f \in K^+(\mathbb{R}_+)$ then there exists $a \in \mathbb{R}_+$ such that $f(t) \leq a\, e^{-t}$ for every $t \in \mathbb{R}_+$ and hence

$$\int_0^\infty \int_0^\infty f(s+t) d\mu(s) d\nu(t) \leq a\, L\mu(1)\, L\nu(1) < \infty .$$

Therefore $\mu * \nu$ exists, and the first equalities show that $L(\mu * \nu) = L\mu \cdot L\nu$. ⌋

5. Coercive Bilinear Forms

In this section we shall prove a general projection theorem which will be used only in section 4 of chapter VIII. Let H be a (real) Hilbert space with scalar product $(\cdot,\cdot)$ and corresponding norm $\|\cdot\|$. A continuous bilinear form $a : H \times H \to \mathbb{R}$ is called *coercive* if there exists a constant $c > 0$ such that $a(u,u) \geq c\|u\|^2$ for every $u \in H$.

5.1. THEOREM (G. STAMPACCHIA). Let a be a coercive bilinear form on $H \times H$, L a continuous linear functional on H and A a non-empty closed convex subset of H. Then there exists a unique $u \in A$ such that $a(u,v) \geq L(v)$ for all $v \in A-u$.

Proof. First of all we note that there exists a unique $w \in H$ such that

$$L(v) = (w,v) \quad \text{for all} \quad v \in H .$$

Let us denote the point u in the statement of the theorem by $G(w,a,A)$.

The *uniqueness* will result from the following inequality. If $w_1, w_2 \in H$ and $u_1 = G(w_1,a,A)$, $u_2 = G(w_2,a,A)$ then $\|u_2 - u_1\| \leq \frac{1}{c}\|w_2 - w_1\|$. Since $a(u_i,v) \geq (w_i,v)$ for all $v \in A-u_i$ $(i=1,2)$ and $u_2 - u_1 \in A-u_1$, $u_1 - u_2 \in A-u_2$

we obtain

$$c\|u_2-u_1\|^2 \le a(u_2-u_1,u_2-u_1) = a(u_2,u_2-u_1) - a(u_1,u_2-u_1)$$
$$\le (w_2,u_2-u_1) - (w_1,u_2-u_1) = (w_2-w_1,u_2-u_1) \le \|w_2-w_1\| \circ \|u_2-u_1\| .$$

The *existence* will be proved in several steps. Let us introduce the following bilinear forms on $H \times H$:

$$\alpha(u,v) = \frac{a(u,v) + a(v,u)}{2}, \quad \beta(u,v) = \frac{a(u,v) - a(v,u)}{2},$$
$$a_t(u,v) = \alpha(u,v) + t\beta(u,v) \quad (0 \le t \le 1) \circ$$

Clearly, every a_t is coercive with the same constant c.

i) The statement is true for $a_0 = \alpha$: Let $I : A \to \mathbb{R}$ be defined by $I(u) = \alpha(u,u) - 2(w,u)$ and set $d := \inf_{u\in A} I(u)$.
Then for every $u \in A$,

$$I(u) \ge c\|u\|^2 - 2\|u\| \cdot \|w\| = c(\|u\|^2 - 2\|u\| \cdot \|\tfrac{1}{c}w\| + \|\tfrac{1}{c}w\|^2) - \tfrac{1}{c}\|w\|^2$$
$$= c(\|u\| - \|\tfrac{1}{c}w\|)^2 - \tfrac{1}{c}\|w\|^2 \ge -\tfrac{1}{c}\|w\|^2 \circ$$

There exists a sequence $(u_n) \subset A$ such that $d \le I(u_n) \le d + \frac{1}{n}$ for every $n \in \mathbb{N}$. Since

$$c\|u_n - u_m\|^2 \le \alpha(u_n - u_m, u_n - u_m) = 2\alpha(u_n,u_n) + 2\alpha(u_m,u_m) - 4\alpha\left(\frac{u_n+u_m}{2}, \frac{u_n+u_m}{2}\right)$$
$$= 2I(u_n) + 2I\left(\frac{u_m+u_n}{2}\right) < 2\left(\frac{1}{n} + \frac{1}{m}\right),$$

(u_n) is a Cauchy sequence in A, hence converges to some $u \in A$ satisfying $I(u) = d$.

Let $v \in A-u$, $v \neq 0$. Then $u + \lambda v \in A$ for some $\lambda > 0$. Since $I(u+\lambda v) \ge I(u)$ we obtain $\alpha(u+\lambda v,u+\lambda v) - 2(w,u+\lambda v) \ge \alpha(u,u) - 2(w,u)$, hence $\lambda\alpha(u,v) + \lambda\alpha(v,u+\lambda v) \ge 2\lambda(w,v)$, which implies $\alpha(u,v) + \alpha(v,u+\lambda v) \ge 2(w,v)$. Letting $\lambda \to 0$, the assertion follows.

ii) If $0 \le s \le 1$ and if the statement is true for a_s, then it is true for all a_t such that $t \le s + t_0$, where $t_0 < \frac{c}{M}$ and $|\beta(u,v)| \le M \cdot \|u\| \cdot \|v\|$ for all $u,v \in H$. For the proof of this statement let us introduce a mapping $T : H \to A$ by $T(y) = G(w',a_s,A)$ where $w' \in H$ is defined by the following continuous linear functional

$$v \mapsto (w',v) = (w,v) - (t-s)\beta(y,v) .$$

Since for every $v \in A-T(y)$

$$a_s(T(y),v) \geq (w',v) = (w,v) - (t-s)\beta(y,v)$$

we obtain for every $y_1, y_2 \in H$ using the inequality of the beginning of the proof

$$\|T(y_2) - T(y_1)\| \leq \frac{1}{c} \| w_2' - w_1' \| \leq \frac{1}{c}(t-s)M \| y_2 - y_1 \| \leq \frac{t_o M}{c} \| y_2 - y_1 \| .$$

Therefore, T is a contraction, and there exists a unique element $u \in A$ such that $u = T(u)$. By the definition of T we obtain for every $v \in A-u$, $a_s(u,v) \geq (w,v) - (t-s)\beta(u,v)$ which implies the desired inequality $a_t(u,v) = a_s(u,v) + (t-s)\beta(u,v) \geq (w,v)$.

iii) By (i) the statement of the theorem is true for $a_o = \alpha$, hence by (ii) it is true for all a_t such that $t < t_o$. Continuing this process the assertion follows after a finite number of steps. ┘

<u>5.2. EXERCISE.</u> Let a be a coercive bilinear form on $H \times H$ and L a continuous linear functional on H. Furthermore, let (A_n) be an increasing (decreasing resp.) sequence of non-empty closed convex subsets of H, $A_o = \overline{\bigcup_{n=1}^{\infty} A_n}$ ($A_o = \bigcap_{n=1}^{\infty} A_n \neq \emptyset$ resp.), and, for every $n \in \mathbb{N}_o$, let u_n be the unique element of A_n such that $a(u_n,v) \geq L(v)$ for all $v \in A_n - u_n$. Then $\lim u_n = u_o$.

II. Excessive Functions

The aim of this chapter is an intrinsic characterization of the set of excessive functions of a sub-Markov semigroup. In section 1 we introduce kernels on X. In section 2 we derive basic properties of functions which are supermedian with respect to a kernel on X. On the one hand these functions form one of our standard examples (discrete potential theory), on the other hand their study allows us to obtain fundamental properties of excessive functions of sub-Markov semigroups and resolvents (section 3) which constitute one major motivation for the notion of balayage spaces (section 4). Introducing the cone $\mathcal{P}$ of continuous potentials on a balayage space (section 5) it is possible to construct associated potential kernels (section 6), resolvents (section 7), and semigroups (section 8). As a consequence we obtain a correspondence between balayage spaces and sub-Markov semigroups having "many" continuous excessive functions.

1. Kernels

A *kernel* V on X is a mapping $V : X \times \mathcal{B}(X) \to \overline{\mathbb{R}}_+$ such that

(K_1) $x \mapsto V(x,B)$ is Borel measurable for every $B \in \mathcal{B}(X)$,

(K_2) $B \mapsto V(x,B)$ is a measure on $(X,\mathcal{B}(X))$ for every $x \in X$.

V is called a *sub-Markov (Markov, bounded resp.)* kernel, if

$$V(x,X) \leq 1 \ (V(x,X) = 1\ ,\ \sup_{x \in X} V(x,X) < +\infty \quad \text{resp.}).$$

If $x \mapsto V(x,K)$ is bounded for every compact set $K \subset X$ then V is called *proper*.

Let V be a kernel on X. For every $f \in \mathcal{B}^+(X)$ define a numerical function

Vf on X by

$$Vf(x) = \int f(y)\, V(x,dy) \qquad (x \in X) .$$

Since f is the limit of an increasing sequence of step functions $\varphi_n \in \mathcal{B}^+(X)$ for which $V\varphi_n \in \mathcal{B}^+(X)$ by (K_1) one obtains $Vf \in \mathcal{B}^+(X)$. The mapping $V : \mathcal{B}^+(X) \to \mathcal{B}^+(X)$ has the following properties

(K_1') $V(f+g) = Vf+Vg \quad (f,g \in \mathcal{B}^+(X))$,

(K_2') $V(\lambda f) = \lambda Vf \quad (f \in \mathcal{B}^+(X) , \lambda \in \mathbb{R}_+)$,

(K_3') $f_n \in \mathcal{B}^+(X) , f_n \uparrow f \Rightarrow Vf = \lim Vf_n$.

Conversely, every mapping $V : \mathcal{B}^+(X) \to \mathcal{B}^+(X)$ such that (K_1') - (K_3') hold defines a kernel (also denoted by V) on X by

$$V(x,B) := V1_B(x) \qquad (x \in X , B \in \mathcal{B}(X)) .$$

Furthermore, if $f \in \mathcal{B}(X)$ such that $\{Vf^+ = +\infty\} \cap \{Vf^- = +\infty\} = \emptyset$ we set

$$Vf := Vf^+ - Vf^- .$$

<u>1.1. EXAMPLES.</u> 1. The identity $I : \mathcal{B}^+(X) \to \mathcal{B}^+(X)$ defines a Markov kernel on X .

2. If $X = \mathbb{R}$ and $t > 0$, then by

$$T_t f(x) = f(x-t) \qquad (x \in \mathbb{R} , f \in \mathcal{B}^+(\mathbb{R}))$$

a Markov kernel T_t is defined on $\mathbb{R}$. T_t is called *translation* by t (to the left).

3. If λ is a measure on X and $g : X \times X \to \overline{\mathbb{R}}_+$ Borel measurable then

$$Vf(x) := \int g(x,y)\, f(y)\, \lambda(dy) \qquad (x \in X , f \in \mathcal{B}^+(X))$$

defines a kernel on X . Especially ($X = \mathbb{R}^n$, λ = Lebesgue measure on $\mathbb{R}^n$) the operators P_t of the Brownian semigroup (see section 0.2) are Markov kernels.

4. If $X = \mathbb{R}^n$ and $U = B_r(a)$ then a kernel H_U on $\mathbb{R}^n$ is defined as follows (see section 0.1):

$$H_U f(x) := \begin{cases} \int f\, P_x\, d\sigma_{a,r} & , \text{ if } x \in U , \\ f(x) & , \text{ if } x \notin U . \end{cases}$$

H_U is called the *harmonic kernel* associated with U .

5. If V,W are kernels on X , then V+W , αV $(\alpha \in \mathbb{R}_+)$, and VW are ker-

nels on X defined by

$$(V+W)f := Vf + Wf ,$$

$$(\alpha V)f := \alpha Vf ,$$

$$(VW)f := V(Wf) = \int Wf(y)\, V(\cdot,dy) \qquad (f \in \mathcal{B}^+(X)) .$$

For every sequence (V_n) of kernels

$$(\sum_{n=1}^{\infty} V_n)f := \sum_{n=1}^{\infty} V_n f \qquad (f \in \mathcal{B}^+(X))$$

defines a kernel on X .

6. If V is a kernel on X and $\widetilde{V}$ is a kernel on $\widetilde{X}$, the *product kernel* $V \otimes \widetilde{V}$ on $X \times \widetilde{X}$ will be defined by

$$(V\otimes\widetilde{V})g(x,\widetilde{x}) = \iint g(y,\widetilde{y})\, V(x,dy)\, \widetilde{V}(\widetilde{x},d\widetilde{y}) \qquad (x \in X ,\ \widetilde{x} \in \widetilde{X} ,\ g \in \mathcal{B}^+(X \times \widetilde{X})) .$$

In particular, for every $t > 0$,

$$(V\otimes T_t)g(x,s) = \int g(y,s-t)\, V(x,dy) .$$

7. If V is a kernel on X and μ is a measure on $(X,\mathcal{B}(X))$ then

$$(\mu V)(B) := \int V(x,B)\, \mu(dx) \qquad (B \in \mathcal{B}(X))$$

defines a measure μV on $(X,\mathcal{B}(X))$.

1.2. PROPOSITION. Let S be a function cone on X and $T : W(S) \to \mathcal{B}_r(X)$ ($T : K(X) \to \mathcal{B}_r(X)$ resp.) such that T is additive, positively homogeneous and increasing. Then there exists a unique kernel V on X such that $Vf = Tf$ for every $f \in W(S)$ ($f \in K(X)$ resp.).

Proof. For every $x \in X$, the mapping $f \mapsto Tf(x)$ is an additive, positively homogeneous and increasing functional on $W(S)$, hence by (I.1.4) there exists a unique measure μ_x on X such that

$$Tf(x) = \int f \, d\mu_x .$$

Let F be the set of all functions $f \in \mathcal{B}^+(X)$ such that the function

$$Vf : x \mapsto \int f \, d\mu_x$$

is Borel measurable on X . Then $K^+(X) \subset F$ by (I.1.2). Hence $F = \mathcal{B}^+(X)$ by the monotone class theorem. Thus V defines a kernel on X which has the desired properties. Evidently, V is uniquely determined. ┘

1.3. COROLLARY. Let V be a bounded kernel on X and let $W : \mathcal{B}_b(X) \to \mathcal{B}_b(X)$ be linear such that, for every $f \in \mathcal{B}_b^+(X)$, $0 \leq Wf \leq Vf$. Then W can be extended to a (bounded) kernel on X.

Proof. By (1.2) there exists a kernel $\tilde{W}$ on X such that $\tilde{W}f = Wf$ for every $f \in \mathcal{K}(X)$. If $(f_n) \subset \mathcal{B}_b^+(X)$ and $f_n \uparrow f \in \mathcal{B}_b^+(X)$ then

$$0 \leq \inf W(f-f_n) \leq \inf V(f-f_n) = 0 ,$$

hence $Wf_n \uparrow Wf$. Therefore $\tilde{W}$ is an extension of W by the monotone class theorem. ┘

A kernel V on X is called a *strong Feller kernel* if $V(\mathcal{B}_b(X)) \subset \mathcal{C}_b(X)$.

1.4. EXAMPLE. The operators P_t, $t > 0$, of Brownian semigroup on $\mathbb{R}^n$ are strong Feller kernels by (0.2.2).

The following proposition shows that strong Feller kernels have in fact a seemingly much stronger property which is very useful when dealing with products (see section (V.2)).

1.5. PROPOSITION. Let V be a strong Feller kernel on X and let (f_n) be a sequence in $\mathcal{B}_b(X)$ which is uniformly bounded and pointwise converging to a function f. Then $\lim_{n\to\infty} Vf_n(x_n) = Vf(x)$ for every sequence (x_n) in X converging to $x \in X$.

Proof. For every $m \in \mathbb{N}$ let $g_m = \sup_{n \geq m} f_n$. If $n,m \in \mathbb{N}$ such that $n \geq m$ then $g_m \geq f_n$, hence $Vg_m \geq Vf_n$. Therefore

$$Vg_m(x) = \lim_{n\to\infty} Vg_m(x_n) \geq \limsup_{n\to\infty} Vf_n(x_n)$$

for every $m \in \mathbb{N}$. Since the sequence (g_m) is decreasing to f, we obtain that

$$Vf(x) = \lim_{m\to\infty} Vg_m(x) \geq \limsup_{n\to\infty} Vf_n(x_n) .$$

An application to the sequence $(-f_n)$ yields that

$$Vf(x) \leq \liminf_{n\to\infty} Vf_n(x_n) .$$

Thus $\lim_{n\to\infty} Vf_n(x_n) = Vf(x)$. ┘

1.6. COROLLARY. Let V be a strong Feller kernel on X and $\widetilde{V}$ a strong Feller kernel on $\widetilde{X}$, $\widetilde{X}$ locally compact with countable base. Then $V \otimes \widetilde{V}$ is a strong Feller kernel on $X \times \widetilde{X}$.

Proof. Let $g \in \mathcal{B}_b(X \times \widetilde{X})$ and let $(x_n, \widetilde{x}_n)$ be a sequence in $X \times \widetilde{X}$ converging to $(x_0, \widetilde{x}_0)$. For every $n \in \mathbb{N} \cup \{0\}$ define $f_n : X \to \mathbb{R}$ by

$$f_n(x) = \int g(x,\widetilde{y})\, \widetilde{V}(\widetilde{x}_n, d\widetilde{y}) = Vg_x(\widetilde{x}_n) .$$

Then (f_n) is a uniformly bounded sequence in $\mathcal{B}_b(X)$ and $\lim_{n\to\infty} f_n(x) = \lim_{n\to\infty} Wg_x(\widetilde{x}_n) = Wg_x(\widetilde{x}_0) = f_0(x)$ for every $x \in X$. Thus we conclude by (1.5) that

$$\lim_{n\to\infty} (V\otimes\widetilde{V})g(x_n,\widetilde{x}_n) = \lim_{n\to\infty} Vf_n(x_n) = Vf_0(x_0) = (V\otimes\widetilde{V})g(x_0,\widetilde{x}_0) .$$

1.7. EXERCISE. A mapping $V : \mathcal{B}^+(X) \to \mathcal{B}^+(X)$ defines a kernel on X if and only if $V(0) = 0$ and $V(\sum_{n=1}^{\infty} f_n) = \sum_{n=o}^{\infty} V(f_n)$ for every sequence (f_n) in $\mathcal{B}^+(X)$.

2. Supermedian Functions

Let P be a kernel on X. A function $s \in \mathcal{B}^+(X)$ is called P-*supermedian* if

$$P s \leq s .$$

Evidently, the set S_P of all P-supermedian functions is a σ-stable convex cone. Furthermore, $\inf s_n \in S_P$ for any sequence (s_n) in S_P. Finally, if $s \in S_P$ and $t \in \mathcal{B}^+(X)$ such that $Ps \leq t \leq s$ then $Pt \leq Ps \leq t$, hence $t \in S_P$.

The following two propositions due to G. MOKOBODZKI contain fundamental properties of reducing functions in potential theory. For every $f \in \mathcal{B}(X)$ let

$$R_f = \inf \{s \in S_P : s \geq f\} .$$

Obviously, $R_f = R_{f^+}$. Note that clearly $R_f \in S_P$ whenever $R_f \in \mathcal{B}(X)$. Surprisingly, this is always true.

2.1. PROPOSITION. Let $f \in \mathcal{B}^+(X)$ and define a sequence (g_n) in $\mathcal{B}^+(X)$ by $g_1 := f$ and $g_{n+1} := \sup (g_n, Pg_n)$. Then $g := \sup g_n \in S_P$ and $R_f = g$.

Proof. The sequence (g_n) is increasing, hence

$$Pg = \sup Pg_n \leq \sup g_{n+1} = g ,$$

i.e. $g \in S_P$. Since $f = g_1 \leq g$ we have $R_f \leq g$. Conversely, let $s \in S_P$ with $s \geq f$. Proceeding by induction we obtain $s \geq g_n$ for every $n \in \mathbb{N}$, i.e. $s \geq g$. Thus $R_f \geq g$. ┘

2.2. PROPOSITION. Let $s, t \in S_P$ and $f \in \mathcal{B}^+(X)$ such that $s = t + f$. Then there exists a function $t' \in S_P$ such that $t' \leq t$ and $s = t' + R_f$.

Proof. Let $h \in \mathcal{B}^+(X)$ such that

$$s = Ps + h$$

and define a sequence (t_n) by

$$t_1 := t , \quad t_{n+1} := \inf(t_n, Pt_n + h) .$$

We note that the decreasing sequence (t_n) is contained in S_P . Indeed, $t_1 = t \in S_P$ and if $t_n \in S_P$ for some $n \in \mathbb{N}$ then $Pt_n \leq t_{n+1} \leq t_n$, hence $t_{n+1} \in S_P$. As in (2.1) we define an increasing sequence (g_n) in $\mathcal{B}^+(X)$ by

$$g_1 = f , \quad g_{n+1} = \sup (g_n, Pg_n) .$$

Then

$$s = t_n + g_n$$

for every $n \in \mathbb{N}$. Indeed, $s = t + f = t_1 + g_1$. Suppose that $s = t_n + g_n$ for some $n \in \mathbb{N}$. Then $Ps = Pt_n + Pg_n$, hence

$$s = Ps + h = Pt_n + h + Pg_n$$

and

$$s = \inf (t_n, Pt_n + h) + \sup (g_n, Pg_n) = t_{n+1} + g_{n+1} .$$

Defining $t' = \inf t_n$ we conclude that $t' \in S_P$ and $s = t' + R_f$. ┘

Because of the importance of the above properties in potential theory we shall say that a convex cone $S \subset \mathcal{B}^+(X)$ is a *potential cone* provided the following holds: Let $u,v \in S$ and $f \in \mathcal{B}^+(X)$ such that $f = (u-v)^+$ on $\{v < \infty\}$ and $f = 0$ on $\{v = \infty\}$. Then $R_f := \inf\{s \in S : s \geq f\} \in S$ and $u = w + R_f$ for some $w \in S$.

2.3. EXERCISE. Let P be a kernel on X. Then $R_f = \sup(f, P R_f)$ for every $f \in B^+(X)$.

3. Semigroups and Resolvents

A family $\mathbb{P} = (P_t)_{t>0}$ of kernels on X is called a *semigroup* if

$$P_{s+t} = P_s P_t \qquad \text{for all } s, t > 0 .$$

$\mathbb{P}$ is called a *sub-Markov* (*Markov* resp.) *semigroup* if $P_t 1 \leq 1$ ($P_t 1 = 1$ resp.) for every $t > 0$.

If $\mathbb{P}$ is a sub-Markov semigroup on X we denote by $S_{\mathbb{P}}$ ($E_{\mathbb{P}}$ resp.) the set of all $\mathbb{P}$-*supermedian* ($\mathbb{P}$-*excessive* resp.) functions, i.e.

$$S_{\mathbb{P}} := \{u \in B^+(X) : P_t u \leq u \text{ for every } t > 0\} ,$$

$$E_{\mathbb{P}} := \{u \in B^+(X) : \sup_{t>0} P_t u = u\} .$$

3.1. EXAMPLES. 1. *Brownian semigroup* on $\mathbb{R}^n$ (0.2).

2. *Translation semigroup* $\mathbb{T}$ on $\mathbb{R}$: $\mathbb{T} = (T_t)_{t>0}$ defined by $T_t(x,\cdot) = \varepsilon_{x-t}$ $(t>0, x \in \mathbb{R})$ is a Markov semigroup; $S_{\mathbb{T}}$ is the set of all positive numerical functions on $\mathbb{R}$ which are increasing and $E_{\mathbb{T}}$ is the set of all left-continuous functions in $S_{\mathbb{T}}$. Note that an increasing function on $\mathbb{R}$ is left-continuous if and only if it is l.s.c. .

3. *Pseudo-Poisson semigroup*. Let P be a sub-Markov kernel on X. For every $t > 0$ define

$$P_t = e^{-t} \sum_{k=0}^{\infty} \frac{t^k}{k!} P^k .$$

Then $\mathbb{P} = (P_t)_{t>0}$ is a sub-Markov semigroup such that

$$E_{\mathbb{P}} = S_{\mathbb{P}} = S_P .$$

Indeed, obviously $S_P \subset S_{\mathbb{P}}$ and $S_{\mathbb{P}} = E_{\mathbb{P}}$ since $e^{-t} f \leq P_t f$ for every $f \in B^+(X)$, $t > 0$.

Finally let $u \in S_{\mathbb{P}}$, $x \in X$ such that $u(x) < \infty$, and $t > 0$. Then $P_t u \leq u$ implies that $e^{-t} \sum_{k=1}^{\infty} \frac{t^k}{k!} P^k u(x) \leq (1-e^{-t}) u(x)$, hence $e^{-t} P u(x) \leq \frac{1-e^{-t}}{t} u(x)$.

Letting t tend to zero we conclude that $Pu(x) \leq u(x)$. Thus $u \in S_P$.

Note that $\mathbb{P}$ is a Markov semigroup if and only if P is a Markov kernel.

If $X = \mathbb{Z}$ and $P(x,\cdot) = \varepsilon_{x+1}$ then $\mathbb{P}$ is the *Poisson semigroup* and $E_{\mathbb{P}}$ is the set of all decreasing positive numerical functions on $\mathbb{Z}$.

In the following let $\mathbb{P}$ be a sub-Markov semigroup on X.

3.2. Lemma. For every $u \in S_{\mathbb{P}}$, the mapping $t \longmapsto P_t u$ is decreasing. For every $u \in E_{\mathbb{P}}$ the mapping $t \longmapsto P_t u$ is decreasing and right continuous.

Proof. Let $u \in S_{\mathbb{P}}$. Then for all $s,t > 0$

$$P_{t+s}u = P_t P_s u \leq P_t u .$$

If even $u \in E_{\mathbb{P}}$ then $P_s u$ is increasing to u as s tends to zero, hence for every $t > 0$

$$\lim_{s\to 0} P_{s+t}u = \lim_{s\to 0} P_t P_s u = P_t u .$$ ┘

3.3. PROPOSITION. $S_{\mathbb{P}}$ and $E_{\mathbb{P}}$ are σ-stable convex cones. Furthermore, $P_t(S_{\mathbb{P}}) \subset S_{\mathbb{P}}$ and $P_t(E_{\mathbb{P}}) \subset E_{\mathbb{P}}$ for every $t > 0$.

Proof. By (3.2),

$$E_{\mathbb{P}} = \{u \in S_{\mathbb{P}} : \lim_{t\to 0} P_t u = u\} .$$

Hence $S_{\mathbb{P}}$ and $E_{\mathbb{P}}$ are convex cones. Let (u_n) be an increasing sequence in $S_{\mathbb{P}}$ and $u = \sup u_n$. Then for every $t > 0$

$$P_t u = \sup_n P_t u_n \leq \sup_n u_n = u ,$$

i.e. $u \in S_{\mathbb{P}}$. If even $u_n \in E_{\mathbb{P}}$ for every $n \in \mathbb{N}$ then

$$\sup_{t>0} P_t u = \sup_{t>0} \sup_n P_t u_n = \sup_n \sup_{t>0} P_t u_n = \sup_n u_n = u ,$$

i.e. $u \in E_{\mathbb{P}}$.

Furthermore, let $v \in S_{\mathbb{P}}$ and $t > 0$. Then for every $s > 0$ by (3.2)

$$P_s P_t v = P_{s+t} v \leq P_t v ,$$

i.e. $P_t v \in S_{\mathbb{P}}$. If even $v \in E_{\mathbb{P}}$ then

$$\lim_{s\to 0} P_s P_t v = \lim_{s\to 0} P_{t+s} v = P_t v .$$ ┘

Let us now suppose that $\mathbb{P}$ is *measurable*, i.e. such that for every $f \in B^+(X)$ the mapping $(x,t) \mapsto P_t f(x)$ of $X \times \mathbb{R}_+^*$ into $\overline{\mathbb{R}}_+$ is measurable. For every $\lambda > 0$ and $f \in B^+(X)$ we may then define a function $V_\lambda f$ by

$$V_\lambda f(x) = \int_0^\infty e^{-\lambda t} P_t f(x)\, dt .$$

3.4. PROPOSITION. The family $\mathbb{V} = (V_\lambda)_{\lambda>0}$ has the following properties:

1. λV_λ is a sub-Markov kernel for every $\lambda > 0$.
2. For every $\lambda,\mu > 0$,

$$V_\lambda = V_\mu + (\mu-\lambda) V_\lambda V_\mu \qquad (\textit{resolvent equation}).$$

Proof. 1. Let $\lambda > 0$. The mapping $f \mapsto V_\lambda f$ from $B^+(X)$ into $B^+(X)$ is apparently additive, positively homogeneous and increasing. If (f_n) is an increasing sequence in $B^+(X)$ and $f = \sup f_n$ then $(P_t f_n(x))$ is increasing to $P_t f(x)$ for every $x \in X$ and $t \in \mathbb{R}_+^*$. Hence $(V_\lambda f_n)$ is increasing to $V_\lambda f$. Thus V_λ is a kernel. Furthermore

$$\lambda V_\lambda 1 = \lambda \int_0^\infty e^{-\lambda t} P_t 1\, dt \le \lambda \int_0^\infty e^{-\lambda t}\, dt = 1 .$$

2. Let $\lambda, \mu > 0$, $\lambda \neq \mu$, and $f \in B_b^+(X)$. Then

$$V_\lambda V_\mu f = \int_0^\infty e^{-\lambda s} P_s V_\mu f\, ds$$

$$= \int_0^\infty e^{-\lambda s} \Big(\int_0^\infty e^{-\mu t} P_s P_t f\, dt\Big) ds = \int_0^\infty e^{-(\lambda-\mu)s} \Big(\int_0^\infty e^{-\mu(s+t)} P_{s+t} f\, dt\Big)\, ds$$

$$= \int_0^\infty e^{-(\lambda-\mu)s} \Big(\int_s^\infty e^{-\mu t} P_t f\, dt\Big)\, ds = \int_0^\infty \Big(\int_0^t e^{-(\lambda-\mu)s}\, ds\Big)\, e^{-\mu t} P_t f\, dt$$

$$= \frac{1}{\lambda-\mu} \int_0^\infty (1 - e^{-(\lambda-\mu)t})\, e^{-\mu t} P_t f\, dt = \frac{1}{\lambda-\mu} (V_\mu f - V_\lambda f) .$$

Since the family $\mathbb{V}$ will be useful when studying $E_{\mathbb{P}}$ we define quite generally: A family $\mathbb{V} = (V_\lambda)_{\lambda>0}$ of kernels on X is a *sub-Markov resolvent* if $\mathbb{V}$ satisfies (1) and (2) of (3.4).

3.5. REMARKS. Let $\mathbb{V} = (V_\lambda)_{\lambda>0}$ be a sub-Markov resolvent on X.

1. The resolvent equation implies that $V_\lambda V_\mu = V_\mu V_\lambda$ for all $\lambda,\mu > 0$.
2. For every $f \in B^+(X)$, the mapping $\lambda \mapsto V_\lambda f$ is decreasing since for all

$0 < \lambda < \mu$

$$V_\lambda f = V_\mu f + (\mu-\lambda) V_\lambda V_\mu f \geq V_\mu f .$$

Hence we may define a kernel V_0 on X by

$$V_0 f := \sup_{\lambda>0} V_\lambda f = \lim_{\lambda\to 0} V_\lambda f \qquad (f \in \mathcal{B}^+(X)) .$$

V_0 is called the *potential kernel* of $\mathbb{V}$. Let $\lambda > 0$. The resolvent equation implies that

$$V_0 V_\lambda = V_\lambda V_0 \text{ and } V_0 = V_\lambda + \lambda V_\lambda V_0 .$$

Furthermore,

$$I + \lambda V_0 = \sum_{n=0}^{\infty} (\lambda V_\lambda)^n .$$

Indeed, let $f \in \mathcal{B}_b^+(X)$ and $\alpha > 0$. Then

$$\lambda V_\alpha f = \lambda V_{\lambda+\alpha} f + (\lambda V_{\lambda+\alpha})(\lambda V_\alpha f)$$

and hence for every $n \in \mathbb{N}$

$$f + \lambda V_\alpha f = \sum_{i=0}^{n} (\lambda V_{\lambda+\alpha})^i f + (\lambda V_{\lambda+\alpha})^n (\lambda V_\alpha f) .$$

Let g be the limit of the decreasing sequence $((\lambda V_{\lambda+\alpha})^n (\lambda V_\alpha f))$. Then evidently $\lambda V_{\lambda+\alpha} g = g$, hence

$$\lambda V_{\lambda+\alpha} V_\alpha g = V_\alpha (\lambda V_{\lambda+\alpha} g) = V_\alpha g = V_{\lambda+\alpha} g + \lambda V_{\lambda+\alpha} V_\alpha g ,$$

i.e. $V_{\lambda+\alpha} g = 0$ and therefore $g = 0$. Thus

$$f + \lambda V_\alpha f = \sum_{n=0}^{\infty} (\lambda V_{\lambda+\alpha})^n f .$$

Letting α tend to zero the assertion follows.

3. For every $\alpha > 0$, $\mathbb{V}^\alpha := (V_{\alpha+\lambda})_{\lambda>0}$ is a sub-Markov resolvent on X. Since $V_\alpha = V_{\alpha+\lambda} + \lambda V_\alpha V_{\alpha+\lambda}$ and $V_\alpha V_{\alpha+\lambda} 1 \leq \frac{1}{\alpha^2}$ for every $\lambda > 0$, the potential kernel of $\mathbb{V}^\alpha$ is the bounded kernel V_α.

4. Suppose now that $\mathbb{V}$ is the resolvent of a measurable sub-Markov semigroup $\mathbb{P}$. Then for every $f \in \mathcal{B}^+(X)$, $x \in X$,

$$V_0 f(x) = \int_0^\infty P_t f(x)\, dt .$$

V_0 is also called the *potential kernel* of $\mathbb{P}$. Again, let $\alpha > 0$. Then $\mathbb{P}^\alpha := (e^{-\alpha t} P_t)_{t>0}$ is a measurable sub-Markov semigroup and $\mathbb{V}^\alpha$ is the resolvent of $\mathbb{P}^\alpha$. In particular, the potential kernel of $\mathbb{P}^\alpha$ is the *bounded* kernel V_α.

5. The potential kernels corresponding to the semigroups of (3.1) are easily calculated, namely:

1. Brownian semigroup on $\mathbb{R}^n$:

$$V_0 f(x) = c_n \int \frac{f(y)\lambda^n(dy)}{\|x-y\|^{n-2}} = c_n\, N^0 * f(x)\ , \quad c_n = \frac{\Gamma(\frac{n}{2}-1)}{2\pi^{n/2}}\ , \text{ if } n \geq 3$$

and

$$V_0 f(x) = \infty \cdot \lambda(\{f>0\}) \text{ if } n \leq 2\ .$$

2. Translation semigroup on $\mathbb{R}$:

$$V_0 f(x) = \int_{-\infty}^{x} f(t)dt = (f\lambda)(]-\infty,x])\ .$$

3. Pseudo-Poisson semigroup:

$$V_0 f(x) = \sum_{k=0}^{\infty} P^k f(x)\ .$$

In the following let $\mathbb{V} = (V_\lambda)_{\lambda>0}$ be a sub-Markov resolvent on X. A function $s \in B^+(X)$ is called $\mathbb{V}$-*supermedian* if for every $\lambda > 0$,

$$\lambda V_\lambda s \leq s\ .$$

The set of all $\mathbb{V}$-supermedian functions will be denoted by $S_\mathbb{V}$. Since $S_\mathbb{V} = \bigcap_{\lambda>0} S_{\lambda V_\lambda}$ it will turn out that $S_\mathbb{V}$ is a potential cone. We shall establish first that $\lambda \mapsto \lambda V_\lambda f$ is increasing for every $f \in S_\mathbb{V}$.

3.6. LEMMA. Let $0 < \lambda < \mu$ and $t \in B^+(X)$ such that $\mu V_\mu t \leq t$. Then $\lambda V_\lambda t \leq \mu V_\mu t$.

Proof. Suppose first that t is bounded by a natural number n. Then $V_\lambda t \leq V_\lambda n \leq \frac{n}{\lambda} < \infty$. By the resolvent equation,

$$V_\lambda t = V_\mu t + (\mu-\lambda)\, V_\lambda\, V_\mu t \leq \frac{t}{\mu} + (\mu-\lambda)\, V_\lambda(\frac{t}{\mu})\ .$$

Hence $\mu V_\lambda t \leq t + (\mu-\lambda)\, V_\lambda t$, i.e.

$$\lambda V_\lambda t \leq t\ .$$

Using the above equality we obtain

$$\lambda V_\lambda t = \lambda V_\mu t + \lambda(\mu-\lambda)\, V_\mu\, V_\lambda t \leq \lambda V_\mu t + (\mu-\lambda)\, V_\mu t = \mu V_\mu t\ .$$

The general statement now follows easily since we may apply the preceding considerations to $t_n := \inf(t,n)$, $n \in \mathbb{N}$, and then let n tend to infinity. ┘

<u>3.7. PROPOSITION.</u> $S_{\mathbb{V}}$ is a potential cone having the following properties:

1. $S_{\mathbb{V}}$ is σ-stable, $V_0(\mathcal{B}^+(X)) \subset S_{\mathbb{V}}$.
2. $\inf s_n \in S_{\mathbb{V}}$ for every sequence (s_n) in $S_{\mathbb{V}}$.

Proof. It is easily verfied that $S_{\mathbb{V}}$ is a σ-stable convex cone and $\inf s_n \in S_{\mathbb{V}}$ for every sequence (s_n) in $S_{\mathbb{V}}$. Furthermore, $V_0(\mathcal{B}^+(X)) \subset S_{\mathbb{V}}$ since for every $f \in \mathcal{B}^+(X)$ and every $\lambda > 0$,

$$\lambda V_\lambda V_0 f \leq V_\lambda f + \lambda V_\lambda V_0 f = V_0 f .$$

Now let $u, v \in S_{\mathbb{V}}$ and $f \in \mathcal{B}^+(X)$ such that $u = v + f$. For every $n \in \mathbb{N}$ let

$$w_n = \inf \{w \in S_{nV_n} : w \geq f\} .$$

By (2.1) and (2.2), $w_n \in S_{nV_n}$ and there exists $v_n \in S_{nV_n}$ such that

$$u = v_n + w_n .$$

By (3.6), the sequence (S_{nV_n}) is decreasing and

$$S_{\mathbb{V}} = \bigcap_{\lambda>0} S_{\lambda V_\lambda} = \bigcap_{n=1}^{\infty} S_{nV_n} .$$

Hence the sequence (w_n) is increasing and $w' := \sup w_n = \sup_{n\geq m} w_n \in S_{mV_m}$ for every $m \in \mathbb{N}$, i.e. $w' \in S_{\mathbb{V}}$. Therefore on one hand $w' \geq R_f$ since $w' \geq f$. If on the other hand $w \in S_{\mathbb{V}}$ such that $w \geq f$ then $w \geq w_n$ for every $n \in \mathbb{N}$ and hence $w \geq w'$, i.e. $R_f \geq w'$. Thus $R_f = w' \in S_{\mathbb{V}}$.

Since (w_n) is increasing, the sequence (v_n) is decreasing on $\{u < \infty\}$. Hence the sequence $(v_n + \frac{1}{n} u)$ is decreasing on X. For every $n \in \mathbb{N}$, $v_n + \frac{1}{n} u \in S_{nV_n}$. Therefore $v' := \inf (v_n + \frac{1}{n} u) \in S_{\mathbb{V}}$ and

$$u = \inf (v_n + \frac{1}{n} u) + \sup w_n = v' + R_f .$$ ┘

Let $s \in S_{\mathbb{V}}$. By (3.6), the mapping $\lambda \mapsto \lambda V_\lambda s$ is increasing, hence

$$\tilde{s} := \sup_{\lambda>0} \lambda V_\lambda s = \lim_{\lambda\to\infty} \lambda V_\lambda s \in \mathcal{B}^+(X) .$$

s is called $\mathbb{V}$-*excessive* if $\tilde{s} = s$. Let $E_{\mathbb{V}}$ denote the set of all $\mathbb{V}$-excessive functions, i.e.

$$E_{\mathbb{W}} = \{s \in \mathcal{B}^+(X) : \sup_{\lambda>0} \lambda V_\lambda s = s\} .$$

3.8. PROPOSITION. 1. For every $s \in S_{\mathbb{W}}$, $\tilde{s}$ is the greatest minorant of s which is contained in $E_{\mathbb{W}}$. The mapping $s \mapsto \tilde{s}$ of $S_{\mathbb{W}}$ on $E_{\mathbb{W}}$ is additive, positively homogeneous and increasing. Furthermore, $\widetilde{\sup s_n} = \sup \tilde{s}_n$ for every increasing sequence (s_n) in $S_{\mathbb{W}}$.

2. $E_{\mathbb{W}}$ is a σ-stable potential cone, $V_0(\mathcal{B}^+(X)) \subset E_{\mathbb{W}}$.

Proof. Let $s \in S_{\mathbb{W}}$. Then $\tilde{s} \leq s$ and for every $\lambda > 0$,

$$\lambda V_\lambda \tilde{s} \leq \lambda V_\lambda s \leq \tilde{s} ,$$

i.e. $\tilde{s} \in S_{\mathbb{W}}$. Hence $s \mapsto \tilde{s}$ is a mapping of $S_{\mathbb{W}}$ into $S_{\mathbb{W}}$. It follows immediately from the definition that for all $s, t \in S_{\mathbb{W}}$ and $\alpha \geq 0$

$$\widetilde{s+t} = \tilde{s} + \tilde{t} , \quad \widetilde{\alpha s} = \alpha \tilde{s}$$

and $\tilde{s} \leq \tilde{t}$ if $s \leq t$. If (s_n) is an increasing sequence in $S_{\mathbb{W}}$ then

$$\sup_n \tilde{s}_n = \sup_n \sup_{\lambda>0} \lambda V_\lambda s_n = \sup_{\lambda>0} \sup_n \lambda V_\lambda s_n$$

$$= \sup_{\lambda>0} \lambda V_\lambda (\sup_n s_n) = \widetilde{\sup s_n} .$$

Let $s \in S_{\mathbb{W}}$ and $t \in E_{\mathbb{W}}$ such that $t \leq s$. Then

$$t = \sup_{\lambda>0} \lambda V_\lambda t \leq \sup_{\lambda>0} \lambda V_\lambda s = \tilde{s} .$$

Hence it remains to show that $\tilde{s} \in E_{\mathbb{W}}$. Let $\lambda > 0$, $n \in \mathbb{N}$ and $s_n = \inf(s,n)$. Then $s_n \in S_{\mathbb{W}}$ and for every $\mu > \lambda$,

$$V_\lambda s_n = V_\mu s_n + (\mu-\lambda) V_\lambda V_\mu s_n$$

where $V_\mu s_n \leq \frac{s_n}{\mu}$ and hence $\lambda V_\lambda V_\mu s_n \leq \lambda V_\lambda (\frac{s_n}{\mu}) \leq \frac{s_n}{\mu}$. Letting μ tend to infinity we obtain that $V_\lambda s_n = V_\lambda \tilde{s}_n$ and hence

$$V_\lambda s = \sup_n V_\lambda s_n = \sup_n V_\lambda \tilde{s}_n = V_\lambda \tilde{s} .$$

Thus

$$\tilde{s} = \sup_{\lambda>0} \lambda V_\lambda s = \sup_{\lambda>0} \lambda V_\lambda \tilde{s} ,$$

i.e. $\tilde{s} \in E_{\mathbb{V}}$.

Finally, if $f \in \mathcal{B}_b^+(X)$ then $V_0 f \in E_{\mathbb{V}}$ since $\lambda V_\lambda V_0 f + V_\lambda f = V_0 f$ and $V_\lambda f \le \frac{1}{\lambda} \|f\|$. Hence for every $f \in \mathcal{B}^+(X)$

$$V_0 f = \sup_n V_0(\inf(f,n)) \in E_{\mathbb{V}} .$$

$\lrcorner$

<u>3.9. REMARK.</u> The proof shows that $V_0 1_{\{\tilde{s}<s\}} = \sup_{\lambda>0} V_\lambda 1_{\{\tilde{s}<s\}} = 0$ for every $s \in S_{\mathbb{V}}$.

<u>3.10. LEMMA.</u> For every $\lambda > 0$, $V_\lambda(S_{\mathbb{V}}) \subset E_{\mathbb{V}}$.

Proof. Let $\lambda > 0$ and $s \in S_{\mathbb{V}}$. Then

$$\sup_{\mu>0} \mu V_\mu V_\lambda s = \sup_{\mu>0} V_\lambda(\mu V_\mu s) = V_\lambda \tilde{s} = V_\lambda s .$$

$\lrcorner$

The following characterization of $\mathbb{V}$-excessive functions is very useful.

<u>3.11. PROPOSITION.</u> Let $\mathbb{V}$ be a sub-Markov resolvent on X such that the potential kernel V_0 is proper. Then

$$E_{\mathbb{V}} = \{s \mid \exists\, f_n \in \mathcal{B}_b^+(X) \text{ with } V_0 f_n \uparrow s\} .$$

Proof. Let $g \in \mathcal{B}^+(X)$ such that $g > 0$ and $V_0 g < \infty$. Then $V_\lambda f = 0$ on $\{V_0 g = 0\}$ for every $\lambda > 0$ and $f \in \mathcal{B}^+(X)$ since

$$0 \le V_\lambda f \le V_0 f = \sup V_0(\inf(mg,f)) \le \sup m\, V_0 g .$$

Consider now $s \in E_{\mathbb{V}}$. Then $s = \sup \lambda V_\lambda s$ and hence

$$s = 0 \text{ on } \{V_0 g = 0\} .$$

Defining

$$s_n = \inf(s,n,nV_0 g)$$

we thus obtain an increasing sequence (s_n) in $S_{\mathbb{V}}$ such that $\sup s_n = s$. In particular, for every $n \in \mathbb{N}$,

$$nV_n s_n \le (n+1)V_{n+1} s_n \le (n+1)V_{n+1} s_{n+1} \le s_{n+1} \le s$$

and

$$\sup_n nV_n s_n \ge \sup_m \sup_n nV_n s_m = \sup_n nV_n s = s .$$

Therefore $nV_n s_n \uparrow s$. Let $n \in \mathbb{N}$ and

$$f_n = n(s_n - nV_n s_n) .$$

We claim that

$$V_o f_n = nV_n s_n .$$

This is immediate from the resolvent equation if V_o is bounded. Since we are only assuming that V_o is proper, we have to be a bit more careful. Let $\lambda > 0$. Then by the resolvent equation

$$V_\lambda f_n = n(V_\lambda s_n - (n-\lambda)V_\lambda V_n s_n) - n\lambda V_\lambda V_n s_n = nV_n s_n - n\lambda V_\lambda V_n s_n$$

where

$$n\lambda V_\lambda V_n s_n \leq \lambda V_\lambda s_n \leq n\lambda V_\lambda V_o g = n(V_o g - V_\lambda g) .$$

Since $\lim_{\lambda\to 0} V_\lambda g = V_o g$ we conclude that $V_o f_n = \lim_{\lambda\to 0} V_\lambda f_n = nV_n s_n$. Thus

$$V_o f_n = nV_n s_n \uparrow s .$$

If conversely $s \in \mathcal{B}^+(X)$ such that $V_o f_n \uparrow s$ for some sequence (f_n) in $\mathcal{B}^+(X)$ then $s \in E_{\mathbb{V}}$ by (3.8). ┘

<u>3.12. PROPOSITION.</u> Let $\mathbb{V} = (V_\lambda)_{\lambda>0}$ be a sub-Markov resolvent on X . Then

$$E_{\mathbb{V}} = \bigcap_{\alpha>0} E_{\mathbb{V}^\alpha} .$$

Proof. Let $s \in E_{\mathbb{V}}$ and $\alpha > 0$. Then for every $\lambda > 0$

$$\lambda V_{\alpha+\lambda}\, s \leq (\alpha+\lambda)\, V_{\alpha+\lambda}\, s \leq s$$

and $\lim_{\lambda\to\infty} \lambda V_{\alpha+\lambda}\, s = \lim_{\lambda\to\infty} (\alpha+\lambda)\, V_{\alpha+\lambda}\, s = s$ since $\lim_{\lambda\to\infty} \frac{\lambda}{\alpha+\lambda} = 1$. Hence $s \in E_{\mathbb{V}^\alpha}$.

Conversely, let $t \in E_{\mathbb{V}^\alpha}$ for every $\alpha > 0$. Then for every $\lambda > 0$ and $0 < \alpha < \lambda$,

$$(\lambda-\alpha)\, V_\lambda t = (\lambda-\alpha)\, V_{\alpha+(\lambda-\alpha)}\, t \leq t ,$$

hence $\lambda V_\lambda t \leq t$. Furthermore,

$$\lim_{\lambda\to\infty} \lambda V_\lambda t = \lim_{\lambda\to\infty} (\lambda-\alpha)\, V_{\alpha+(\lambda-\alpha)} t = t .$$

Therefore $t \in E_{\mathbb{V}}$. ┘

<u>3.13. COROLLARY.</u> Let $\mathbb{P}$ be a measurable sub-Markov semigroup on X and let $\mathbb{V}$ be the resolvent of $\mathbb{P}$. Then $S_{\mathbb{P}} \subset S_{\mathbb{V}}$ and $E_{\mathbb{P}} = E_{\mathbb{V}}$.

Proof. Let $s \in S_{\mathbb{P}}$. Then for every $\lambda > 0$,

$$\lambda V_\lambda s = \lambda \int_0^\infty e^{-\lambda t} P_t s \, dt \leq \lambda \int_0^\infty e^{-\lambda t} s \, dt = s ,$$

hence $s \in S_{\mathbb{V}}$. If even $s \in E_{\mathbb{P}}$ then

$$\lim_{\lambda\to\infty} \lambda V_\lambda s = \lim_{\lambda\to\infty} \int_0^\infty e^{-t} P_{t/\lambda} s \, dt = \int_0^\infty e^{-t} s \, dt = s ,$$

hence $s \in E_{\mathbb{V}}$. Thus $S_{\mathbb{P}} \subset S_{\mathbb{V}}$ and $E_{\mathbb{P}} \subset E_{\mathbb{V}}$.

Let $f \in \mathcal{B}^+(X)$. Then for every $s > 0$,

$$P_s V_0 f = \int_0^\infty P_s P_t f \, dt = \int_0^\infty P_{s+t} f \, dt = \int_s^\infty P_t f \, dt$$

and hence

$$\sup_{s>0} P_s V_0 f = V_0 f ,$$

i.e. $V_0 f \in E_{\mathbb{P}}$. If V_0 is bounded the inclusion $E_{\mathbb{V}} \subset E_{\mathbb{P}}$ then follows by (3.11) and (3.3). In the general case we hence conclude by (3.5.4) that $E_{\mathbb{V}^\alpha} \subset E_{\mathbb{P}^\alpha}$ for every $\alpha > 0$ and therefore

$$E_{\mathbb{V}} = \bigcap_{\alpha>0} E_{\mathbb{V}^\alpha} \subset \bigcap_{\alpha>0} E_{\mathbb{P}^\alpha} = E_{\mathbb{P}} .$$ ┘

3.14. EXAMPLE. Translation semigroup. If $\mathbb{P}$ is the translation semigroup $\mathbb{T}$ on $\mathbb{R}$ (see 3.1.2), then $S_{\mathbb{P}} \neq S_{\mathbb{V}}$ since e.g. $1_{\{0\}} \in S_{\mathbb{V}} \setminus S_{\mathbb{P}}$.

3.15. LEMMA. Let $g : \mathbb{R}_+^* \times X \to \mathbb{R}$ such that for every $x \in X$ the function $t \mapsto g(t,x)$ is right continuous and integrable on $\mathbb{R}_+^*$. Then the following statements are equivalent:

(1) $x \longmapsto g(t,x)$ is Borel measurable for every $t \geq 0$.

(2) g is $\mathcal{B}(\mathbb{R}_+^*) \otimes \mathcal{B}(X)$-measurable.

(3) $x \longmapsto \int_0^\infty e^{-\alpha t} g(t,x) \, dt$ is Borel measurable for every $\alpha > 0$.

Proof. (1) ⇒ (2): By the right continuity of $t \mapsto g(t,x)$ we have for every $t > 0$ and $x \in X$

$$g(t,x) = \lim_{n\to\infty} \sum_{i=1}^{\infty} 1_{[\frac{i-1}{n}, \frac{i}{n}[}(t) g(\tfrac{i}{n}, x) .$$

(2) ⇒ (3): Obvious.

(3) ⇒ (1): The Stone-Weierstrass theorem yields that $x \mapsto \int \varphi(t) g(t,x) \, dt$ is Borel measurable for every $\varphi \in \mathcal{K}(\mathbb{R}_+^*)$. Let $s > 0$ and let (φ_n) be a sequence in $\mathcal{K}(\mathbb{R}_+^*)$ such that $\int \varphi_n(t) \, dt = 1$ and $S(\varphi_n) \subset [s, s + \frac{1}{n}]$. Using the

right continuity of $t \mapsto g(t,x)$ it is easy to check that for every $x \in X$

$$\lim_{n\to\infty} \int \varphi_n(t)g(t,x)\,dt = g(s,x)\ .$$

This shows that $x \mapsto g(s,x)$ is Borel measurable. ┘

3.16. PROPOSITION. Let $\mathbb{V}$ be a sub-Markov resolvent on X, $\mathbb{P}$ a sub-Markov semigroup on X, and let P be a min-stable function cone on X.

1. If $E_{\mathbb{V}} = S(P)$ then, for every $f \in K(X)$, $\lambda V_\lambda f$ tends locally uniformly to f as λ tends to infinity.

2. If $E_{\mathbb{P}} = S(P)$ then, for every $f \in K(X)$, $P_t f$ tends locally uniformly to f as t tends to zero. In particular, $\mathbb{P}$ is measurable.

Proof. 1. Let $p \in P$ and $\lambda > 0$. Then $V_\lambda p \in E_{\mathbb{V}}$, hence $V_\lambda p$ is l.s.c. As λ tends to infinity $\lambda V_\lambda p$ tends to p. Hence the convergence is locally uniform. Since by (I.1.2) $(P-P) \cap K(X)$ is dense in $K(X)$ with respect to uniform convergence and $\lambda V_\lambda 1 \le 1$ for every $\lambda > 0$ the statement follows.

2. The first part follows analogously since $P_t p \in E_{\mathbb{P}}$ and $P_t 1 \le 1$ for every $t > 0$. Furthermore if $f \in K(X)$ and $t > 0$ then

$$\lim_{s\downarrow 0} P_{t+s} f = \lim_{s\downarrow 0} P_t P_s f = P_t f\ ,$$

i.e. $t \mapsto P_t f(x)$ is right continuous for every $x \in X$. Applying (3.15) to the function $g : (t,x) \mapsto e^{-t} P_t f(x)$ we obtain that $(t,x) \mapsto P_t f(x)$ is $\mathcal{B}(\mathbb{R}_+^*) \otimes \mathcal{B}(X)$-measurable. Thus $\mathbb{P}$ is measurable. ┘

3.17. PROPOSITION. Let $\mathbb{P}$ be a sub-Markov semigroup on X. Then the following statements are equivalent:

(1) $\lim_{t\to 0} P_t f = f$ for every $f \in K(X)$.

(2) $\lim_{t\to 0} P_t f = f$ for every $f \in C_b(X)$.

(3) $1 \in E_{\mathbb{P}}$ and $\lim_{t\to 0} P_t f = 0$ on $\complement S(f)$ for every $f \in C_b(X)$.

Proof. (1) ⇒ (2): Let $f \in C_b(X)$ such that $|f| \le 1$ and $x \in X$. There exists $g \in K(X)$ such that $g \le 1$ and $g(x) = 1$. Then $f = fg + f(1-g)$ where $fg \in K(X)$, $|f(1-g)| \le 1 - g$ and $\lim_{t\to 0} P_t(1-g)(x) \le 1 - \lim_{t\to 0} P_t g(x) = 0$, hence

$\lim_{t\to 0} P_t f(x) = f(x)$.

(2) ⇒ (3): Trivial.

(3) ⇒ (1): Let $f \in K(X)$, $x \in X$ and $\varepsilon > 0$. Then there exist $f_1, f_2 \in C_b(X)$ such that $f = f(x) + f_1 + f_2$, $|f_1| \le \varepsilon$, $f_2 = 0$ in a neighborhood of x . Hence for all $t > 0$

$$|P_t f - f(x)| \le f(x)(1 - P_t 1) + P_t|f_1| + P_t|f_2|$$

where $P_t|f_1| \le \varepsilon$, $\lim_{t\to 0} P_t 1 = 1$ and $\lim_{t\to 0} P_t|f_2|(x) = 0$. Hence $\lim_{t\to 0} P_t f(x) = f(x)$. ┘

3.18. EXERCISES. 1. Let $(V_\lambda)_{\lambda>0}$ be a sub-Markov resolvent on X and $f \in B_b(X)$. Then $V_\lambda f = \int_\lambda^\infty V_t^2 f \, dt$ for every $\lambda > 0$.

2. Let $\mathbb{P} = (P_t)_{t>0}$ be a family of sub-Markov kernels on $\mathbb{R}^n$, $n \ge 1$. Then $\mathbb{P}$ is translation invariant, i.e. $P_t(x+a, B+a) = P_t(x,B)$ for all $t > 0$, $B \in B(\mathbb{R}^n)$ and $x,a \in \mathbb{R}^n$, if and only if there exists a family $(\mu_t)_{t>0}$ of measures on $\mathbb{R}^n$ such that $\mu_t(\mathbb{R}^n) \le 1$ and $P_t f = \mu_t * f$ for all $t > 0$ and $f \in B^+(\mathbb{R}^n)$. Moreover, $\mathbb{P}$ is a semigroup of kernels if and only if $(\mu_t)_{t>0}$ is a convolution semigroup of measures, i.e. $\mu_s * \mu_t = \mu_{s+t}$ for all $s,t > 0$. If $\mathbb{P}$ is a semigroup then $\mathbb{P}$ is measurable if and only if the function $t \mapsto \mu_t(\varphi)$ is measurable for every $\varphi \in K(\mathbb{R}^n)$, and in this case the potential kernel V of $\mathbb{P}$ is given by $Vf = \kappa * f$ where $\kappa = \int_0^\infty \mu_t \, dt$.

4. Balayage Spaces

In the sequel we shall restrict our considerations to semigroups and resolvents whose excessive functions are l.s.c. as in the classical case of Brownian semigroup. The natural desire to deal only with continuous functions leads to the introduction of another topology on X , the fine topology. This topology suggested in classical potential theory by H. CARTAN in 1944 has many pathologic properties. But every point has a fundamental system of neighborhoods which are compact in the initial topology. This implies at once that X is a Baire space with respect to the fine topology which has been considered a mere curiosity at that time. However, the consequence that two finely continuous functions are identical provided they coincide on the complement of a finely meager set is extremely useful.

Let $\mathcal{W}$ be a convex cone of positive l.s.c. numerical functions on X. The coarsest topology on X which is finer than the initial topology and for which all functions of $\mathcal{W}$ are continuous will be called the $(\mathcal{W}\text{-})$ *fine topology*. Topological notions with respect to the fine topology are distinguished by the term "fine(ly)" or affix "f" from those pertaining to the initial topology on X.

4.1. LEMMA. Let $x \in X$ and let V be a fine neighborhood of x. Then there exist a compact neighborhood K of x, a function $u \in \mathcal{W}$, and $\alpha \in \mathbb{R}_+^*$ such that $u(x) < \alpha$ and $\{z \in K : u(z) \leq \alpha\} \subset V$. In particular, x has a fundamental system of fine neighborhoods which are compact in the initial topology.

Proof. By definition of the fine topology there exist an open set U of X, functions $u_i \in \mathcal{W}$, and real numbers α_i $(1 \leq i \leq n)$, such that

$$x \in U \cap \bigcap_{i=1}^{n} \{u_i < \alpha_i\} \subset V .$$

Let $\varepsilon > 0$ such that

$$u_i(x) + n\,\varepsilon < \alpha_i \quad (1 \leq i \leq n) .$$

Since every function of $\mathcal{W}$ is l.s.c., there exists a compact neighborhood K of x in U such that $u_i(x) - \varepsilon < u_i(z)$ for all $z \in K$ and $1 \leq i \leq n$. Define

$$u := \sum_{i=1}^{n} u_i \quad \text{and} \quad \alpha := u(x) + \varepsilon .$$

Then $u \in \mathcal{W}$ and $u(x) < \alpha < +\infty$. Let $z \in K$ such that $u(z) \leq \alpha$. Then for every $1 \leq j \leq n$

$$u_j(z) = u(z) - \sum_{\substack{i=1\\ i\neq j}}^{n} u_i(z) \leq \alpha - \sum_{\substack{i=1\\ i\neq j}}^{n} (u_i(x) - \varepsilon) = u_j(x) + n\varepsilon < \alpha_j ,$$

thus

$$W := \{z \in K : u(z) \leq \alpha\} \subset V .$$

Evidently, W is a fine neighborhood of x which is compact in the initial topology. ┘

4.2. COROLLARY. X endowed with the fine topology is a Baire space. More generally, if (U_n) is a sequence of finely open subsets of X then $A = \bigcap_{n=1}^{\infty} U_n$ is a Baire space with respect to the fine topology.

Proof. Let (G_n) be a sequence of finely open subsets of X such that A is contained in the fine closure of $G_n \cap A$. Furthermore, let G be a finely open subset of X such that $G \cap A \neq \emptyset$. Using (4.1) we may easily construct by induction a sequence (K_n) of non-empty compact subsets of X such that K_{n+1} is contained in the fine interior of K_n and $K_n \subset G \cap U_n \cap G_n$ for every $n \in \mathbb{N}$. The assertion then follows from

$$\emptyset \neq \bigcap_{n=1}^{\infty} K_n \subset G \cap A \cap \bigcap_{n=1}^{\infty} G_n .$$ ┘

The following definition of a balayage space $(X,\mathcal{W})$ combining elementary properties of excessive functions with the concepts of function cone and fine topology leads to an intrinsic characterization of cones of excessive functions of "nice" sub-Markov semigroups. Moreover, chapter III will show that $(X,\mathcal{W})$ is a balayage space if and only if $\mathcal{W}$ is the set of positive hyperharmonic functions of a family of harmonic kernels on X. These equivalences would certainly be sufficient to justify the introduction of the concept of a balayage space. However, as the name already indicates its main importance lies in the fact that it contains precisely the tools necessary for the development of balayage theory (see chapter VI).

$(X,\mathcal{W})$ is called a *balayage space* if the following axioms are satisfied:

(B_1) $\mathcal{W}$ is σ-stable.

(B_2) $\widehat{\inf \mathcal{V}}^f \in \mathcal{W}$ for every subset $\mathcal{V}$ of $\mathcal{W}$.

(B_3) If $u, v', v'' \in \mathcal{W}$ such that $u \leq v' + v''$ there exist $u', u'' \in \mathcal{W}$ such that $u = u' + u''$, $u' \leq v'$, $u'' \leq v''$.

(B_4) There exists a function cone $\mathcal{P} \subset \mathcal{C}^+(X)$ such that $\mathcal{W} = \mathcal{S}(\mathcal{P})$.

Let us note that (B_1) implies by (I.1.7) that $\sup \mathcal{V} \in \mathcal{W}$ for every increasingly filtered subset $\mathcal{V}$ of $\mathcal{W}$.

4.3. PROPOSITION. Suppose that $(X,\mathcal{W})$ satisfies (B_2) and let $\mathcal{V} \subset \mathcal{W}$. Then $\widehat{\inf \mathcal{V}} = \widehat{\inf \mathcal{V}}^f \in \mathcal{W}$ and $\{\widehat{\inf \mathcal{V}} < \inf \mathcal{V}\}$ is finely meager.

Proof. Let $u = \inf \mathcal{V}$. Since obviously $\hat{u} \leq \hat{u}^f \leq u$ and $\hat{u}^f \in \mathcal{W}$ is l.s.c., we have $\hat{u} = \hat{u}^f$. For every $n \in \mathbb{N}$ let

$$E_n := \{\hat{u} \leq \inf (n, u - \tfrac{1}{n})\} .$$

Then every E_n is finely closed and

$$\bigcup_{n=1}^{\infty} E_n = \{\hat{u} < u\} .$$

Hence it suffices to show that the fine interior of every E_n is empty.

Let $n \in \mathbb{N}$ and $x_0 \in E_n$. Then $\hat{u}(x_0) \leq n < \infty$. Furthermore, there exists a neighborhood V of x_0 such that

$$\hat{u} > \hat{u}(x_0) - \frac{1}{2n} \quad \text{on } V .$$

Hence for every $x \in V \cap E_n$

$$u(x) - \frac{1}{n} \geq \hat{u}(x) > \hat{u}(x_0) - \frac{1}{2n} ,$$

i.e.

$$u(x) > \hat{u}(x_0) + \frac{1}{2n} .$$

This shows that x_0 is not a finely interior point of E_n since otherwise $V \cap E_n$ would be a fine neighborhood of x_0 and we would obtain the contradiction

$$\hat{u}(x_0) > \inf u(V \cap E_n) \geq \hat{u}(x_0) + \frac{1}{2n} .$$ ┘

For every numerical function f on X let

$$R_f = \inf\{u \in \mathcal{W} : u \geq f\} .$$

If (B_2) holds then $\hat{R}_f \in \mathcal{W}$. If in addition f is finely l.s.c. then $f \leq \hat{R}_f$, hence $R_f \leq \hat{R}_f$ and therefore $R_f = \hat{R}_f \in \mathcal{W}$.

4.4. PROPOSITION. Suppose that $(X,\mathcal{W})$ satisfies (B_2). Then (B_3) holds if and only if $\mathcal{W}$ is a potential cone.

Proof. Assume first that (B_3) holds. Since f is finely l.s.c., we have $R_f \in \mathcal{W}$. Since $u \leq v + R_f$, there exist $u', u'' \in \mathcal{W}$ such that

$$u = u' + u'', \ u' \leq v \ \text{ and } \ u'' \leq R_f \ ,$$

thus $u \leq v + u''$ and therefore $f \leq u''$ which implies $R_f \leq u''$. Hence $R_f + u' = u'' + u' = u$. So $\mathcal{W}$ is a potential cone.

Suppose now that $\mathcal{W}$ is a potential cone. Let $u, v', v'' \in \mathcal{W}$ such that $u \leq v' + v''$ and define

$$f := \begin{cases} (u-v')^+ & \text{on } \{v' < +\infty\} \\ 0 & \text{on } \{v' = +\infty\} \end{cases} .$$

Then $R_f \in \mathcal{W}$ and $R_f + w = u$ where $w \in \mathcal{W}$ which implies

$$u = \min(v' + R_f \, , w + R_f) = \min(v', w) + R_f \ .$$

Since $u \leq v' + v''$, we have $f \leq v''$ and therefore $R_f \leq v''$. Thus $u' := \min(v', w)$ and $u'' := R_f$ satisfy the desired relations $u = u' + u''$, $u' \leq v'$, $u'' \leq v''$. ┘

<u>4.5. PROPOSITION.</u> Suppose that $(X,\mathcal{W})$ satisfies (B_1) and (B_2) . Furthermore let $\mathcal{P}$ be a function cone on X such that $\mathcal{W} = S(\mathcal{P})$ and $p - R_{p-q} \in \mathcal{W}$ for all $p,q \in \mathcal{P}$. Then $(X,\mathcal{W})$ is a balayage space.

Proof. By (4.4), it suffices to show that $\mathcal{W}$ is a potential cone.

First let $p \in \mathcal{P}$ and $v \in \mathcal{W}$. We claim that $p - R_{p-v} \in \mathcal{W}$. Indeed, there exists an increasing sequence (q_n) in $\mathcal{P}$ such that $\sup q_n = v$. For every $n \in \mathbb{N}$,

$$w_n := p - R_{p-q_n} \in \mathcal{W} \ .$$

Evidently, the sequence (w_n) is increasing and hence $\sup w_n \in \mathcal{W}$,

$$p = \sup w_n + \inf R_{p-q_n} \ .$$

Obviously, $\inf R_{p-q_n} \geq R_{p-v}$. Conversely, let $w \in \mathcal{W}$ such that $w > p-v$, i.e. $w + v > p$. Let $x \in X$ and $\varepsilon > 0$. Since $\mathcal{P}$ is a function cone, there exist $p' \in \mathcal{P}$ and a compact subset K of X such that $p'(x) < \varepsilon$ and $p' \geq p$ on $\complement K$. There exists $n \in \mathbb{N}$ such that $w + q_n > p$ on K . Then $w + q_n + p' \geq p$, $w + p' \geq p - q_n$ and hence $w + p' \geq R_{p-q_n}$. In particular, $R_{p-q_n}(x) \leq w(x) + \varepsilon$. Therefore $\inf R_{p-q_n} \leq R_{p-v}$. Thus

$$p - R_{p-v} = p - \inf R_{p-q_n} = \sup w_n \in \mathcal{W} \ .$$

Consider now $u, v \in \mathcal{W}$ and let (p_n) , (q_n) be increasing sequences in $\mathcal{P}$

such that

$$\sup p_n = u \quad , \quad \sup q_n = v \ .$$

By the preceding considerations there exist $u_n \in \mathcal{W}$ such that

$$u_n + R_{p_n - v} = p_n \ .$$

Since the sequence (R_{p_n-v}) is increasing, $w' := \lim_{n\to\infty} R_{p_n-v} \in \mathcal{W}$. Let

$$f = \begin{cases} (u-v)^+ & \text{on } \{v<\infty\} \\ 0 & \text{on } \{v=\infty\} \end{cases} .$$

Since $p_n - v \leq f$, we have $w' \leq R_f$. On the other hand $w' \geq \lim_{n\to\infty} (p_n - v)^+ = f$ and hence $w' \geq R_f$. Thus

$$R_f = \lim_{n\to\infty} R_{p_n-v} \ .$$

For every $n \in \mathbb{N}$ let

$$w_n = \inf_{m\geq n} u_m \ .$$

Then (w_n) is an increasing sequence such that

$$\sup w_n = \lim_{n\to\infty} u_n = u - R_f \quad \text{on} \quad \{R_f < \infty\} \ .$$

Evidently, $R_f \leq u$ and hence $u = \infty$ on $\{R_f = \infty\}$. Therefore

$$\sup w_n + R_f = u \ .$$

Furthermore, $(\hat{w}_n)$ is an increasing sequence in $\mathcal{W}$, hence $w := \sup \hat{w}_n \in \mathcal{W}$. The set $E := \bigcup_{n=1}^{\infty} \{\hat{w}_n < w_n\}$ is finely meager by (4.3) and

$$w + R_f = \sup w_n + R_f = u \quad \text{on} \quad \complement E \ .$$

Hence $w + R_f = u$. Thus $\mathcal{W}$ is a potential cone, and $(X,\mathcal{W})$ is a balayage space.⌋

4.6. PROPOSITION. Suppose that $(X,\mathcal{W})$ satisfies (B_1), (B_2), and (B_3) . Then the following statements are equivalent:

(1) $(X,\mathcal{W})$ is a balayage space.

(2) The set $\mathcal{P}$ of all $p \in \mathcal{W} \cap C(X)$ such that $\frac{p}{v} \in C_o(X)$ for some strictly positive $v \in \mathcal{W} \cap C(X)$ is a min-stable function cone satisfying $S(\mathcal{P}) = \mathcal{W}$.

(3) $\mathcal{W}$ is linearly separating, there exist strictly positive functions $u_o, v_o \in \mathcal{W} \cap C(X)$ such that $\frac{u_o}{v_o} \in C_o(X)$, and $w = \sup\{u \in \mathcal{W} \cap C(X) : u \leq w\}$ for every $w \in \mathcal{W}$.

Proof. Obviously, (2) implies (1) and (1) implies (3). So assume that (3) holds and define $\mathcal{P}$ as in (2). Then $u_0 \in \mathcal{P}$ and it is immediately verified that $\mathcal{P}$ is a min-stable convex cone. If $u \in \mathcal{W} \cap C(X)$ then $u_n = \inf(u, n\,u_0) \in \mathcal{P}$ for every $n \in \mathbb{N}$ and $\sup u_n = u$. Hence for every $w \in \mathcal{W}$,

$$w = \sup\{p \in \mathcal{P} : p \leq w\} .$$

Since $\mathcal{W}$ is linearly separating, we conclude that $\mathcal{P}$ is linearly separating. Moreover, given $w \in \mathcal{W}$, the set $F = \{p \in \mathcal{P} : p \leq w\}$ is increasingly filtered and hence $\sup F = \sup p_n$ for some increasing sequence (p_n) in F, i.e. $w \in S(\mathcal{P})$. Indeed, let $p_1, p_2 \in F$ and define $p = R_{\sup(p_1,p_2)}$. Then $\sup(p_1, p_2) \leq p \leq w$. Moreover,

$$p = R_{p_1+p_2} - \inf(p_1, p_2) ,$$

where the functions $p_1 + p_2$ and $\inf(p_1, p_2)$ are contained in the subset $\mathcal{P}$ of $\mathcal{W}$. By (4.4), $p \in \mathcal{W}$ and there exists a function $w' \in \mathcal{W}$ such that $p + w' = p_1 + p_2$. This shows that p is a continuous minorant of $p_1 + p_2$ and hence $p \in \mathcal{P}$.

In order to prove that (2) holds it remains to show that $\mathcal{P}$ is adapted. So let $p \in \mathcal{P}$ and choose a strictly positive function $v \in \mathcal{W} \cap C(X)$ such that $\frac{p}{v} \in C_0(X)$. Define $q = \sum_{n=1}^{\infty} \inf(p, 2^{-n} v)$. Then $q \in \mathcal{W}$ and $\frac{q}{v}$ is the sum of the uniformly convergent series $\sum_{n=1}^{\infty} \inf(\frac{p}{v}, 2^{-n})$ in $C_0(X)$, hence $\frac{q}{v} \in C_0(X)$. Therefore $q \in \mathcal{P}$. Given $n \in \mathbb{N}$, there exists a compact subset K of X such that $p \leq 2^{-n} v$ on $\complement K$, and we obtain that $q \geq n\,p$ on $\complement K$. So $p \in o(q)$ finishing the proof. ⌋

<u>4.7. THEOREM.</u> Let $\mathbb{V} = (V_\lambda)_{\lambda > 0}$ be a sub-Markov resolvent on X. Then the following statements are equivalent:

(1) $(X, E_{\mathbb{V}})$ is a balayage space.

(2) For every $f \in \mathcal{K}(X)$, $\lim_{\lambda \to \infty} \lambda V_\lambda f = f$. Moreover, $E_{\mathbb{V}}$ is linearly separating, there exist strictly positive functions $u_0, v_0 \in E_{\mathbb{V}} \cap C(X)$ such that $\frac{u_0}{v_0} \in C_0(X)$, and, for every $w \in E_{\mathbb{V}}$, $w = \sup\{v \in E_{\mathbb{V}} \cap C(X) : v \leq w\}$.

Furthermore, if $(X, E_{\mathbb{V}})$ is a balayage space, then $\lim_{\lambda \to \infty} \lambda V_\lambda(x, U) = 1$ for every finely open set $U \in \mathcal{B}(X)$ and $x \in U$. In particular, $1 \in E_{\mathbb{V}}$ and $E_{\mathbb{V}}$ is the set of all finely l.s.c. functions in $S_{\mathbb{V}}$.

Proof. (1) ⇒ (2): (3.16) and (4.6).

(2) ⇒ (1): By the last assumption, $E_{\mathbb{V}}$ is a convex cone of l.s.c. functions. Let $U \in \mathcal{B}(X)$ be an $E_{\mathbb{V}}$-finely open set and $x \in U$. By (4.1), there exists a compact neighborhood K of x, a function $u \in E_{\mathbb{V}}$ and $\alpha \in \mathbb{R}_+$ such that $0 < u(x) < \alpha$ and $K \cap \{u \leq \alpha\} \subset U$. Let $0 \leq \beta < u(x)$. Then there exists a compact neighborhood L of x such that $L \subset K$ and $u > \beta$ on L. For every $\lambda > 0$

$$\begin{aligned} u(x) &\geq \lambda V_\lambda u(x) \geq \alpha\, \lambda V_\lambda(x, K \cap \{u > \alpha\}) + \beta\, \lambda V_\lambda(x, L \cap \{u \leq \alpha\}) \\ &= (\alpha-\beta)\lambda V_\lambda\, (x, K \cap \{u > \alpha\}) + \beta\lambda V_\lambda(x, (K \cap \{u > \alpha\}) \cup (L \cap \{u \leq \alpha\})) \\ &\geq (\alpha - u(x))\, \lambda V_\lambda(x, K \cap \{u > \alpha\}) + \beta\lambda V_\lambda(x, L)\ . \end{aligned}$$

Choosing $f \in K(X)$ such that $0 \leq f \leq 1_L$ and $f(x) = 1$ we see that $\lim_{\lambda\to\infty} \lambda V_\lambda(x,L) = 1$. Hence

$$\lim_{\lambda\to\infty} \lambda V_\lambda(x, K \cap \{u > \alpha\}) = 0$$

and

$$1 = \lim_{\lambda\to\infty} \lambda V_\lambda(x,K) = \lim_{\lambda\to\infty} \lambda V_\lambda(x, K \cap \{u \leq \alpha\}) \leq \liminf_{\lambda\to\infty} \lambda V_\lambda(x,U)\ .$$

Thus $\lim_{\lambda\to\infty} \lambda V_\lambda(x,U) = 1$.

By (4.4) and (4.6), it remains to show that $E_{\mathbb{V}}$ is a potential cone satisfying (B_1) and (B_2). If (s_n) is an increasing sequence in $E_{\mathbb{V}}$ then $\sup s_n \in E_{\mathbb{V}}$ by (3.8), i.e. (B_1) holds.

Now let $\mathcal{V}$ be a non-empty subset of $S_{\mathbb{V}}$. By the topological lemma of Choquet (I.1.8), there exists a countable subset $\mathcal{V}_0$ of $\mathcal{V}$ such that $\widehat{\inf \mathcal{V}_0} = \widehat{\inf \mathcal{V}}$. Let $v_0 = \inf \mathcal{V}_0$, $v = \inf \mathcal{V}$. Then $v_0 \in S_{\mathbb{V}}$, hence $\tilde{v}_0 = \lim_{\lambda\to\infty} \lambda V_\lambda v_0 \in E_{\mathbb{V}}$ and $\tilde{v}_0 \leq v_0$. In particular,

$$\tilde{v}_0 \leq \hat{v}_0 = \hat{v} \leq \hat{v}^f \leq \hat{v}_0^f\ .$$

On the other hand by the first part of the proof

$$\hat{v}_0^f \leq \lim_{\lambda\to\infty} \lambda V_\lambda\, \hat{v}_0^f \leq \lim_{\lambda\to\infty} \lambda V_\lambda\, v_0 = \tilde{v}_0\ .$$

Thus $\hat{v}^f = \tilde{v}_0 \in E_{\mathbb{V}}$, i.e. (B_2) holds.

Finally, let $u,v \in E_{\mathbb{V}}$ and $f \in \mathcal{B}^+(X)$ defined by $f = (u-v)^+$ on $\{v < \infty\}$ and $f = 0$ on $\{v = \infty\}$. Then $u = \inf(u,v) + f$. Hence by (3.7)

$$w = \inf\{s \in S_{\mathbb{W}} : s \geq f\} \in S_{\mathbb{W}}$$

and there exists $w' \in S_{\mathbb{W}}$ such that $u = w + w'$. By (3.8), $\tilde{w} = \lim_{\lambda\to\infty} \lambda V_\lambda w \in E_{\mathbb{W}}$, $\tilde{w}' = \lim_{\lambda\to\infty} \lambda V_\lambda w' \in E_{\mathbb{W}}$ and

$$u = \tilde{w} + \tilde{w}' .$$

Since $u \leq v + f \leq v + w$ we conclude that $u = \tilde{u} \leq \widetilde{v+w} = \tilde{v} + \tilde{w} = v + \tilde{w}$, i.e. $f \leq \tilde{w}$. Using $E_{\mathbb{W}} \subset S_{\mathbb{W}}$ we thus obtain

$$w \leq \inf\{s \in E_{\mathbb{W}} : s \geq f\} \leq \tilde{w} \leq w ,$$

i.e.

$$\inf\{s \in E_{\mathbb{W}} : s \geq f\} = \tilde{w} \in E_{\mathbb{W}}$$

and $E_{\mathbb{W}}$ is a potential cone. ┘

4.8. REMARK. If the potential kernel V of $\mathbb{W}$ is proper and $\lim_{\lambda\to\infty} \lambda V_\lambda f = f$ for every $f \in K(X)$ then $E_{\mathbb{W}}$ is linearly separating. Indeed, let $x,y \in X$, $x \neq y$, and $a \geq 0$. Then there exists $f \in K^+(X)$ such that $f(x) \neq a\,f(y)$ (e.g. $f(x) = 1$, $f(y) = 0$). Since $\lim_{\lambda\to\infty} \lambda V_\lambda f = f$ and $V_\lambda f + \lambda V\, V_\lambda f = Vf$ there exists $\lambda > 0$ such that $V(V_\lambda f)(x) \neq aV(V_\lambda f)(y)$ or $Vf(x) \neq aVf(y)$. By (3.8), $V(V_\lambda f) \in E_{\mathbb{W}}$ and $Vf \in E_{\mathbb{W}}$.

4.9. COROLLARY. Let $\mathbb{P} = (P_t)_{t>0}$ be a sub-Markov semigroup on X. Then the following two statements are equivalent:

(1) $(X,E_{\mathbb{P}})$ is a balayage space.

(2) For every $f \in K(X)$, $\lim_{t\to 0} P_t f = f$. Moreover, $E_{\mathbb{P}}$ is linearly separating, there exist strictly positive functions $u_0, v_0 \in E_{\mathbb{P}} \cap C(X)$ such that $\frac{u_0}{v_0} \in C_0(X)$, and, for every $w \in E_{\mathbb{P}}$, $w = \sup\{v \in E_{\mathbb{P}} \cap C(X) : v \leq w\}$.

Furthermore, if $(X,E_{\mathbb{P}})$ is a balayage space, then $\lim_{t\to 0} P_t(x,U) = 1$ for every finely open set $U \in \mathcal{B}(X)$ and $x \in U$. In particular, $1 \in E_{\mathbb{P}}$ and $E_{\mathbb{P}}$ is the set of all finely l.s.c. functions in $S_{\mathbb{P}}$.

In section 8 we shall prove that for every balayage space $(X,\mathcal{W})$ with $1 \in \mathcal{W}$ there exists a sub-Markov semigroup $\mathbb{P}$ such that $\mathcal{W} = E_{\mathbb{P}}$. An application of (4.9) to (3.1) yields fundamental examples of balayage spaces.

4.10. EXAMPLES. 1. Classical potential theory. Let $\mathbb{P}$ be the Brownian semigroup on $\mathbb{R}^n$, $n \geq 3$. Then $(\mathbb{R}^n, E_{\mathbb{P}})$ is a balayage space having the following properties:

(a) $\lambda^n(U) > 0$ for every non-empty finely open $U \in \mathcal{B}(\mathbb{R}^n)$.

(b) Every countable subset of $\mathbb{R}^n$ is finely closed. In particular, only finite subsets of $\mathbb{R}^n$ are finely compact.

Indeed, let $P = E_{\mathbb{P}} \cap C_0(\mathbb{R}^n)$. Then by (0.3.2), (0.1.6), (I.1.1.5) P_σ is a min-stable function cone and $E_{\mathbb{P}} = S(P) = S(P_\sigma)$, hence $(\mathbb{R}^n, E_{\mathbb{P}})$ is a balayage space satisfying (a) by (4.9). Finally, let $A = \{x_m : m \in \mathbb{N}\}$ be a countable subset of $\mathbb{R}^n$ and $x \in \mathbb{R}^n \setminus A$. Define $u \in E_{\mathbb{P}}$ by

$$u = \sum_{m=1}^{\infty} (2^m N(x_m, x))^{-1} N^{x_m} .$$

Then $u(x) = 1$ and $u(x_m) = \infty$ for every $m \in \mathbb{N}$. Hence $V = \{u < \infty\}$ is a finely open neighborhood of x such that $V \cap A = \emptyset$. Thus (b) holds.

2. Translation on $\mathbb{R}$. Let $\mathbb{T}$ be the translation semigroup on $\mathbb{R}$. Then $(\mathbb{R}, E_{\mathbb{T}})$ is a balayage space. Indeed, let P be the set of all increasing functions $p \in C^+(\mathbb{R})$ such that $\lim_{x\to-\infty} p(x) = 0$. Then $P = P_\sigma$ is a min-stable function cone by (I.1.1.3). Furthermore, if $u \in E_{\mathbb{T}}$ and (φ_n) is an increasing sequence in $K^+(\mathbb{R})$ such that $\sup \varphi_n = u$ then defining $p_n(x) := \sup_{y \leq x} \varphi_n(y)$ we obtain an increasing sequence (p_n) in P such that $u = \sup p_n$.
Since $E_{\mathbb{T}}$ is the set of all left continuous increasing functions $u \geq 0$ on $\mathbb{R}$, the fine topology is generated by the semi-open intervals $]a,b]$.

3. Discrete potential theory. A balayage space $(X,\mathcal{W})$ is called *discrete* if the topology of X is discrete, i.e. if X is a countable set with the discrete topology. If X is discrete and $\mathbb{P}$ is a pseudo-Poisson semigroup on X such that $E_{\mathbb{P}}$ separates the points of X then $(X,E_{\mathbb{P}})$ is a (discrete) balayage space and $1 \in E_{\mathbb{P}}$. This will follow from (III.1.1.3) and (III.6.11). Conversely, if $(X,\mathcal{W})$ is a discrete balayage space and $1 \in \mathcal{W}$ then there exists a pseudo-Poisson semigroup $\mathbb{P}$ on X such that $\mathcal{W} = E_{\mathbb{P}}$. This will follow from (3.1.3) and (5.7).

If X is discrete then $(X,\mathcal{B}^+(X))$ is certainly a balayage space, called *trivial balayage space*. An associated semigroup is for example given by $P_t = I$, $t > 0$.

4.11. EXERCISES. 1. The $\mathcal{W}$-fine topology is always completely regular.

2. Suppose that $(X,\mathcal{W})$ satisfies (B_1) and that for any pair of points $x,y \in X$ with $x \neq y$ there exists $v \in \mathcal{W}$ such that $v(x) < \infty$ and $v(y) = \infty$. Then each point $x \in X$ is finely isolated or does not admit a countable fundamental system of fine neighborhoods.

3. Suppose that $(X,\mathcal{W})$ satisfies (B_1) and (B_2) . Then for every $u \in \mathcal{W}$ the sets $\{u = 0\}$ and $\overline{\{u < \infty\}}^f$ are finely open and closed.

4. Suppose that $(X,\mathcal{W})$ satisfies (B_1), (B_2) and let (v_n) be a sequence in $\mathcal{W}$ which converges pointwise to a finely l.s.c. function v . Then $v \in \mathcal{W}$.

5. Let $(X,\mathcal{W})$ be a balayage space such that $1 \in \mathcal{W}$ and $\mathcal{W} = S(\mathcal{W} \cap C_0(X))$. Let $X_\Delta = X \cup \{\Delta\}$ be the one-point compactification of X and let $\mathcal{W}_\Delta$ be the set of all functions $u : X_\Delta \to \overline{\mathbb{R}}_+$ such that $(u - u(\Delta))|_X \in \mathcal{W}$ if $u(\Delta) < \infty$ and $u = \infty$ if $u(\Delta) = \infty$. Then $(X_\Delta,\mathcal{W}_\Delta)$ is a balayage space such that $\mathcal{W}_\Delta|_X = \mathcal{W}$ and $R_1^{\{\Delta\}} = 1$.

5. Continuous Potentials

Let $(X,\mathcal{W})$ be a balayage space. A function $p \in \mathcal{W} \cap C(X)$ such that $\frac{p}{v} \in C_0(X)$ for some strictly positive $v \in \mathcal{W} \cap C(X)$ will be called a continuous *potential* on X . In the following let $\mathcal{P}$ denote the set of all continuous potentials on X . We know already from (4.6) that $\mathcal{P}$ is a min-stable function cone and $S(\mathcal{P}) = \mathcal{W}$.

5.1. EXAMPLES. 1. Classical potential theory on $\mathbb{R}^n$, $n \geq 3$. Every hyperharmonic function $p \in C_0^+(\mathbb{R}^n)$ is a potential. In section V.4 we shall give a complete characterization of $\mathcal{P}$.

2. Translation on $\mathbb{R}$. An increasing function $p \in C^+(\mathbb{R})$ is a potential if and only if $\lim_{x\to-\infty} p(x) = 0$, i.e. $\mathcal{P}$ is the set of all functions $x \mapsto \mu(]-\infty,x])$ where $\mu \in M_+(\mathbb{R})$ is non-atomic such that $\mu(]-\infty,0]) < \infty$.

For every $A \subset X$ and $u \in \mathcal{W}$ we define the *reduit* of u on A by

$$R_u^A = \inf\{v \in \mathcal{W} : v \leq u , v = u \text{ on } A\}$$
$$= \inf\{v \in \mathcal{W} : v \geq u \text{ on } A\} = R_{u1_A} .$$

Obviously,

$$R_u^A \leq u , R_u^A = u \text{ on } A ,$$
$$u \leq v , A \subset B \Rightarrow R_u^A \leq R_v^B ,$$
$$R_{u+v}^A \leq R_u^A + R_v^A ,$$
$$R_{\alpha u}^A = \alpha R_u^A \qquad (\alpha \in \mathbb{R}_+) .$$

5.2. PROPOSITION. $\mathcal{P}$ is the largest function cone contained in $\mathcal{W}$ and $\mathcal{P} = \mathcal{P}_\sigma = \mathcal{W} \cap C_p(X)$. $\mathcal{P}$ is the set of all $p \in \mathcal{W} \cap C(X)$ such that $R_p^{\complement K}$ decreases locally uniformly to zero as the compact set K increases to X . Furthermore, $R_f \in \mathcal{P}$ for every $f \in C_p^+(X)$.

Proof. If $\mathcal{P}'$ is a function cone contained in $\mathcal{W}$ and $p \in \mathcal{P}'$, then there exists a strictly positive function $q \in \mathcal{P}' \subset \mathcal{W}$ such that $p \in o(q)$ and hence $p \in \mathcal{P}$. So $\mathcal{P}$ is the largest function cone contained in $\mathcal{W}$. Moreover, it follows immediately from the definition of $\mathcal{P}$ that $\mathcal{P} = \mathcal{W} \cap C_p(X)$.

Let $\mathcal{P}'$ denote the set of all $p \in \mathcal{W} \cap C(X)$ such that $R_p^{\complement K}$ decreases locally uniformly to zero as the compact set K increases to X . Then $\mathcal{P}'$ is a convex cone. Let $p \in \mathcal{P}$ and $v \in \mathcal{W} \cap C(X)$ such that $p \in o(v)$. Fix a compact subset L of X and $\varepsilon > 0$. Then there exists $\delta > 0$ and a compact subset K of X such that $\delta v \leq \varepsilon$ on L and $p \leq \delta v$ on $\complement K$, and hence $R_p^{\complement K} \leq \delta v \leq \varepsilon$ on L . Therefore $\mathcal{P} \subset \mathcal{P}'$.

Let $\varphi \in K^+(X)$ and let $p_0 \in \mathcal{P} , p_0 > 0$. Consider $u \in \mathcal{W}$ such that $u \geq \varphi$ and let $\varepsilon > 0$. Then $u + \varepsilon p_0 > \varphi$. Since $u \in S(\mathcal{P})$ there exists a potential $p \in \mathcal{P}$ such that $p \leq u$ and $p + \varepsilon p_0 > \varphi$ on $S(\varphi)$ and hence $u + \varepsilon p_0 \geq p + \varepsilon p_0 \geq R_\varphi$. This shows that

$$R_\varphi = \inf\{p \in \mathcal{P} : p \geq \varphi\} .$$

In particular, R_φ is u.s.c.. Since we already know that $R_\varphi \in \mathcal{W}$, we obtain $R_\varphi \in \mathcal{P}$.

Now let $f \in C^+_{P'}(X)$ and $p \in P'$ such that $f \leq p$. Let L be a compact subset of X and $\varepsilon > 0$. Then there exists a compact subset K of X such that $R_p^{\complement K} \leq \varepsilon$ on L. Let $\varphi \in K^+(X)$ such that $\varphi \leq f$ and $\varphi = f$ on K. Then evidently

$$R_\varphi \leq R_f \leq R_\varphi + R_p^{\complement K} .$$

In particular,

$$0 \leq R_f - R_\varphi \leq \varepsilon \quad \text{on } L .$$

Since $R_\varphi \in C^+(X)$ we conclude that $R_f \in C^+(X)$. Thus $R_f \in P'$ since $R_f \in W$ and $R_f \leq p$.

Fix $p \in P'$, $x \in X$ and $\varepsilon > 0$. There exists a compact subset K of X such that $R_p^{\complement K}(x) < \varepsilon$ and a function $\varphi \in K^+(X)$ such that $\varphi \leq p$ and $\varphi = p$ on K. Then $q := R_{p-\varphi} \in P'$, $q(x) \leq R_p^{\complement K}(x) < \varepsilon$ and $q = p$ on $\complement S(\varphi)$. Hence P'_σ is a function cone by (I.1.1.3).

Since $P \subset P' \subset P'_\sigma \subset W$ and P is the largest function cone contained in W we finally conclude that $P = P' = P'_\sigma = P_\sigma$. ┘

An important consequence of (B_3) is the additivity of the reduit $u \mapsto R_u^A$. This property yields the existence of *reduced measures* $\varepsilon_x^{\varrho A}$, $x \in X$, leading to a characterization of functions in W by integral inequalities. An extensive study of such measures is postponed to chapter VI.

<u>5.3. PROPOSITION.</u> Let $A \subset X$ and $u_1, u_2 \in W$ such that $u_1 + u_2 < \infty$ on A. Then

$$R^A_{u_1+u_2} = R^A_{u_1} + R^A_{u_2} .$$

Proof. Let $v \in W$ such that $v \leq u_1 + u_2$ and $v = u_1 + u_2$ on A. By (B_3) there exist $v_1, v_2 \in W$ such that $v = v_1 + v_2$ and $v_1 \leq u_1, v_2 \leq u_2$. Then $v_1 = u_1$ on A and $v_2 = u_2$ on A, hence $R^A_{u_1} \leq v_1$ and $R^A_{u_2} \leq v_2$, i.e. $v = v_1 + v_2 \geq R^A_{u_1} + R^A_{u_2}$ and therefore $R^A_{u_1+u_2} \geq R^A_{u_1} + R^A_{u_2}$. The converse inequality is trivial. ┘

<u>5.4. COROLLARY.</u> Let $A \subset X$. Then there exists a unique measure $\varepsilon_x^{\varrho A}$ on X

which is supported by $\bar{A}$ such that

$$\int p \, d\overset{\circ}{\varepsilon}{}_x^A = R_p^A(x) \qquad \text{for every } p \in P .$$

Proof. The mapping $p \mapsto R_p^A(x)$ from P into $\mathbb{R}_+$ is additive by (5.3). Furthermore, it is obviously positively homogeneous and increasing. Hence by (I.1.4) there exists a unique measure $\overset{\circ}{\varepsilon}{}_x^A$ on X such that

$$\int p \, d\overset{\circ}{\varepsilon}{}_x^A = R_p^A(x) \qquad \text{for every } p \in P .$$

Let $f \in K^+(X)$ and $f = 0$ on A. Take $p_0 \in P$, $p_0 > 0$. By (I.1.2) there exist for every $\varepsilon > 0$ functions $p, q \in P$ such that

$$0 \le p - q \le f \le p - q + \varepsilon \, p_0 .$$

In particular $p = q$ on A, hence $R_p^A = R_q^A$ and therefore

$$0 \le \int f \; d\overset{\circ}{\varepsilon}{}_x^A \le \int (p - q + \varepsilon p_0) \; d\overset{\circ}{\varepsilon}{}_x^A = R_{\varepsilon p_0}^A(x) \le \varepsilon p_0(x) ,$$

thus $\int f \; d\overset{\circ}{\varepsilon}{}_x^A = 0$. ⌋

5.5. PROPOSITION. For every lower P-bounded l.s.c. numerical function u on X the following statements are equivalent:

(1) $u \in W$.

(2) $\int u \; d\overset{\circ}{\varepsilon}{}_x^{\complement U} \le u(x)$ for all $x \in X$ and every open neighborhood U of x.

(3) For every $x \in X$ and every open neighborhood U of x there exists a subset V of U such that $\overset{\circ}{\varepsilon}{}_x^{\complement V} \neq \varepsilon_x$ and $\int u \; d\overset{\circ}{\varepsilon}{}_x^{\complement V} \le u(x)$.

Proof. (1) ⇒ (2): By (B_4) there exists an increasing sequence $(p_n) \subset P$ such that $u = \sup p_n$. Hence for every open set U and $x \in U$ by (5.4)

$$\int u \; d\overset{\circ}{\varepsilon}{}_x^{\complement U} = \sup \int p_n \; d\overset{\circ}{\varepsilon}{}_x^{\complement U} = \sup R_{p_n}^{\complement U}(x) \le \sup p_n(x) = u(x) .$$

(2) ⇒ (3): If $x \in X$ and U is an open neighborhood of x then $\overset{\circ}{\varepsilon}{}_x^{\complement U} \neq \varepsilon_x$ by (5.4), hence we may choose $V = U$.

(3) ⇒ (1): Let $f \in C_P(X)$, $f \le u$ and $A := \{f = R_f\}$. Defining

$$F := P + \mathbb{R}_+ u - \mathbb{R}_+ R_f$$

it suffices to show $Ch_F X \subset A$. Indeed, $u - R_f \ge f - R_f = 0$ on A , hence by the minimum principle (I.2.2) $u - R_f \geq 0$, i.e. $u \geq R_f \geq f$. This yields $u \in S(P) = \mathcal{W}$.

Let $x \in \complement A$. By assumption, there exists a subset V of $\complement A$ such that $\varepsilon_x^{\complement V} \neq \varepsilon_x$ and $\int u \, d\varepsilon_x^{\complement V} \leq u(x)$. Since by (I.2.5) $Ch_{P-\mathbb{R}_+ R_f} X \subset A \subset \complement V$ we have $R_{R_f}^{\complement V} = R_f$, hence $\int R_f \, d\varepsilon_x^{\complement V} = R_f(x)$. Therefore $\varepsilon_x^{\complement V} \in M_x(F)$, $x \notin Ch_F X$. ┘

5.6. COROLLARY. $(X,\mathcal{W})$ has the *truncation property*, i.e. for every open set U in X and every $u,v \in \mathcal{W}$ such that $u \geq v$ on $\complement U$ the function w defined by

$$w = \begin{cases} \inf(u,v) & \text{on } U \\ v & \text{on } \complement U \end{cases}$$

is contained in $\mathcal{W}$.

5.7. EXAMPLE. Discrete potential theory. If P is the kernel defined by $P(x,\cdot) = \varepsilon_x^{\complement\{x\}}$ then $\mathcal{W} = S_P$.

5.8. EXERCISES. 1. Let U be a finely open subset of X, $x \in X$, and $v \in \mathcal{W}$. Then $\int v \, d\varepsilon_x^U = R_v^U(x)$.

2. Let $(\mathbb{R},\mathcal{W})$ be the balayage space associated with translation on $\mathbb{R}$. Let $A \subset \mathbb{R}$ and $x \in \mathbb{R}$. Then the measure ε_x^A is supported by $\bar{A} \cap]-\infty,x]$.

3. Let $(\mathbb{Z},\mathcal{W})$ be the balayage space associated with the Poisson semigroup. Let $A \subset \mathbb{Z}$ and $n \in \mathbb{Z}$. Then the measure ε_n^A is supported by $\{m \in A : m \geq n\}$.

6. Construction of Kernels

As before let $(X,\mathcal{W})$ be a balayage space and let P be the function cone of all continuous real potentials on X . We shall assign to every $p \in P$ a kernel $(x,B) \mapsto p_B(x)$ having similar properties as the kernels

$$(x,B) \mapsto \int_B \frac{\mu(dy)}{\|x - y\|^{n-2}}$$

in classical potential theory.

Given $u,v \in \mathcal{W}$ we shall write $u \prec v$ if there exists a function $w \in \mathcal{W}$ such that $u + w = v$. The relation " $\prec$ " is called *specific order* on $\mathcal{W}$. We note that $u \prec v$ and $v \in P$ implies that u and $v - u$ are contained in

$\mathcal{W} \cap C_P(X)$, i.e. $u, v - u \in \mathcal{P}$.

For every $f \in C_P(X)$ we define the *fine support* of f by

$$\delta(f) := Ch_{\mathcal{P}-\mathbb{R}_+ f} X ,$$

i.e. $\delta(f)$ is the set of all $x \in X$ such that ε_x is the only measure $\mu \in M_x(\mathcal{P})$ satisfying $\mu(f) \geq f(x)$.

6.1. LEMMA. If $p, q \in \mathcal{P}$, and $p \prec q$ then $\delta(p) \subset \delta(q)$.

Proof. Let $x \in \delta(p)$ and $\mu \in M_x(\mathcal{P})$ such that $\mu(q) \geq q(x)$. Since $\mu(p) \leq p(x)$ and $\mu(q-p) \leq (q-p)(x)$, we have $\mu(p) = p(x)$, hence $\mu = \varepsilon_x$. Therefore $x \in \delta(q)$. ┘

6.2. LEMMA. Let $p \in \mathcal{P}$ and let F be an increasingly filtered subset of $\mathcal{P}$ such that $q \prec p$ for every $q \in F$. Then $\sup F \in \mathcal{P}$ and $\sup F \prec p$.

Proof. Let $G = \{p - q : q \in F\}$. Then $G \subset \mathcal{P}$ and $p = \sup F + \inf G$ where $\sup F \in \mathcal{W}$ by (5.6). Therefore $p = \sup F + \widehat{\inf G}$ and hence $\sup F \in \mathcal{W} \cap C_P(X) = \mathcal{P}$ (and $\inf G = \widehat{\inf G} \in \mathcal{P}$) . ┘

For every $p \in \mathcal{P}$ we define the *carrier* of p by

$$C(p) = \overline{\delta(p)} .$$

6.3. PROPOSITION. For every $p \in \mathcal{P}$, $R_p^{\delta(p)} = p$. Moreover, $C(p)$ is the smallest closed subset A of X such that $R_p^A = p$.

Proof. If $u \in \mathcal{W}$ and $u \geq p$ on $\delta(p)$ then $u \geq p$ by (I.2.2). Hence $R_p^{\delta(p)} = p$. Let $A \subset X$ be closed and $R_p^A = p$. For every $x \in \complement A$ we have $\varepsilon_x^A \in M_x(\mathcal{P}) \setminus \{\varepsilon_x\}$, hence $x \in \complement\, \delta(p)$. Therefore $\delta(p) \subset A$, i.e. $C(p) \subset A$. ┘

6.4. COROLLARY. For all $p, q \in \mathcal{P}$,

$$C(p + q) = C(p) \cup C(q) .$$

Proof. By (6.1) , $C(p) \cup C(q) \subset C(p + q)$. Furthermore, by (5.3)

$$R_{p+q}^{C(p) \cup C(q)} = R_p^{C(p) \cup C(q)} + R_q^{C(p) \cup C(q)} = p + q .$$

Hence $C(p+q) \subset C(p) \cup C(q)$ by (6.3). ┘

6.5. LEMMA. Let A be a closed set in X and let F be a subset of P such that $q' := \sup F \in P$ and $C(q) \subset A$ for every $q \in F$. Then $C(q') \subset A$.

Proof. For every $q \in F$, $R^A_{q'} \geq R^A_q = q$, i.e. $R^A_{q'} \geq q'$. Hence $C(q') \subset A$ by (6.3). ┘

Given $p \in P$ we define for every $B \in \mathcal{B}(X)$

$$p_B = \sup \{q \in P : q \prec p, C(q) \subset B\} .$$

We note that $p_\emptyset = 0$, $p_X = p$.

6.6. PROPOSITION. For every $B \in \mathcal{B}(X)$, $p_B \in P$, $p_B \prec p$ and $C(p_B) \subset \bar{B}$.

Proof. Let $B \in \mathcal{B}(X)$ and let F be the set of all $q \in P$ such that $q \prec p$ and $C(q) \subset B$. By (6.2) and (6.5) it suffices to show that F is increasingly filtered. So let $q_1, q_2 \in F$. Defining

$$q = q_1 + R_{q_2 - q_1}$$

we certainly have $q \geq q_1$, $q \geq q_2$. Furthermore, $R_{q_2-q_1} \prec q_2$, hence $C(R_{q_2-q_1}) \subset C(q_2)$ by (6.4) and therefore

$$C(q) = C(q_1) \cup C(R_{q_2-q_1}) \subset C(q_1) \cup C(q_2) \subset B .$$

Finally, $q \prec p$ since

$$p - q = p - q_1 - R_{q_2-q_1} = (p - q_1) - R_{(p-q_1)-(p-q_2)}$$

where $p - q_1, p - q_2 \in P$. ┘

6.7. COROLLARY. For every $B \in \mathcal{B}(X)$, $p_B = \sup \{p_A : A \text{ closed}, A \subset B\}$.

Proof. Let $q \in P$ such that $q \prec p$ and $C(q) \subset B$. Then $q \leq p_{C(q)}$ by definition of $p_{C(q)}$. Hence $p_B \leq \sup \{p_A : A \text{ closed}, A \subset B\}$. The converse inequality is trivial. ┘

6.8. LEMMA. Let $p_1, p_2 \in P$ such that $p_1 \prec p$, $p_2 \prec p$ and $C(p_1) \cap C(p_2) = \emptyset$.

Then $p_1 + p_2 \prec p$.

Proof. Let $x \in \complement C(p_1)$ and let U be an open neighborhood of x in $\complement C(p_1)$. Then $C(p_1) \subset \complement U$, hence $R_{p_1}^{\complement U} = p_1$ and in particular $\varepsilon_x^{\complement U}(p_1) = p_1(x)$. Therefore

$$\varepsilon_x^{\complement U}(p - (p_1 + p_2)) = \varepsilon_x^{\complement U}(p - p_2) - \varepsilon_x^{\complement U}(p_1)$$
$$\leq (p - p_2)(x) - p_1(x) = (p - (p_1 + p_2))(x) .$$

Similarly for every $x \in \complement C(p_2)$. Since $\complement C(p_1) \cup \complement C(p_2) = X$ we conclude by (5.5) that $p - (p_1 + p_2) \in \mathcal{W} \cap C_P(X) = P$. ┘

<u>6.9. LEMMA.</u> Let B_1 , $B_2 \in \mathcal{B}(X)$ such that $B_1 \cap B_2 = \emptyset$. Then $p_{B_1} + p_{B_2} \leq p_{B_1 \cup B_2}$.

Proof. Let $q_1, q_2 \in P$ such that $q_i \prec p$ and $C(q_i) \subset B_i$, $i = 1, 2$. Then $C(q_1) \cap C(q_2) = \emptyset$, hence $q_1 + q_2 \prec p$ by (6.8). Furthermore $C(q_1 + q_2) = C(q_1) \cup C(q_2) \subset B_1 \cup B_2$. Therefore $q_1 + q_2 \leq p_{B_1 \cup B_2}$. Thus $p_{B_1} + p_{B_2} \leq p_{B_1 \cup B_2}$. ┘

<u>6.10 LEMMA.</u> For every open subset U of X , $C(p - p_U) \subset \complement U$.

Proof. Let U be an open subset of X and $q = p - p_U$. By (6.3), it suffices to show that $R_q^{\complement U} = q$. Let $x \in X$ and $\varepsilon > 0$. There exists a closed subset A of U such that $q_1 = p - p_A$ satisfies

$$R_{q_1}^{\complement U}(x) < R_q^{\complement U}(x) + \varepsilon .$$

Furthermore, there exists $q_2 \in P$ such that $q_2 > q_1$ on $\complement U$ and $q_2(x) < R_{q_1}^{\complement U}(x) + \varepsilon$. Choosing $q' = R_{q_1 - q_2}$ we have $q' \in P, q' \prec q_1$, hence $p_A + q' \prec p$ and by (I.2.5)

$$C(q') \subset \{q_1 - q_2 = q'\} \subset \{q_1 - q_2 \geq 0\} \subset U .$$

Hence $p_A + q' \leq p_U$ and

$$q = p - p_U \leq p - (p_A + q') = q_1 - q' \leq q_2 .$$

In particular,

$$q(x) \leq q_2(x) < R_{q_1}^{\complement U}(x) + \varepsilon < R_q^{\complement U}(x) + 2\varepsilon .$$

Thus $R_q^{\complement U} = q$. ┘

6.11. PROPOSITION. For every open subset U of X, $p_U + p_{\complement U} = p$.

Proof. By (6.6) and (6.10), $p - p_U \leq p_{\complement U}$, i.e. $p \leq p_U + p_{\complement U}$. By (6.9), $p_U + p_{\complement U} \leq p$. ┘

Let $\mathcal{A}$ denote the family of all $B \in \mathcal{B}(X)$ such that $p_B + p_{\complement B} = p$. We already know by (6.11) that $\mathcal{A}$ contains every open subset of X and we shall see very soon that $\mathcal{A} = \mathcal{B}(X)$.

6.12. LEMMA. For every $B \in \mathcal{A}$, $p_B = \inf\{p_U : U \text{ open}, U \supset B\}$.

Proof. (6.7) and (6.11). ┘

6.13. PROPOSITION. Let $B_1 \in \mathcal{A}$ and $B_2 \in \mathcal{B}(X)$ such that $B_1 \cap B_2 = \emptyset$. Then $p_{B_1 \cup B_2} = p_{B_1} + p_{B_2}$.

Proof. Let A be a closed subset of $B_1 \cup B_2$ and let U be an open neighborhood of B_1. Applying (6.11) to $q := p_A$ we obtain $q = q_U + q_{\complement U}$. Since $q \prec p$ we have $q_U \leq p_U$ by definition of p_U. Furthermore, $q_{\complement U} \prec q$, hence

$$C(q_{\complement U}) \subset C(q) \cap \complement U \subset A \setminus U \subset B_2$$

by (6.4) and (6.6), and therefore $q_{\complement U} \leq p_{B_2}$. So we conclude that $q \leq p_U + p_{B_2}$ and hence $p_{B_1 \cup B_2} \leq p_{B_1} + p_{B_2}$ by (6.7) and (6.12). Thus $p_{B_1 \cup B_2} = p_{B_1} + p_{B_2}$ by (6.9). ┘

6.14. LEMMA. For every $B \in \mathcal{B}(X)$, $p_B = \sup\{p_K : K \text{ compact}, K \subset B\}$.

Proof. Let A be a closed subset of B. Let $q \in P$ such that $p \in o(q)$ and $\varepsilon > 0$. Then there exists a compact subset L of X such that $p \leq \varepsilon q$ on $\complement L$. By (6.11) and (6.13), $p_A = p_{A \cap L} + p_{A \setminus L}$ where

$$p_{A \setminus L} = R_{p_{A \setminus L}}^{\overline{\complement L}} \leq R_p^{\overline{\complement L}} \leq \varepsilon q$$

since $C(p_{A \setminus L}) \subset \overline{\complement L}$. Thus

$$\sup\{p_K : K \text{ compact}, K \subset B\} \geq p_A$$

and the statement follows by (6.7). ┘

6.15. PROPOSITION. Let (B_n) be an increasing sequence in A and $\bigcup_{n=1}^{\infty} B_n = B$. Then $B \in A$ and $\sup p_{B_n} = p_B$.

Proof. Let $x \in X$ and $\varepsilon > 0$. By (6.12), there exist open neighborhoods V_n of B_n such that

$$p_{V_n}(x) \leq p_{B_n}(x) + \frac{\varepsilon}{2^n}$$

for every $n \in \mathbb{N}$. Defining $U_n = V_1 \cup \ldots \cup V_n$ we have

$$p_{U_n}(x) \leq p_{B_n}(x) + \sum_{i=1}^{n} \frac{\varepsilon}{2^i}$$

for every $n \in \mathbb{N}$. This is trivial if $n = 1$ and the induction step follows easily from

$$p_{U_{n+1}} + p_{B_n} \leq p_{U_{n+1}} + p_{U_n \cap V_{n+1}}$$

$$= p_{U_n} + p_{U_{n+1} \setminus U_n} + p_{U_n \cap V_{n+1}} = p_{U_n} + p_{V_{n+1}} .$$

Now let K be a compact subset of $U = \bigcup_{n=1}^{\infty} U_n$. Then there exists $n \in \mathbb{N}$ such that $K \subset U_n$ and hence

$$p_K(x) \leq p_{U_n}(x) \leq \sup p_{B_n}(x) + \varepsilon .$$

Thus by (6.14), $p_U(x) \leq \sup p_{B_n}(x) + \varepsilon$. Since U is an open set containing B we conclude in particular that

$$p_B(x) \leq \sup p_{B_n}(x) + \varepsilon$$

and hence $p_B = \sup p_{B_n}$. Furthermore,

$$p_B(x) + p_{\complement B}(x) \geq (p_U(x) - \varepsilon) + p_{\complement U}(x) = p(x) - \varepsilon$$

and hence $p_B + p_{\complement B} = p$ by (6.9). Thus $B \in A$. ┘

6.16. PROPOSITION. $A = \mathcal{B}(X)$.

Proof. By (6.11), every open subset of X is contained in A. Clearly, $B \in A$ if and only if $\complement B \in A$. If $B_1, B_2 \in A$ such that $B_1 \cap B_2 = \emptyset$ then by (6.13)

$$p_{B_1 \cup B_2} + p_{\complement(B_1 \cup B_2)} = p_{B_1} + p_{B_2} + p_{\complement(B_1 \cup B_2)} = p_{B_1} + p_{\complement B_1} = p ,$$

i.e. $B_1 \cup B_2 \in A$. If (B_n) is an increasing sequence in A then $\bigcup_{n=1}^{\infty} B_n \in A$ by (6.15). Thus $A = \mathcal{B}(X)$. ⌋

6.17. THEOREM. For every $p \in P$, the mapping $V : (x,B) \mapsto p_B(x)$ is a kernel on X such that

(1) $V1 = p$,

(2) $Vf \in P$ and $C(Vf) \subset S(f)$ for every $f \in \mathcal{B}_b^+(X)$.

The kernel V is uniquely determined by (1) and (2). V is called the *potential kernel associated with* p .

Proof. By (6.16), (6.13) and (6.15), V is a kernel on X such that $V1 = p$. If $f \in \mathcal{B}_b^+(X)$ is a step function then $Vf \in P$ and $C(Vf) \subset S(f)$ by (6.6), and $Vf + V(\|f\| - f) = \|f\|p$. Hence property (2) follows from (6.2) and (6.5). Suppose now that W is a kernel on X satisfying (1) and (2) and let $B \in \mathcal{B}(X)$. Then

$$W\,1_B = \sup\{W1_K : K \subset B \text{ compact}\}$$
$$\leq \sup\{q \in P : C(q) \subset B\,,\, q \prec p\} = p_B\ .$$

Similarly, $W\,1_{\complement B} \leq p_{\complement B}$. Since $W\,1_B + W\,1_{\complement B} = W\,1 = p = p_B + p_{\complement B}$ we conclude that $W\,1_B = p_B$. ⌋

6.18. COROLLARY. 1. For all $p,q \in P$, $\alpha,\beta \in \mathbb{R}_+$ and $B \in \mathcal{B}(X)$, $(\alpha p + \beta q)_B = \alpha p_B + \beta q_B$.

2. For all $p \in P$ and $A,B \in \mathcal{B}(X)$, $(p_A)_B = p_{A \cap B}$.

Proof. 1. The mapping $(x,B) \mapsto \alpha p_B(x) + \beta q_B(x)$ has the properties characterizing the kernel $(x,B) \mapsto (p+q)_B(x)$.

2. The mapping $(x,B) \mapsto p_{A \cap B}(x)$ has the properties characterizing the kernel $(x,B) \mapsto (p_A)_B(x)$. ⌋

6.19. REMARK. Let P_c denote the convex cone of all $q \in P$ such that $C(q)$ is compact. Then by (6.17) every $p \in P$ is the sum of a sequence (q_n) in P_c . Furthermore, (6.3) implies $R_f \in P_c$ for every $f \in K^+(X)$, hence $W = S(P_c)$.

In the next section we shall see that the fine support $\delta(p)$ is always a Borel set. Hence the following result amounts to the statement that $p_{\complement\delta(p)} = 0$ for every $p \in P$.

6.20. PROPOSITION. Let $p \in P$ and $A \in \mathcal{B}(X)$ such that $A \cap \delta(p) = \emptyset$. Then $p_A = 0$.

Proof. Let K be a compact subset of A. Then $\delta(p_K) \subset K \cap \delta(p) = \emptyset$, hence $p_K = 0$ by (6.3). Thus $p_A = 0$. ⌟

6.21. EXERCISES. 1. If V is the potential kernel associated to $p \in P_b$ then $V : C_b(X) \to C_b(X)$ is injective if and only if $C(p) = X$.

2. Let $p \in P_b$ be a strict potential and let V be the associated potential kernel. For every $x \in X$ let $\mathcal{U}_x$ denote the system of all open neighborhood of x. Then for every $f \in C_b(X)$ and every $x \in X$

$$\lim_{U,\mathcal{U}_x} \frac{Vf(x) - \varepsilon_x^{\complement U}(Vf)}{p(x) - \varepsilon_x^{\complement U}(p)} = f(x) .$$

(Hint: Given $\delta > 0$ choose $U_0 \in \mathcal{U}_x$ and $f_1, f_2 \in C_b(X)$ such that $f_1 = 0$ on U_0, $|f_2| \leq \delta$ and $f = f(x) + f_1 + f_2$.)

7. Construction of Resolvents

Let V be a kernel on X. A function $u \in \mathcal{B}^+(X)$ is called V-*dominant* if for all $f \in \mathcal{B}_b(X)$ such that $Vf \leq u$ on $\{f > 0\}$ we have $Vf \leq u$. We shall say that V satisfies the *complete maximum principle* if the constant 1 is V-dominant.

7.1. PROPOSITION. If there exists a sub-Markov resolvent $\mathbb{V}$ on X such that V is the potential kernel of $\mathbb{V}$ then every $\mathbb{V}$-supermedian function is V-dominant. In particular, V satisfies the complete maximum principle.

Proof. Let $\mathbb{V} = (V_\lambda)_{\lambda>0}$ be a sub-Markov resolvent on X such that

$\sup_{\lambda>0} V_\lambda = V$. Let $f \in \mathcal{B}_b(X)$ and $u \in S_{\mathbb{V}}$ such that

$$Vf \leq u \quad \text{on} \quad \{f > 0\} ,$$

i.e. $Vf^+ \leq Vf^- + u$ on $\{f > 0\} = \{f^+ > 0\}$. Since $Vf^- \in S_{\mathbb{V}}$ by (3.7) we may replace u by $u + Vf^-$, i.e. we may suppose that $f \geq 0$. Fixing a $\lambda > 0$ we have by (3.5.2)

$$\sum_{n=0}^{\infty} (\lambda V_\lambda)^n f = f + \lambda Vf ,$$

hence

$$(*) \qquad \sum_{n=0}^{\infty} (\lambda V_\lambda)^n f \leq \|f\| + \lambda u \quad \text{on} \quad \{f > 0\} .$$

We claim that for every $n \in \mathbb{N}_0$

$$\sum_{i=0}^{n} (\lambda V_\lambda)^i f \leq \|f\| + \lambda u .$$

This inequality is trivial if $n = 0$. Suppose that it holds for some $n \in \mathbb{N}$. Using $\lambda V_\lambda(\|f\| + \lambda u) \leq \|f\| + \lambda u$ we obtain

$$\sum_{i=1}^{n+1} (\lambda V_\lambda)^i f \leq \|f\| + \lambda u ,$$

hence

$$\sum_{i=0}^{n+1} (\lambda V_\lambda)^i f = f + \sum_{i=1}^{n+1} (\lambda V_\lambda)^i f \leq \|f\| + \lambda u \quad \text{on} \quad \{f = 0\} .$$

However by (*) this inequality holds on $\{f > 0\}$ as well. Thus

$$f + \lambda Vf = \lim_{n\to\infty} \sum_{i=0}^{n} (\lambda V_\lambda)^i f \leq \|f\| + \lambda u .$$

Dividing by λ and letting λ tend to infinity we conclude that $Vf \leq u$. ┘

In the following we suppose conversely that the kernel V is bounded and satisfies the complete maximum principle. We shall see that these conditions are in fact sufficient to establish the existence of a corresponding sub-Markov resolvent.

<u>7.2. LEMMA.</u> Let $\lambda > 0$, $f,g \in \mathcal{B}_b(X)$ and let u be a V-dominant function such that $f \leq u$ and $g + \lambda Vg = Vf$. Then $\lambda g \leq u$.

Proof. We have $V(\lambda(f - \lambda g)) = \lambda g \leq f \leq u$ on $\{f - \lambda g \geq 0\}$, hence $\lambda g = V(\lambda(f - \lambda g)) \leq u$. ┘

7.3. COROLLARY. For every $\lambda > 0$, the linear operator $I + \lambda V : \mathcal{B}_b(X) \to \mathcal{B}_b(X)$ is injective. If $f \in \mathcal{B}_b(X)$ and $g = (I + \lambda V)^{-1}(Vf)$ then $\lambda \| g \| \leq \| f \|$.

Proof. Let $\lambda > 0$ and $f, g \in \mathcal{B}_b(X)$ such that $g + \lambda Vg = Vf$. Since $\pm f \leq \| f \|$ and the constant $\| f \|$ is V-dominant we obtain by (7.2) that $\pm \lambda g \leq \| f \|$, i.e. $\lambda \| g \| \leq \| f \|$. Choosing $f = 0$ we see that $I + \lambda V$ is injective. ┘

7.4. PROPOSITION. For every $\lambda \geq 0$, $I + \lambda V : \mathcal{B}_b(X) \to \mathcal{B}_b(X)$ is an algebraic isomorphism.

Proof. The statement is almost trivial for every $\lambda \geq 0$ such that $\lambda \| V \| < 1$ for then

$$(I + \lambda V)^{-1} = \sum_{n=0}^{\infty} (-\lambda V)^n .$$

Let $\lambda_0 > 0$ and suppose that $I + \lambda_0 V : \mathcal{B}_b(X) \to \mathcal{B}_b(X)$ is an algebraic isomorphism and let $0 \leq \lambda < 2\lambda_0$. Then by (7.3)

$$\| (\lambda - \lambda_0)(I + \lambda_0 V)^{-1} V \| \leq \frac{|\lambda - \lambda_0|}{\lambda} < 1 ,$$

hence $I + (\lambda - \lambda_0)(I + \lambda_0 V)^{-1} V : \mathcal{B}_b(X) \to \mathcal{B}_b(X)$ an isomorphism. Therefore

$$I + \lambda V = (I + \lambda_0 V)(I + (\lambda - \lambda_0)(I + \lambda_0 V)^{-1} V)$$

is an isomorphism as well. ┘

For every $\lambda \geq 0$, we define $V_\lambda : \mathcal{B}_b(X) \to \mathcal{B}_b(X)$ by

$$V_\lambda := (I + \lambda V)^{-1} V .$$

In particular, $V_0 = V$.

7.5. PROPOSITION. For every $\lambda > 0$, the linear mapping $V_\lambda : \mathcal{B}_b(X) \to \mathcal{B}_b(X)$ is positive and $\| \lambda V_\lambda \| \leq 1$. For all $\lambda, \mu \geq 0$,

$$V_\lambda - V_\mu = (\mu - \lambda) V_\lambda V_\mu .$$

Proof. By (7.3), $\| \lambda V_\lambda \| \leq 1$. Furthermore, if $f \in \mathcal{B}_b(X)$ and $f \leq 0$ then $V_\lambda f \leq 0$ by (7.2) since 0 is V-dominant. Hence V_λ is positive. Finally let $\lambda, \mu \geq 0$. Then

$$(I + \lambda V)(V_\lambda - V_\mu) = (I + \lambda V)V_\lambda - (I + \mu V)V_\mu + (\mu - \lambda)VV_\mu$$
$$= (\mu - \lambda)(I + \lambda V)V_\lambda V_\mu ,$$

hence

$$V_\lambda - V_\mu = (\mu - \lambda)V_\lambda V_\mu .$$ ⌟

Let $\lambda > 0$. Then for every $f \in \mathcal{B}_b^+(X)$,

$$Vf - V_\lambda f = \lambda V_\lambda Vf \geq 0 .$$

Thus by (1.3), V_λ may be extended to a kernel on X which we shall again denote by V_λ . For every $f \in \mathcal{B}_b^+(X)$,

$$0 \leq Vf - V_\lambda f = \lambda V_\lambda Vf \leq \lambda V Vf ,$$

hence $\lim_{\lambda \to 0} \| Vf - V_\lambda f \| = 0$. Consider now $f \in \mathcal{B}^+(X)$ and take $f_n := \inf(f,n)$. Then

$$\sup_{\lambda>0} V_\lambda f = \sup_{\lambda>0} \sup_n V_\lambda f_n = \sup_n \sup_{\lambda>0} V_\lambda f_n = \sup_n Vf_n = Vf .$$

Thus we have obtained the following result.

<u>7.6. PROPOSITION.</u> $\mathbb{V} = (V_\lambda)_{\lambda>0}$ is a sub-Markov resolvent on X and V is the potential kernel of $\mathbb{V}$.

<u>7.7. THEOREM.</u> Let V be a bounded kernel on X . Then there exists a sub-Markov resolvent $\mathbb{V} = (V_\lambda)_{\lambda>0}$ on X such that $\sup_{\lambda>0} V_\lambda = V$ if and only if V satisfies the complete maximum principle. The resolvent $\mathbb{V}$ is uniquely determined by V .

Proof. In view of (7.1) and (7.6) it suffices to show the uniqueness of $\mathbb{V}$. So let $(V'_\lambda)_{\lambda>0}$ be a second resolvent such that $\sup_{\lambda>0} V'_\lambda = V$ and let $\lambda > 0$ and $f \in \mathcal{B}_b(X)$. Then

$$V_\lambda f + \lambda V V_\lambda f = Vf = V'_\lambda f + \lambda V V'_\lambda f ,$$

defining $g := V_\lambda f - V'_\lambda f$ we hence have $g + \lambda Vg = 0$, i.e. $g = 0$ by (7.4). Thus $V_\lambda f = V'_\lambda f$. ⌟

We shall now apply the previous considerations to a balayage space $(X,\mathcal{W})$.

7.8. THEOREM. Let $1 \in \mathcal{W}$ and $p \in \mathcal{P}_b$. Then the following holds:

1. There exists a unique sub-Markov resolvent $\mathbb{V} = (V_\lambda)_{\lambda>0}$ such that $E_{\mathbb{V}} \subset \mathcal{W} \subset S_{\mathbb{V}}$ and $V_0 1 = p$. V_0 is the potential kernel associated with p.

2. For every strict $q_0 \in \mathcal{P}_b$, $\bigcap_{u \in \mathcal{W}} \{\sup \lambda V_\lambda u = u\} = \{\sup \lambda V_\lambda q_0 = q_0\} = \delta(p)$. In particular, $\delta(p)$ is a finely closed Borel set.

3. $E_{\mathbb{V}} = \mathcal{W} \Leftrightarrow \delta(p) = X \Leftrightarrow p$ is strict.

Proof. 1. Let V be the potential kernel associated with p. We claim first that every function $u \in \mathcal{W}$ is V-dominant. Indeed, let $u \in \mathcal{W}$ and $f \in \mathcal{B}_b(X)$ such that $Vf \leq u$ on $\{f > 0\}$. Let K be a compact subset of $\{f > 0\}$. Then $f1_K \leq f^+$, hence $V(f1_K) \leq Vf^+ \leq Vf^- + u$ on $\{f > o\}$. We have $q := V(f1_K) \in \mathcal{P}$, $C(q) \subset K$ and $Vf^- + u \in \mathcal{W}$. Therefore we conclude by (6.3) that $q \leq Vf^- + u$. Hence $Vf^+ \leq Vf^- + u$, i.e. $Vf \leq u$. Thus u is V-dominant.

In particular, the bounded kernel V satisfies the complete maximum principle. Hence by (7.7) there exists a (unique) sub-Markov resolvent $\mathbb{V} = (V_\lambda)_{\lambda>0}$ such that $V_0 = V$. Then of course $V_0 1 = V1 = p$. Since $V(\mathcal{B}_b^+(X)) \subset \mathcal{P}$ we conclude by (3.11) that $E_{\mathbb{V}} \subset \mathcal{W}$. Let $u \in \mathcal{W}$ and $\lambda > 0$. Then for every $n \in \mathbb{N}$ the bounded function $u_n := \inf(u,n) \in \mathcal{W}$ is V-dominant, hence $\lambda V_\lambda u_n \leq u_n$ by (7.2) and therefore $\lambda V_\lambda u \leq u$. Thus $\mathcal{W} \subset S_{\mathbb{V}}$.

Suppose now that $V' = (V'_\lambda)_{\lambda>0}$ is an arbitrary resolvent on X such that $E_{\mathbb{V}'} \subset \mathcal{W} \subset S_{\mathbb{V}'}$ and $V'_0 1 = p$. We want to prove that $\mathbb{V}' = \mathbb{V}$. By (7.7), it suffices to show that $V'_0 = V$.

So let $f \in \mathcal{B}_b^+(X)$. Then $V'_0 f \in E_{\mathbb{V}'}$ by (3.11), hence $V'_0 f \in \mathcal{W}$. Similarly $V'_0(\|f\| - f) \in \mathcal{W}$. Since

$$V'_0 f + V'_0(\|f\| - f) = V'_0(\|f\|) = \|f\| p$$

we have in fact $V'_0 f \in \mathcal{P}$. If $u \in \mathcal{W}$ such that $u \geq V'_0 f$ on $S(f)$ then $u \geq V'_0 f$ on X by (7.1). Hence $C(V'_0 f) \subset S(f)$ by (6.3). This shows that V'_0 is the potential kernel V associated with p.

2. Let $x \in \delta(p)$. The mapping

$$T : q \mapsto \lim_{\lambda\to\infty} \lambda V_\lambda q(x)$$

of P into $\mathbb{R}_+$ is additive, increasing and positively homogeneous (see (3.8)). By (I.1.4) there exists a measure τ on X such that

$$T(q) = \int q \, d\tau$$

for every $q \in P$. Evidently, $\int q \, d\tau \leq q(x)$ for every $q \in P$. Moreover,

$$\int p \, d\tau = \lim_{\lambda \to \infty} \lambda V_\lambda V1(x) = V1(x) = p(x) .$$

Thus $\tau = \varepsilon_x$, i.e. $\lim_{\lambda \to \infty} \lambda V_\lambda q(x) = q(x)$ for every $q \in P$ and hence $\lim_{\lambda \to \infty} \lambda V_\lambda u(x) = u(x)$ for every $u \in S(P) = W$.

Consider now $\mu, \nu \in M_+(X)$ such that $\mu(p) = \nu(p) < \infty$ and $\mu(q) \leq \nu(q)$ for every $q \in P$. Suppose furthermore that $q_o \in P_b$ is strict and

$$\lim_{\lambda \to \infty} \lambda V_\lambda q_o = q_o \qquad \nu\text{-a.e. .}$$

Let $f \in B_b^+(X)$. Then Vf, $V(\|f\| - f) \in P$, hence $\mu(Vf) \leq \nu(Vf)$ and $\mu(V(\|f\| - f)) \leq \nu(V(\|f\| - f))$. Since $V1 = p$ and $\mu(p) = \nu(p)$ we therefore have $\mu(Vf) = \nu(Vf)$. Since $\lambda V_\lambda q_o = V(\lambda(q_o - \lambda V_\lambda q_o))$ we conclude in particular that $\mu(\lambda V_\lambda q_o) = \nu(\lambda V_\lambda q_o)$ and hence

$$\mu(\lim_{\lambda \to \infty} \lambda V_\lambda q_o) = \nu(\lim_{\lambda \to \infty} \lambda V_\lambda q_o) = \nu(q_o) .$$

On the other hand

$$\mu(\lim_{\lambda \to \infty} \lambda V_\lambda q_o) \leq \mu(q_o) \leq \nu(q_o)$$

since $W \subset S_W$. Hence $\mu(q_o) = \nu(q_o)$ and therefore $\mu = \nu$ since q_o is strict.

Choosing $\nu = \varepsilon_x$ this shows in particular that $x \in \delta(p)$ if $\lim_{\lambda \to \infty} \lambda V_\lambda q_o(x) = q_o(x)$.

3. Knowing that $\delta(p)$ is the set of all $x \in X$ such that $\sup_{\lambda > 0} \lambda V_\lambda u = u$ for every $u \in W$ the first equivalence is obvious. Moreover the preceding considerations show that p is strict if $\lim_{\lambda \to \infty} \lambda V_\lambda q_o = q_o$, i.e. if $\delta(p) = X$. Conversely, if p is strict then of course $\delta(p) = X$. ⌟

7.9. REMARK. The assumption $1 \in W$ is not very restrictive. Indeed, let (X, W) be an arbitrary balayage space and choose a strictly positive $f \in C(X)$. Then $(X, \frac{1}{f} W)$ is evidently a balayage space having the same fine topology and such that for every numerical function g on X

$$\frac{1}{f}{}^{W}R_g = \inf\{\tfrac{u}{f} : u \in W, \tfrac{u}{f} \geq g\} = \tfrac{1}{f} \inf\{u \in W : u \geq f\,g\} = \tfrac{1}{f} \cdot {}^{W}R_{fg} .$$

If $f \in W$ then obviously $1 \in \frac{1}{f} W$.

Let $\mu, \nu \in M_+(X)$ and $g \in B^+(X)$. Then $\mu(g) \leq \nu(g)$ if and only if $(f\mu)\,(\frac{g}{f}) \leq (f\nu)\,(\frac{g}{f})$, and $\mu = \nu$ if and only if $f\mu = f\nu$. This implies that for every $p \in P$ the fine support $\delta(p)$ coincides with the fine support of the potential $\frac{p}{f}$ with respect to $(X, \frac{1}{f} W)$.

Fix $p \in P$ and $v_0 \in W \cap C(X)$ such that $v_0 > 0$ and take $f = p + v_0$. Then $1 \in \frac{1}{f} W$ and $\frac{p}{f}$ is a bounded potential with respect to $(X, \frac{1}{f} W)$. Applying (6.20), (7.8), and the previous considerations we obtain that $\delta(p)$ is a finely closed Borel set, $p_{\complement\delta(p)} = 0$, and that p is strict if and only if $\delta(p) = X$.

The characterization of the fine support of $p \in P$ obtained in (7.8) allows the reduit on $\delta(p)$ to be given in a very nice way from below.

7.10. PROPOSITION. Let $p \in P$, $u \in W$, and let F denote the set of all $q \in P$ such that $q \leq u$ and $q \prec \lambda p$ for some $\lambda \geq 0$. Then F is increasingly filtered and $\sup F = R_u^{\delta(p)}$.

Proof. Let $q_1, q_2 \in F$. Then $q := R_{\sup(q_1,q_2)} \in P$ and $\sup(q_1,q_2) \leq q \leq u$. If $\lambda_1, \lambda_2 \in \mathbb{R}_+$ such that $q_1 \prec \lambda_1 p$, $q_2 \prec \lambda_2 p$, then $q = R_{q_1+q_2-\inf(q_1,q_2)} \prec q_1+q_2 \prec (\lambda_1+\lambda_2)p$. This shows that $q \in F$. So F is increasingly filtered.

If $q \in F$ and $\lambda > 0$ such that $q \prec \lambda p$ then $\delta(q) \subset \delta(\lambda p) = \delta(p)$ and hence $R_u^{\delta(p)} \geq R_q^{\delta(q)} = q$ by (6.3). Therefore $R_u^{\delta(p)} \geq \sup F$.

The converse inequality is a consequence of the preceding theorem. Indeed, by (7.9) we may assume without loss of generality that $1 \in W$ and that p is bounded. Then by (7.8) there exists a resolvent $(V_\lambda)_{\lambda>0}$ on X such that V_0 is the potential kernel associated to p . For every $n \in \mathbb{N}$ let $u_n = \inf(u,n)$. Then $n\,V_n u_n \leq u_n \leq u$, $n\,V_n u_n = n\,V_0(u_n - n\,V_n u_n) \in P$, and $n\,V_n u_n \prec n\,V_0 n = n^2 p$, i.e. $n\,V_n u_n \in F$ for every $n \in \mathbb{N}$. Since $\sup n\,V_n u_n = u$ on $\delta(p)$ by (7.8), we obtain that $\sup F = u$ on $\delta(p)$. The set F being increasingly filtered we know that $\sup F \in \mathcal{W}$. Thus we finally conclude that $\sup F \geq R_u^{\delta(p)}$. ┘

The following characterization of specific minorants of a potential $p \in \mathcal{P}$ will allow us to sharpen the result of (7.10). This will be useful in section (VI.12).

7.11. PROPOSITION. Let $p \in \mathcal{P}$ and let V denote the potential kernel associated with p . Then for every $q \in \mathcal{P}$ the following statements are equivalent:

(1) $q \prec p$.

(2) There exists $f \in \mathcal{B}^+(X)$ such that $f \leq 1$ and $Vf = q$.

Proof. (1) ⇒ (2): By (7.9) we may assume without loss of generality that $1 \in \mathcal{W}$ and that p is bounded. Then by (7.8) there exists a sub-Markov resolvent $\mathbb{V} = (V_\lambda)_{\lambda>0}$ such that $E_{\mathbb{V}} \subset \mathcal{W} \subset S_{\mathbb{V}}$ and $V_0 = V$. Given $q \in \mathcal{P}$ satisfying $q \prec p$ we define functions $f_n \in \mathcal{B}^+(X)$ by

$$f_n = n(q - n\, V_n q) \qquad (n \in \mathbb{N}) .$$

Since $p - q \in \mathcal{W} \subset S_{\mathbb{V}}$ we obtain that, for every $n \in \mathbb{N}$, $n V_n(p-q) \leq p-q$ and hence

$$q - n\, V_n q \leq p - n\, V_n p = V1 - n V_n V1 = V_n 1 \leq \frac{1}{n} .$$

Therefore, $f_n \leq 1$ for every $n \in \mathbb{N}$.

Let (x_n) be a dense sequence in X and define

$$\mu = \sum_{n=1}^{\infty} \frac{1}{2^n} V(x_n, \cdot) .$$

Then μ is a finite measure on X since $\mu(X) = \sum_{n=1}^{\infty} 2^{-n} p(x_n) \leq \|p\|$. By the weak compactness of the unit ball in $\mathcal{B}_b(X)$ there exists a function $f \in \mathcal{B}^+(X)$ such that $f \leq 1$ and

$$\lim_{n\to\infty} \int f_n g \, d\mu = \int fg \, d\mu$$

for every μ-integrable function $g \in \mathcal{B}^+(X)$.

Now fix $x \in X$. If $A \in \mathcal{B}(X)$ such that $\mu(A) = 0$ then $V1_A(x_n) = 0$ for every $n \in \mathbb{N}$ and hence $V1_A(x) = 0$ since $V1_A$ is continuous. So by the Radon-Nikodym theorem the finite measure $V(x,\cdot)$ has an integrable density $g \in \mathcal{B}^+(X)$ with respect to μ . Thus

$$Vf(x) = \int fg\, d\mu = \lim_{n\to\infty} \int f_n g\, d\mu = \lim_{n\to\infty} Vf_n(x) = q(x) .$$

(2) ⇒ (1): If $f \in \mathcal{B}^+(X)$ and $f \leq 1$ then $V(1-f) \in P$ and $Vf + V(1-f) = V1 = p$. ┘

7.12. COROLLARY. Let $p \in P$ and $A \in \mathcal{B}(X)$ such that $p_A = p$, and let V be the potential kernel associated with p. Then for every $u \in W$ there exists a sequence (f_n) in $\mathcal{B}_b^+(X)$ such that (Vf_n) is increasing to $R_u^{\delta(p)}$ and $S(f_n)$ is a compact subset of A for every $n \in \mathbb{N}$.

Proof. We shall improve the result of (7.10) and then apply (7.11). Let $u \in W$, define F as in (7.10) and let F_A be the set of all $q \in F$ such that $C(q)$ is a compact subset of A. Given $q \in F$ we obviously have $q_K \in F_A$ for every compact subset of A, and $q = q_A = \sup\{q_K : K \text{ compact} \subset A\}$. Hence $\sup F_A = \sup F = R_u^{\delta(p)}$. If $q_1, q_2 \in F_A$ then $q := R_{\sup(q_1,q_2)} \in F_A$, since $q \in F$ and $q \prec q_1 + q_2$, hence $C(q) \subset C(q_1) \cup C(q_2)$. So F_A is increasingly filtered.

By (I.1.7), there exists an increasing sequence (q_n) in F_A such that $\sup q_n = R_u^{\delta(p)}$. Fix $n \in \mathbb{N}$. By (7.11), there exists a function $g_n \in \mathcal{B}_b^+(X)$ such that $Vg_n = q_n$. If L is a compact subset of $\complement C(q_n)$ then $C(V(g_n 1_L)) \subset C(q_n) \cap L = \emptyset$, i.e. $V(g_n 1_L) = 0$. Defining $f_n = g_n 1_{C(q_n)}$ we thus have $Vf_n = q_n$ and $S(f_n) \subset C(q_n)$ finishing the proof. ┘

7.13. EXERCISES. 1. Let V be a bounded kernel on X satisfying the complete maximum principle, let $u \in \mathcal{B}_b^+(X)$ be V-dominant, and let $\lambda > 0$.

a) $0 \leq (I + \lambda V)^{-1} u \leq u$ and $\{(I + \lambda V)^{-1} u > 0\} = \{u > 0\}$.

b) If $f \in \mathcal{B}_b^+(X)$ such that $\{f > 0\} \subset \{u > 0\}$ then $\{Vf > 0\} \subset \{u > 0\}$.

2. Let $p \in P$ and $\mu, \nu \in M_+(X)$ such that $\mu(q) \leq \nu(q)$ for every $q \in P$, $\mu(p) = \nu(p) < \infty$ and $\nu(\complement \delta(p)) = 0$. Then $\mu = \nu$.

8. Construction of Semigroups

Assume for a moment that $(P_t)_{t>0}$ is a sub-Markov semigroup on X with resolvent $(V_\lambda)_{\lambda>0}$ and bounded potential kernel V. Let $f \in \mathcal{B}_b^+(X)$ and $x \in X$.

Then

$$P_t\, Vf(x) = \int_t^\infty P_s f(x)\, ds \ ,$$

i.e. if we denote by μ the measure on $\mathbb{R}_+$ having density $s \longmapsto P_s f(x)$ with respect to Lebesgue measure we have

$$P_t\, Vf(x) = \mu(]t,\infty[) \ .$$

On the other hand the Laplace transform of μ is given by

$$L\mu(\alpha) = \int_0^\infty e^{-\alpha s}\, \mu(ds) = \int_0^\infty e^{-\alpha s}\, P_s f(x)\, ds$$

$$= V_\alpha f(x) = Vf(x) - \alpha V_\alpha Vf(x) \ ,$$

and in order to show that a function $g :]0,\infty[\longrightarrow \mathbb{R}$ is a Laplace transform we have to prove that g is completely monotone, i.e. that $(-1)^n g^{(n)} \geq 0$ for all $n \geq 0$. These observations are the motivation for (8.1) and the construction of semigroups given in (8.2).

<u>8.1. PROPOSITION.</u> Let $\mathbb{V} = (V_\lambda)_{\lambda>0}$ be a sub-Markov resolvent on X and $f \in \mathcal{B}^+(X)$ such that $f \leq u < \infty$ for some $u \in S_{\mathbb{V}}$. Then for each $x \in X$, the function $\alpha \longmapsto V_\alpha f(x)$ is infinitely differentiable on $]0,\infty[$ and if $D = \frac{d}{d\alpha}$,

$$D^n V_\alpha f = (-1)^n n! (V_\alpha)^{n+1} f \ ,$$

$$D^n(\alpha V_\alpha f) = (-1)^{n+1} n! (V_\alpha)^n (I - \alpha V_\alpha) f \ .$$

Proof. Let G be the set of all $g \in \mathcal{B}^+(X)$ such that $g \leq c\, u$ for some real $c \geq 0$. Since $\alpha V_\alpha u \leq u$ we clearly have $V_\alpha(G) \subset G$ for every $\alpha > 0$.

Let $g \in G$, $g \leq c\, u$ and $\alpha, \beta > 0$, $\alpha \neq \beta$. Then by the resolvent equation

$$\frac{V_\beta g - V_\alpha g}{\beta - \alpha} = - V_\beta V_\alpha g \ .$$

In particular, for every $i,j \in \mathbb{N}$

$$|(V_\beta - V_\alpha)(V_\beta)^j (V_\alpha)^i g| = |\beta - \alpha| (V_\beta)^{j+1} (V_\alpha)^{i+1} g \leq \frac{|\beta - \alpha|}{\alpha^{i+1} \beta^{j+1}}\, cu$$

Therefore the identity

$$(V_\beta)^n g - (V_\alpha)^n g = (V_\beta - V_\alpha)[(V_\beta)^{n-1} g + (V_\beta)^{n-2} V_\alpha g + \ldots + (V_\alpha)^{n-1} g]$$

shows that $\lim\limits_{\beta \to \alpha} (V_\beta)^n g = (V_\alpha)^n g$. Furthermore, it implies that

$$\lim_{\beta\to\alpha} \frac{(V_\beta)^n g - (V_\alpha)^n g}{\beta - \alpha} = \lim_{\beta\to\alpha} (-V_\beta V_\alpha[\ \dots\ ]) = -n(V_\alpha)^{n+1} g\ ,$$

i.e.

$$D(V_\alpha)^n g = -\ n(V_\alpha)^{n+1} g\ .$$

Proceeding by induction we obtain

$$D^n V_\alpha g = (-1)^n n!(V_\alpha)^{n+1} g\ .$$

Moreover

$$D(\alpha V_\alpha g) = V_\alpha g + \alpha D V_\alpha g = V_\alpha g - \alpha (V_\alpha)^2 g$$

and, for every $n \in \mathbb{N}$,

$$D[(V_\alpha)^{n-1}(I - \alpha V_\alpha)g] = -(n - 1)(V_\alpha)^n g - (V_\alpha)^n g + n\alpha(V_\alpha)^{n+1} g$$

$$= -n(V_\alpha)^n [I - \alpha V_\alpha] g\ .$$

Thus we finally conclude that for every $n \in \mathbb{N}$

$$D^n(\alpha V_\alpha g) = (-1)^{n+1} n!(V_\alpha)^n [I - \alpha V_\alpha] g\ .$$ ⌋

For every $F \subset B^+(X)$ let $\bar{S}(F)$ denote the smallest set $\bar{F} \subset B^+(X)$ such that $F \subset \bar{F}$ and $\bar{F}$ is closed with respect to increasing limits, i.e. $S(\bar{F}) \subset \bar{F}$. Obviously $F \subset S(F) \subset S(S(F)) \subset \ldots \subset \bar{S}(F)$.

<u>8.2. THEOREM.</u> Let $\mathbb{V} = (V_\lambda)_{\lambda>0}$ be a sub-Markov resolvent on X and $P \subset S_{\mathbb{V}}$ a min-stable function cone such that $1 \in S(P)$ and $V_\lambda(P) \subset \bar{S}(P)$ for every $\lambda > 0$. Then there exists a unique sub-Markov semigroup $\mathbb{P} = (P_t)_{t>0}$ on X such that $t \longmapsto P_t q$ is right continuous for every $q \in P$ and $\mathbb{V}$ is the resolvent of $\mathbb{P}$. In fact, $t \longmapsto P_t v$ is right continuous for every $v \in \bar{S}(P)$.

Proof. Let us first suppose that V_0 is bounded $V_0 q < \infty$ for every $q \in P$. Let $q \in P, x \in X$ and define $f : \mathbb{R}_+ \to \mathbb{R}_+$ by

$$f(\alpha) = \tilde{q}(x) - \alpha V_\alpha q(x)\ .$$

Then $f(0) = \lim_{\alpha\downarrow 0} f(\alpha) = \tilde{q}(x)$ and

$$(-1)^n D^n f(\alpha) = n!(V_\alpha)^n (I - \alpha V_\alpha) q(x) \geq 0\ ,$$

i.e. f is completely monotone. Consequently by the theorem of Bernstein (I.4.2) there exists a positive measure λ_x^q on $\mathbb{R}_+$ such that

$$\tilde{q}(x) - \alpha V_\alpha q(x) = \int_{\mathbb{R}_+} e^{-\alpha t}\ \lambda_x^q(dt)$$

for every $\alpha \geq 0$. By the very definition of $\tilde{q}$ we have

$$\lambda_x^q(\{0\}) = \lim_{\alpha\to\infty} \int_{\mathbb{R}_+} e^{-\alpha t}\, \lambda_x^q(dt) = \tilde{q}(x) - \lim_{\alpha\to\infty} \alpha V_\alpha q(x) = 0 ,$$

and so λ_x^q is carried by $]0,\infty[$. For every $t \geq 0$ we define

$$P_t q(x) = \lambda_x^q(]t,\infty[) .$$

Then the mapping $t \mapsto P_t q(x)$ is decreasing, right continuous and $\lim_{t\downarrow 0} P_t q(x) = \tilde{q}(x)$ since λ_x^q does not charge $\{0\}$. Moreover,

$$\int_0^\infty e^{-\alpha t}\, P_t q(x)\, dt = \int_0^\infty e^{-\alpha t} \left(\int_{]t,\infty[} \lambda_x^q(ds)\right) dt$$

$$= \int_0^\infty \left(\int_0^s e^{-\alpha t}\, dt\right) \lambda_x^q(ds) = \frac{1}{\alpha}\int_0^\infty (1 - e^{-\alpha s})\, \lambda_x^q(ds)$$

$$= \frac{1}{\alpha}(\tilde{q}(x) - [\tilde{q}(x) - \alpha V_\alpha q(x)]) = V_\alpha q(x) .$$

By the uniqueness theorem for Laplace transforms the mapping $q \mapsto \lambda_x^q$ is additive and positively homogeneous. Hence, for every $t \geq 0$, the mapping $q \mapsto P_t q(x)$ is additive and positively homogeneous. Let $p,q \in P$ such that $q \leq p$ and $f = p - q$. Then, for every $n \in \mathbb{N}$,

$$(-1)^n D^n (V_\alpha f) = n!(V_\alpha)^{n+1} f \geq 0 .$$

Hence, by the theorem of Bernstein $\alpha \mapsto V_\alpha f(x)$ is the Laplace transform of a positive measure. On the other hand,

$$V_\alpha f(x) = V_\alpha p(x) - V_\alpha q(x) = \int_0^\infty e^{-\alpha t}(P_t p(x) - P_t q(x))\, dt .$$

Consequently by the uniqueness of the Laplace transform $P_t p(x) - P_t q(x) \geq 0$ for every $t \geq 0$. Applying (I.1.4) we thus obtain, for every $t \geq 0$ and every $x \in X$, a unique measure $P_t(x,\cdot)$ on X such that

$$P_t q(x) = \int q(y) P_t(x,dy)$$

for every $q \in P$. Since $t \mapsto P_t q(x)$ is right continuous and $V_\alpha q(x) = \int_0^\infty e^{-\alpha t} P_t q(x) dt$. for every $q \in P$ we conclude that $t \mapsto P_t f(x)$ is measurable and

$$V_\alpha f(x) = \int_0^\infty e^{-\alpha t} P_t f(x) dt$$

for every $f \in B^+(X)$.

Let $q \in P$. Then

$$(-1)^n D^n(\frac{q}{\alpha} - V_\alpha q) = \frac{n!}{\alpha^{n+1}}(q - (\alpha V_\alpha)^{n+1} q) \geq 0 .$$

Since $\alpha \mapsto \frac{q}{\alpha} - V_\alpha q$ is the Laplace transform of $t \mapsto q - P_t q$ and the mapping $t \mapsto q - P_t q$ is right continuous we conclude that $q - P_t q \geq 0$ for every $t \geq 0$.

By assumption, there exists a sequence (q_n) in P such that $\sup q_n = 1$. Therefore

$$P_t 1 = \sup P_t q_n \leq \sup q_n \leq 1 .$$

By (3.15), the mapping $(t,x) \mapsto P_t f(x)$ is measurable for every $f \in P$ and hence for every $f \in B^+(X)$.

Let F be the set of all $f \in B^+(X)$ such that $t \mapsto P_t f(x)$ is decreasing and right continuous. Then $P \subset F$ and $S(F) \subset F$, hence $\bar{S}(F) \subset F$. In particular, $t \mapsto P_t V_\alpha q(x)$ is decreasing and right continuous for every $q \in P$ and $\alpha > 0$. To complete the proof of existence under the present hypotheses it remains to show that $(P_t)_{t \geq 0}$ forms a semigroup. Let $q \in P$ and $t \geq 0$. Then both $s \mapsto P_t P_s q$ and $s \mapsto P_{t+s} q$ are right continuous in s , and so it suffices to show that they have the same Laplace transform. That is, after an obvious change of integration,

$$\text{(i)} \quad P_t V_\alpha q(x) = \int_0^\infty e^{-\alpha s} P_{t+s} q \, ds = e^{\alpha t} \int_t^\infty e^{-\alpha s} P_s q \, ds .$$

But for fixed $\alpha > 0$ both sides of (i) are right continuous in t . Thus (i) will hold identically in t provided both sides have the same Laplace transform. Thus the proof of the semigroup property reduces to verifying

$$\text{(ii)} \quad V^\beta V^\alpha q(x) = \int_0^\infty e^{-\beta t} e^{\alpha t} (\int_t^\infty e^{-\alpha s} P_s q \, ds) \, dt .$$

The right hand side of (ii) is equal to

$$\int_0^\infty e^{-\alpha s} P_s q \, (\int_0^s e^{-(\beta-\alpha)t} \, dt) \, ds = (\beta - \alpha)^{-1} \int_0^\infty (e^{-\alpha s} - e^{-\beta s}) \, P_s q \, ds$$

$$= (\beta - \alpha)^{-1}(V_\alpha q - V_\beta q) = V_\beta V_\alpha q$$

provided $\beta \neq \alpha$. If $\beta = \alpha$ the same computation shows that the right side of (ii) equals

$$\int_0^\infty s e^{-\alpha s} P_s q \, ds = -D(\int_0^\infty e^{-\alpha s} P_s q \, ds) = -DV_\alpha q = (V_\alpha)^2 q$$

which establishes (ii) when $\beta = \alpha$.

For the general case let $\beta > 0$ and consider $\mathbb{V}^\beta = (V_{\beta+\lambda})_{\lambda>0}$. By what has already been proved there exists a unique sub-Markov semigroup $(P_t^\beta)_{t\geq 0}$ on X such that $t \longmapsto P_t^\beta q$ is right continuous for every $q \in P$ and

$$V_{\alpha+\beta} q = \int_0^\infty e^{-\alpha t} P_t^\beta q \, dt$$

for every $\alpha > 0$. For all $\alpha > \beta > \gamma > 0$ and $q \in P$

$$\int_0^\infty e^{-\alpha t} e^{\beta t} P_t^\beta q \, dt = V_\alpha q = \int_0^\infty e^{-\alpha t} e^{\gamma t} P_t^\gamma q \, dt$$

and hence $e^{\beta t} P_t^\beta q = e^{\gamma t} P_t^\gamma q$ for every $t \geq 0$ by the uniqueness of the Laplace transform. Defining

$$P_t = e^{\beta t} P_t^\beta \quad , \quad \beta > 0$$

we thus obtain a semigroup $(P_t)_{t\geq 0}$ on X such that

$$V_\alpha = \int_0^\infty e^{-\alpha t} P_t \, dt$$

for every $\alpha > 0$. The mapping $t \longmapsto P_t v$ is right continuous for every $v \in \bar{S}(P)$ and $P_t 1 \leq 1$ for every $t \geq 0$ since $P_t 1 = e^{\beta t} P_t^\beta 1 \leq e^{\beta t}$ for every $\beta > 0$. The uniqueness of (P_t) follows from the uniqueness theorem of Laplace transforms. ┘

8.3. COROLLARY. Let $\mathbb{V} = (V_\lambda)_{\lambda>0}$ be a sub-Markov resolvent on X such that $E_\mathbb{V} \subset S(P) \subset S_\mathbb{V}$ for some function cone P . Then there exists a unique sub-Markov semigroup $\mathbb{P} = (P_t)_{t>0}$ on X such that $\mathbb{V}$ is the resolvent of $\mathbb{P}$ and $t \mapsto P_t q$ is right continuous for every $q \in S(P)$.

Proof. $P' = S_\mathbb{V} \cap C_P(X)$ is a min-stable function cone on X , $1 \in S(P')$ and by (3.10), $V_\lambda(P') \subset V_\lambda(S_\mathbb{V}) \subset E_\mathbb{V} \subset S(P')$. Thus the statement follows from (8.2). ┘

8.4. COROLLARY. Let X be compact and let $\mathbb{V} = (V_\lambda)_{\lambda>0}$ be a sub-Markov resolvent on X such that $V_\lambda(C(X)) \subset C(X)$ for every $\lambda > 0$ and $S_\mathbb{V} \cap C(X)$ separates the points of X . Then there exists a unique sub-Markov semigroup $\mathbb{P} = (P_t)_{t>0}$ on X such that $t \longmapsto P_t f$ is right continuous for every $f \in C(X)$

and $\mathbb{V}$ is the resolvent of $\mathbb{P}$.

The preceding corollary is basically the theorem of the existence of semigroups associated with Ray resolvents (see exercise 8.7.2).

8.5. LEMMA. Let $(X,\mathcal{W})$ be a balayage space such that $1 \in \mathcal{W}$, let $\mathbb{P}$ be a sub-Markov semigroup on X such that $E_{\mathbb{P}} = \mathcal{W}$ and let $\mathbb{V}$ be the resolvent of $\mathbb{P}$. Then the following statements are equivalent:

(1) For every $t > 0$, $P_t(\mathcal{P}_c) \subset \mathcal{C}_b(X)$.

(2) For every $t > 0$, $P_t(\mathcal{P}) \subset \mathcal{P}$.

(3) For every $t > 0$, $P_t(\mathcal{C}_{\mathcal{P}}(X)) \subset \mathcal{C}_{\mathcal{P}}(X)$.

(4) For every $\lambda > 0$, $V_\lambda(\mathcal{P}_c) \subset \mathcal{C}_b(X)$.

(5) For every $\lambda > 0$, $V_\lambda(\mathcal{P}) \subset \mathcal{P}$.

(6) For every $\lambda > 0$, $V_\lambda(\mathcal{C}_{\mathcal{P}}(X)) \subset \mathcal{C}_{\mathcal{P}}(X)$.

Proof. (1) ⇒ (2): Fix $p \in \mathcal{P}$ and choose an increasing sequence (K_n) of compact sets such that $X = \bigcup_{n=1}^{\infty} K_n$. For every $n \in \mathbb{N}$ let

$$p_n := p_{K_n}, \qquad q_n := p_{\complement K_n}.$$

Then $p_n \in \mathcal{P}_c$, $q_n \in \mathcal{P}$ and $p_n + q_n = p$. Since (q_n) is decreasing and $\lim_{n\to\infty} q_n = 0$, the sequence (q_n) converges locally uniformly to zero. Fix $t > 0$. Then for every $n \in \mathbb{N}$

$$P_t p = P_t p_n + P_t q_n$$

where $P_t p_n \in \mathcal{C}_b(X)$ and $P_t q_n \leq q_n$. Hence $P_t p$ is continuous. Furthermore, $P_t p \in \mathcal{W}$ by (3.3) and $P_t p \leq p$. Hence $P_t p \in \mathcal{W} \cap \mathcal{C}_{\mathcal{P}}(X) = \mathcal{P}$.

(2) ⇒ (3): Let $f \in \mathcal{C}_{\mathcal{P}}(X)$ and $p_o \in \mathcal{P}$ such that $|f| \leq p_o$. Choose $q_o \in \mathcal{P}$ such that $p_o \in o(q_o)$. Given $\varepsilon > 0$, we obtain from (I.1.3) that there exist $p, q \in \mathcal{P}$ satisfying

$$p - q \leq f \leq p - q + \varepsilon q_o .$$

Then, for every $t > 0$,

$$P_t p - P_t q \leq P_t f \leq P_t p - P_t q + \varepsilon P_t q_o$$

where $P_t q_o \leq q_o$. Therefore $P_t(\mathcal{C}_{\mathcal{P}}(X)) \subset \mathcal{C}_{\mathcal{P}}(X)$.

(3) ⇒ (1): Trivial since every $p \in \mathcal{P}$ is a bounded function in $\mathcal{C}_{\mathcal{P}}(X)$ and

P_t is a sub-Markov kernel.

(4) ⇒ (5) ⇒ (6) ⇒ (4): Similarly using the fact that $V_\lambda(\mathcal{W}) \subset \mathcal{W}$ by (3.10) and $\lambda V_\lambda 1 \leq 1$.

(1) ⇒ (4): Immediate consequence of Lebesgue's convergence theorem.

(4) ⇒ (1): Let $q \in \mathcal{P}_c$, $t > 0$, $x \in X$ and suppose that $P_t q$ is not continuous at x. Since $P_t q \in \mathcal{W}$ is l.s.c., there hence exists an $\varepsilon > 0$ and a sequence (x_n) in X such that $\lim_{n\to\infty} x_n = x$ and

$$P_t q(x_n) \geq P_t q(x) + 2\varepsilon$$

for every $n \in \mathbb{N}$. There exists $\eta > 0$ such that $P_\eta q + \varepsilon > q$ on $C(q)$ and hence $P_\eta q + \varepsilon \geq q$. If $t < s < t + \eta$ then $P_t q \leq P_{t+\eta} q + P_t \varepsilon \leq P_s q + \varepsilon$ and hence for every $n \in \mathbb{N}$

$$P_s q(x) \leq P_t q(x) \leq P_t q(x_n) - 2\varepsilon \leq P_s q(x_n) - \varepsilon .$$

Thus for every $\lambda > 0$,

$$\liminf_{n\to\infty} V_\lambda q(x_n) \geq \int_0^\infty e^{-\lambda s} \liminf_{n\to\infty} P_s q(x_n)\, ds \geq \int_0^\infty e^{-\lambda s} P_s q(x)\, ds + \varepsilon\eta e^{-\lambda(t+\eta)} > V_\lambda q(x)$$

and hence $V_\lambda q$ is not continuous at x. ┘

8.6. THEOREM. Let $(X,\mathcal{W})$ be a balayage space such that $1 \in \mathcal{W}$, and let $p \in \mathcal{P}_b$ be a strict potential. Then there exists a unique sub-Markov semigroup $\mathbb{P} = (P_t)_{t>0}$ on X such that $E_{\mathbb{P}} = \mathcal{W}$ and the potential kernel V of $\mathbb{P}$ satisfies $V1 = p$. V is the potential kernel associated with p. Furthermore, $P_t(\mathcal{P}) \subset \mathcal{P}$, $P_t(\mathcal{K}(X)) \subset \mathcal{C}_b(X)$, and $P_t(\mathcal{C}_\mathcal{P}(X)) \subset \mathcal{C}_\mathcal{P}(X)$ for every $t > 0$.

Proof. By (7.8) there exists a sub-Markov resolvent $\mathbb{V}$ on X such that $E_{\mathbb{V}} = \mathcal{W}$ and the potential kernel V of $\mathbb{V}$ is the potential kernel associated with p. By (8.3), there exists a sub-Markov semigroup $\mathbb{P} = (P_t)_{t>0}$ on X such that $\mathbb{V}$ is the resolvent of $\mathbb{P}$. In particular, $E_{\mathbb{P}} = E_{\mathbb{V}} = \mathcal{W}$.

By the resolvent equation we have $V_\lambda(\mathcal{P}_b) \subset V(\mathcal{B}_b(X)) \subset \mathcal{C}_b(X)$ for every $\lambda > 0$, hence $P_t(\mathcal{P}) \subset \mathcal{P}$ and $P_t(\mathcal{C}_\mathcal{P}(X)) \subset \mathcal{C}_\mathcal{P}(X)$ for every $t > 0$ by (8.5). Moreover, $P_t(\mathcal{K}(X)) \subset \mathcal{C}_b(X)$ for every $t > 0$ since P_t is sub-Markov and $\mathcal{K}(X) \subset \mathcal{C}_\mathcal{P}(X)$.

Suppose now that $\mathbb{P}'$ is an arbitrary semigroup on X such that $E_{\mathbb{P}'} = \mathcal{W}$ and the potential kernel V' of $\mathbb{P}'$ satisfies $V'1 = 1$. By (3.16), $\mathbb{P}'$ is measurable. Let $\mathbb{V}'$ be the resolvent of $\mathbb{P}'$. Then $\mathbb{V}' = \mathbb{V}$ by (7.8) and hence $\mathbb{P}' = \mathbb{P}$ by (8.6) and (3.2). ┘

8.7. EXERCISES. 1. Let X be compact and $\mathbb{V} = (V_\lambda)_{\lambda>0}$ a sub-Markov resolvent on X such that

(R_1) $V_\lambda(C(X)) \subset C(X)$ for every $\lambda > 0$.

Then the following properties are equivalent:

(R_2) $\bigcup_{\lambda>0} S_{\mathbb{V}^\lambda} \cap C(X)$ separates the points of X.

(R_2') $S_{\mathbb{V}^\lambda} \cap C(X)$ separates the points of X for some $\lambda > 0$.

(R_2'') $S_{\mathbb{V}^\lambda} \cap C(X)$ separates the points of X for all $\lambda > 0$.

2. Let X be compact and let $\mathbb{V} = (V_\lambda)_{\lambda>0}$ be a *Ray resolvent* on X, i.e. the sub-Markov resolvent $\mathbb{V}$ satisfies (R_1) and (R_2). Then there exists a unique sub-Markov semigroup $\mathbb{P} = (P_t)_{t>0}$ on X such that $t \mapsto P_t f$ is right continuous for every $f \in C(X)$ and $\mathbb{V}$ is the resolvent of $\mathbb{P}$. (Hint: Use (8.4) and the idea finishing the proof of (8.2).)

3. Let X be the subset of $\mathbb{R}^2$ consisting of all points $x_s = (s,0)$, $s \in [0,1]$, and $x_s^\pm = (s, \pm 1)$, $s \in [-1,0]$. For every $s < -1$ let $x_s^\pm = x_{-1}^\pm$. Given $t > 0$, define

$$P_t(x_s,\cdot) = \begin{cases} \varepsilon_{x_{s-t}} & , \quad s - t \geq 0 \\ \frac{1}{2}\varepsilon_{x_{s-t}^+} + \frac{1}{2}\varepsilon_{x_{s-t}^-} & , \quad s - t < 0 \end{cases}$$

if $s \geq 0$, and

$$P_t(x_s^\pm,\cdot) = \varepsilon_{x_{s-t}^\pm}$$

if $s < 0$. Then $\mathbb{P} = (P_t)_{t>0}$ is a measurable Markov semigroup on X such that the resolvent $\mathbb{V}$ of $\mathbb{P}$ is a Ray resolvent. However, $(X, E_{\mathbb{P}1})$ is no balayage space.

III. Hyperharmonic Functions

The definition of hyperharmonic functions in classical potential theory and the properties of Poisson integrals suggest a general definition of a family of harmonic kernels and its (hyper)harmonic functions. The actual choice of the axioms in section 1 will guarantee a one-to-one correspondence between these families and balayage spaces.

In section 2 we show that reducing of unit masses on the complement of open subsets of a balayage space defines an associated family of harmonic kernels. Conversely, the study of convergence properties of harmonic functions (section 3), of sheaf properties and minimum principles (section 4), of regularizations (section 5), and finally of potentials leads to the result that every family of harmonic kernels is associated with a balayage space (section 6).

A balayage space having no absorbing points (section 7) and satisfying the local truncation property (section 8) is a harmonic space, it can be characterized by corresponding properties of its associated family of harmonic kernels and sub-Markov semigroups.

1. Harmonic Kernels

Throughout this chapter X will denote a locally compact space with countable base. For every family $\mathcal{U}$ of open subsets of X, every $x \in X$ and open subset V of X we define

$$\mathcal{U}_x := \{U \in \mathcal{U} : x \in U\} \ , \ \mathcal{U}(V) := \{U \in \mathcal{U} : \overline{U} \subset V\} \ .$$

Let U be an open subset of X. A kernel H_U on X will be called a *sweeping kernel* (relative to U) if $H_U(x,U) = 0$ for every $x \in U$ and $H_U(x,\cdot) = \varepsilon_x$ for every $x \in \complement U$. Let F be a filter on U converging to a point $z \in U^*$. We shall say that F is *regular* (with respect to H_U) if $\lim_{x,F} H_U(x,\cdot) = \varepsilon_z$, i.e. $\lim_F H_U f = f(z)$ for every $f \in K(X)$. Otherwise F is called *non-regular*.

Now let U be a base of relatively compact open subsets of X and let $(H_U)_{U \in U}$ be a family of corresponding sweeping kernels. For every open subset V of X let $^*H^+(V)$ denote the set of all positive *hyperharmonic* functions on V, i.e.

$$^*H^+(V) := \{v \in B^+(X) : v_{|V} \text{ l.s.c.}, H_U v \leq v \,\forall U \in U(V)\},$$

let $S^+(V)$ denote the set of all positive *superharmonic* functions on V, i.e.

$$S^+(V) := \{s \in {}^*H^+(V) : H_U s_{|U} \in C(U) \;\forall\; U \in U(V)\}$$

and let $H^+(V)$ denote the set of all positive *harmonic* functions on V, i.e.

$$\begin{aligned} H^+(V) &:= \{h \in S^+(V) : H_U h = h \;\forall\; U \in U(V)\} \\ &= \{h \in B^+(X) : h_{|V} \in C(V), H_U h = h \;\forall\; U \in U(V)\}. \end{aligned}$$

For every function $f : X \to \overline{\mathbb{R}}$ the *reduit* R_f is defined by

$$R_f := \inf\{u \in {}^*H^+(X) : u \geq f\}.$$

For every $A \subset X$ and $f : X \to \overline{\mathbb{R}}_+$ we define

$$R_f^A := R_{f1_A} = \inf\{u \in {}^*H^+(X) : u \geq f \text{ on } A\}.$$

$(H_U)_{U \in U}$ will be called a *family of harmonic kernels* if the following axioms are satisfied:

(H_1) For every $x \in X$, $\lim_{U,U_x} H_U(x,\cdot) = \varepsilon_x$ or $R_1^{\{x\}}$ is l.s.c. at x.

(H_2) $H_V H_U = H_U$ for every $V, U \in U$ with $\overline{V} \subset U$.

(H_3) $H_U f \in C_b(U)$ for every $U \in U$ and every $f \in B_b(X)$ with compact support.

(H_4) For every $x \in U \in U$ there exists an Evans function w, i.e. a function $w \in {}^*H^+(U)$ such that $w(x) < \infty$ and $\lim_F w = \infty$ for every non-regular ultrafilter F on U.

(H_5) $^*H^+(X)$ is linearly separating, and there exists a strictly positive function $s_0 \in S^+(X) \cap C(X)$.

In the next section we shall see that for every balayage space the reducing of measures on the complement of open subsets gives a family of harmonic kernels. Sections 3 - 6 are devoted to the converse. Treating these problems we have to prove properties which prepare a deeper study of potential theory starting in chapter VI. But first of all we look at our examples.

1.1. EXAMPLES

1. Classical potential theory.

Let X be an open subset of $\mathbb{R}^n$, $n \geq 1$, and suppose that X is relatively compact if $n \leq 2$. Let $\mathcal{U}$ be the family of all open balls $B_r(a)$ such that $\overline{B}_r(a) \subset X$. For every $U = B_r(a) \in \mathcal{U}$ we define a sweeping kernel H_U by

$$H_U f(x) = \int f \, P_x^{a,r} \, d\sigma_{a,r} \qquad (f \in \mathcal{B}^+(X)\ ,\ x \in U)\ .$$

Then $(H_U)_{U \in \mathcal{U}}$ is a family of harmonic kernels on X such that for every open subset V of X the set $\mathcal{H}^+(V)$ (resp. $^*\mathcal{H}^+(V)$) defined by $(H_U)_{U \in \mathcal{U}}$ is the set of all positive functions on V which are harmonic (resp. hyperharmonic) in the classical sense.

Indeed it suffices to recall the results of chapter 0. For every $x \in U \in \mathcal{U}$, $H_U(x, U^*) = 1$, $H_U(x, \complement U^*) = 0$. Hence (H_1) is satisfied. Let $V, U \in \mathcal{U}$ such that $\overline{V} \subset U$ and let $f \in \mathcal{B}_b^+(X)$. By (0.1.2), $H_U f$ is harmonic on U (in the classical sense) and hence $H_V H_U f = H_U f$ by (0.1.3). Hence (H_2) and (H_3) hold. Furthermore, let $\mathcal{F}$ be an ultrafilter on U converging to $z \in U^*$. Then $\mathcal{F}$ is regular by (0.1.2). Therefore (H_4) is trivially satisfied: we may choose $w = 0$. The restriction of every affinely linear function $u : \mathbb{R}^n \to \mathbb{R}$ on X is contained in $\mathcal{H}(X)$ and hence in $\mathcal{S}(X)$. Therefore (H_5) is satisfied if X is relatively compact. Suppose now that $n \geq 3$ and let $x, y \in X$, $x \neq y$, and $\lambda > 0$. Then $N^x \in {}^*\mathcal{H}^+(X)$, $N^x(x) = \infty > \lambda N^x(y)$. Furthermore, $1 \in \mathcal{S}^+(X) \cap \mathcal{C}(X)$. Thus (H_5) holds as well.

2. Translation on $\mathbb{R}$.

Let $X = \mathbb{R}$ and $\mathcal{U} = \{]\alpha,\beta[\ : \alpha,\beta \in \mathbb{R}, \alpha < \beta\}$. For every $U =]\alpha,\beta[$ we define a sweeping kernel H_U by

$$H_U(x,\cdot) = \varepsilon_\alpha \qquad (x \in]\alpha,\beta[) .$$

Then $(H_U)_{U \in \mathcal{U}}$ is a family of harmonic kernels on $\mathbb{R}$. Indeed, (H_1), (H_2) and (H_3) are easily verified. Furthermore, let $U =]\alpha,\beta[$ and define $w : X \to \mathbb{R}_+$ by

$$w(x) = \begin{cases} \frac{1}{\beta - x} & , x \in U \\ 0 & , x \in \complement U . \end{cases}$$

Then $w \in {}^*\mathcal{H}^+(U)$ and $\lim\limits_{x \to \beta} w(x) = \infty$. Since trivially $\lim\limits_{x \to \alpha} H_U(x,\cdot) = \varepsilon_\alpha$, we conclude that (H_4) holds. (H_5) is satisfied since e.g. the functions $1, e^x \in \mathcal{S}^+(\mathbb{R})$ linearly separate the points of $\mathbb{R}$.

We finally note that, for every open subset V of $\mathbb{R}$, $\mathcal{H}^+(V)$ (${}^*\mathcal{H}^+(V)$ resp.) is the set of all positive functions on V which are constant and real-valued (increasing and left-continuous resp.) on every interval contained in V . In particular, ${}^*\mathcal{H}^+(\mathbb{R}) = E_{\mathbb{T}}$.

3. Discrete potential theory.

Let X be discrete and let P be a sub-Markov kernel on X . For every $x \in X$ we define a sweeping kernel $H_{\{x\}}$ by

$$H_{\{x\}}(x,A) = \begin{cases} \frac{P(x,A \setminus \{x\})}{1-P(x,\{x\})} & , \quad P(x,\{x\}) < 1 \\ 0 & , \quad P(x,\{x\}) = 1 . \end{cases}$$

Then the family $(H_{\{x\}})_{x \in X}$ trivially satisfies (H_1) - (H_4) . Since $P(x,\complement\{x\}) = 0$ if $P(x,\{x\}) = 1$ we have for every positive numerical function v on X

$$(*) \qquad \begin{aligned} Pv(x) &= P(x,\{x\})v(x) + P(v1_{\complement\{x\}})(x) \\ &= P(x,\{x\})v(x) + (1 - P(x,\{x\}))H_{\{x\}}v(x) . \end{aligned}$$

This shows that $Pv \leq v$ if and only if $H_{\{x\}}v \leq v$ for every $x \in X$, i.e. $S_P = {}^*\mathcal{H}^+(X)$. Moreover, $\{v \in S_P : Pv < \infty\} = \mathcal{S}^+(X)$, hence $1 \in \mathcal{S}^+(X)$, and (H_5) is satisfied if and only if S_P separates the points of X .

This condition holds e.g. if there exists a function $v \in S_P$ such that $\inf P^n v < v$. Indeed, let $x,y \in X$, $x \neq y$. Since $(P^n v)$ is a decreasing sequence in S_P it remains to consider the case where $P^n v(x) = P^n v(y)$ for every $n \in \mathbb{N}$. Let $n \in \mathbb{N}$ such that $P^n v(x) < P^{n-1} v(x)$ and define

$$w(z) = \begin{cases} P^{n-1}v(z) & , \quad z \neq x , \\ P^n v(x) & , \quad z = x . \end{cases}$$

Then $P^n v \leq w \leq P^{n-1} v$, hence $w \in S_P$ and

$$w(x) = P^n v(x) < P^{n-1} v(x) = P^{n-1} v(y) = w(y) .$$

We finally note that in case $P(x,\{x\}) < 1$ for every $x \in X$ the equation (*) implies that a real function $h \geq 0$ on X is harmonic if and only if $Ph = h$.

For example, if $X = \mathbb{Z}$ and $P(x,\cdot) = \varepsilon_{x+1}$, then $Pv < v$ for every strictly decreasing function $v : \mathbb{Z} \to \mathbb{R}_+$, hence

$$H_{\{x\}}(y,\cdot) = \begin{cases} \varepsilon_{x+1} & , \; y = x \\ \varepsilon_y & , \; y \neq x \end{cases} \quad (x \in \mathbb{Z})$$

defines a family of harmonic kernels on $\mathbb{Z}$.

In the following let $(H_U)_{U \in \mathcal{U}}$ *be a family of harmonic kernels on* X .

<u>1.2. LEMMA.</u> Let $U \in \mathcal{U}$ and $f \in \mathcal{B}^+(X)$. Then $H_U f \in {}^*\mathcal{H}^+(U)$. If there exists a function $g \in \mathcal{B}^+(X)$ such that $f \leq g$ and $H_U g|_U \in \mathcal{C}(U)$ then $H_U f \in \mathcal{H}^+(U)$.

Proof. There exist $f_n \in \mathcal{B}_b^+(X)$ with compact support such that $f_n \uparrow f$. Hence

$$H_U f = \sup H_U f_n$$

is l.s.c. on U by (H_3). Thus $H_U f \in {}^*\mathcal{H}^+(U)$ by (H_2). Now let $g \in \mathcal{B}^+(X)$ such that $f \leq g$ and $H_U g|_U \in \mathcal{C}(U)$. There exists $f' \in \mathcal{B}^+(X)$ with $f + f' = g$. Hence

$$H_U f + H_U f' = H_U g$$

where $H_U f$ and $H_U f'$ are l.s.c. on U . Therefore $H_U f|_U \in \mathcal{C}(U)$ and even $H_U f \in \mathcal{H}^+(U)$ by (H_2). ┘

The following proposition is an immediate consequence of the definitions and of (1.2).

<u>1.3. PROPOSITION.</u> Let V be an open subset of X . Then the following statements hold:

1. $^*\mathcal{H}^+(V)$ is a min-stable convex cone. If (v_n) is an increasing sequence in $^*\mathcal{H}^+(V)$ then $\sup v_n \in {}^*\mathcal{H}^+(V)$.

2. $\mathcal{S}^+(V)$ is a convex subcone of $^*\mathcal{H}^+(V)$. If $s \in \mathcal{S}^+(V)$ and $v \in {}^*\mathcal{H}^+(V)$ then $\inf(s,v) \in \mathcal{S}^+(V)$.

3. $\mathcal{H}^+(V)$ is a convex subcone of $\mathcal{S}^+(V)$ which is hereditary to the left, i.e. if $s,t \in \mathcal{S}^+(V)$ such that $s + t \in \mathcal{H}^+(V)$ then $s,t \in \mathcal{H}^+(V)$.

4. If W is an open subset of V then $^*\mathcal{H}^+(V) \subset {}^*\mathcal{H}^+(W)$, $\mathcal{S}^+(V) \subset \mathcal{S}^+(W)$ and $\mathcal{H}^+(V) \subset \mathcal{H}^+(W)$.

1.4. PROPOSITION. Let V be an open subset of X and let $\sum_{n=1}^{\infty} s_n$ be a series in $\mathcal{S}^+(V)$ which is locally uniformly convergent in V . Then $s = \sum_{n=1}^{\infty} s_n \in \mathcal{S}^+(V)$.

Proof. By (1.3), $s \in {}^*\mathcal{H}^+(V)$. Let $U \in \mathcal{U}(V)$. Then $H_U s = \sum_{n=1}^{\infty} H_U s_n$ where $H_U s_n \leq s_n$. Hence the series $\sum_{n=1}^{\infty} H_U s_n$ is locally uniformly convergent on U , i.e. $H_U s|_U \in C(U)$ since $H_U s_n|_U \in C(U)$ for every $n \in \mathbb{N}$. ┘

1.5. EXERCISES. 1. Let $x_0 \in \mathbb{R}^2$, $\varepsilon > 0$, and $X = \mathbb{R}^2 \setminus \overline{B}_\varepsilon(x_0)$. Then the sweeping kernels $H_{B_r(a)}$, $\overline{B_r(a)} \subset X$, given by the Poisson integral form a family of harmonic kernels on X . How about the case $X = \mathbb{R}^2 \setminus \{x_0\}$, $x_0 \in \mathbb{R}^2$?

2. Let $x_0 \in \mathbb{R}$ and $X = \mathbb{R} \setminus \{x_0\}$. Then the sweeping kernels $H_{]a,b[}$, $[a,b] \subset X$, given by the Poisson integral form a family of harmonic kernels on X .

3. Let $(X,\mathcal{W})$ be a balayage space and let Y be a discrete subspace of X . Then $H_{\{y\}}(y,\cdot) = \overset{\circ}{\varepsilon}{}_y^{Y \setminus \{y\}}$, $y \in Y$, defines a family of harmonic kernels on Y .

2. Harmonic Structure of a Balayage Space

Let $(X,\mathcal{W})$ be a balayage space and $\mathcal{P}$ the convex cone of all continuous real potentials. For any open subset U of X and $x \in X$ let

$$H_U(x,\cdot) := \overset{\circ}{\varepsilon}{}_x^{\complement U} .$$

Furthermore, denote by $\mathcal{U}$ the set of all relatively compact open subsets of X . The aim of this section is to show that $(H_U)_{U \in \mathcal{U}}$ is a family of harmonic kernels on X .

For every open subset U of X let $*\mathcal{H}(U)$ denote the set of all functions $u \in \mathcal{B}(X)$ such that u is l.s.c. in U , and $-\infty < H_V u(x) \leq u(x)$ for every $V \in \mathcal{U}(U)$ and every $x \in V$.

An immediate consequence of (II.5.5) is the following property of $\mathcal{W}$.

2.1. COROLLARY. Let $(U_i)_{i \in I}$ be a family of open subsets of X such that $\bigcup_{i \in I} U_i = X$. Then $\bigcap_{i \in I} *\mathcal{H}^+(U_i) = \mathcal{W}$. In particular, $*\mathcal{H}^+(X) = \mathcal{W}$.

2.2. PROPOSITION. Let U be an open subset of X , $V \in \mathcal{U}(U)$ and $c \in \mathbb{R}_+$. Then there exists a potential $q \in \mathcal{P}$ and an increasing mapping $v \mapsto q_v$ from $\{v \in *\mathcal{H}^+(U) : v \leq c \text{ on } V\}$ into $\mathcal{W}$ such that

$$q_v = v + q \quad \text{on} \quad V .$$

For every v which is continuous on U we may choose $q_v \in \mathcal{P}$.

Proof. Let K be a compact subset of U such that $\overline{V} \subset \overset{\circ}{K}$. By (I.1.2) there exist $q, q' \in \mathcal{P}$ such that $q \leq q'$, $q = q'$ on $\complement K$ and $q' - q \geq c$ on $\overline{V}$. For any $v \in *\mathcal{H}(U)$ such that $0 \leq v \leq c$ on U define

$$q_v := \inf(v + q, q')$$

Evidently, $q_v \in *\mathcal{H}^+(U)$. Furthermore, $q_v \leq q'$ and $q_v = q'$ on $\complement K$, hence $q_v \in *\mathcal{H}^+(\complement K)$. Since $U \cup \complement K = X$, we obtain by (2.1) that $q_v \in \mathcal{W}$. The stated properties of the map $v \mapsto q_v$ immediately follow from the definition of q_v . ┘

For every open set U in X define

$$\mathcal{H}(U) = *\mathcal{H}(U) \cap (-*\mathcal{H}(U)) = \{h \in \mathcal{B}(X) : h|_U \in \mathcal{C}(U) , H_V h = h \ \forall V \in \mathcal{U}(U)\} .$$

2.3. PROPOSITION. Let U be an open subset of X and let (h_n) be an increasing sequence in $\mathcal{H}(U)$ which is bounded by some function $w \in \mathcal{W} \cap \mathcal{C}(X)$. Then $\sup h_n \in \mathcal{H}(U)$.

Proof. We may suppose without loss of generality that $(h_n) \subset \mathcal{H}^+(U)$. Indeed, let $w \in \mathcal{W} \cap \mathcal{C}(X)$ such that $|h_n| \leq w$ for every $n \in \mathbb{N}$. Then $0 \leq h_n - h_1 \leq 2w$, knowing that $\sup(h_n - h_1) \in H^+(U)$ we conclude that $\sup h_n = \sup(h_n - h_1) + h_1 \in H(U)$. Let us note furthermore that we only have to show that $h := \sup h_n$ is u.s.c.

Let V be a relatively compact open set such that $\overline{V} \subset U$ and let $c = \sup w(\overline{V})$. We choose $q \in P$ and a mapping $v \mapsto q_v$ according to the preceding proposition (2.2). Then (q_{h_n}) is an increasing sequence in W and for every $n \in \mathbb{N}$

$$h_n + q = q_{h_n} \quad \text{on } V,$$

hence

$$h + q = \sup q_{h_n} \quad \text{on } V.$$

Since $\sup q_{h_n} \in W$ we conclude that h is finely continuous on V. On the other hand, for every $n \in \mathbb{N}$, $v_n := w - h_n \in {}^*H(U)$, $0 \leq v_n \leq w$ and

$$v_n + q = q_{v_n} \quad \text{on } V,$$

hence

$$w - h + q = \inf v_n + q = \inf q_{v_n} \quad \text{on } V.$$

Since $w - h + q$ is finely continuous we have $\widehat{\inf q_{v_n}}^f = \inf q_{v_n}$ on V. But being a function in W the function $\widehat{\inf q_{v_n}}^f$ is l.s.c. Thus $\inf q_{v_n}$ is l.s.c. on V. Since w and q are continuous, we finally conclude that h is u.s.c. on V. ┘

2.4. PROPOSITION. Let f be a l.s.c. function on X which is majorized by some function $w \in W \cap C(X)$. Then $R_f \in H^+(\complement S(f))$.

Proof. We consider first a function $\varphi \in K^+(X)$. By (II.5.2), $R_\varphi \in P$. Let $V \in U$ such that $\overline{V} \cap S(\varphi) = \emptyset$. If $u \in W$ such that $u \geq R_\varphi$ on $\complement V$ then evidently $u \geq \varphi$. Hence $R_{R_\varphi}^{\complement V} \geq R_\varphi$, i.e. $R_{R_\varphi}^{\complement V} = R_\varphi$. Thus $R_\varphi \in H^+(\complement S(\varphi))$.

In order to treat the general case we choose an increasing sequence (φ_n) in $K^+(X)$ such that $\sup \varphi_n = f$. Then (R_{φ_n}) is an increasing sequence in $H^+(\complement S(f))$ majorized by w. Evidently $\sup R_{\varphi_n} \leq R_f$, $\sup R_{\varphi_n} \in W$, and $\sup R_{\varphi_n} \geq \sup \varphi_n = f$, hence $\sup R_{\varphi_n} \geq R_f$. Thus $R_f = \sup R_{\varphi_n} \in H^+(\complement S(f))$. ┘

2.5. COROLLARY. Let U be an open subset of X, $w \in W \cap C(X)$ and $v \in W$ such that $v \leq w$. Then $R_v^U \in H^+(\complement \overline{U})$.

2.6. COROLLARY. Let $v \in \mathcal{W} \cap C(X)$ and let A be a closed subset of X. Then there exists a decreasing sequence (U_n) of open sets containing A such that $R_v^A = \inf R_v^{U_n}$. In particular, $R_v^A \in H^+(\complement A)$.

Proof. Let (V_m) be a decreasing sequence of open subsets of X containing A such that $A = \bigcap_{m=1}^{\infty} \overline{V}_m$. Fix $m \in \mathbb{N}$ and let $\mathcal{V}_m$ denote the family of all open sets U such that $A \subset U \subset V_m$. If $w \in \mathcal{W}$ such that $w \geq v$ on A and if $u_o \in \mathcal{W}$ is strictly positive then $U := V_m \cap \{w + u_o > v\} \in \mathcal{V}_m$ and $R_v^U \leq w + u_o$. This shows that $R_v^A \leq \inf \{R_v^U : U \in \mathcal{V}_m\}$ and hence $R_v^A = \inf \{R_v^U : U \in \mathcal{V}_m\}$.

Since every function R_v^U, $U \in \mathcal{V}_m$, is continuous on $\complement \overline{V}_m$ by (2.5), we may choose a decreasing sequence $(U_{mn})_{n \in \mathbb{N}}$ in $\mathcal{V}_m$ such that $R_v^A = \inf_n R^{U_{mn}}$ on $\complement \overline{V}_m$. For every $n \in \mathbb{N}$ let $U_n = \bigcap_{m=1}^{n} U_{mn}$. Then (U_n) is a decreasing sequence of open sets containing A such that $R_v^A = \inf R_v^{U_n}$. Since $R_v^{U_n} \in H(\complement \overline{V}_n)$ by (2.5) and the sequence $(-R_v^{U_n})$ is increasing, we conclude by (2.3) that $\inf R_v^{U_n} \in H^+(\complement \overline{V}_n)$ for every $n \in \mathbb{N}$, i.e. $R_v^A \in H^+(\complement A)$. ⌟

2.7. PROPOSITION. For every $x \in X$, $\lim_{U,\mathcal{U}_x} H_U(x,\cdot) = \overset{\circ}{\varepsilon}{}_x^{\complement\{x\}}$. In particular, $\lim_{U,\mathcal{U}_x} H_U(x,\cdot) = \varepsilon_x$ if x is not finely isolated.

Proof. Let (U_n) be a sequence in $\mathcal{U}_x$ such that $\overline{U}_{n+1} \subset U_n$ and $\bigcap_{n=1}^{\infty} U_n = \{x\}$ and let $p \in P$. Then $(R_p^{\complement \overline{U}_n})$ is an increasing sequence in $\mathcal{W}$, hence $w = \sup R_p^{\complement \overline{U}_n} \in \mathcal{W}$ such that $w \geq p$ on $\complement\{x\}$. Therefore $\sup R_p^{\complement U_n} = \sup R_p^{\complement \overline{U}_n} = R_p^{\complement\{x\}}$, i.e. $\lim_{U,\mathcal{U}_x} R_p^{\complement U} = R_p^{\complement\{x\}}$. If x is not finely isolated then of course $R_p^{\complement\{x\}} = p$. Thus an application of (I.1.2) finishes the the proof. ⌟

2.8. THEOREM. The family $(H_U)_{U \in \mathcal{U}}$ associated with a balayage space $(X,\mathcal{W})$ is a family of harmonic kernels on X whose set of positive hyperharmonic functions on X coincides with $\mathcal{W}$. Furthermore, $\mathcal{W} \cap C(X) = S^+(X) \cap C(X)$ and

$$P = \{p \in \mathcal{W} \cap C(X) : \inf \{R_p^{\complement K} : K \text{ compact} \subset X\} = 0\}$$
$$= \{p \in \mathcal{W} \cap C(X) : h \in H^+(X) \text{ , } h \leq p \Rightarrow h = 0\} .$$

Proof. By (II.5.4), (2.6) and (I.1.2) every H_U, $U \in \mathcal{U}$, is a sweeping kernel.

It remains to show that the axioms (H_1) - (H_5) are satisfied.

(H_1): Follows from (2.7) remarking that for a finely isolated point $x \in X$ one has by (B_2)

$$R_1^{\{x\}} = \hat{R}_1^{\{x\}} \in \mathcal{W} .$$

(H_2): Let $V,U \in \mathcal{U}, \overline{V} \subset U$. For every $p \in \mathcal{P}$ we have by (2.6)

$$H_U p = R_p^{\complement U} \in \mathcal{H}^+(U) ,$$

hence $H_V H_U p = H_U p$ which implies $H_V H_U = H_U$ by (I.1.2).

(H_3): Let $f \in \mathcal{B}_b(X)$ with compact support and let $U \in \mathcal{U}$.

Case 1: $f \in \mathcal{K}^+(X)$. Choose $p_0 \in \mathcal{P}, p_0 > 0$. Then for every $\varepsilon > 0$ there exist by (I.1.2) $p,q \in \mathcal{P}$ such that $0 \leq p - q \leq f \leq p - q + \varepsilon p_0$. Hence

$$|H_U f - (H_U p - H_U q)| \leq \varepsilon p_0 .$$

This implies the continuity of $H_U f$ on U by (2.6). Furthermore, $H_U f$ is bounded on U since $H_U f \leq p + q + \varepsilon p_0$.

Case 2: Let $f \geq 0$ be l.s.c. Choose $p \in \mathcal{P}$ and an increasing sequence $(f_n) \subset \mathcal{K}^+(X)$ such that $\sup f_n = f \leq p$. Then $(H_U f_n)$ is an increasing sequence in $\mathcal{H}^+(U)$ such that $\sup H_U f_n = H_U f \leq H_U p = R_p^{\complement U} \leq p$. Therefore $H_U f \in \mathcal{H}^+(U)$ by (2.3).

Case 3: $f \geq 0$. Choose $\varphi \in \mathcal{K}^+(X)$ such that $f \leq \varphi$. Then

$$H_U f = \inf \{H_U g : g \text{ l.s.c.}, f \leq g \leq \varphi\}$$

is u.s.c. on U by case 2. Similarly $H_U(\varphi - f)$ is u.s.c. on U. Since $H_U f + H_U(\varphi - f) = H_U \varphi$ is continuous and bounded on U we therefore conclude that $H_U f|_U \in \mathcal{C}_b(U)$.

Case 4: The general case follows from case 3 since $H_U f = H_U f^+ - H_U f^-$.

(H_4): Let $U \in \mathcal{U}$ and choose a sequence (U_n) of open sets such that $\overline{U}_n \subset U_{n+1}$ and $\bigcup_{n=1}^{\infty} U_n = U$. Let $p \in \mathcal{P}$ be strict. For every $n \in \mathbb{N}$ let $f_n \in \mathcal{C}(X)$ such that $0 \leq f_n \leq p, f_n = 0$ on $\overline{U}_n$ and $f_n = p$ on $\complement U_{n+1}$. Then $p_n := R_{f_n} \in \mathcal{P}$ and

$$H_U p = R_p^{\complement U} \leq p_n \leq R_p^{\complement U_n} .$$

Hence $\lim p_n = H_U p$ by (2.6).

Let $x \in U$. We may assume $p_n(x) - H_U p(x) \leq \frac{1}{2^n}$. If

$$w := \sum_{n=1}^{\infty} (p_n - H_U p)$$

then $w \in {}^*\mathcal{H}^+(U)$ by (2.6) and $w(x) \leq 1$. If $\mathcal{F}$ is a non-regular ultrafilter on U such that $\lim_{\mathcal{F}} =: z \in U^*$ then $a := p(z) - \lim_{\mathcal{F}} H_U p > 0$ since p is strict. For every $n \in \mathbb{N}$ we have $p_n = p$ on $\complement U_{n+1}$, hence for every $m \in \mathbb{N}$

$$w \geq \sum_{n=1}^{m} (p_n - H_U p) = m(p - H_U p) \text{ on } \complement U_{m+1}$$

which implies $\lim_{\mathcal{F}} w \geq m a$ and thus $\lim_{\mathcal{F}} w = \infty$ since $m \in \mathbb{N}$ was arbitrary and $a > 0$.

(H_5): By (2.6), $\mathcal{P} \subset \mathcal{S}^+(X) \cap \mathcal{C}(X)$. Consider now $v \in \mathcal{W} \cap \mathcal{C}(X)$ and let $U \in \mathcal{U}$. There exists an increasing sequence (q_n) in $\mathcal{P}$ such that $\sup q_n = v$. Then

$$H_U v = \sup H_U q_n = \sup R_{q_n}^{\complement U} \leq R_v^{\complement U} .$$

By (2.6), $R_v^{\complement U} \in \mathcal{H}^+(U)$, hence for every open V such that $\overline{V} \subset U$

$$H_U v = H_V H_U v \in \mathcal{H}^+(V)$$

by (1.2). This shows that $H_U v \in \mathcal{H}^+(U)$ and therefore $v \in \mathcal{S}^+(X) \cap \mathcal{C}(X)$.

Let $p \in \mathcal{W} \cap \mathcal{C}(X)$. If $p \in \mathcal{P}$ then of course $\inf \{R_p^{\complement K} : K \text{ compact } \subset X\} = 0$. Suppose now that $\inf \{R_p^{\complement K} : K \text{ compact } \subset X\} = 0$ and let $h \in \mathcal{H}^+(X)$ such that $h \leq p$. Let K be a compact subset of X and let $U \in \mathcal{U}$ such that $K \subset U$. Then

$$h = H_U h \leq H_U p = R_p^{\complement U} \leq R_p^{\complement K} .$$

Therefore $h = 0$.

Suppose finally that 0 is the greatest minorant of p in $\mathcal{H}^+(X)$. Let (K_n) be an exhaustion of X and let $h = \inf R_p^{\complement K_n}$. For every $n \in \mathbb{N}$, $R_p^{\complement K_n} \in \mathcal{H}^+(\mathring{K}_n)$ by (2.4). Applying (2.3) we obtain $h \in \mathcal{H}^+(\mathring{K}_n)$ for every $n \in \mathbb{N}$, i.e. $h \in \mathcal{H}^+(X)$. Therefore $h \in \mathcal{H}^+(X)$, $h = 0$. Furthermore, $(R_p^{\complement K_n})$ converges to zero locally uniformly. Thus $p \in \mathcal{P}$. ┘

2.9. REMARK. The proof of (2.8) shows that (H_4) holds for every open *not necessarily* relatively compact subset U of X.

2.10. EXAMPLE. Discrete potential theory. Let X be discrete and P a sub-Markov kernel on X such that S_P is linearly separating and $P(x,\cdot) \neq \varepsilon_x$ for every $x \in X$. Let $V = \sum_{n=0}^{\infty} P^n$ (see (II.3.5.5)). Defining $(H_{\{x\}})_{x \in X}$ as in (1.1.3) we obtain

$$P = \{q \in S_P : q < \infty , \lim_{n\to\infty} P^n q = 0\}$$
$$= \{Vf : f \in \mathcal{B}^+(X) , Vf < \infty\} .$$

Indeed, by (1.1.3) a real function $h \geq 0$ on X is harmonic if and only if $Ph = h$. Hence the first equality follows from (2.8) and the fact that for every $q \in S_P$ the function $h = \lim_{n\to\infty} P^n q$ is the greatest minorant h of q such that $Ph = h$.

If $f \in \mathcal{B}^+(X)$ such that $Vf < \infty$, then $P^n Vf = \sum_{m=n}^{\infty} P^m f \leq Vf$ for every $n \in \mathbb{N}$, hence $Vf \in S_P$ and $\lim_{n\to\infty} P^n Vf = 0$.

For the converse, let $q \in P$. Then $f := q - Pq \in \mathcal{B}^+(X)$ and, for every $n \in \mathbb{N}$,

$$\sum_{m=0}^{n} P^m f = q - P^{n+1} q ,$$

hence $Vf = q$.

2.11. EXERCISE. For every $n \in \mathbb{N}$, let $(X_n, \mathcal{W}_n)$ be a balayage space such that $1 \in \mathcal{W}_n$. Let X be the topological sum of the spaces X_n, $n \in \mathbb{N}$, and for every $n \in \mathbb{N}$ let $\tau_n \in M_+(X)$ such that $\tau_n(X_n) = 0$ and $\tau_n(X) \leq 1$. If U is a relatively compact open subset of X_n with harmonic kernel H_U^n define a kernel H_U on X by

$$H_U(x,B) = \begin{cases} H_U^n(x, B\cap X_n) + (1 - H_U^n(x,X_n))\tau_n(B) , & x \in U , \\ \varepsilon_x(B) , & x \in \complement U . \end{cases}$$

Then we have a family (H_U) of harmonic kernels on X such that $1 \in {}^*\mathcal{H}^+(X)$. If $\tau_n(X) = 1$ for every $n \in \mathbb{N}$ then $1 \in \mathcal{H}(X)$.

3. Convergence Properties

Throughout the sections 3 - 6 let $(H_U)_{U \in \mathcal{U}}$ be a family of harmonic kernels on X . Let V be an open subset of X .

3.1. PROPOSITION. Let (h_n) be a decreasing sequence in $\mathcal{H}^+(V)$. Then $\inf h_n \in \mathcal{H}^+(V)$.

Proof. Let $h = \inf h_n$ and $U \in \mathcal{U}(V)$. Then $h \in \mathcal{B}^+(X)$, $H_U h|_U \in \mathcal{C}(U)$ by (1.2) and

$$H_U h = \inf H_U h_n = \inf h_n = h \quad .$$

Hence $h \in \mathcal{H}^+(V)$. ⌟

3.2. PROPOSITION. Let F be a decreasingly filtered subset of $\mathcal{H}^+(V)$. Then there exists a decreasing sequence (h_n) in F such that $\inf h_n = \inf F$ on V . In particular, $\inf F$ is continuous on V .

Proof. By (I.1.7), there exists a decreasing sequence (h_n) in F such that $\inf h_n = \inf F$ on V . By (3.1), $\inf h_n$ is continuous on V . ⌟

3.3. COROLLARY 1. Let F be a decreasingly filtered subset of $\mathcal{H}^+(V)$ containing a sequence (h_n) such that $\inf h_n = \inf F$ on $\complement V$. Then $\inf F \in \mathcal{H}^+(V)$.

3.4. COROLLARY 2. Let $U \in \mathcal{U}$ and let f be a positive numerical function on X . Then the function

$$H_U^* f : x \mapsto \int^* f(y) H_U(x,dy)$$

is l.s.c. on U .

Proof. If $g \in \mathcal{B}_b^+(X)$ with compact support then $H_U g \in \mathcal{H}^+(U)$ by (H_2) and (H_3). Let (K_n) be an exhaustion of X . For every $n \in \mathbb{N}$ let

$$f_n = \inf (f , n1_{K_n}) \quad .$$

Then

$$F_n := \{H_U g : g \in \mathcal{B}_b^+(X) \ , \ S(g) \subset K_n \ , \ g \geq f_n\}$$

is a decreasingly filtered subset of $\mathcal{H}^+(U)$, hence

$$H_U^* f_n = \inf F_n$$

is continuous on U by (3.2). Thus $H_U^* f = \sup H_U^* f_n$ is l.s.c. on U. ┘

3.5. EXERCISE. Let $(H_U)_{U \in \mathcal{U}}$ be a family of sweeping kernels on X satisfying (H_2). Let $U \in \mathcal{U}$ such that $H_U \varphi \in C(U)$ for every $\varphi \in \mathcal{K}(X)$ and $\sup h_n \in C(U)$ for every bounded increasing sequence (h_n) in $\mathcal{H}^+(U)$. Then $H_U f \in \mathcal{H}(U)$ for every $f \in \mathcal{B}_b(X)$ with compact support.

4. Minimum Principle and Sheaf Properties

Let V be an open subset of X. For every $x \in V$ let $\mathcal{V}(x)$ be a fundamental system of neighborhoods $U \in \mathcal{U}(V)$ of x. We define

$$^*\mathcal{H}_{\mathcal{V}}(V) = \{v \in \mathcal{B}(X) : v|_V \text{ l.s.c.}, -\infty < H_U v(x) \leq v(x) \ \forall x \in V \ \forall U \in \mathcal{V}(x)\}$$

and

$$\begin{aligned} \mathcal{H}_{\mathcal{V}}(V) &= {}^*\mathcal{H}_{\mathcal{V}}(V) \cap (-{}^*\mathcal{H}_{\mathcal{V}}(V)) \\ &= \{h \in \mathcal{B}(X) : h|_V \in C(V), \ H_U h(x) = h(x) \ \forall x \in V \ \forall U \in \mathcal{V}(x)\} . \end{aligned}$$

Evidently, $^*\mathcal{H}(V) \subset {}^*\mathcal{H}_{\mathcal{V}}(V)$ and $\mathcal{H}(V) \subset \mathcal{H}_{\mathcal{V}}(V)$. We intend to show that $^*\mathcal{H}^+(V) = {}^*\mathcal{H}^+_{\mathcal{V}}(V)$ and $\mathcal{H}^+(V) = \mathcal{H}^+_{\mathcal{V}}(V)$. We start by a useful minimum principle.

4.1. PROPOSITION. Let V be relatively compact, $v \in {}^*\mathcal{H}_{\mathcal{V}}(V)$ such that $v \geq 0$ on $\complement V$ and $\liminf_{x \to z} v(x) \geq 0$ for every $z \in V^*$. Then $v \geq 0$.

Proof. Let $s \in S^+(X) \cap C(X)$ be strictly positive, let $\varepsilon > 0$ and

$$w = v + \varepsilon s .$$

Obviously w is l.s.c. on V and for every $z \in V^*$

$$\liminf_{x \to z} w(x) \geq \liminf_{x \to z} v(x) + \varepsilon \liminf_{x \to z} s(x) \geq \varepsilon s(z) > 0 .$$

Hence

$$K := \{x \in V : w(x) \leq 0\}$$

is a compact subset of V. Let $x \in K$, $U \in \mathcal{V}(x)$ and

$$\mu = H_U(x, \cdot)|_K .$$

Then $\mu \neq \varepsilon_x$ since $H_U(x,U) = 0$. Furthermore $w \geq 0$ on $\complement K$ and hence

$$\mu(w|_K) \leq H_U w(x) = H_U v(x) + \varepsilon H_U s(x) \leq v(x) + \varepsilon s(x) = w(x) .$$

Obviously, $\mu(u|_K) \leq H_U u(x) \leq u(x)$ for every $u \in {}^*\mathcal{H}^+(V)$. Defining

$$\mathcal{F} = ({}^*\mathcal{H}^+(V) + \mathbb{R}^+ w)|_K$$

we thus have $\mu \in M_x(\mathcal{F})$. Hence $\mathrm{Ch}_{\mathcal{F}} K = \emptyset$. Applying (I.2.1) we therefore conclude that $w \geq 0$ on K and hence $w \geq 0$. Thus $v \geq 0$. ┘

4.2. LEMMA. Let $U \in \mathcal{U}$ and $z \in U^*$.

1. If $v \in \mathcal{B}^+(X)$ is l.s.c. at z then $\lim_{\mathcal{F}} H_U v \geq v(z)$ for every regular ultrafilter $\mathcal{F}$ on U converging to z .

2. If g is a numerical function on U and $a \in \mathbb{R}$ such that $\liminf_{x \to z, x \in U} g(x) > -\infty$ and $\lim_{\mathcal{F}} g \geq a$ for every regular ultrafilter $\mathcal{F}$ on U converging to z then $\liminf_{x \to z, x \in U} (g+w)(x) \geq a$ for every Evans function w on U .

Proof. 1. Let $\mathcal{F}$ be a regular ultrafilter on U converging to z and let $a < v(z)$. Then there exists a function $f \in \mathcal{K}^+(X)$ such that $f \leq v$ and $a < f(z)$. Hence

$$\lim_{\mathcal{F}} H_U v \geq \lim_{\mathcal{F}} H_U f = f(z) > a$$

and therefore $\lim_{\mathcal{F}} H_U v \geq v(z)$.

2. Let w be an Evans function on U . There exists an ultrafilter $\mathcal{F}$ on U such that $\lim \mathcal{F} = z$ and

$$\liminf_{x \to z, x \in U} (g+w)(x) = \lim_{\mathcal{F}}(g+w) .$$

If $\mathcal{F}$ is regular then

$$\lim_{\mathcal{F}} (g+w) \geq \lim_{\mathcal{F}} g \geq a .$$

If $\mathcal{F}$ is non-regular then

$$\lim_{\mathcal{F}} (g+w) = \lim_{\mathcal{F}} w = \infty \geq a$$

since $\lim_{\mathcal{F}} g \geq \liminf_{x \to z, x \in U} g(x) > -\infty$. Thus the statement follows. ┘

4.3. COROLLARY. Let $V \in \mathcal{U}$ and $v \in {}^*\mathcal{H}_V(V)$ such that $\inf v(V) > -\infty$, $v \geq 0$ on $\complement V$ and $\lim_{\mathcal{F}} v \geq 0$ for every regular ultrafilter $\mathcal{F}$ on V . Then

$v \geq 0$.

Proof. Let w be an Evans function on V . By (4.2), $\liminf_{x \to z} (v+w)(x) \geq 0$ for every $z \in V^*$. Hence $v + w \geq 0$ by (4.1). Applying (H_4) we finally obtain $v \geq 0$. ┘

4.4. PROPOSITION. $^*\mathcal{H}^+_{\mathcal{V}}(V) = {}^*\mathcal{H}^+(V)$ and $\mathcal{H}^+_{\mathcal{V}}(V) = \mathcal{H}^+(V)$.

Proof. Let $v \in {}^*\mathcal{H}^+_{\mathcal{V}}(V)$, $U \in \mathcal{U}(V)$ and $f \in \mathcal{B}^+_b(X)$ with compact support such that $f \leq v$ and f is continuous on V . Then $h := H_U f \in \mathcal{H}^+(U)$, $v - H_U f \in {}^*\mathcal{H}_{\mathcal{V}}(U)$ and $H_U f = f \leq v$ on $\complement U$. Furthermore for every regular ultrafilter $\mathcal{F}$ on U converging to a point $z \in U^*$

$$\lim_{\mathcal{F}} (v - H_U f) \geq v(z) - \lim_{\mathcal{F}} H_U f = v(z) - f(z) \geq 0 .$$

Hence $v \geq H_U f$ by (4.3). Since v is the limit of an increasing sequence of such functions f we thus conclude that $v \geq H_U v$. Therefore $v \in {}^*\mathcal{H}^+(V)$.

Consider now $h \in \mathcal{H}^+_{\mathcal{V}}(V)$. Then $h \in {}^*\mathcal{H}^+_{\mathcal{V}}(V) = {}^*\mathcal{H}^+(V)$. Choosing $U \in \mathcal{U}(V)$ we therefore have $H_U h \leq h$. On the other hand $H_U h \in {}^*\mathcal{H}^+(U)$ by (1.2) and $H_U h = h$ on $\complement U$. Let $g \in \mathcal{B}(X)$ such that $h + g = H_U h$. Then $g \in {}^*\mathcal{H}_{\mathcal{V}}(V)$. Let $\mathcal{F}$ be a regular ultrafilter on U converging to $z \in U^*$. There exists $f \in \mathcal{K}^+(X)$ such $f \leq h$ and $f(z) = h(z)$. Hence

$$\lim_{\mathcal{F}} g \geq \lim_{\mathcal{F}} (H_U f - h) = f(z) - h(z) = 0 .$$

Applying (4.3) again we obtain $H_U h \geq h$. Thus $H_U h = h$, $h \in \mathcal{H}^+(V)$. ┘

4.5. COROLLARY 1. Let $(V_i)_{i \in I}$ be a family of open subsets of X . Then

$$^*\mathcal{H}^+\Big(\bigcup_{i \in I} V_i\Big) = \bigcap_{i \in I} {}^*\mathcal{H}^+(V_i)$$

and

$$\mathcal{H}^+\Big(\bigcup_{i \in I} V_i\Big) = \bigcap_{i \in I} \mathcal{H}^+(V_i) .$$

Proof. Let $V = \bigcup_{i \in I} V_i$. Defining

$$\mathcal{V}(x) = \bigcup_{i \in I} \mathcal{U}_x(V_i) \qquad (x \in V)$$

every $\mathcal{V}(x)$ is a fundamental system of neighborhoods of x . Obviously,

$$\bigcap_{i \in I} {}^*\mathcal{H}^+(V_i) = {}^*\mathcal{H}^+_{\mathcal{V}}(V) \ , \quad \bigcap_{i \in I} \mathcal{H}^+(V_i) = \mathcal{H}^+_{\mathcal{V}}(V) \ .$$

Thus the statement follows immediately from (4.4). ┘

4.6. COROLLARY 2. Let V,W be open subsets of X such that $W \subset V$. Let $v \in {}^*\mathcal{H}^+(V)$ and $w \in {}^*\mathcal{H}^+(W)$ such that $w \geq v$ on $\complement W$ and $\liminf_{x \to z} w(x) \geq v(z)$ for every $z \in W^* \cap V$. Then $\inf(v,w) \in {}^*\mathcal{H}^+(V)$.

Proof. Obviously, $u := \inf(v,w) \in {}^*\mathcal{H}^+(W)$. Furthermore, u is l.s.c. on V and $u = v$ on $\complement W$. Hence for every $x \in V \setminus W$ and $U \in \mathcal{U}_x(V)$

$$H_U u(x) \leq H_U v(x) \leq v(x) = u(x) \ .$$

Defining

$$\mathcal{V}(x) = \begin{cases} \mathcal{U}_x(W) \ , & x \in W \\ \mathcal{U}_x(V) \ , & x \in V \setminus W \end{cases}$$

we thus have $u \in {}^*\mathcal{H}^+_{\mathcal{V}}(V)$, i.e. $u \in {}^*\mathcal{H}^+(V)$. ┘

4.7. REMARK. Let us note that the preceding results (see the proof of (4.1)) did not require that ${}^*\mathcal{H}^+(X)$ is linearly separating. For every $V \in \mathcal{U}$ a linear separation of the points of V by ${}^*\mathcal{H}^+(V)$ would have been sufficient.

4.8. EXAMPLE. Classical potential theory. Let U be an open subset of $\mathbb{R}^n$, $n \geq 1$. Then (4.4) implies that a l.s.c. function $v : U \to \overline{\mathbb{R}}_+$ is hyperharmonic on U if (and only if) for every $x \in U$ there exist arbitrarily small $r > 0$ such that

$$\int u \, d\sigma_{x,r} \leq u(x) \ .$$

4.9. EXERCISES. Let $p \in \mathcal{P}_b$ be a strict potential on a balayage space $(X,\mathcal{W})$. For every $x \in X$, let $\mathcal{V}_x \subset \mathcal{U}_x$ be a fundamental system of neighborhoods of x. Given $f \in \mathcal{B}(X)$ such that $f \geq -v$ for some $v \in \mathcal{W} \cap \mathcal{C}(X)$, define $\overline{A}f, \underline{A}f : X \to \overline{\mathbb{R}}$ by

$$\overline{A}f(x) = \limsup_{U, \mathcal{V}_x} \frac{f(x) - H_U f(x)}{p(x) - H_U p(x)} \ ,$$

$$\underline{A}f(x) = \liminf_{U, \mathcal{V}_x} \frac{f(x) - H_U f(x)}{p(x) - H_U p(x)}$$

if $f(x) < \infty$, and $\overline{A}f(x) = \underline{A}f(x) = \infty$ if $f(x) = \infty$. If $\overline{A}f(x) = \underline{A}f(x) \in \mathbb{R}$, define $Af(x) = \overline{A}f(x) = \underline{A}f(x)$.

1. Let $f \in \mathcal{B}^+(X)$ and let U be an open subset of X such that f is l.s.c. on U. Then: $f \in {}^*\mathcal{H}^+(U) \Leftrightarrow \underline{A}f \geq 0$ on $U \Leftrightarrow \overline{A}f \geq 0$ on U. (Hint: In order to prove that $f \in {}^*\mathcal{H}^+(U)$ if $\overline{A}f \geq 0$ on U consider $f + \varepsilon p$, $\varepsilon > 0$.)

2. Let $f \in \mathcal{B}^+(X)$ and let U be an open subset of X such that f is continuous and real on U. Then: $f \in \mathcal{H}^+(U) \Leftrightarrow \underline{A}f = 0$ on $U \Leftrightarrow \overline{A}f = 0$ on $U \Leftrightarrow Af = 0$ on U.

3. In classical potential theory, $Af = \frac{\Delta f}{\Delta p}$ if p and f are twice continuously differentiable and $\mathcal{V}_x = \{B_r(x) : r > 0\}$. (Hint: Use Taylor's formula to show that $\lim_{r\to 0} \frac{2n}{r^2}(f(x) - H_{x,r}f(x)) = -\Delta f(x)$.)

5. Regularizations

Let X_0 denote the set of all $x \in X$ such that $\lim_{U,\mathcal{U}_x} H_U(x,\cdot) = \varepsilon_x$.

5.1. LEMMA. Let $v \in {}^*\mathcal{H}^+(X)$ and $x \in X_0$. Then

$$v(x) = \sup_{U \in \mathcal{U}_x} H_U v(x) = \lim_{U,\mathcal{U}_x} H_U v(x) .$$

Proof. Let $\alpha < v(x)$. Then there exists a function $f \in \mathcal{K}^+(X)$ such that $f \leq v$ and $f(x) > \alpha$. There exists $U_0 \in \mathcal{U}$ such that $H_U f(x) > \alpha$ for every $U \in \mathcal{U}$ contained in U_0. Then for every such $U \in \mathcal{U}$

$$\alpha < H_U f(x) \leq H_U v(x) \leq v(x) .$$

Thus the statement follows. ┘

5.2. PROPOSITION. Let $\mathcal{V} \subset {}^*\mathcal{H}^+(X)$, $v = \inf \mathcal{V}$. Then $\hat{v} \in {}^*\mathcal{H}^+(X)$, $\hat{v} = v$ on $X \setminus X_0$ and for every $x \in X_0$,

$$\hat{v}(x) = \sup_{U \in \mathcal{U}_x} H_U^* v(x) = \lim_{U,\mathcal{U}_x} H_U^* v(x) .$$

Proof. Let $U \in \mathcal{U}$. Then

$$H_U^* \hat{v} \leq H_U^* v \leq \inf_{v' \in \mathcal{V}} H_U v' \leq \inf_{v' \in \mathcal{V}} v' = v$$

where $H_U^* v$ is l.s.c. on U by (3.4), hence

$$H_U^* v \leq \hat{v} \quad .$$

Therefore $\hat{v} \in {}^*H^+(X)$.

Consider now $x \in X \setminus X_0$. Then $v' \geq v(x) R_1^{\{x\}}$ for every $v' \in V$ and hence

$$v \geq v(x) R_1^{\{x\}} \quad .$$

By (H_1), $\liminf_{y \to x} R_1^{\{x\}}(y) \geq 1$ and hence $\hat{v}(x) \geq v(x)$. The rest of the statement follows by (5.1). ∎

<u>5.3. COROLLARY.</u> Let $V_i \subset {}^*H^+(X)$, $v_i = \inf V_i$, $i = 1,2$. Then $\widehat{v_1 + v_2} = \hat{v}_1 + \hat{v}_2$.

Proof. It is clear that $\hat{v}_1 + \hat{v}_2 \leq v_1 + v_2$ since $\hat{v}_1 + \hat{v}_2$ is a l.s.c. minorant of $v_1 + v_2$. Let $V = V_1 + V_2$. Then $V \subset {}^*H^+(X)$ and $\inf V = v_1 + v_2$. Hence for every $x \in X \setminus X_0$,

$$\widehat{v_1 + v_2}(x) = (v_1 + v_2)(x) = \hat{v}_1(x) + \hat{v}_2(x) \quad .$$

Finally, let $x \in X_0$. Then

$$\widehat{v_1 + v_2}(x) = \lim_{U,\mathcal{U}_x} H_U^*(v_1 + v_2)(x)$$
$$\leq \lim_{U,\mathcal{U}_x} (H_U^* v_1(x) + H_U^* v_2(x)) = \hat{v}_1(x) + \hat{v}_2(x) \quad .$$ ∎

<u>5.4. COROLLARY.</u> Let $v \in {}^*H^+(X)$, $U \in \mathcal{U}$ and $u \in {}^*H^+(U)$ such that $H_U v \leq u \leq v$. Then $\hat{u} \in {}^*H^+(X)$.

Proof. Let w be an Evans function on U and $u' = u + w$. Let $z \in U^*$ and let F be a regular ultrafilter on U converging to z . Then

$$\lim_F u \geq \lim_F H_U v \geq v(z)$$

by (4.2) and hence

$$\liminf_{x \to z, x \in U} u'(x) \geq v(z)$$

again by (4.2). Since $u' \in {}^*H^+(U)$ and $u' \geq v$ on $\complement U$ we conclude by (4.6) that $\inf(u',v) \in {}^*H^+(X)$.

By (H_4), u is the infimum of the family of all $\inf(u+w,v)$ where w is an Evans function on U. Thus $\hat{u} \in {}^*\mathcal{H}^+(X)$ by (5.2). ┘

5.5. REMARK. We note that $\hat{u} = u$ on $\complement U^*$ and $\hat{u}(z) = \liminf_{x\to z, x\in U} u(x)$ for every $z \in U^*$.

If $f : X \to \overline{\mathbb{R}}_+$ then $\hat{R}_f$ is hyperharmonic by (5.2). If f is l.s.c. then $f \le R_f$ implies $f \le \hat{R}_f$, hence $R_f \le \hat{R}_f$, i.e. $R_f = \hat{R}_f \in {}^*\mathcal{H}^+(X)$.

5.6. PROPOSITION. Let f be a function in $C^+(X)$ which is bounded by a superharmonic function. Then $R_f \in \mathcal{S}^+(X) \cap C(X)$. Furthermore $H_U R_f = R_f$ for every $U \in \mathcal{U}$ such that $H_U f \ge f$. In particular, R_f is harmonic on $\complement S(f)$.

Proof. By the precedings remarks and (1.3.2) we have $s := R_f \in \mathcal{S}^+(X)$. If $U \in \mathcal{U}$ such that $H_U f \ge f$, then $H_U s \ge H_U f \ge f$, hence $\widehat{H_U s} \ge f$. Thus $\widehat{H_U s} \ge s$, i.e. $H_U s = s$. It remains to show that s is u.s.c. and real.

Fix $x \in X$, $\varepsilon > 0$, and $a,b \in \mathbb{R}$ such that $a \le b$, $a \le \sup_{U\in\mathcal{U}_x} H_U s(x)$, $f(x) \le b \le s(x)$. By (H_5) there exists a function $t \in \mathcal{S}^+(X) \cap C(X)$ such that

$$b - a < t(x) < b - a + \varepsilon \ .$$

Then $f(x) - t(x) \le b - t(x) < a$, hence there exists $U \in \mathcal{U}_x$ such that

$$f(x) - t(x) < H_U s(x) \ .$$

So there exists $V \in \mathcal{U}_x$ such that $\overline{V} \subset U$ and $f - t < H_U s$ on V and hence

$$f \le t + \widehat{H_V s}$$

since $H_U s = H_V H_U s \le H_V s$ and $H_V s = s \ge f$ on $\complement V$. Therefore $s = R_f \le t + \widehat{H_V s}$. In particular, $s(x) < \infty$. Choosing $a = \sup_{U\in\mathcal{U}_x} H_U s(x)$, $b = s(x)$ and using the continuity of $t + \widehat{H_V s}$ on V we finally conclude that

$$\limsup_{y\to x} s(y) \le t(x) + \widehat{H_V s}(x) < s(x) - a + \varepsilon + \widehat{H_V s}(x) \le s(x) + \varepsilon \ .$$

Thus s is u.s.c. at x. ┘

5.7. COROLLARY. Let $s \in {}^*\mathcal{H}^+(X)$. Then there exists an increasing sequence (s_n) in $\mathcal{S}^+(X) \cap C(X)$ such that $s = \sup s_n$.

Proof. Let (φ_n) be an increasing sequence in $K^+(X)$ such that $s = \sup \varphi_n$. Defining $s_n = R_{\varphi_n}$ we have by (5.6) and (H_5), $s_n \in S^+(X) \cap C(X)$, $\varphi_n \leq s_n \leq s_{n+1} \leq s$ and hence $\sup s_n = s$. ┘

In the following the $*H^+(X)$-fine topology on X will simply be called *fine topology* and we shall adopt the notations of (II.4).

5.8. LEMMA. Let $x \in X_0$ and let K be a compact fine neighborhood of x. Then $\lim_{U,\mathcal{U}_x} H_U(x,K) = 1$.

Proof. By (II.4.1), there exists a relatively compact open neighborhood V of x, a function $u \in *H^+(X)$ and $\alpha \in \mathbb{R}$ such that $u(x) < \alpha$ and

$$V \cap \{u < \alpha\} \subset K .$$

Let $s \in S^+(X) \cap C(X)$ such that $s(x) = u(x)$ and define $u' := \inf(u,s)$. Let $u(x) < \beta < \alpha$ and $W = V \cap \{s < \beta\}$. Then W is an open neighborhood of x. Furthermore,

$$u - u' \geq \alpha - \beta > 0 \quad \text{on} \quad W \setminus K$$

and hence

$$(\alpha - \beta) \limsup_{U,\mathcal{U}_x} H_U(x, W \setminus K) \leq \lim_{U,\mathcal{U}_x} H_U(u - u')(x) = u(x) - u'(x) = 0 ,$$

i.e. $\lim_{U,\mathcal{U}_x} H_U(x, W \setminus K) = 0$. Since $\lim_{U,\mathcal{U}_x} H_U(x,W) = 1$ by definition of X_0 the statement follows. ┘

5.9. COROLLARY. For every $\mathcal{V} \subset *H^+(X)$,

$$\widehat{\inf \mathcal{V}} = \widehat{\inf \mathcal{V}}^f .$$

Proof. Let $\mathcal{V} \subset *H^+(X)$, $v = \inf \mathcal{V}$. Evidently

$$\hat{v} \leq \hat{v}^f \leq v .$$

In particular, $\hat{v} = \hat{v}^f = v$ on $X \setminus X_0$ by (5.2). So let $x \in X_0$ and $\alpha < \hat{v}^f(x)$. Then there exists a compact fine neighborhood K of x such that $\alpha \leq v$ on K. Hence for every $U \in \mathcal{U}_x$,

$$\alpha H_U(x,K) \leq H_U^* v(x) \leq \hat{v}(x)$$

and therefore $\alpha \leq \hat{v}(x)$ by (5.8). Thus $\hat{v}^f(x) \leq \hat{v}(x)$. ┘

5.10. EXERCISES. 1. Let $(H_U)_{U \in \mathcal{U}}$ be a family of sweeping kernels on X such that $\widehat{\inf \mathcal{V}} = \widehat{\inf \mathcal{V}}^f$ for every $\mathcal{V} \subset \mathcal{H}^+(X)$. Then $\inf h_n \in \mathcal{H}^+(X)$ for every decreasing sequence (h_n) in $\mathcal{H}^+(X)$

2. Let $X = [0,1[$, $\mathcal{U} = \{]a,b[\; : 0 < a < b < 1\} \cup \{[0,b[\; : 0 < b < 1\}$ and define

$$H_{]a,b[}(x,\cdot) = \frac{x-a}{b-a}\varepsilon_b + \frac{b-x}{b-a}\varepsilon_a$$

if $0 < a < x < b < 1$ and

$$H_{[0,b[}(x,\cdot) = \frac{x}{b}\varepsilon_b$$

if $0 \leq x < b < 1$, and $H_U(x,\cdot) = \varepsilon_x$ for every $U \in \mathcal{U}$ and $x \in \complement U$. Then $(H_U)_{U \in \mathcal{U}}$ is a family of harmonic kernels on X such that ${}^*\mathcal{H}^+(X)$ is the set of all l.s.c. positive numerical functions on X which are concave on $]0,1[$. In particular, the point 0 is finely isolated and hence the function $v := 1_{\{0\}}$ is finely continuous. We have $H_U v = 0 \leq v$ for every $U \in \mathcal{U}$ and $\hat{v}(0) = 0 < 1 = \hat{v}^f(0)$.

6. Potentials

Looking at (2.8) we are led to the following definition. A function $p \in \mathcal{S}^+(X)$ is called a *potential* if the constant zero is the only positive harmonic minorant of p . Let $\mathcal{P}(X)$ denote the set of all potentials on X . If $p \in \mathcal{P}(X)$ and $u \in {}^*\mathcal{H}^+(X)$ then obviously $\inf(p,u) \in \mathcal{P}(X)$.

6.1. PROPOSITION. Let (U_n) be a sequence in $\mathcal{U}$ covering X and let (k_n) be a sequence in $\mathbb{N}$ such that for every $m \in \mathbb{N}$ the set $\{n \in \mathbb{N} : k_n = m\}$ is infinite. Let $s \in \mathcal{S}^+(X)$ and define (s_n) by $s_0 = s$ and $s_n = \widehat{H_{U_{k_n}} s_{n-1}}$. Then (s_n) is a decreasing sequence in $\mathcal{S}^+(X)$, $h_s := \lim_{n\to\infty} s_n$ is the greatest harmonic minorant of s and $p_s := s - h_s$ is a potential.

Proof. By (5.4), (s_n) is a decreasing sequence in $S^+(X)$. Let $m \in \mathbb{N}$ and let (l_n) be a strictly increasing sequence in $\mathbb{N}$ such that $k_{l_n} = m$. Then for every $n \geq 2$,

$$s_{l_n} \leq H_{U_m} s_{l_n - 1} \leq s_{l_n - 1} .$$

Letting n tend to infinity we conclude that

$$h_s \leq H_{U_m} h_s \leq h_s .$$

Hence $h_s \in H^+(U_m)$ by (1.2). Applying (4.5) we obtain $h_s \in H^+(X)$. Thus h_s is a harmonic minorant of s.

Let $h \in H^+(X)$ such that $h \leq s$. Since $H_{U_m} h = h$ for every $m \in \mathbb{N}$ induction by n yields immediately that $h \leq s_n$ for every n and hence $h \leq h_s$.

Obviously, $p_s = s - h_s \in S^+(X)$. Let $g \in H^+(X)$ such that $g \leq p_s$. Then $g + h_s \in H^+(X)$ and $g + h_s \leq s$, hence $g + h_s \leq h_s$, i.e. $g = 0$. Thus $p_s \in P(X)$. ┘

6.2. COROLLARY. $P(X)$ is a convex cone. $S^+(X)$ is the direct sum of $P(X)$ and $H^+(X)$.

Proof. By (6.1) and (5.3), the mapping $s \mapsto h_s$ is additive and s is a potential if and only if $h_s = 0$.

Let $q_1, q_2 \in P(X)$. Then $q_1 + q_2 \in S^+(X)$ and

$$h_{q_1 + q_2} = h_{q_1} + h_{q_2} = 0 + 0 = 0 .$$

Hence $q_1 + q_2 \in P(X)$. It is obvious that $\alpha q \in P(X)$ for every $q \in P(X)$ and $\alpha \geq 0$. Therefore $P(X)$ is a convex cone.

Let $s \in S^+(X)$. By (6.1), $s = h_s + p_s$ where $h_s \in H^+(X)$ and $p_s \in P(X)$. Consider now $g \in H^+(X)$ and $q \in P(X)$ such that $s = g + q$. Then

$$h_s = h_g + h_q = g + 0 = g ,$$

and hence $p_s = s - h_s = s - g = q$. ┘

6.3. PROPOSITION. Let (p_n) be a sequence of potentials on X such that $p = \sum_{n=1}^{\infty} p_n$ is superharmonic. Then p is a potential.

Proof. Let $x \in X$, $U \in \mathcal{U}_x$ and $\varepsilon > 0$. Since $\sum_{n=1}^{\infty} H_U p_n(x) = H_U p(x) < \infty$ there exists $n \in \mathbb{N}$ such that $\sum_{i=n+1}^{\infty} H_U p_i(x) < \varepsilon$. Let

$$q_1 = \sum_{i=1}^{n} p_i \ , \quad q_2 = \sum_{i=n+1}^{\infty} p_i \quad .$$

Then $q_1, q_2 \in S^+(X)$ and

$$h_p(x) = h_{q_1}(x) + h_{q_2}(x) = H_U h_{q_2}(x) \leq H_U q_2(x) < \varepsilon \quad .$$

Hence $h_p = 0$, i.e. $p \in P(X)$. ┘

6.4. PROPOSITION. Let (V_i) be an open covering of X and $u \in \bigcap_{i \in I} {}^*H(V_i)$ such that $u \geq -p$ for some $p \in P(X)$. Then $u \in {}^*H^+(X)$.

Proof. There exist sequences (U_n) and (W_n) in $\mathcal{U}$ such that $X = \bigcup_{n=1}^{\infty} W_n$, $\overline{W}_n \subset U_n$ and $\overline{U}_n \subset V_{i_n}$ for some $i_n \in I$. Let (k_n) be a sequence as in (6.1) and define (p_n) by $p_o = p$ and $p_n = \widehat{H_{U_{k_n}} p_{n-1}}$. Then by (6.1), (p_n) is a decreasing sequence in $P(X)$ such that $\inf p_n = 0$. We claim that $u \geq -p_n$ for every $n \in \mathbb{N}$ and hence $u \geq 0$. Indeed, $u \geq -p = -p_o$ and if $n \in \mathbb{N}$ such that $u \geq -p_{n-1}$ then

$$u \geq H_{U_{k_n}} u \geq - H_{U_{k_n}} p_{n-1} \geq - \widehat{H_{W_{k_n}} p_{n-1}} = - p_n \quad .$$

Thus by (4.5) $u \in \bigcap_{i \in I} {}^*H^+(V_i) = {}^*H^+(X)$. ┘

6.5. COROLLARY. $P(X) = \{s \in S^+(X) : u \in {}^*H(X) \ , \ u + s \geq 0 \Rightarrow u \geq 0\}$.

Proof. Let $s \in S^+(X)$ such that $u \geq 0$ for every $u \in {}^*H(X)$ satisfying $u + s \geq 0$. Then $-h_s \in {}^*H(X)$ and $(-h_s) + s = p_s \geq 0$, hence $-h_s \geq 0$, i.e. $h_s = 0$, $s = p_s \in P(X)$. This shows one inclusion. The inverse inclusion follows from (6.4). ┘

6.6. COROLLARY (Minimum principle). Let U be an open subset of X and $v \in {}^*H(U)$ such that $\liminf_{x \to z} v(x) \geq 0$ for every $z \in U^*$, $v \geq 0$ on $\complement U$ and $v \geq -p$ for some $p \in P(X)$. Then $v \geq 0$.

Proof. Let $s \in S^+(X) \cap C(X)$ be strictly positive, $\varepsilon > 0$ and $u = \inf(v + \varepsilon s, 0)$. Then obviously $u \in {}^*H(U)$. Moreover, $u \leq 0$ on X and

$u = 0$ on an open set V containing $\complement U$ since $\liminf_{x \to z} (v + \varepsilon s)(x) \geq \varepsilon s(z) > 0$ for every $z \in U^*$. Since $u \geq -p$ we therefore conclude by (6.4) that $u \geq 0$, i.e. $v + \varepsilon s \geq 0$. Thus $v \geq 0$. ┘

Let $\mathcal{P} = \mathcal{P}(X) \cap \mathcal{C}(X)$. Then $\mathcal{P}$ is a min-stable convex cone. We want to show that $\mathcal{P}$ is a function cone. We already know by (1.4) and (6.3) that $\mathcal{P} = \mathcal{P}_\sigma$.

<u>6.7. PROPOSITION.</u> Let $p \in \mathcal{P}$, $x \in X$, and $\varepsilon > 0$. Then there exists a potential $q \in \mathcal{P}$ such that $q(x) < \varepsilon$ and $q \geq p$ outside of a compact set.

Proof. Let (K_n) be a sequence of compact subsets of X such that $K_n \subset \overset{\circ}{K}_{n+1}$ and $\bigcup_{n=1}^{\infty} K_n = X$. For every $n \in \mathbb{N}$ let $\varphi_n \in \mathcal{C}(X)$ such that $0 \leq \varphi_n \leq 1$, $\varphi_n = 0$ on K_n and $\varphi_n = 1$ on $\complement K_{n+1}$ and define

$$p_n = R_{\varphi_n p} \quad .$$

By (5.6), $p_n \in \mathcal{P}$, p_n is harmonic on $\overset{\circ}{K}_n$ and p_n is equal to p on $\complement K_{n+1}$. The sequence (φ_n) being decreasing the sequence (p_n) is decreasing as well. So we conclude by (3.1) that $\inf p_n \in \mathcal{H}^+(\overset{\circ}{K}_m)$ for every $m \in \mathbb{N}$ and hence $\inf p_n \in \mathcal{H}^+(X)$. Since $\inf p_n \leq p$ and p is a potential we obtain $\inf p_n = 0$ which implies the assertion. ┘

<u>6.8. PROPOSITION.</u> Let $U \in \mathcal{U}$ and $x \in U$. Then there exists $p \in \mathcal{P}$ such that $H_U p(x) < p(x)$, in particular $p(x) > 0$.

Proof. By (6.2), it suffices to find a function $s \in \mathcal{S}^+(X) \cap \mathcal{C}(X)$ such that $H_U s(x) < s(x)$. By (H_5) there exists a function $s_0 \in \mathcal{S}^+(X) \cap \mathcal{C}(X)$ such that $s_0(x) > 0$. If $H_U(x,\cdot) = 0$ then $H_U s_0(x) = 0 < s_0(x)$. So let us suppose that $H_U(x,\cdot) \neq 0$ and let y be a point in the support of $H_U(x,\cdot)$. By (H_5) and (5.7) there exist $s_1, s_2 \in \mathcal{S}^+(X) \cap \mathcal{C}(X)$ such that $s_1(x)s_2(y) < s_1(y)s_2(x)$. Defining

$$s = \inf\,(s_1(x)s_2, s_2(x)s_1) \quad \text{and} \quad t = s_2(x)s_1$$

we have $s,t \in \mathcal{S}^+(X) \cap \mathcal{C}(X)$, $s \leq t$, $s(y) < t(y)$ and $s(x) = t(x)$. Thus

$$H_U s(x) < H_U t(x) \leq t(x) = s(x) \quad .$$

┘

6.9. COROLLARY. For every $x \in X$ and every neighborhood V of x there exist $p,q \in \mathcal{P}$ such that $q \le p$, $q(x) < p(x)$ and $q = p$ on $\complement V$. In particular, $\mathcal{P}$ is linearly separating.

Proof. Let $U \in \mathcal{U}_x$ such that V is a neighborhood of $\overline{U}$ and let $p \in \mathcal{P}$ such that $H_U p(x) < p(x)$. There exists a function $\varphi \in C^+(X)$ such that $\varphi \le H_U p$ and $\varphi = p$ on $\complement V$. Then $q := R_\varphi \in \mathcal{P}(X) \cap C(X)$, $q = p$ on $\complement V$ and $q \le \widehat{H_U p}$ hence $q(x) \le H_U p(x) < p(x)$.

6.10. COROLLARY. $\mathcal{P}$ is a function cone such that $*\mathcal{H}^+(X) = S(\mathcal{P})$.

Proof. By (6.8), (6.9) and (6.7) and (I.1.1) $\mathcal{P}$ is a function cone. Let $p_o \in \mathcal{P}$, $p_o > 0$ and let $u \in *\mathcal{H}^+(X)$. There exists an increasing sequence (φ_n) in $\mathcal{K}^+(X)$ such that $\sup \varphi_n = u$. Every φ_n is bounded by a multiple of p_o , hence $p_n = R_{\varphi_n} \in \mathcal{P}$ by (5.6). Obviously the sequence (p_n) is increasing and $\sup p_n = u$.

6.11. THEOREM. $(X, *\mathcal{H}^+(X))$ is a balayage space. For every $x \in U \in \mathcal{U}$,

$$H_U(x,\cdot) = \varepsilon_x^{\circ\complement U} .$$

Proof. By (1.3), $*\mathcal{H}^+(X)$ is a convex cone of l.s.c. functions on X satisfying (B_1). (5.2) and (5.9) yield (B_2). By (6.10), $\mathcal{P}$ is a function cone such that $*\mathcal{H}^+(X) = S(\mathcal{P})$, in particular (B_4) holds. In order to prove (B_3) it suffices by (II.4.5) to show that $p - R_{p-q} \in *\mathcal{H}^+(X)$ for all $p,q \in \mathcal{P}$. So let $p,q \in \mathcal{P}$. Then $s := R_{p-q} \in S^+(X) \cap C^+(X)$ by (5.6) (in fact $s \in \mathcal{P}$ since $s \le p$). Hence $p - s \in C^+(X)$.

Let $U \in \mathcal{U}$. We want to show that $H_U(p - s) \le p - s$. Let $g \in \mathcal{K}^+(X)$ such that $g \le p$, let w be an Evans function on U , $\varepsilon > 0$ and define

$$s' = \inf (s, H_U s + p - H_U g + \varepsilon w) .$$

Since $H_U g \le H_U p \le p$ we have $H_U s + p - H_U g + \varepsilon w \in *\mathcal{H}^+(U)$. Furthermore

$$H_U s + p - H_U g + \varepsilon w = s + p - g + \varepsilon w \ge s \text{ on } \complement U .$$

Let $z \in U^*$ and let $\mathcal{F}$ be a regular ultrafilter on U converging to z . Then by (4.2)

$$\lim_F (H_U s + p - H_U g) \geq s(z) + p(z) - g(z) \geq s(z)$$

and hence

$$\liminf_{x \to z, x \in U} (H_U s + p - H_U g + \varepsilon w)(x) \geq s(z) \quad .$$

By (4.6), we conclude that $s' \in {}^*\mathcal{H}^+(X)$. We have

$$H_U g \leq H_U p \leq H_U q + H_U s \leq q + H_U s \quad ,$$

$$p = H_U g + p - H_U g \leq q + H_U s + p - H_U g$$

and hence $p - q \leq s'$. Therefore $s \leq s'$. We deduce that

$$s \leq H_U s + p - H_U g + \varepsilon w \quad ,$$

therefore

$$s \leq H_U s + p - H_U g$$

and

$$H_U(p - s) \leq p - s \quad .$$

Thus $p - s \in {}^*\mathcal{H}^+(X)$. So $(X, {}^*\mathcal{H}^+(X))$ is a balayage space.

Given $p \in \mathcal{P}$ and $U \in \mathcal{U}$ we finally have to show that $H_U p = R_p^{\complement U}$ on U . If $v \in {}^*\mathcal{H}^+(X)$ and $v \geq p$ on $\complement U$ then $v \geq H_U v \geq H_U p$. Hence $R_p^{\complement U} \geq H_U p$. In order to prove the converse inequality we take an Evans function w on U . Then $u := H_U p + w \in {}^*\mathcal{H}^+(U)$ and we conclude from (4.2) that $\liminf_{x \to z, x \in U} u(x) \geq p(z)$ for every $z \in U^*$. Hence $v := \inf (H_U p + w, p) \in {}^*\mathcal{H}^+(X)$ by (4.6). Since $v \geq p$ on $\complement U$ we deduce that $v \geq R_p^{\complement U}$. In particular, $H_U p \geq R_p^{\complement U}$ on U . ⌟

6.12. PROPOSITION. For every $p \in \mathcal{P}$, $C(p)$ is the smallest closed subset A of X such that $p \in \mathcal{H}^+(\complement A)$.

Proof. By (II.6.3) and (2.6), $p = R_p^{C(p)} \in \mathcal{H}^+(\complement C(p))$. Let A be any closed subset of X such that $p \in \mathcal{H}^+(\complement A)$. If $u \in {}^*\mathcal{H}^+(X)$ such that $u \geq p$ on A , then $u \geq p$ by (6.6). Hence $R_p^A = p$, i.e. $C(p) \subset A$ by (II.6.3) . ⌟

Let us note that by (4.5) for every $u \in \mathcal{B}^+(X)$ there exists a smallest closed subset A of X such that $u \in \mathcal{H}(\complement A)$. In view of (6.12) it will be denoted by $C(u)$.

6.13. REMARK. Let $\mathcal{U}'$ be the set of all relatively compact open subsets of X. Combining (6.11) and (2.8) we may extend $(H_U)_{U\in\mathcal{U}}$ to a family $(H_U)_{U\in\mathcal{U}'}$ of harmonic kernels. However, in view of (4.4) we know that for every open subset V of X the set of all positive hyperharmonic (harmonic resp.) functions on V is not changed if we replace $(H_U)_{U\in\mathcal{U}}$ by $(H_U)_{U\in\mathcal{U}'}$. By (1.2) this holds as well for the positive superharmonic functions. Indeed, let $s \in \mathcal{B}^+(X)$ be superharmonic on V with respect to $(H_U)_{U\in\mathcal{U}}$. Then s is hyperharmonic on V with respect to $(H_U)_{U\in\mathcal{U}'}$. Let $U \in \mathcal{U}'$ and $W \in \mathcal{U}$ such that $\overline{W} \subset U$. Then $H_W s|_W \in C(W)$ and hence $H_U s = H_W H_U s$ is continuous on W by (1.2). Thus s is superharmonic with respect to $(H_U)_{U\in\mathcal{U}'}$. We finally conclude that the convex cone of all potentials on X is not changed by the extension of the family of harmonic kernels.

Of course the same is true if $\mathcal{U}'' \subset \mathcal{U}$ is a base of X and we replace $(H_U)_{U\in\mathcal{U}}$ by $(H_U)_{U\in\mathcal{U}''}$.

We are now able to extend some of the results of section 5.

6.14. PROPOSITION. Suppose that for every $x \in X\smallsetminus X_0$ there exists a neighborhood U such that $U\smallsetminus\{x\} \in \mathcal{U}$. Let $v : X \to \overline{\mathbb{R}}_+$ such that $H_U^* v \le v$ for every $U \in \mathcal{U}$. Then $\hat{v} = \hat{v}^f \in {}^*\mathcal{H}^+(X)$, the set $\{\hat{v} < v\}$ is finely meager, and for every $x \in X_0$,

$$\hat{v}(x) = \sup_{U\in\mathcal{U}_x} H_U^* v(x) = \lim_{U,\mathcal{U}_x} H_U^* v(x) .$$

Proof. For every $U \in \mathcal{U}$, $H_U\hat{v} \le H_U^* v \le v$ where $H_U^* v$ is l.s.c. on U by (3.4), hence $H_U^* v \le \hat{v}$ on U. Therefore $\hat{v} \in {}^*\mathcal{H}^+(X)$. Obviously $\hat{v} \le \hat{v}^f \le v$.

First let $x \in X_0$. Then for every $\alpha < \hat{v}^f(x)$ there exists a compact fine neighborhood K of x such that $\alpha \le v$ on K and hence by (5.8)

$$\alpha = \alpha \lim_{U,\mathcal{U}_x} H_U(x,K) \le \liminf_{U,\mathcal{U}_x} H_U^* v(x) \le \hat{v}(x) .$$

Therefore

$$\hat{v}^f(x) = \hat{v}(x) = \sup_{U\in\mathcal{U}_x} H_U^* v(x) = \lim_{U,\mathcal{U}_x} H_U^* v(x) .$$

Consider now $x \in X\smallsetminus X_0$. Let $U \in \mathcal{U}_x$ and $V = U\smallsetminus\{x\}$. By (6.9) there exist $p,q \in \mathcal{P}$ such that $q \le p$, $q(x) < p(x)$ and $q = p$ on $\complement U$. Since $R_p^{\complement V} \ge R_p^{\{x\}}$

we obtain by (H_1) that $\lim_{y\to x} R_p^{\complement V}(y) = p(x)$. Likewise, $\lim_{y\to x} R_q^{\complement V} = q(x)$. Since $q = p$ on $\complement U = (\complement V)\setminus\{x\}$ we deduce that $\lim_{y\to x} \varepsilon_y^{\complement V}(\{x\}) = 1$, i.e. $\lim_{y\to x} H_V(y,\{x\}) = 1$. Therefore

$$\hat{v}(x) = \liminf_{y\to x} v(y) \geq \liminf_{y\to x} H_V^* v(y) \geq v(x) \liminf_{y\to x} H_V(y,\{x\}) = v(x) .$$

Thus $\hat{v} = \hat{v}^f = v$ on $X\setminus X_0$. As in the proof of (II.4.3) we finally conclude that the set $\{\hat{v} < v\}$ is finely meager. ┘

6.15. REMARK. Of course the assumption on $\mathcal{U}$ holds if $\mathcal{U}$ is the set of all relatively compact open subsets of X . Note, however, that the statement on the behavior of v on $X\setminus X_0$ may be wrong if $\mathcal{U}$ is only a base of X (see exercise 5.10.2).

6.16. EXERCISES. 1. If X is compact then $^*\mathcal{H}(X) \cap \mathcal{C}(X) = \mathcal{P}$ (in particular, $\mathcal{H}(X) = \{0\}$) .

2. Let $s,t \in \mathcal{S}^+(X)$ such that $\{s \neq t\}$ is relatively compact. Then $h_s = h_t$.

7. Absorbing and Finely Isolated Points

Let $(X,\mathcal{W})$ be a balayage space and $(H_U)_{U\in\mathcal{U}}$ a family of harmonic kernels such that $^*\mathcal{H}^+(X) = \mathcal{W}$. For every $x \in X$ let α_x denote the measure $\overset{\circ}{\varepsilon}{}_x^{\complement\{x\}}$. A point $x \in X$ is called *absorbing* if $\alpha_x = 0$. Let A_0 denote the set of all absorbing points in X . We recall that the set X_0 defined in (III.5) is by (2.7) the set of all $x \in X$ such that $\alpha_x = \varepsilon_x$ and hence $A_0 \subset \complement X_0$.

7.1. PROPOSITION. For every $x \in X$ the following statements are equivalent:

(1) x is absorbing.

(2) $\overset{\circ}{\varepsilon}{}_x^B = 0$ for every $B \subset \complement\{x\}$.

(3) $H_U(x,\cdot) = 0$ for every $U \in \mathcal{U}_x$.

(4) $h(x) = 0$ for every function $h \geq 0$ which is harmonic in a neighborhood of x .

Proof. (1) ⇒ (2): Let $p \in P$, $p > 0$. Then, for every $B \subset \complement\{x\}$, $\overset{\circ B}{\varepsilon_x}(p) = R_p^B(x) \leq R_p^{\complement\{x\}}(x) = \overset{\circ\complement\{x\}}{\varepsilon_x}(p) = 0$, hence $\overset{\circ B}{\varepsilon_x} = 0$.

(2) ⇒ (3): Obvious since $H_U(x,\cdot) = \overset{\circ\complement U}{\varepsilon_x}$.

(3) ⇒ (4): If $h \in \mathcal{B}^+(X)$ is harmonic in an open neighborhood V of x we choose $U \in \mathcal{U}$ such that $\overline{U} \subset V$ and obtain that $h(x) = H_U h(x) = 0$.

(4) ⇒ (3): Let $U \in \mathcal{U}_x$ and $p \in P$, $p > 0$. Then $H_U p = R_p^{\complement U} \in \mathcal{H}^+(U)$ by (6.8) and (2.6), hence $H_U p(x) = 0$, i.e. $H_U(x,\cdot) = 0$.

(3) ⇒ (1): (2.7) . ┘

7.2. PROPOSITION. 1. The set $A_0 \subset \complement X_0$ is a closed, finely open, and discrete subspace of X .

2. A point $x \in X$ is finely isolated if and only if $\alpha_x \neq \varepsilon_x$. The set $\complement X_0$ of all finely isolated points is countable.

3. If $u \in \mathcal{W}$ and $v : X \to \overline{\mathbb{R}}_+$ such that $\alpha_x(u) \leq v(x) \leq u(x)$ for every $x \in X$ then $v \in \mathcal{W}$.

Proof. Let $x \in X$. If x is not finely isolated, then $\alpha_x = \varepsilon_x$ by (2.7). If x is finely isolated then $\alpha_x \neq \varepsilon_x$, since otherwise by (5.8) $\lim_{U,\mathcal{U}_x} H_U(x,\{x\}) = 1$ which is impossible since $H_U(x,U) = 0$ for every $U \in \mathcal{U}_x$. Thus $\complement X_0$ is the set of all finely isolated points. Let $p \in P$ and for every $n \in \mathbb{N}$ let

$$A_n = \{x \in X : \alpha_x(p) \leq (1-\tfrac{1}{n})p(x)\} \quad .$$

Then A_n is a closed subset of $\complement X_0$. Indeed, let $x \in \complement A_n$, i.e. $R_p^{\complement\{x\}}(x) > (1-\frac{1}{n})p(x)$. Then by (2.7) there exists an open neighborhood U of x such that $R_p^{\complement U}(x) > (1-\frac{1}{n})p(x)$ and by (2.6) there exists a neighborhood V of x such that $R_p^{\complement U} > (1-\frac{1}{n})p$ on V . Therefore $R_p^{\complement\{y\}}(y) > (1-\frac{1}{n})p(y)$ for every $y \in V$, i.e. $V \subset \complement A_n$.

Consider now $v : X \to \mathbb{R}_+$ such that $\alpha_x(p) \leq v(x) \leq p(x)$ for every $x \in X$. Let $z \in \complement X_0$ and define $v_z : X \to \mathbb{R}_+$ by

$$v_z(x) = \begin{cases} p(x) , & x \neq z , \\ v(x) , & x = z . \end{cases}$$

Then v is l.s.c. If $x \in U \in \mathcal{U}$ then $\varepsilon_x^{\complement U}(v_z) \leq \varepsilon_x^{\complement U}(p) \leq p(x) = v_z(x)$ if $x \neq z$ and $\varepsilon_x^{\complement U}(v_z) = \varepsilon_x^{\complement U}(p) \leq \alpha_x(p) \leq v(x) = v_z(x)$ if $x = z$. Therefore $v_z \in \mathcal{W}$ by (II.5.5).

For every $n \in \mathbb{N}$ let

$$v_n = \inf \{v_z : z \in A_n\} .$$

Since A_n is closed and the points of A_n are finely isolated we obtain by (B_2) that $v_n = \hat{v}_n^f \in \mathcal{W}$. Obviously

$$v = \inf \{v_z : z \in \complement X_0\}$$

and $v \leq v_n \leq v + \frac{1}{n}p$ for every $n \in \mathbb{N}$. Therefore v is l.s.c. and $v = \hat{v}^f \in \mathcal{W}$ by (B_2). This proves (3).

Suppose now that we have chosen a strict potential $p \in \mathcal{P}$. Then $A_1 = A_0$ and $\bigcup_{n=1}^{\infty} A_n = \complement X_0$. Let $n \in \mathbb{N}$, $B \subset A_n$ and define $v : X \to \mathbb{R}_+$ by

$$v(x) = \begin{cases} p(x) , & x \in \complement B , \\ \alpha_x(p), & x \in B . \end{cases}$$

Then $v \in \mathcal{W}$ implies that v is l.s.c. and hence B is closed since $\alpha_x(p) \leq (1 - \frac{1}{n})p(x)$ for every $x \in B$. Thus A_n is a discrete subspace of X and hence countable. Consequently, $\complement X_0$ is countable as well. ┘

7.3. PROPOSITION. Let $\mathbb{P}$ be a sub-Markov semigroup on X such that $(X,E_{\mathbb{P}})$ is a balayage space. Then for every $x \in X$ the following statements are equivalent:

(1) x is absorbing.

(2) $P_t(x,\complement\{x\}) = 0$ for every $t > 0$.

(3) There exists $\lambda \in \mathbb{R}_+$ such that $P_t(x,\cdot) = e^{-\lambda t}\varepsilon_x$ for every $t > 0$.

Proof. (1) ⇒ (2): If x is absorbing then $1_{\complement\{x\}} \in E_{\mathbb{P}}$ by (7.1), hence for every $t > 0$

$$P_t 1_{\complement\{x\}}(x) \leq 1_{\complement\{x\}}(x) = 0 ,$$

i.e. $P_t(x,\complement\{x\}) = 0$.

(2) ⇒ (3): For every $t > 0$ there exists $\lambda_t \in [0,1]$ such that $P_t(x,\cdot) = \lambda_t \varepsilon_x$. By the semigroup property of $\mathbb{P}$ we have for all $s,t > 0$

$$\lambda_{s+t} = \lambda_s \lambda_t \quad .$$

Hence there exists $\lambda \in \mathbb{R}_+$ such that $\lambda_t = e^{-\lambda t}$ for every $t > 0$.

(3) ⇒ (1): Obviously $1_{\complement\{x\}}$ is a l.s.c. function in $S_{\mathbb{P}}$, hence $1_{\complement\{x\}} \in E_{\mathbb{P}}$. Thus for every $U \in \mathcal{U}_x$

$$H_U 1_{\complement\{x\}}(x) \leq 1_{\complement\{x\}}(x) = 0 \quad ,$$

i.e. $H_U(x,\cdot) = 0$. Therefore x is absorbing by (7.1). ┘

7.4. PROPOSITION. Let $\mathbb{P}$ be a sub-Markov semigroup on X such that $(X, E_{\mathbb{P}})$ is a balayage space having no absorbing points and let $u \in E_{\mathbb{P}} \cap C(X)$. If $P_t u = u$ for some $t > 0$ then u is harmonic. If u is a potential then $\lim_{t\to\infty} P_t u = 0$.

Proof. By (7.3), for every $x \in X$, there exists $t > 0$ such that $P_t(x,\cdot) \neq \varepsilon_x$. If $p \in E_{\mathbb{P}} \cap C(X)$ is a potential such that $P_t p = p$ for every $t > 0$, we therefore have $\delta(p) = \emptyset$, i.e. $p = 0$.

Suppose now that $u \in C(X)$ such that $P_{t_0} u = u$ for some $t_0 > 0$. Then $P_{nt_0} u = u$ for every $n \in \mathbb{N}$, hence $P_t u = u$ for every $t > 0$ since $t \mapsto P_t u$ is decreasing. u is superharmonic by (2.8). The potential part p of u obviously satisfies $P_t p = p$ for every $t > 0$, hence $p = 0$, i.e. u is harmonic.

Finally consider a potential $u \in E_{\mathbb{P}} \cap C(X)$ and let $p = \lim_{t\to\infty} P_t u$. Then $P_t p = p$ and hence $P_t(u-p) \leq u - p$ for every $t > 0$. Moreover, $\lim_{t\to\infty} P_t(u-p) = u-p$. Therefore $p \in E_{\mathbb{P}}$ and $u - p \in E_{\mathbb{P}}$, hence $p \in E_{\mathbb{P}} \cap C(X)$. The inequality $p \leq u$ implies that p is a potential. Thus $p = 0$. ┘

The following example shows that in general the converse relations in (7.4) do not hold.

7.5. EXAMPLE. Consider the sub-Markov semigroup $\mathbb{T}^*$ on $\mathbb{R}_+^*$ defined by $T_t^*(x,\cdot) = \varepsilon_{x-t}$ if $t < x$ and $T_t^*(x,\cdot) = 0$ if $t \geq x$ (translation semigroup on $\mathbb{R}_+^*$). Then $E_{\mathbb{T}^*}$ is the set of all left-continuous increasing positive numerical functions on $\mathbb{R}_+^*$. $(\mathbb{R}_+^*, E_{\mathbb{T}^*})$ is a harmonic space (it is the restriction of $(\mathbb{R}, E_{\mathbb{T}})$ on $\mathbb{R}_+^*$) such that the constant 1 is harmonic in spite of the fact that $\lim_{t\to\infty} T_t^* 1 = 0$ on $\mathbb{R}_+^*$.

7.6. COROLLARY. Let $\mathbb{P}$ be a Markov semigroup on X such that $(X, E_{\mathbb{P}})$ is a balayage space having no absorbing points. Then

$$P_b = \{q \in E_{\mathbb{P}} \cap C_b(X) : \lim_{t\to\infty} P_t q = 0\} \quad ,$$

$$H_b^+(X) = \{h \in E_{\mathbb{P}} \cap C_b(X) : P_t h = h \quad \text{for every} \quad t > 0\}$$

$$= \{h \in E_{\mathbb{P}} \cap C_b(X) : P_t h = h \quad \text{for some} \quad t > 0\} \quad .$$

Proof. By (7.4), $1 \in H(X)$. Let $h \in H^+(X)$, $0 \leq h \leq 1$. Fix $t > 0$. Since h , $1 - h \in H^+(X) \subset E_{\mathbb{P}}$ we obtain that $P_t h \leq h$ and

$$P_t(-h) = P_t(1-h) - P_t 1 \leq 1 - h - 1 = - h \; ,$$

i.e. $P_t h = h$.

Consider now $q \in E_{\mathbb{P}} \cap C_b(X)$ such that $\lim_{t\to\infty} P_t q = 0$ and let $h \in H^+(X)$ such that $h \leq q$. Then $\lim_{t\to\infty} P_t h = 0$ and therefore $h = 0$ by the above considerations. Thus $q \in P_b$.

An application of (7.4) finishes the proof. ┘

7.7. REMARK. The situation may change completely if X has absorbing points: If X is discrete and $\mathbb{P}$ is the trivial semigroup given by $P_t = I$, $t > 0$, then $P_b = B_b^+(X) = \{f \in B_b^+(X) : P_t f = f$ for every $t > 0\}$ and $H_b^+(X) = \{0\} = \{f \in B_b^+(X) : \lim_{t\to\infty} P_t f = 0\}$!

7.8. EXERCISE. Let A be a non-empty discrete subset of $\mathbb{R}$ and let W be the set of all l.s.c. positive numerical functions w on $\mathbb{R}$ such that w is concave on each interval contained in $\mathbb{R} \setminus A$. Then $(\mathbb{R}, W)$ is a balayage space such that every point $a \in A$ is absorbing.

8. Harmonic Spaces

Let (X,W) be a balayage space and $(H_U)_{U \in U}$ a family of harmonic kernels such that $*H^+(X) = W$. We shall say that (X,W) has the *local truncation property* if the following holds: For every open U in X and every $u,v \in W$ such that $u \geq v$ on U^* the function w defined by

$$w = \begin{cases} \inf(u,v) & \text{on } U \\ v & \text{on } \complement U \end{cases}$$

is containd in $\mathcal{W}$.

8.1. PROPOSITION. The following statements are equivalent:

(1) $(X,\mathcal{W})$ has the local truncation property.

(2) For every open V in X and every $x \in V$, $\overset{o}{\varepsilon}{}_x^{\complement V}(\complement V^*) = 0$.

(3) For every $x \in U \in \mathcal{U}$, $H_U(x,\complement U^*) = 0$.

(4) For every open V in X , every $u \in {}^*\mathcal{H}(V)$ and $v \in \mathcal{B}(X)$ such that $v = u$ on V we have $v \in {}^*\mathcal{H}(V)$.

Proof. (1) ⇒ (2): Let V be an open subset of X . Let $p,q \in \mathcal{P}$ such that $q \leq p$ and $q = p$ on V^* . Let $u \in \mathcal{W}$ such that $u \geq q$ on $\complement V$. Then $u \geq q = p$ on V^* , hence by the local truncation property the function w defined by

$$w = \begin{cases} \inf(u,p) & \text{on } V \\ p & \text{on } \complement V \end{cases}$$

is contained in $\mathcal{W}$. Therefore $w \geq R_p^{\complement V}$. In particular, $u \geq R_p^{\complement V}$ on V . So we obtain that $R_q^{\complement V} \geq R_p^{\complement V}$ on V , i.e. $R_q^{\complement V} = R_p^{\complement V}$ on V since $q \leq p$. Thus for every $x \in V$,

$$\overset{o}{\varepsilon}{}_x^{\complement V}(p-q) = R_p^{\complement V}(x) - R_q^{\complement V}(x) = 0 \quad .$$

By (6.9), this yields $\overset{o}{\varepsilon}{}_x^{\complement V}(\complement V^*) = 0$.

(2) ⇒ (3) ⇒ (4): Trivial.

(4) ⇒ (1): Let U be an open set in X and $u,v \in \mathcal{W}$ such that $u \geq v$ on U^* . Define w by

$$w = \begin{cases} \inf(u,v) & \text{on } U \\ v & \text{on } \complement U \text{ .} \end{cases}$$

Since $\inf(u,v) \in {}^*\mathcal{H}^+(X)$ and $\inf(u,v) = w$ on U we know by (4) that $w \in {}^*\mathcal{H}^+(U)$. Since w is l.s.c. and $w = v$ on $\complement U$ we conclude by (4.6) that $w = \inf(v,w) \in {}^*\mathcal{H}^+(X)$. ┘

8.2. PROPOSITION. If $1 \in \mathcal{W}$ the following statements are equivalent:

(1) $(X,\mathcal{W})$ has the local truncation property.

(2) There exists a sub-Markov semigroup $\mathbb{P} = (P_t)_{t>0}$ on X such that $E_{\mathbb{P}} = \mathcal{W}$ and $\lim_{t\downarrow 0} \frac{1}{t} P_t f = 0$ locally uniformly on $\complement S(f)$ for every $f \in \mathcal{K}^+(X)$.

Proof. (1) ⇒ (2): Let $p \in \mathcal{P}_b$ be a strict potential. Let A be a countable base of relatively compact open subsets of X which is stable under finite unions and let $\{(U_n, V_n) : n \in \mathbb{N}\}$ be the set of all pairs $(U,V) \in A \times A$ such that $\overline{U} \subset V$. Given $n \in \mathbb{N}$ we choose $f_n \in \mathcal{K}^+(X)$ such that $f_n \leq 1$, $f_n = 1$ on $\overline{U}_n$ and $f_n = 0$ on $\complement \overline{V}_n$ and define

$$p_n = R_{f_n p} \quad , \quad q_n = R_{(1-f_n)p_n} \quad .$$

Then $p_n, q_n \in \mathcal{P}$, $q_n \leq p_n \leq p$, q_n is harmonic on U_n by (2.4), and p_n is equal to the strict potential p on U_n . Hence $q_n < p_n$ on U_n . Moreover $q_n = p_n$ on $\complement \overline{V}_n$ and p_n is harmonic on $\complement \overline{V}_n$ by (2.4). Using (8.1) we conclude that q_n is harmonic on $\complement \overline{V}_n$.

We now define a strict potential $q \in \mathcal{P}_b$ by

$$q = p + \sum_{n=1}^{\infty} \frac{1}{2^n}(p_n + q_n) \quad .$$

Then by (II.8.9) there exists a sub-Markov semigroup $\mathbb{P} = (P_t)_{t>0}$ on X such that $E_{\mathbb{P}} = \mathcal{W}$ and the potential kernel V of $\mathbb{P}$ is the bounded kernel $(x,B) \mapsto q_B(x)$.

Let $f \in \mathcal{K}^+(X)$ and let K be a compact subset of $\complement S(f)$. There exists $n \in \mathbb{N}$ such that $S(f) \subset U_n$ and $K \cap \overline{V}_n = \emptyset$. We choose a function $g \in \mathcal{K}^+(X)$ such that $g = 0$ on K , $g = 2^n$ on $\overline{V}_n$ and claim that $Vg - q_n \in \mathcal{P}$. Indeed, if W denotes the kernel $(x,B) \mapsto (q_n)_B$ then it follows from the definition of q and (II.6.18.1) that $Vg - \frac{1}{2^n} Wg \in \mathcal{P}$. Furthermore, $C(q_n) \subset \overline{V}_n$ since q_n is harmonic on $\complement \overline{V}_n$ and hence $Wg = 2^n q_n$ by (II.6.18.2).

Let $t > 0$. Then

$$P_t(Vg - q_n) \leq Vg - q_n \quad ,$$

i.e.

$$q_n - P_t q_n \leq Vg - P_t Vg = \int_0^t P_s g \, ds \quad .$$

There exists $\alpha > 0$ such that $f \leq \alpha(p_n - q_n)$ and hence on K

$$P_tf \le \alpha(P_tp_n - P_tq_n) = \alpha(P_tp_n - P_tq_n + q_n - p_n)$$
$$\le \alpha(q_n - P_tq_n) \le \alpha\int_0^t P_sg\,ds \quad .$$

By (II.3.16) P_sg converges to g locally uniformly as s tends to zero. Since $g = 0$ on K we conclude that $\frac{1}{t}P_tf$ converges to 0 uniformly on K as t tends to zero.

(2) ⇒ (1): Let $\mathbb{P} = (P_t)_{t>0}$ be a sub-Markov semigroup on X such that $E_{\mathbb{P}} = \mathcal{W}$ and $\lim_{t\downarrow 0}\frac{1}{t}P_tf = 0$ locally uniformly on $\complement S(f)$ for every $f \in \mathcal{K}^+(X)$. Let U be an open subset of X, $u \in \mathcal{W}$ and $q \in \mathcal{P}$ such that $u \ge q$ on U^*. There exists $q' \in \mathcal{P}$ such that $q' > 0$ and $q \in o(q')$. We take $\varepsilon > 0$ and define a function u_ε by

$$u_\varepsilon = \begin{cases} \inf(u + \varepsilon q', q) & \text{on } U \\ q & \text{on } \complement U \quad . \end{cases}$$

We claim that $u_\varepsilon \in \mathcal{W}$. We note first that u_ε is l.s.c. Let

$$v_\varepsilon = \inf(u + \varepsilon q', q) \quad .$$

Then $v_\varepsilon = q$ in a neighborhood of U^* and the set $\{v_\varepsilon < q\}$ is relatively compact. Hence there exists a function $f \in \mathcal{K}^+(X)$ such that $u_\varepsilon - v_\varepsilon \le f$ and $S(f) \cap \overline{U} = \emptyset$.

Let $\eta > 0$. Since $\{u_\varepsilon < q\}$ is a relatively compact subset of U there exists $t_0 > 0$ such that

$$\frac{1}{t}P_tf(x) \le \eta$$

for every $x \in \{u_\varepsilon < q\}$ and every $0 < t \le t_0$. Let $0 < t \le t_0$. Then for every $x \in \{u_\varepsilon < q\}$

$$P_tu_\varepsilon(x) \le P_tv_\varepsilon(x) + P_tf(x) \le v_\varepsilon(x) + t\eta = u_\varepsilon(x) + t\eta$$

whereas for every $x \in \{u_\varepsilon \ge q\}$

$$P_tu_\varepsilon(x) \le P_tq(x) \le q(x) = u_\varepsilon(x) \le u_\varepsilon(x) + t\eta \quad .$$

Proceeding by induction and using the fact that $P_t1 \le 1$ we obtain that

$$P_tu_\varepsilon \le u_\varepsilon + t\eta$$

for every $n \in \mathbb{N}$ and every $0 < t \le nt_0$, i.e. for every $t > 0$. Since $\eta > 0$

is arbitrary we deduce that

$$P_t u_\varepsilon \leq u_\varepsilon$$

for every $t > 0$ and hence $u_\varepsilon \in \mathcal{W}$ by (II.4.6). Letting ε tend to zero we conclude that the function w defined by

$$w = \begin{cases} \inf(u,q) & \text{on } U \\ q & \text{on } \complement U \end{cases}$$

is contained in $\mathcal{W}$.

Suppose finally that $v \in \mathcal{W}$ such that $u \geq v$ on U^*. Taking an increasing sequence (q_n) in $\mathcal{P}$ such that $\sup q_n = v$ and defining corresponding functions w_n as above we obtain an increasing sequence (w_n) in $\mathcal{W}$ such that

$$\sup w_n = \begin{cases} \inf(u,v) & \text{on } U \\ q & \text{on } \complement U . \end{cases}$$

Since $\sup w_n \in \mathcal{W}$ we have shown that $(X,\mathcal{W})$ has the local truncation property.

<u>8.3. PROPOSITION.</u> If $(X,\mathcal{W})$ has the local truncation property then $X_0 \cup A_0 = X$, i.e. a point is finely isolated if and only if it is absorbing.

Proof. Let $x \in X$ be finely isolated. By (II.4.1) there exists an open subset U of X, a function $v \in \mathcal{W}$ and $\alpha \in \mathbb{R}$ such that $\{x\} = U \cap \{v < \alpha\}$. Suppose that there exists an open neighborhood V of x and a function $h \in \mathcal{H}^+(V)$ such that $h(x) = 1$. Let $\frac{v(x)}{\alpha} < \beta < 1$ and let $W \in \mathcal{U}$ such that $x \in W$ and $\overline{W} \subset U \cap \{y \in V : \beta h(y) < 1\}$. Then we obtain the contradiction

$$v(x) < \alpha\beta = \alpha\beta H_W h(x) \leq \alpha H_W(x,W^*) \leq H_W v(x) \leq v(x) .$$

So x is absorbing by (7.1).

We shall say that $(X,\mathcal{W})$ is a *harmonic space* if $(X,\mathcal{W})$ has the local truncation property and if X has no finely isolated points or, equivalently, if $(X,\mathcal{W})$ has the local truncation property and if X has no absorbing points. *)

---

*) The reader who is familiar with the standard monograph of C. CONSTANTINESCU - A. CORNEA [4] will easily verify that our notion of a harmonic space corresponds to a $\mathcal{P}$-harmonic space in their terminology.

8.4. EXAMPLES. The balayage spaces of our examples "classical potential theory" and "translation on $\mathbb{R}$" are harmonic spaces since $H_U(x,\complement U^*) = 0$ and $H_U(x,U^*) = 1$ for every $x \in U \in \mathcal{U}$. Let us note that the condition of (8.2) is satisfied by the Brownian semigroup on $\mathbb{R}^n$ ((0.2.4)) and trivially by the translation semigroup on $\mathbb{R}$. On the other hand a balayage space $(X,\mathcal{W})$ clearly cannot be a harmonic space if X has isolated points as in example (III.1.1.3). Furthermore, the balayage spaces $(X,\mathcal{W})$ associated with pseudo-Poisson semigroups do not satisfy the local truncation property unless $\mathcal{W}$ consists of all positive numerical functions on X .

Historically, the notion of a harmonic space is closely connected to linear elliptic and parabolic partial differential equations of second order. We shall study this relation in chapter VIII.

Summarizing results of chapters II and III we obtain at once the following characterization of harmonic spaces.

8.5. THEOREM. Let $\mathcal{W}$ be a convex cone of numerical functions on X such that $1 \in \mathcal{W}$. Then the following properties are equivalent:

(1) $(X,\mathcal{W})$ is a harmonic space.

(2) There exists a family $(H_U)_{U \in \mathcal{U}}$ of harmonic kernels on X such that:

(a) $^*\mathcal{H}^+(X) = \mathcal{W}$.

(b) $H_U(x,\complement U^*) = 0$ for every $x \in U \in \mathcal{U}$.

(c) $H_U(x,U^*) > 0$ for every $x \in U \in \mathcal{U}$.

(3) There exists a sub-Markov semigroup $\mathbb{P} = (P_t)_{t>0}$ on X and a min-stable function cone $\mathcal{P} \subset \mathcal{C}^+(X)$ such that:

(a) $E_{\mathbb{P}} = S(\mathcal{P}) = \mathcal{W}$.

(b) $\lim_{t\to 0} \frac{1}{t} P_t f = 0$ locally uniformly on $\complement S(f)$ for every $f \in \mathcal{K}(X)$.

(c) For every $x \in X$, $P_t(x,\complement\{x\}) > 0$ for some $t > 0$.

8.6. EXERCISES. 1. X is locally connected if $(X,\mathcal{W})$ is a harmonic space. (Hint: Let U be an open neighborhood of $x \in X$, let $h \in \mathcal{H}(U)$ such that $h > 0$ on U and consider the family $\{h 1_V : x \in V \subset U , V$ and $U \setminus V$ open$\}$.)

2. Let X be the subspace $\{0\} \cup \{\frac{1}{n} : n \in \mathbb{N}\}$ of $\mathbb{R}$ and let $\mathcal{W}$ be the set

of all decreasing functions $v : X \to \overline{\mathbb{R}}_+$ such that $\lim_{n \to \infty} v(\frac{1}{n}) = v(0)$. Then $(X,\mathcal{W})$ is a balayage space. Note that X is not locally connected.

3. Characterize the discrete balayage spaces having the local truncation property.

IV. Markov Processes

This chapter is devoted to a construction of a probabilistic model of balayage spaces. Having collected some basic material from probability theory (section 1) we define Markov processes (section 2) which are the stochastic processes corresponding to sub-Markov semigroups (section 3). To overcome measurability problems in later sections it is necessary to modify processes by completion of various σ-algebras (section 4). Our applications in chapter VI will require a Markov property where the constant times are replaced by certain random times (e.g. hitting times). Therefore we introduce stopping times (section 5) and study the strong Markov property (section 6). It turns out that Markov processes associated with balayage spaces are equivalent to Hunt processes (section 7) having all the regularity properties needed later. A global presentation of the equivalence of our four views of potential theory (balayage space, family of harmonic kernels, sub-Markov semigroup, Hunt process) finishes this chapter.

1. Stochastic Processes

Let T and Λ be index sets. A *stochastic process* with values in a measurable space $(E,\mathcal{E})$ is a collection $\mathfrak{X} = (\Omega, \mathfrak{M}, (X_t)_{t\in T}, (P^\lambda)_{\lambda\in\Lambda})$ such that

(SP_1) For every $\lambda \in \Lambda$, $(\Omega, \mathfrak{M}, P^\lambda)$ is a probability space.

(SP_2) For every $t \in T$, $X_t : \Omega \to E$ is $\mathfrak{M}$-$\mathcal{E}$-measurable.

For every $\omega \in \Omega$, the map $t \mapsto X_t(\omega)$ from T into E is called the *path* corresponding to ω. The space $(E,\mathcal{E})$ is called *state space* of $\mathfrak{X}$.

Two stochastic processes $\mathfrak{X} = (\Omega, \mathfrak{M}, (X_t)_{t\in T}, (P^\lambda)_{\lambda\in\Lambda})$ and $\overline{\mathfrak{X}} = (\overline{\Omega}, \overline{\mathfrak{M}}, (\overline{X}_t)_{t\in T}, (\overline{P}^\lambda)_{\lambda\in\Lambda})$ with the same state space $(E,\mathcal{E})$ are called *equivalent* if for every $\lambda \in \Lambda$, every finite subset $\{t_1,\dots,t_n\}$ of T and $A_1\dots,A_n \in E$ the following equality holds:

$$P^\lambda[X_{t_1} \in A_1,\dots,X_{t_n} \in A_n] = \overline{P}^\lambda\ [\overline{X}_{t_1} \in A_1,\dots,\overline{X}_{t_n} \in A_n]\,.$$

<u>1.1. REMARKS.</u> Let $\mathfrak{X} = (\Omega, \mathfrak{M}, (X_t)_{t\in T}, (P^\lambda)_{\lambda\in\Lambda})$ be a stochastic process with values in (E,E).

1. Define (E^T,E^T) to be usual product measurable space. For each $t \in T$ let $\pi_t : E^T \to E$ the coordinate map and let $\Phi : \Omega \to E^T$ be defined as $\Phi(\omega) = (X_t(\omega))_{t\in T}$. Since Φ is $\mathfrak{M}$-E^T-measurable, for every $\lambda \in \Lambda$ the image measure $\Phi(P^\lambda)$ of P^λ on E^T is defined. Therefore,

$$\widetilde{\mathfrak{X}} = (E^T, E^T, (\pi_t)_{t\in T}, (\Phi(P^\lambda))_{\lambda\in\Lambda})$$

is a stochastic process with values in (E,E), which is certainly equivalent to $\mathfrak{X}$. $\widetilde{\mathfrak{X}}$ is called the (first) *canonical process* (associated with $\mathfrak{X}$). It follows at once from the definitions that two processes are equivalent if and only if they have the same first canonical process.

2. Let $\mathcal{F}(T)$ be the set of all non-empty finite subsets of T and define for $I \in \mathcal{F}(T)$ the product measurable space (E^I, E^I), the canonical projection $\pi_I : E^T \to E^I$. If $I = \{t\}$ then $\pi_I = \pi_t$. If $I, J \in \mathcal{F}(T)$ such that $I \subset J$ we will denote the natural projection of E^J on E^I by π_I^J. If $\lambda \in \Lambda$ and $I = \{t_1,\dots,t_n\} \in \mathcal{F}(T)$ let P_I^λ be the common distribution of $X_{t_1},\dots,X_{t_n}$ with respect to P^λ, i.e. for every $A \in E^I$

$$P_I^\lambda(A) = P^\lambda[(X_{t_1},\dots,X_{t_n}) \in A]\,.$$

The set $(P_I^\lambda)_{\lambda\in\Lambda, I\in\mathcal{F}(T)}$ is called the family of *finite-dimensional distributions* of $\mathfrak{X}$. Hence two processes are equivalent if and only if they have the same family of finite-dimensional distributions.

3. For every $\lambda \in \Lambda$ the family $(P_I^\lambda)_{I\in\mathcal{F}(T)}$ is *projective* on (E,E), i.e. for all $I, J \in \mathcal{F}(T)$ such that $I \subset J$

$$\pi_I^J(P_J^\lambda) = P_I^\lambda\,,$$

having $\Phi(P^\lambda)$ as *projective limit*, i.e. for all $I \in \mathcal{F}(T)$

$$\pi_I(\Phi(P^\lambda)) = P_I^\lambda .$$

The following proposition which is a special case of Kolmogorov's theorem gives an answer to the converse problem, namely whether a given projective family of probability measures on the measurable spaces $(E^I, \mathcal{E}^I)$, $I \in \mathcal{F}(T)$, is the family of finite-dimensional distributions of a stochastic process.

1.2. PROPOSITION. Let E be a compact metrizable space. Then every projective family $(P_I)_{I \in \mathcal{F}(T)}$ of probability measures P_I on $(E^I, \mathcal{B}(E)^I)$ has a projective limit P on $(E^T, \mathcal{B}(E)^T)$. In particular, $(E^T, \mathcal{B}(E)^T, (\pi_t)_{t \in T}, P)$ is a stochastic process with values in $(E, \mathcal{B}(E))$ having $(P_I)_{I \in \mathcal{F}(T)}$ as its family of finite-dimensional distributions.

Proof. Let H be the vector space of real continuous functions on the compact space E^T depending only on a finite number of variables. Then the approximation theorem of Stone-Weierstrass yields at once that H is dense in $C(E^T)$ with respect to the topology of uniform convergence. For every $h \in H$ there exist $I \in \mathcal{F}(T)$ and $h_I \in C(E^I)$ such that $h = h_I \circ \pi_I$. Furthermore, $\int_{E^I} h_I \, dP_I$ is independent of the representation of h. Extending the positive linear functional $h \mapsto \int_{E^I} h_I \, dP_I$ to a positive linear functional on $C(E^T)$ we obtain a probability measure P on $(E^T, \mathcal{B}(E)^T)$ such that for every $I \in \mathcal{F}(T)$ and $h_I \in C(E^I)$

$$\int_{E^T} h_I \circ \pi_I \, dP = \int_{E^I} h_I \, dP_I$$

which implies the assertions. ∎

Let $(\Omega, \mathfrak{M}, P)$ be a probability space. If Y is a numerical random variable, i.e. $Y : \Omega \to \overline{\mathbb{R}}$ is $\mathfrak{M}$-$\mathcal{B}(\overline{\mathbb{R}})$-measurable, then the *expectation* of Y is defined by

$$E(Y) = \int Y \, dP$$

provided the integral makes sense.

If $\mathfrak{M}_0$ is a σ-algebra contained in $\mathfrak{M}$ and if Y is a numerical random variable then $E(Y|\mathfrak{M}_0)$ denotes the *conditional expectation* of Y given $\mathfrak{M}_0$, i.e. $E(Y|\mathfrak{M}_0)$ is used to denote any numerical function $f \in \mathfrak{M}_0$ such that

$\int_A f\,dP = \int_A Y\,dP$ for every $A \in \mathfrak{m}_0$. Since such a function f is only determined up to a set in $\mathfrak{m}_0$ of P-measure zero relations involving conditional expectations hold only "almost surely with respect to P". We mostly omit this phrase.

If $X = 1_A$, $A \in \mathfrak{m}$ we write $P(A|\mathfrak{m}_0)$ instead of $E(1_A|\mathfrak{m}_0)$. Finally, if the σ-algebra $\mathfrak{m}_0$ is of the form $\sigma(\{f_i \in E_i,\ i \in I\})$, we might shortly write $E(X|f_i\ ;\ i \in I)$.

Now we shall consider a special class of stochastic processes. Let $(\Omega, \mathfrak{m}, P)$ be a probability space and $(\mathfrak{m}_t)_{t \in \mathbb{R}_+}$ be an increasing family of sub-σ-algebras of $\mathfrak{m}$. A family $(X_t)_{t \in \mathbb{R}_+}$ of P-integrable functions $X_t : \Omega \to \overline{\mathbb{R}}_+$ is called an $(\mathfrak{m}_t)$-*supermartingale* (with respect to $(\Omega, \mathfrak{m}, P)$) if for all $0 \leq s \leq t$,

$$X_t \in \mathfrak{m}_t \quad \text{and} \quad E(X_t|\mathfrak{m}_s) \leq X_s .$$

It is called *right continuous* if $t \mapsto X_t$ is right continuous P-a.s.

In the sequel we shall use the following facts on supermartingales (see P.A, MEYER [1] and R.K. GETOOR [1]).

1.3. PROPOSITION. Let $(X_t)_{t \in \mathbb{R}_+}$ be an $(\mathfrak{m}_t)$-supermartingale and D a countable dense subset of $\mathbb{R}_+$. Then:

(1) For almost all $\omega \in \Omega$, $\lim_{s \uparrow t, s \in D} X_s(\omega)$ and $\lim_{s \downarrow t, s \in D} X_s(\omega)$ exist and are finite for all $t \in \mathbb{R}_+$.

(2) For each $t \in \mathbb{R}_+$, $P[X_t > 0,\ \inf_{s<t, s \in D} X_s = 0] = 0$.

1.4. PROPOSITION. Let $((X_t^n)_{t \in \mathbb{R}_+})_{n \in \mathbb{N}}$ be an increasing sequence of right continuous $(\mathfrak{m}_t)$-supermartingales. Then $(\sup_n X_t^n)_{t \in \mathbb{R}_+}$ is a right continuous $(\mathfrak{m}_t)$-supermartingale.

2. Markov Processes

As usual let X be a locally compact space with a countable base, and let $X_\Delta := X \cup \{\Delta\}$ where Δ is the point at infinity if X is noncompact and an isolated point if X is compact. Let W denote the set of all maps $w : \overline{\mathbb{R}}_+ \to X_\Delta$

such that $w(\infty) = \Delta$ and $w(s+t) = \Delta$ if $w(s) = \Delta$. Let $\pi_t : W \to X_\Delta$ denote the coordinate map $w \mapsto w(t)$ and define in W the σ-algebras

$$M := \sigma(\pi_t : t \in \overline{\mathbb{R}}_+),\ M_t := \sigma(\pi_s : s \in \overline{\mathbb{R}}_+ ,\ s \leq t),\ t \in \overline{\mathbb{R}}_+ .$$

Finally, let $\varphi_t : W \to W$ be the *shift operator* defined by $(\varphi_t w)(s) := w(t+s)$.

A (normal) *Markov process of function space type* is given by a family $(P^x)_{x \in X_\Delta}$ of probability measures P^x on (W,M) having the following properties:

(MF_1) For every $x \in X_\Delta$, $P^x[\pi_0 = x] = 1$.

(MF_2) For every $t \in \mathbb{R}_+$ and every $B \in \mathcal{B}(X)$, the function $x \mapsto P^x[\pi_t \in B]$ is Borel measurable on X.

(MF_3) *(Markov Property)* For every $x \in X$, $B \in \mathcal{B}(X)$ and $s,t \in \mathbb{R}_+$,
$P^x[\pi_{t+s} \in B | M_t] = P^{\pi_t}[\pi_s \in B]$.

2.1. REMARK. Note that condition (MF_3) implies that for every $B \in \mathcal{B}(X)$ and $s,t \in \mathbb{R}_+$ the function

$$w \mapsto P^{\pi_t(w)}[\pi_s \in B]$$

is M_t-measurable on W. Hence (MF_3) may be replaced by the following condition.

(MF_3') For every $x \in X$, $B,B_1,\dots,B_n \in \mathcal{B}(X)$ and $0 \leq t_1 < \dots < t_n < \infty$, $s \in \mathbb{R}_+$,

$$P^x(\{w \in W : w(t_1) \in B_1,\dots,w(t_n) \in B_n,\ w(t_n+s) \in B\})$$
$$= \int_{\{w \in W: w(t_1) \in B_1,\dots,w(t_n) \in B_n\}} P^{w(t_n)}[\pi_s \in B]\, P^x(dw)$$

The intuitive meaning is the following: We have a particle moving in X (or more generally, we have a system such that the states of this system are represented by points of X). Having moved in X from time 0 to some random time ζ the particle will die (or at least leave X). This is formalized by letting the particle jump to Δ. For every starting point x we have a probability measure P^x on the space W of possible motions. Then the Markov property states the following: Knowing the motion up to time t, the position at time $t+s$ can be obtained by making a new start at the position reached at time t and waiting for a time s. In particular, given the position at time t the future movement is independent of the history up to time t. This may be expressed by saying that the particle has no memory.

In many situations it is convenient to distinguish between elementary events and the corresponding paths. This leads to the following definition: A (normal) *Markov process* $\mathfrak{X}$ with *state space* X is a collection $(\Omega, \mathfrak{M}, \mathfrak{M}_t, X_t, \Theta_t, P^x)$ such that the following holds:

(M_0)(a) For every $x \in X_\Delta$, $(\Omega, \mathfrak{M}, P^x)$ is a probability space.

(b) For every $t \in \overline{\mathbb{R}}_+$, $\mathfrak{M}_t$ is a sub-σ-algebra of $\mathfrak{M}$ such that $\mathfrak{M}_t \subset \mathfrak{M}_{t+s}$ for every $s \in \overline{\mathbb{R}}_+$.

(c) For every $t \in \overline{\mathbb{R}}_+$, $X_t : \Omega \to X_\Delta$ is $\mathfrak{M}_t$-$\mathcal{B}(X_\Delta)$-measurable such that $X_\infty = \Delta$ and $X_{t+s}(\omega) = \Delta$ whenever $X_t(\omega) = \Delta$, $\omega \in \Omega$.

(d) For every $t \in \overline{\mathbb{R}}_+$, $\Theta_t : \Omega \to \Omega$ such that $X_s \circ \Theta_t = X_{t+s}$ for all $s \in \overline{\mathbb{R}}_+$.

(M_1) For every $x \in X_\Delta$, $P^x[X_0 = x] = 1$.

(M_2) For every $t \in \mathbb{R}_+$ and $B \in \mathcal{B}(X)$, the function $x \mapsto P^x[X_t \in B]$ is Borel measurable on X.

(M_3) (*Markov Property*) For every $x \in X$, $B \in \mathcal{B}(X)$ and $s, t \in \mathbb{R}_+$, $P^x[X_{t+s} \in B | \mathfrak{M}_t] = P^{X_t}[X_s \in B]$.

2.2. REMARKS. 1. Evidently, every Markov process of function space type is a Markov process.

2. In view of (M_0) and (M_1), condition (M_2) is equivalent to

(M_2') For every $t \in \overline{\mathbb{R}}_+$ and $B \in \mathcal{B}(X_\Delta)$, the function $x \mapsto P^x[X_t \in B]$ is Borel measurable on X_Δ.

3. In view of (M_0) - (M_2), the Markov property (M_3) is equivalent to each of the following properties:

(M_3') For every $x \in X_\Delta$, $B \in \mathcal{B}(X_\Delta)$ and $s, t \in \overline{\mathbb{R}}_+$,

$$P^x[X_{t+s} \in B | \mathfrak{M}_t] = P^{X_t}[X_s \in B]$$

(M_3'') For every $x \in X_\Delta$, $f \in \mathcal{B}_b(X_\Delta)$ and $s, t \in \overline{\mathbb{R}}_+$

$$E^x[f(X_{t+s}) | \mathfrak{M}_t] = E^{X_t}[f(X_s)] .$$

2.3. EXAMPLE. $\mathfrak{X}^0 = (\Omega, \mathfrak{M}, \mathfrak{M}_t, X_t, \Theta_t, P^x)$ is called *uniform motion to the left on* $\mathbb{R}$ (with constant speed 1) if $\Omega = \mathbb{R}_\Delta$, $\mathfrak{M} = \mathfrak{M}_t = \mathcal{B}(\mathbb{R}_\Delta)$, $X_t(x) = \Theta_t(x) = x - t$, $X_\infty(x) = X_t(\Delta) = X_\infty(\Delta) = \Delta$ $(x \in \mathbb{R},\ t \in \mathbb{R}_+)$, and $P^x = \varepsilon_x$. Then (M_0) - (M_2) are clearly satisfied. The Markov property (M_3) holds since $[X_t \in B] = B + t$ and P^x - a.s.

$$P^x[X_{t+s} \in B | \mathfrak{M}_t] = 1_{[X_{t+s} \in B]}(x) = 1_{B+t+s}(x) = \varepsilon_{x-t}(B+s) = P^{X_t}[X_s \in B] .$$

2.4. EXERCISE. Let $\mathfrak{X} = (\Omega, \mathfrak{M}, \mathfrak{M}_t, X_t, \Theta_t, P^x)$ be a Markov process on X and let $\mathcal{F}$ be a σ-algebra on Ω such that $\mathcal{F} \subset \mathfrak{M}$ and $\mathcal{F}$ and $\mathfrak{M}_t$ are independent for every t. Then $(\Omega, \mathfrak{M}, \sigma(\mathfrak{M}_t \cup \mathcal{F}), X_t, \Theta_t, P^x)$ is a Markov process.

3. Transition Functions

Let $\mathfrak{X} = (\Omega, \mathfrak{M}, \mathfrak{M}_t, X_t, \Theta_t, P^x)$ be a Markov process with state space X and define for every $x \in X_\Delta$, $B \in \mathcal{B}(X_\Delta)$ and $t \in \overline{\mathbb{R}}_+$

$$P_t^\Delta(x,B) := P^x[X_t \in B] .$$

Then (M_1) and (M_2') imply that P_t^Δ is a Markov kernel on X_Δ such that $P_0^\Delta = I$ and $P_\infty^\Delta(x,.) = \varepsilon_\Delta$ for every $x \in X_\Delta$. Thus for every $t \in \mathbb{R}_+^*$,

$$P_t := P_t^\Delta|_X$$

is a sub-Markov kernel on X. The family $\mathbb{P} := (P_t)_{t>0}$ is called the *transition function* of $\mathfrak{X}$. The name will be motivated by the following proposition.

3.1. EXAMPLE. The transition function of uniform motion to the left on $\mathbb{R}$ is just the translation semigroup $\mathbb{T}$ on $\mathbb{R}$ since

$$P^x[X_t \in B] = \varepsilon_x(B+t) = \varepsilon_{x-t}(B) = T_t(x,B) .$$

3.2. PROPOSITION. $(P_t^\Delta)_{t\in\overline{\mathbb{R}}_+}$ and $(P_t)_{t\in\mathbb{R}_+^*}$ are semigroups on X_Δ and X respectively such that for every $x \in X_\Delta$, $t \in \mathbb{R}_+^*$, and $B \in \mathcal{B}(X_\Delta)$,

$$(1) \qquad P_t^\Delta(x,B) = \begin{cases} P_t(x,B\setminus\{\Delta\}) + (1 - P_t(x,X))\,\varepsilon_\Delta(B) , & x \in X , \\ \varepsilon_\Delta(B) , & x = \Delta . \end{cases}$$

Furthermore, for every $x \in X_\Delta$, $0 \leq t_1 \leq \dots \leq t_n \leq \infty$ and $B_1,\dots,B_n \in \mathcal{B}(X_\Delta)$,

$$(2)\quad P^x[X_{t_1} \in B_1,\dots, X_{t_n} \in B_n] = \int_{B_1} P^\Delta_{t_1}(x,dx_1) \int_{B_2} P^\Delta_{t_2-t_1}(x_1,dx_2)\dots \int_{B_n} P^\Delta_{t_n-t_{n-1}}(x_{n-1},dx_n) .$$

Proof. a) We note first, that (2) is equivalent to

$$\int f(X_{t_1},\dots,X_{t_n})dP^x = \int P^\Delta_{t_1}(x,dx_1)\int P^\Delta_{t_2-t_1}(x_1,dx_2)\dots\int P^\Delta_{t_n-t_{n-1}}(x_{n-1},dx_n)f(x_1,\dots,x_n)$$

for every $f \in \mathcal{B}_b(X^n_\Delta)$.

Let us prove (2) by induction. For $n = 1$ it reduces to the definition of P_{t_1}. Suppose that (2) holds for some $n \in \mathbb{N}$. Let $0 \leq t_1 \leq \dots \leq t_{n+1} \leq \infty$, $B_1,\dots,B_{n+1} \in \mathcal{B}(X_\Delta)$. Then the Markov property yields:

$$P^x[X_{t_1} \in B_1,\dots,X_{t_n} \in B_n, X_{t_{n+1}} \in B_{n+1}] = \int_{[X_{t_1} \in B_1,\dots,X_{t_n} \in B_n]} P^x[X_{t_{n+1}} \in B_{n+1} | \mathfrak{M}_{t_n}]\, dP^x$$

$$= \int_{[X_{t_1} \in B_1,\dots,X_{t_n} \in B_n]} P^{X_{t_n}}[X_{t_{n+1}-t_n} \in B_{n+1}]\, dP^x = \int_{[X_{t_1} \in B_1,\dots,X_{t_n} \in B_n]} P^\Delta_{t_{n+1}-t_n}(X_{t_n}, B_{n+1})\, dP^x .$$

By induction hypothesis and our first observation, the last expression is equal to

$$\int_{B_1} P^\Delta_{t_1}(x,dx_1)\dots \int_{B_n} P^\Delta_{t_n-t_{n-1}}(x_{n-1},dx_n) P^\Delta_{t_{n+1}-t_n}(x_n, B_{n+1})$$

$$= \int_{B_1} P^\Delta_{t_1}(x,dx_1)\dots \int_{B_n} P^\Delta_{t_n-t_{n-1}}(x_{n-1}, dx_n) \int_{B_{n+1}} P^\Delta_{t_{n+1}-t_n}(x_n, dx_{n+1}) .$$

Therefore, (2) holds for all $n \in \mathbb{N}$.

b) If in particular $x \in X_\Delta$, $B \in \mathcal{B}(X_\Delta)$ and $s,t \in \overline{\mathbb{R}}_+$, the first part implies

$$P^\Delta_{s+t}(x,B) = P^x[X_{s+t} \in B] = P^x[X_s \in X_\Delta, X_{s+t} \in B]$$
$$= \int P^\Delta_s(x,dx_1) \int_B P^\Delta_t(x_1,dx_2) = P^\Delta_s P^\Delta_t (x,B) .$$

Thus, $(P^\Delta_t)_{t\in\overline{\mathbb{R}}_+}$ is a semigroup on X_Δ.

c) The first equality of (1) is obvious by the definition of P_t. Since $X_0 = \Delta$ P^Δ-a.s. by (M_1), we conclude from (M_0), that $X_t = \Delta$ P^Δ-a.s. for all $t \in \mathbb{R}^*_+$, hence $P^\Delta_t(\Delta,.) = \varepsilon_\Delta$.

d) It remains to show that $\mathbb{P} = (P_t)_{t\in\mathbb{R}_+^*}$ is a semigroup on X. For all $s,t\in\mathbb{R}_+^*$, $x \in X$, and $B \in \mathcal{B}(X)$ we obtain by the second and third part of the proof

$$P_{s+t}(x,B) = P^{\Delta}_{s+t}(x,B) = \int_{X_\Delta} P^{\Delta}_s(x,dx_1)\, P^{\Delta}_t(x_1,B)$$
$$= \int_X P^{\Delta}_s(x,dx_1)\, P^{\Delta}_t(x_1,B) = \int P_s(x,dx_1)\, P_t(x_1,B) = P_sP_t(x,B) .$$

3.3. REMARK. The equalities (1) and (2) in (3.2) show that the finite-dimensional distributions of $\mathfrak{X}$ are uniquely determined by the transition function $\mathbb{P}$ of $\mathfrak{X}$. Therefore two Markov processes are equivalent if and only if they have the same transition function.

3.4. PROPOSITION. Let $\mathbb{P} = (P_t)_{t\in\mathbb{R}_+^*}$ be a sub-Markov semigroup on X. Then there exists a Markov process $\mathfrak{X}$ with state space X having $\mathbb{P}$ as transition function.

Proof. For every $t \in \mathbb{R}_+^*$ we define a Markov kernel P^{Δ}_t on X_Δ by

$$P^{\Delta}_t(x,B) := \begin{cases} P_t(x,B\setminus\{\Delta\}) + (1-P_t(x,X))\varepsilon_\Delta(B) , & x \in X \\ \varepsilon_\Delta(B) & , x = \Delta . \end{cases}$$

In addition we define $P^{\Delta}_0 = I$ and $P^{\Delta}_\infty(x,.) = \varepsilon_\Delta$ for all $x \in X_\Delta$. It is easily checked that $(P^{\Delta}_t)_{t\in\overline{\mathbb{R}}_+}$ is a semigroup on X_Δ.

Let $x \in X_\Delta$. Let $\mathcal{F}$ be the family of all finite subsets of $\overline{\mathbb{R}}_+$. Given $I = \{t_1,\dots,t_n\} \in \mathcal{F}$ such that $0 \leq t_1 < \dots < t_n \leq \infty$ we define a probability measure P^x_I on $(X^I_\Delta,\mathcal{B}(X_\Delta)^I)$ by

$$P^x_I(B_1\times B_2\times\dots\times B_n) = \int_{B_1} P^{\Delta}_{t_1}(x,dx_1) \int_{B_2} P^{\Delta}_{t_2-t_1}(x_1,dx_2)\dots \int_{B_n} P^{\Delta}_{t_n-t_{n-1}}(x_{n-1},dx_n).$$

We claim that the family $(P_I)_{I\in\mathcal{F}}$ is projective, i.e. $\pi^{I'}_I(P_{I'}) = P_I$ whenever $I \subset I' \in \mathcal{F}$ where $\pi^{I'}_I$ denotes the canonical projection of $X^{I'}_\Delta$ on X^I_Δ. Let $I = \{t_1,\dots t_n\}$ be as above. It evidently suffices to consider the case where $I'\setminus I = \{t'\}$ for some $t' \in \overline{\mathbb{R}}_+$. If $t_i < t' < t_{i+1}$ for some $1 \leq i \leq n-1$ then for all $B_1,\dots,B_n \in \mathcal{B}(X_\Delta)$

$$\pi^{I'}_I(P^x_{I'})(B_1\times\dots\times B_n) = P^x_{I'}(B_1\times\dots\times B_{i-1}\times X_\Delta\times B_{i+1}\times\dots\times B_n)$$

$$= \int_{B_1} P_{t_1}(x,dx_1)\dots \int_{X_\Delta} P_{t'-t_i}(x_i,dx') \int_{B_{i+1}} P_{t_{i+1}-t'}(x',dx_{i+1})\dots \int_{B_n} P_{t_n-t_{n-1}}(x_{n-1},dx_n)$$

$$= P_I^X(B_1\times\dots\times B_n) ,$$

hence $\pi_I^{I'}(P_{I'}^X) = P_I^X$. The case $t' < t_1$ is treated similarly, the case $t_n < t'$ is trivial.

Let $\Omega := X_\Delta^{\overline{\mathbb{R}}_+}$, $\mathfrak{M} := \mathcal{B}(X_\Delta)^{\overline{\mathbb{R}}_+}$ and let π_t denote the coordinate mapping $\omega \mapsto \omega(t)$. Then by the theorem of Kolmogorov (1.2) there exists a probability measure P^X on $(\Omega,\mathfrak{M})$ such that

$$P^X[\pi_{t_1} \in B_1,\dots,\pi_{t_n} \in B_n] = P^X_{\{t_1,\dots,t_n\}}(B_1\times\dots\times B_n)$$

for all $0 \le t_1 < \dots < t_n \le \infty$ and $B_1,\dots,B_n \in \mathcal{B}(X)$.

We shall now replace the coordinate mappings π_t by random variables X_t such that $X_t = \pi_t$ P^X-a.s. and such that the trajectories $\omega \mapsto X_t(\omega)$ belong to W (see section 2).

First let $X_\infty = \Delta$. Then $X_\infty = \pi_\infty$ P^X-a.s. since $P^X[\pi_\infty = \Delta] = P^\Delta_\infty(x,\{\Delta\}) = 1$ by definition of P^Δ_∞. We define $\zeta : \Omega \to \overline{\mathbb{R}}_+$ by

$$\zeta(\omega) := \inf \{s \in \mathbb{Q}_+ : \pi_s(\omega) = \Delta\} .$$

Evidently ζ is $\mathfrak{M}$-measurable. For all $s,t \in \overline{\mathbb{R}}_+$ such that $t < s$

$$P^X[\pi_t = \Delta, \pi_s \neq \Delta] = P^\Delta_t(x,\{\Delta\})\, P^\Delta_{s-t}(\Delta,X) = 0 .$$

Hence for all $t \in \mathbb{R}_+$

$$P^X[\pi_t = \Delta, \zeta > t] = P^X[\pi_t \neq \Delta, \zeta < t] = 0$$

since

$$A_1 := [\pi_t = \Delta, \zeta > t] \subset \bigcup_{s\in\mathbb{Q}, s>t} [\pi_t = \Delta, \pi_s \neq \Delta]$$

and

$$A_2 := [\pi_t \neq \Delta, \zeta < t] = \bigcup_{s\in\mathbb{Q}, s<t} [\pi_s = \Delta, \pi_t \neq \Delta] .$$

Let $a \in X$. Defining $X_t : \Omega \to X_\Delta$, $t \in \mathbb{R}_+$, by

$$X_t(\omega) = \begin{cases} \pi_t(\omega) , & \omega \in \complement(A_1 \cup A_2) \\ a , & \omega \in A_1 \\ \Delta , & \omega \in A_2 \end{cases}$$

we thus have

$$X_t = \pi_t \ P^x\text{-a.s.}$$

By definition, $X_t(\omega) \in X$ if $t < \zeta(\omega)$ and $X_t(\omega) = \Delta$ if $t > \zeta(\omega)$ or $t = \infty$, hence the "trajectory" $\tilde{\omega} : t \mapsto X_t(\omega)$ is contained in the subset W of Ω. We note that $W \cap (A_1 \cup A_2) = \emptyset$, hence $X_s(\omega) = \pi_s(\omega) = \omega(s)$ for all $\omega \in W$ and $s \in \overline{\mathbb{R}}_+$. Let $t \in \overline{\mathbb{R}}_+$. Using the shift operator $\varphi_t : W \to W$ we define $\Theta_t : \Omega \to \Omega$ by $\Theta_t(\omega) = \varphi_t(\tilde{\omega})$, i.e. $\Theta_t(\omega)(s) = \tilde{\omega}(t+s) = X_{t+s}(\omega)$ for every $s \in \overline{\mathbb{R}}_+$. Since $\Theta_t(\Omega) \subset W$ we have, for every $s \in \overline{\mathbb{R}}_+$ and $\omega \in \Omega$, $X_s(\Theta_t(\omega)) = \Theta_t(\omega)(s)$, hence $X_s \circ \Theta_t = X_{t+s}$. Finally defining $\mathfrak{m}_t = \sigma(X_s : s \in \overline{\mathbb{R}}_+, s \leq t)$ the property (M_0) of a Markov process is satisfied.

For every $x \in X_\Delta$,

$$P^x[X_0 = x] = P^x[\pi_0 = x] = P_0^\Delta(x,\{x\}) = 1 ,$$

hence (M_1) is satisfied.

Let $t \in \mathbb{R}_+$ and $B \in \mathcal{B}(X)$. Then $x \mapsto P^x[X_t \in B] = P_t(x,B)$ is Borel measurable on X, i.e. (M_2) holds.

It remains to show the Markov property. Since $X_t = \pi_t \ P^x$-a.s. we have by definition of P^x for every $x \in X_\Delta$, $0 \leq t_1 < \dots < t_n \leq \infty$ and $B_1,\dots,B_n \in \mathcal{B}(X_\Delta)$

$$(*) \qquad P^x[X_{t_1} \in B_1,\dots,X_{t_n} \in B_n] = \int_{B_1} P_{t_1}^\Delta(x,dx_1) \dots\dots \int_{B_n} P_{t_n-t_{n-1}}^\Delta(x_{n-1},dx_n) .$$

Now let $x \in X_\Delta$, $B \in \mathcal{B}(X_\Delta)$ and $s,t \in \mathbb{R}_+$. If $s = 0$ then

$$P^x[X_{t+s} \in B|\mathfrak{m}_t] = 1_{[X_t \in B]} = P^{X_t}[X_0 \in B] .$$

So we may assume that $s > 0$. Let $0 \leq t_1 < \dots < t_n = t$ and let $B_1,\dots,B_n \in \mathcal{B}(X_\Delta)$. Then by (*)

$$\int_{[X_{t_1} \in B_1,\dots,X_{t_n} \in B_n]} P_s^\Delta(X_{t_n},B)\, dP^x = \int_{B_1} P_{t_1}^\Delta(x,dx_1)\dots\dots \int_{B_n} P_{t_n-t_{n-1}}^\Delta(x_{n-1},dx_n)\, P_s(x_n,B)$$

$$= P^x[X_{t_1} \in B_1,\dots\dots,X_{t_n} \in B_n, X_{t_n+s} \in B] .$$

Thus

$$P^x[X_{t+s} \in B|\,\mathfrak{m}_t] = P_s^\Delta(X_t,B) = P^{X_t}[X_s \in B] .$$

We finally note that $P^x[X_t \in B] = P_t^\Delta(x,B)$. Hence $\mathbb{P}$ is the transition function of $(\Omega, \mathfrak{M}, \mathfrak{M}_t, X_t, \Theta_t, P^x)$. ┘

3.5. COROLLARY. Let $\mathbb{P} = (P_t)_{t\in\mathbb{R}_+^*}$ be a sub-Markov semigroup on X . Then there exists a unique Markov process of function space type on X having $\mathbb{P}$ as transition function.

Proof. The uniqueness is an immediate consequence of the fact that the finite-dimensional distributions of a Markov process can be calculated from the transition function. So we have to show the existence.

By (3.4), there exists a Markov process $\mathfrak{X} = (\Omega, \mathfrak{M}, \mathfrak{M}_t, X_t, \Theta_t, P^x)$ having $\mathbb{P}$ as transition function. Let $\Phi : \Omega \to W$ be the mapping which to every $\omega \in \Omega$ associates the trajectory

$$\tilde{\omega} : t \mapsto X_t(\omega) .$$

In other words, Φ is given by

$$\pi_t \circ \Phi = X_t \qquad (t \in \overline{\mathbb{R}}_+) .$$

This implies that Φ is $\mathfrak{M}$-M-measurable.*) We note that Φ is even $\mathfrak{M}_t$-M_t-measurable for every $t \in \overline{\mathbb{R}}_+$. Hence we may define probability measures $\tilde{P}^x$ on (W,M) by

$$\tilde{P}^x := \Phi(P^x) \qquad (x \in X_\Delta) .$$

We claim that $\tilde{\mathfrak{X}} := (W,M,M_t,\pi_t,\varphi_t,\tilde{P}^x)$ is a Markov process. Evidently, for every $x \in X_\Delta$,

$$\tilde{P}^x[\pi_0 = x] = P^x(\Phi^{-1}[\pi_0 = x]) = P^x[X_0 = x] = 1 .$$

Furthermore, for every $t \in \mathbb{R}_+$ and $B \in \mathcal{B}(X)$, the function $x \mapsto \tilde{P}^x[\pi_t \in B] = P^x[X_t \in B]$ is Borel measurable on X . Finally, let $x \in X$, $B \in \mathcal{B}(X)$ and $s,t \in \mathbb{R}_+$. Then for every $A \in M_t$, we have $\Phi^{-1}(A) \in \mathfrak{M}_t$, hence using the Markov property of $\mathfrak{X}$ we obtain

$$\int_A \tilde{P}^{\pi_t}[\pi_s \in B] \, d\tilde{P}^x = \int_{\Phi^{-1}(A)} \tilde{P}^{\pi_t \circ \Phi}[\pi_s \in B] \, dP^x$$

---

*) As in section 2 we denote $M = \sigma(\pi_t : t\in\overline{\mathbb{R}}_+)$ and $M_t = \sigma(\pi_s : s\in\overline{\mathbb{R}}_+, s \leq t)$.

$$= \int_{\Phi^{-1}(A)} P^{X_t}(\Phi^{-1}[\pi_s \in B])dP^x = \int_{\Phi^{-1}(A)} P^{X_t}[X_s \in B]\, dP^x$$

$$= P^x(\Phi^{-1}(A) \cap [X_{t+s} \in B]) = P^x(\Phi^{-1}(A \cap [\pi_{t+s} \in B]))$$

$$= \widetilde{P}^x(A \cap [\pi_{t+s} \in B]) \ ,$$

thus

$$\widetilde{P}^x[\pi_{t+s} \in B|\mathfrak{M}_t] = \widetilde{P}^{\pi_t}[\pi_s \in B] \ .$$

The equality

$$\widetilde{P}^x[\pi_t \in B] = P^x[X_t \in B] = P_t(x,B)$$

shows that $\mathbb{P}$ is the transition function of $\widetilde{\mathfrak{X}}$. ┘

3.6. REMARK. (3.2) and (3.5) show that we have a one-to-one correspondence between sub-Markov semigroups on X and Markov processes of function space type on X .

3.7. EXERCISES. 1. Let $\mathbb{P} = (P_t)_{t > 0}$ be defined on $\mathbb{R}$ by

$$P_t(x,\cdot) = \begin{cases} \varepsilon_{x-t} , & x \le 0 \ \text{ or } \ x - t > 0 , \\ \frac{1}{2}\varepsilon_{x-t}, & x > 0 \ \text{ and } \ x - t \le 0 . \end{cases}$$

Then $\mathbb{P}$ is a sub-Markov semigroup on $\mathbb{R}$. Describe a Markov process having $\mathbb{P}$ as transition function.

2. Let $\mathbb{P} = (P_t)_{t>0}$ be defined on $]-\infty,0]$ by

$$P_t f(x) = \begin{cases} f(x-t) & , x < 0 , \\ e^{-t}f(0) + \int_0^t f(s-t)\, e^{-s}\, ds, & x = 0 . \end{cases}$$

Then $\mathbb{P}$ is a Markov semigroup on $]-\infty,0]$. There exists a Markov process $\mathfrak{X} = (\Omega, \mathfrak{M}, \mathfrak{M}_t, X_t, \Theta_t, P^x)$ on $]-\infty,0]$ having $\mathbb{P}$ as transition function such that $\Omega = \mathbb{R}$ and $X_t(s) = \inf(s-t\ , 0)$.

4. Modifications

Let $\mathfrak{X} = (\Omega, \mathfrak{M}, \mathfrak{M}_t, X_t, \Theta_t, P^x)$ be a Markov process with state space X and define

$$\mathcal{F}^0 := \sigma(X_s : s \in \overline{\mathbb{R}}_+)$$
$$\mathcal{F}^0_t := \sigma(X_s : s \in \overline{\mathbb{R}}_+,\ s \leq t)\ ,\ t \in \overline{\mathbb{R}}_+\ .$$

Then $\mathcal{F}^0$ ($\mathcal{F}^0_t$ resp.) is a sub-σ-algebra of $\mathfrak{m}(\mathfrak{m}_t$ resp.). It is evident that $\mathfrak{X}^0 = (\Omega, \mathcal{F}^0, \mathcal{F}^0_t, X_t, \Theta_t, P^x|_{\mathcal{F}^0})$ is a Markov process. We note that $\Theta_t \in \mathcal{F}_0 / \mathcal{F}_0$ and $\Theta_t \in \mathcal{F}^0_{s+t}/\mathcal{F}^0_s$ for all $s,t \in \overline{\mathbb{R}}_+$ since $X_s \circ \Theta_t = X_{s+t}$. If

$$\zeta(\omega) := \inf \{t \in \mathbb{R}_+ : X_t(\omega) = \Delta\}$$

denotes the *life time* $\zeta : \Omega \to \overline{\mathbb{R}}_+$ of the process then, for every $t \in \mathbb{R}_+$, $[\zeta < t] = \bigcup_{r<t,\, r\in\mathbb{Q}_+} [X_r = \Delta] \in \mathcal{F}^0_t$.

4.1. PROPOSITION. Let Y be an $\mathcal{F}^0$-measurable bounded real function on Ω. Then $x \mapsto E^x(Y)$ is a Borel measurable function on X_Δ such that for every $x \in X_\Delta$ and $t \in \overline{\mathbb{R}}_+$

$$E^x[Y \circ \Theta_t \mid \mathfrak{m}_t] = E^{X_t}(Y)\ .$$

Proof. It suffices to consider the case

$$Y := 1_{[X_{s_1} \in B_1, \ldots, X_{s_n} \in B_n]}$$

where $0 \leq s_1 \leq \ldots \leq s_n \leq \infty$ and $B_1, \ldots, B_n \in \mathcal{B}(X_\Delta)$. By (3.2), $x \mapsto E^x(Y) = P^x[X_{s_1} \in B_1, \ldots, X_{s_n} \in B_n]$ is $\mathcal{B}(X_\Delta)$-measurable on X_Δ. Furthermore, for every $t \in \overline{\mathbb{R}}_+$ and $B \in \mathcal{B}(X_\Delta)$

$$\int_{[X_t \in B]} E^{X_t}(Y)\, dP^x$$
$$= \int_{[X_t \in B]} dP^x \int_{B_1} P^\Delta_{s_1}(X_t, dx_1) \ldots \int_{B_n} P^\Delta_{s_n - s_{n-1}}(x_{n-1}, dx_n)$$
$$= \int_B P^\Delta_t(x, dx_0) \int_{B_1} P^\Delta_{s_1}(x_0, dx_1) \ldots \int_{B_n} P^\Delta_{s_n - s_{n-1}}(x_{n-1}, dx_n)$$
$$= P^x[X_t \in B, X_{t+s_1} \in B_1, \ldots, X_{t+s_n} \in B_n]$$
$$= \int_{[X_t \in B]} Y \circ \Theta_t\, dP^x\ ,$$

hence

$$E^x[Y \circ \Theta_t \mid X_t] = E^{X_t}(Y)\ .$$

Thus it will be sufficient to show that for every $x \in X_\Delta$, $0 \leq t \leq t_1 \leq \dots \leq t_n \leq \infty$ and $B_1, \dots B_n \in \mathcal{B}(X_\Delta)$

$$(*) \quad \begin{aligned} &P^x[X_{t_1} \in B_1, \dots, X_{t_n} \in B_n \mid \mathfrak{m}_t] \\ &= P^x[X_{t_1} \in B_1, \dots, X_{t_n} \in B_n \mid X_t] . \end{aligned}$$

If $n = 1$ then $(*)$ is an immediate consequence of $P^x[X_{t_1} \in B_1 \mid \mathfrak{m}_t] = P^{X_t}[X_{t_1-t} \in B_1]$ whenever $t < t_1$. Suppose that $(*)$ holds for some $n \in \mathbb{N}$. Let $x \in X_\Delta$, $0 \leq t \leq t_1 \leq \dots \leq t_{n+1} \leq \infty$, $B_1, \dots, B_{n+1} \in \mathcal{B}(X_\Delta)$ and $A' := [X_{t_1} \in B_1]$, $A'' := [X_{t_2} \in B_2, \dots, X_{t_{n+1}} \in B_{n+1}]$, $A := A' \cap A''$. Then by induction hypothesis $P^x[A'' \mid \mathfrak{m}_{t_1}] = P^x[A'' \mid X_{t_1}]$, hence

$$P^x[A \mid \mathfrak{m}_{t_1}] = 1_{A'} P^x[A'' \mid \mathfrak{m}_{t_1}] = 1_{A'} P^x[A'' \mid X_{t_1}] = P^x[A \mid X_{t_1}] .$$

The case $n = 1$ yields

$$E^x(P^x[A \mid X_{t_1}] \mid \mathfrak{m}_t) = E^x(P^x[A \mid X_{t_1}] \mid X_t) .$$

Therefore

$$\begin{aligned} P^x[A \mid \mathfrak{m}_t] &= E^x(P^x[A \mid \mathfrak{m}_{t_1}] \mid \mathfrak{m}_t) = E^x(P^x[A \mid X_{t_1}] \mid \mathfrak{m}_t) \\ &= E^x(P^x[A \mid X_{t_1}] \mid X_t) = E^x(P^x[A \mid \mathfrak{m}_{t_1}] \mid X_t) = P^x[A \mid X_t] . \end{aligned}$$

Thus $(*)$ holds for $n + 1$. ┘

Let $\mu \in M_+^1(X_\Delta)$. Since $x \mapsto P^x(A)$ is $\mathcal{B}(X_\Delta)$-measurable for every $A \in \mathcal{F}^0$ we may define a probability measure P^μ on $\mathcal{F}^0$ by

$$P^\mu(A) := \int P^x(A)\mu(dx) \quad (A \in \mathcal{F}^0) .$$

Let $t \in \overline{\mathbb{R}}_+$. Then for all $B \in \mathcal{B}(X_\Delta)$

$$P^\mu[X_t \in B] = \int P^x[X_t \in B]\mu(dx) = \int P_t^\Delta(x,B)\mu(dx) ,$$

hence

$$X_t(P^\mu) = \mu P_t^\Delta .$$

In particular $X_0(P^\mu) = \mu$.

Furthermore we have for every $\mathcal{F}^0$-measurable bounded real function Y on Ω and every $A \in \mathcal{F}_t^0$

$$\int_A E^{X_t}(Y)dP^\mu = \int_{X_\Delta} (\int_A E^{X_t}(Y)dP^x)\mu(dx) = \int_{X_\Delta} (\int_A Y \circ \Theta_t \, dP^x)\, \mu(dx) = \int_A Y \circ \Theta_t \, dP^\mu ,$$

hence

$$E^{\mu}[Y\circ\Theta_t \mid \mathcal{F}_t^0] = E^{X_t}(Y)$$

if $E^{\mu}(\cdot)$ means integration with respect to P^{μ}. In particular,

$$E^{\mu}(Y\circ\Theta_t) = \int E^{X_t}(Y)dP^{\mu} = \int E^x(Y)X_t(P^{\mu})(dx) .$$

Thus for every $A \in \mathcal{F}^0$

$$P^{\mu}(\Theta_t^{-1}(A)) = P^{X_t(P^{\mu})}(A) ,$$

i.e.

$$\Theta_t(P^{\mu}) = P^{X_t(P^{\mu})} .$$

Let (Ω, A, P) be a probability space and let A_1 be a sub-σ-algebra of A. Then we denote the P- *completion* of A_1 (with respect to A) by A_1^P, i.e. A_1^P is the smallest sub-σ-algebra of A containing A_1 and all subsets of P-null sets in A. Obviously

$$A_1^P = \{A \subset \Omega : \exists\, A_1 \in A_1 \text{ and } A_0 \in A \text{ such that } A \,\Delta\, A_1 \subset A_0 \text{ and } P(A_0) = 0\} .$$

<u>4.2. LEMMA.</u> Let (Ω, A, P) be a probability space and let A_1 be a sub-σ-algebra of A. Let (Ω', A') be a measurable space and let A_1' be a sub-σ-algebra of A'. Suppose that Y is a mapping of Ω into Ω' such that $Y \in A/A'$ and $Y \in A_1/A_1'$. Then $Y \in A_1^P / A_1'^{Y(P)}$.

Proof. Let $A' \in A_1'^{Y(P)}$. Then there exist $A_1' \in A_1'$ and $A_0' \in A'$ such that $A' \,\Delta\, A_1' \subset A_0'$ and $Y(P)(A_0') = 0$. It follows that $Y^{-1}(A') \,\Delta\, Y^{-1}(A_1') \subset Y^{-1}(A_0')$ and $P(Y^{-1}(A_0')) = 0$ where $Y^{-1}(A_1') \in A_1$ and $Y^{-1}(A_0') \in A$. Thus $Y^{-1}(A') \in A_1^P$. ⌟

Let

$$\mathcal{B}(X_\Delta)^* := \bigcap_{\mu \in M_+^1(X_\Delta)} \mathcal{B}(X_\Delta)^{\mu} .$$

Evidently, $\mathcal{B}(X_\Delta)^*$ is a σ-algebra on X_Δ. The elements of $\mathcal{B}(X_\Delta)^*$ are called *universally measurable* subsets of X_Δ. $\mathcal{B}(X)^* := \{B \in \mathcal{B}(X_\Delta)^* : B \subset X\}$ is the set of all universally measurable subsets of X. Let

$$\mathcal{F} := \bigcap_{\mu \in M_+^1(X_\Delta)} (\mathcal{F}^0)^{P^{\mu}} , \quad \mathcal{F}_t := \bigcap_{\mu \in M_+^1(X_\Delta)} (\mathcal{F}_t^0)^{P^{\mu}} .$$

Then $(\mathcal{F}_t)_{t\in\bar{\mathbb{R}}_+}$ is an increasing family of sub-σ-algebras of the σ-algebra $\mathcal{F}$ on Ω. If $A \in \mathcal{F}^0$ such that $P^x(A) = 0$ for every $x \in X_\Delta$ then $P^\mu(A) = 0$ for every $\mu \in M_+^1(X_\Delta)$ and hence every subset of A is contained in $\mathcal{F}_0$. For every $\mu \in M_+^1(X_\Delta)$ ($x \in X_\Delta$ resp.) the unique extension of P^μ ($P^x \mid_{\mathcal{F}^0}$ resp.) on $\mathcal{F}$ will again be denoted by P^μ (P^x resp.).

<u>4.3. PROPOSITION.</u> $(\Omega, \mathcal{F}, \mathcal{F}_t, X_t, \Theta_t, P^x)$ is a Markov process.

Proof. Let $x \in X$, $B \in \mathcal{B}(X)$ and $A \in \mathcal{F}_t$. Then there exist $A_1 \in \mathcal{F}_t^0$ and $A' \in \mathcal{F}^0$ such that $A \,\Delta\, A_1 \subset A'$ and $P^x(A') = 0$. We obtain

$$P^x(A \cap [X_{t+s} \in B]) = P^x(A_1 \cap [X_{t+s} \in B]) = \int_{A_1} P^{X_t}[X_s \in B]\, dP^x = \int_A P^{X_t}[X_s \in B]\, dP^x .$$

Thus

$$P^x[X_{t+s} \in B \mid \mathcal{F}_t] = P^{X_t}[X_s \in B] .$$

<u>4.4. PROPOSITION.</u> Let $t \in \bar{\mathbb{R}}_+$. Then $X_t \in \mathcal{F}_t / \mathcal{B}(X_\Delta)^*$, $\Theta_t \in \mathcal{F}/\mathcal{F}$ and $\Theta_t \in \mathcal{F}_{s+t}/\mathcal{F}_s$ for every $s \in \mathbb{R}_+$.

Proof. Let $B \in \mathcal{B}(X_\Delta)^*$ and $\mu \in M_+^1(X_\Delta)$. We have $X_t \in \mathcal{F}_t^0 / \mathcal{B}(X_\Delta)$ and $B \in \mathcal{B}(X_\Delta)^{X_t(P^\mu)}$.

Hence $X_t^{-1}(B) \in (\mathcal{F}_t^0)^{P^\mu}$ by (4.2). This implies that $X_t^{-1}(B) \in \mathcal{F}_t$. Now let $A \in \mathcal{F}$. We have $\Theta_t \in \mathcal{F}^0/\mathcal{F}^0$ and $A \in (\mathcal{F}^0)^{\Theta_t(P^\mu)}$ since $\Theta_t(P^\mu) = P^{X_t(P^\mu)}$. Hence $\Theta_t^{-1}(A) \in (\mathcal{F}^0)^{P^\mu}$ by (4.2). This implies $\Theta_t^{-1}(A) \in \mathcal{F}$. If $s \in \mathbb{R}_+$ and $A \in \mathcal{F}_s$ then similarly $\Theta_t^{-1}(A) \in \mathcal{F}_{s+t}$ since $\Theta_t \in \mathcal{F}_{s+t}^0/\mathcal{F}_s^0$.

<u>4.5. PROPOSITION.</u> Let Y be an $\mathcal{F}$-measurable bounded real function on Ω. Then $x \mapsto E^x(Y)$ is universally measurable on X_Δ and

$$\int Y\, dP^\mu = \int E^x(Y)\mu(dx) \quad {}^{*)}$$

for every $\mu \in M_+^1(X_\Delta)$. Furthermore, for every $\mu \in M_+^1(X_\Delta)$ and $t \in \bar{\mathbb{R}}_+$

$$E^\mu[Y \circ \Theta_t \mid \mathcal{F}_t] = E^{X_t}(Y) .$$

---

*) We are integrating with respect to the extension of μ on $\mathcal{B}(X_\Delta)^*$.

Proof. Let $\mu \in M^1_+(X_\Delta)$. Then there exist bounded $Y_1, Y_2 \in \mathcal{F}^0$ such that $Y_1 \leq Y \leq Y_2$ and $Y_1 = Y_2$ P^μ-a.s. The functions $x \mapsto E^x(Y_i)$ are $\mathcal{B}(X_\Delta)$-measurable by (4.1), $E^x(Y_1) \leq E^x(Y) \leq E^x(Y_2)$ and

$$\int E^x(Y_1)\mu(dx) = \int Y_1\, dP^\mu = \int Y_2\, dP^\mu = \int E^x(Y_2)\mu(dx)\,.$$

This implies that $x \mapsto E^x(Y)$ is universally measurable and

$$\int Y\, dP^\mu = \int E^x(Y)\mu(dx)\,.$$

Now let $t \in \overline{\mathbb{R}}_+$, $x \in X_\Delta$, $\mu = P^x_{X_t}$ and take Y_1, Y_2 as above. Then

$$\int E^{X_t}(Y_2 - Y_1)dP^x = \int E^z(Y_2 - Y_1)\mu(dz) = \int (Y_2 - Y_1)dP^\mu = 0\,,$$

hence

$$E^{X_t}(Y_1) = E^{X_t}(Y_2) \qquad P^x\text{-a.s.}$$

Furthermore, $Y \circ \Theta_t \in \mathcal{F}$ by (4.4) and

$$Y_1 \circ \Theta_t = Y_2 \circ \Theta_t \qquad P^x\text{-a.s.}$$

since $P^\mu = \Theta_t(P^x)$. Given $A \in \mathcal{F}_t$ there exist $A_1 \in \mathcal{F}^0_t$ and $A_0 \in \mathcal{F}$ such that $A_1 \Delta A \subset A_0$ and $P^x(A_0) = 0$ and we obtain by (4.1)

$$\int_A Y \circ \Theta_t\, dP^x = \int_{A_1} Y_1 \circ \Theta_t\, dP^x = \int_{A_1} E^{X_t}(Y_1)dP^x = \int_A E^{X_t}(Y)\, dP^x\,.$$

Finally, let $\nu \in M^1_+(X_\Delta)$. Then for every $A \in \mathcal{F}_t$

$$\begin{aligned}\int_A Y \circ \Theta_t\, dP^\nu &= \int \Big(\int_A Y \circ \Theta_t\, dP^x\Big)\, \nu(dx)\\ &= \int \Big(\int_A E^{X_t}(Y)\, dP^x\Big)\, \nu(dx) = \int_A E^{X_t}(Y)\, dP^\nu\,,\end{aligned}$$

i.e. $E^\nu[Y \circ \Theta_t \mid \mathcal{F}_t] = E^{X_t}(Y)$. ∎

For every $t \in \mathbb{R}_+$ let

$$\mathcal{F}^0_{t+} = \bigcap_{s>t} \mathcal{F}^0_s\,, \quad \mathcal{F}_{t+} = \bigcap_{s>t} \mathcal{F}_s\,.$$

Furthermore, let $\mathcal{F}^0_{\infty+} = \mathcal{F}^0_\infty = \mathcal{F}^0$, $\mathcal{F}_{\infty+} = \mathcal{F}_\infty = \mathcal{F}$. Then $(\mathcal{F}^0_{t+})_{t\in\overline{\mathbb{R}}_+}$ ($(\mathcal{F}_{t+})_{t\in\overline{\mathbb{R}}_+}$ resp.) is an increasing family of sub-σ-algebras of $\mathcal{F}^0$ ($\mathcal{F}$ resp.).

<u>4.6. PROPOSITION.</u> If $(\Omega, \mathcal{F}^0, \mathcal{F}^0_{t+}, X_t, \Theta_t, P^x)$ is a Markov process then $\mathcal{F}_{t+} = \mathcal{F}_t$ for every $t \in \overline{\mathbb{R}}_+$.

Proof. Let $t \in \mathbb{R}_+$ and $\mu \in M_+^1(X_\Delta)$. Furthermore, let $0 \leq s_1 < \ldots < s_n \leq \infty$, $f_1, \ldots, f_n \in \mathcal{B}_b(X_\Delta)$ and

$$Y = \prod_{i=1}^{n} f_i \circ X_{s_i}, \quad Y_1 = \prod_{s_i \leq t} f_i \circ X_{s_i}, \quad Y_2 = \prod_{s_i > t} f_i \circ X_{s_i - t}.$$

Then Y_1 is $\mathcal{F}_t^0$-measurable, Y_2 is $\mathcal{F}^0$-measurable and $Y = Y_1 \cdot Y_2 \circ \Theta_t$. Hence by (4.1)

$$E^x[Y \mid \mathcal{F}_{t+}^0] = Y_1 E^{X_t}(Y_2) \qquad P^x\text{-a.s.}$$

for every $x \in X_\Delta$ and therefore

$$E^\mu[Y \mid \mathcal{F}_{t+}^0] = Y_1 E^{X_t}(Y_2) = E^\mu[Y \mid \mathcal{F}_t^0] \qquad P^\mu\text{-a.s.}$$

By the usual argument we conclude that for every $\mathcal{F}$-measurable bounded real function Y on Ω

$$E^\mu[Y \mid \mathcal{F}_{t+}] = E^\mu[Y \mid \mathcal{F}_t] \qquad P^\mu\text{-a.s.}$$

Now let $A \in \mathcal{F}_{t+}$. Choosing $Y = 1_A$ we obtain $1_A = E^\mu[Y \mid \mathcal{F}_t]$ P^μ-a.s. and hence $A \in \mathcal{F}_t^{P^\mu}$. Thus $A \in \mathcal{F}_t$. ┘

The following result is known as *Blumenthal's* 0-1-*law*.

4.7. PROPOSITION. For every $A \in \mathcal{F}_0$ and every $x \in X_\Delta$, $P^x(A) = 0$ or $P^x(A) = 1$.

Proof. Consider first $A_0 \in \mathcal{F}_0^0$. Then $A_0 = [X_0 \in B]$ for some $B \in \mathcal{B}(X_\Delta)$, hence $P^x(A_0) = 1$ if $x \in B$ and $P^x(A_0) = 0$ if $x \notin B$. If $A \in \mathcal{F}_0$ then there exists $A_0 \in \mathcal{F}_0^0$ such that $P^x(A \Delta A_0) = 0$ and hence $P^x(A) = P^x(A_0) \in \{0,1\}$. ┘

4.8. EXAMPLE. If X^0 is uniform motion to the left on $\mathbb{R}$ then $\mathcal{F}^0 = \mathcal{F}_t^0 = \mathcal{B}(\mathbb{R}_\Delta)$ and $\mathcal{F} = \mathcal{F}_t = \mathcal{B}(\mathbb{R}_\Delta)^*$ since $P^\mu = \mu$ for every $\mu \in M_+^1(\mathbb{R}_\Delta)$. Furthermore, $\zeta(x) = \infty$ for every $x \in \mathbb{R}$ and $\zeta(\Delta) = 0$.

4.9. EXERCISE. Describe the σ-algebras $\mathcal{F}^0$, $\mathcal{F}_t^0$, $\mathcal{F}$, $\mathcal{F}_t$, and the life time ζ for the Markov process of (3.7).

5. Stopping Times

Let $(\Omega, \mathfrak{M})$ be a measurable space and let $(\mathfrak{M}_t)_{t\in\overline{\mathbb{R}}_+}$ be an increasing family of sub-σ-algebras of $\mathfrak{M}$. A mapping $T : \Omega \to \overline{\mathbb{R}}_+$ is called *stopping time* with respect to $(\mathfrak{M}_t)$ if $[T \leq t] \in \mathfrak{M}_t$ for every $t \in \mathbb{R}_+$.

5.1. LEMMA. Let T be a stopping time with respect to $(\mathfrak{M}_t)$ and let $t \in \overline{\mathbb{R}}_+$. Then the sets $[T < t], [T = t], [T > t]$ are contained in $\mathfrak{M}_t$.

Proof. Suppose first that $t < \infty$. Then $[T \leq t - \frac{1}{n}] \in \mathfrak{M}_{t-\frac{1}{n}} \subset \mathfrak{M}_t$ for every $n \in \mathbb{N}$, hence

$$[T < t] = \bigcup_{n=1}^{\infty} [T \leq t - \frac{1}{n}] \in \mathfrak{M}_t .$$

Furthermore, $[T = t] = [T \leq t] \setminus [T < t] \in \mathfrak{M}_t$ and $[T > t] = \complement[T \leq t] \in \mathfrak{M}_t$. Finally $[T < \infty] = \bigcup_{n=1}^{\infty} [T \leq n] \in \mathfrak{M}_\infty$ and $[T = \infty] = \complement[T < \infty] \in \mathfrak{M}_\infty$. ┘

5.2. LEMMA. Let S and T be stopping times with respect to $(\mathfrak{M}_t)$. Then $S \vee T$, $S \wedge T$ and $S+T$ are stopping times with respect to $(\mathfrak{M}_t)$.*)

Proof. Let $t \in \mathbb{R}_+$. Then

$$[S \vee T \leq t] = [S \leq t] \cap [T \leq t] \in \mathfrak{M}_t$$

and

$$[S \wedge T \leq t] = [S \leq t] \cup [T \leq t] \in \mathfrak{M}_t .$$

Furthermore,

$$[S+T>t] = [0<S<t,\ S+T>t] \cup [S=0,\ T>t] \cup [S>t,\ T=0] \cup [S \geq t,\ T>0]$$

where

$$[0<S<t,\ S+T>t] = \bigcup_{r\in\mathbb{Q}_+, r<t} [r<S<t,\ T>t-r] .$$

Hence $[S+T \leq t] = \complement[S+T>t] \in \mathfrak{M}_t$. ┘

For every $t \in \overline{\mathbb{R}}_+$ let

$$\mathfrak{M}_{t+} := \bigcap_{\varepsilon>0} \mathfrak{M}_{t+\varepsilon} .$$

$(\mathfrak{M}_t)$ is called *right continuous* if $\mathfrak{M}_{t+} = \mathfrak{M}_t$ for all $t \in \overline{\mathbb{R}}_+$.

---

*) $S \vee T = \sup(S,T)$, $S \wedge T = \inf(S,T)$.

5.3. LEMMA. Let $T : \Omega \to \bar{\mathbb{R}}_+$. Then T is a stopping time with respect to $(\mathfrak{m}_{t+})$ if and only if $[T < t] \in \mathfrak{m}_t$ for every $t \in \mathbb{R}_+$.

Proof. Let T be a stopping time with respect to $(\mathfrak{m}_{t+})$ and let $t \in \mathbb{R}_+$. Then $[T<t] = \bigcup_{n=1}^{\infty} [T \le t - \frac{1}{n}] \in \mathfrak{m}_t$. If conversely $[T<t] \in \mathfrak{m}_t$ for every $t \in \mathbb{R}_+$ then $[T \le t] = \bigcap_{n=m}^{\infty} [T < t + \frac{1}{n}] \in \mathfrak{m}_{t+\frac{1}{m}}$ for every $t \in \mathbb{R}_+$ and $m \in \mathbb{N}$. ┘

5.4. PROPOSITION. Let (T_n) be a sequence of stopping times with respect to $(\mathfrak{m}_t)$. Then $\sup T_n$ is a stopping time with respect to $(\mathfrak{m}_t)$ whereas $\inf T_n$, $\limsup T_n$ and $\liminf T_n$ are stopping times with respect to $(\mathfrak{m}_{t+})$.

Proof. Let $t \in \mathbb{R}_+$. Then

$$[\sup T_n \le t] = \bigcap_{n=1}^{\infty} [T_n \le t] \in \mathfrak{m}_t$$

and

$$[\inf T_n < t] = \bigcup_{n=1}^{\infty} [T_n < t] \in \mathfrak{m}_t .$$

Hence $\sup T_n$ and $\inf T_n$ are stopping times with respect to $(\mathfrak{m}_t)$ and $(\mathfrak{m}_{t+})$ respectively. Now it suffices to recall that $\limsup T_n = \inf_m \sup_{n \ge m} T_n$ and $\liminf T_n = \sup_m \inf_{n \ge m} T_n$. ┘

Given a stopping time T with respect to $(\mathfrak{m}_t)$ we define

$$\mathfrak{m}_T := \{A \in \mathfrak{m} : A \cap [T \le t] \in \mathfrak{m}_t \text{ for every } t \in \mathbb{R}_+\} .$$

We note that $\mathfrak{m}_T$ is a σ-algebra: It is obvious that $\Omega \in \mathfrak{m}_T$ and that $\mathfrak{m}_T$ is closed with respect to countable unions. If $A \in \mathfrak{m}_T$ then for every $t \in \mathbb{R}_+$

$$\complement A \cap [T \le t] = \complement(A \cap [T \le t]) \cap [T \le t] \in \mathfrak{m}_t ,$$

hence $\complement A \in \mathfrak{m}_T$.

Furthermore, no confusion may arise if T is a constant stopping time t_o since then

$$A \cap [T \le t] = \begin{cases} A , & t_o \le t \\ \emptyset , & t_o > t \end{cases}$$

and hence $\mathfrak{m}_T = \mathfrak{m}_{t_o}$.

5.5. PROPOSITION. 1. Every stopping time T with respect to $(\mathfrak{m}_t)$ is $\mathfrak{m}_T$-measurable.

2. If S and T are stopping times with respect to $(\mathfrak{m}_t)$ such that $S \le T$

then $\mathfrak{M}_S \subset \mathfrak{M}_T$.

3. Let (T_n) be a sequence of stopping times with respect to $(\mathfrak{M}_t)$ and suppose that $(\mathfrak{M}_t)$ is right continuous. Then $\mathfrak{M}_{\inf T_n} = \bigcap_{n=1}^{\infty} \mathfrak{M}_{T_n}$.

Proof. 1. Let $s \in \mathbb{R}_+$. Then for every $t \in \mathbb{R}_+$

$$[T \leq s] \cap [T \leq t] = [T \leq s \wedge t] \in \mathfrak{M}_{s \wedge t} \subset \mathfrak{M}_t ,$$

hence $[T \leq s] \in \mathfrak{M}_T$.

2. Let $A \in \mathfrak{M}_S$. Then for every $t \in \mathbb{R}_+$

$$A \cap [T \leq t] = (A \cap [S \leq t]) \cap [T \leq t] \in \mathfrak{M}_t ,$$

hence $A \in \mathfrak{M}_T$.

3. Let $T := \inf T_n$. By (5.4), T is a stopping time with respect to $(\mathfrak{M}_t)$. By (2), $\mathfrak{M}_T \subset \bigcap_{n=1}^{\infty} \mathfrak{M}_{T_n}$. Take now $A \in \bigcap_{n=1}^{\infty} \mathfrak{M}_{T_n}$. Then for every $t \in \mathbb{R}_+$

$$A \cap [T < t] = \bigcup_{n=1}^{\infty} (A \cap [T_n < t]) \in \mathfrak{M}_t ,$$

hence

$$A \cap [T \leq t] = \bigcap_{n=1}^{\infty} A \cap [T < t + \tfrac{1}{n}] \in \mathfrak{M}_{t+} = \mathfrak{M}_t$$

and therefore $A \in \mathfrak{M}_T$. ┘

<u>5.6. PROPOSITION.</u> Let S and T be stopping times with respect to $(\mathfrak{M}_t)$ and let $A \in \mathfrak{M}_T$. Then $A \cap [T \leq S] \in \mathfrak{M}_S$.

Proof . Let $t \in \mathbb{R}_+$. By (5.5), $S \wedge t$ and $T \wedge t$ are $\mathfrak{M}_t$-measurable and hence

$$(A \cap [T \leq S]) \cap [S \leq t] = (A \cap [T \leq t]) \cap [S \leq t] \cap [T \wedge t \leq S \wedge t] \in \mathfrak{M}_t .$$ ┘

<u>5.7. COROLLARY.</u> Let S and T be stopping times with respect to $(\mathfrak{M}_t)$. Then the sets $[S < T]$, $[S = T]$ and $[S > T]$ are contained in $\mathfrak{M}_S \cap \mathfrak{M}_T$.

Proof. Applying (5.6) to $A = \Omega$ we obtain $[T \leq S] \in \mathfrak{M}_S$ and $[S < T] = \complement[T \leq S] \in \mathfrak{M}_S$. Let $R := S \wedge T$. By (5.2) and (5.5), R is $\mathfrak{M}_S$-measurable. Hence $[S \leq T] = [R = S] \in \mathfrak{M}_S$ and $[S > T] = \complement[S \leq T] \in \mathfrak{M}_S$, $[S = T] = [S \leq T] \setminus [S < T] \in \mathfrak{M}_S$. Because of the symmetry of the statement, the proof is complete. ┘

<u>5.8. LEMMA.</u> Let T be a stopping time with respect to $(\mathfrak{M}_t)$ and let

$S : \Omega \to \overline{\mathbb{R}}_+$ be $\mathfrak{m}_T$-measurable such that $S \geq T$. Then S is a stopping time with respect to $(\mathfrak{m}_t)$.

Proof. For every $t \in \overline{\mathbb{R}}_+$, $[S \leq t] \in \mathfrak{m}_T$ and $[S \leq t] = [S \leq t] \cap [T \leq t] \in \mathfrak{m}_t$. ┘

5.9. LEMMA. Let T be a stopping time with respect to $(\mathfrak{m}_{t+})$. Then there exists a sequence (T_n) of stopping times with respect to $(\mathfrak{m}_t)$ such that $T_n(\Omega) \subset \mathbb{Q}_+ \cup \{\infty\}$, $T_n \downarrow T$ and $T_n > T$ on $[T < \infty]$.

Proof. For every $n \in \mathbb{N}$, we define $T_n : \Omega \to \mathbb{Q}_+ \cup \{\infty\}$ by

$$T_n := \begin{cases} \dfrac{k+1}{2^n} & \text{on } [\frac{k}{2^n} \leq T < \frac{k+1}{2^n}], \; k \in \mathbb{N}_0 \\ \infty & \text{on } [T = \infty]. \end{cases}$$

If $n \in \mathbb{N}$, $t \in \mathbb{R}_+$ and $k \in \mathbb{N}_0$ such that $\frac{k}{2^n} \leq t < \frac{k+1}{2^n}$, then

$$[T_n \leq t] = [T < \frac{k}{2^n}] \in \mathfrak{m}_{k2^{-n}} \subset \mathfrak{m}_t.$$

Hence (T_n) is a sequence of stopping times with respect to $(\mathfrak{m}_t)$. The rest of the statement is obvious by definition of (T_n). ┘

5.10. LEMMA. Suppose that $(\mathfrak{m}_t)$ is right continuous. Let S and T be stopping times with respect to $(\mathfrak{m}_t)$ and $\Theta : \Omega \to \Omega$ such that $\Theta \in \mathfrak{m}_{T+s}/\mathfrak{m}_s$ for every $s \in \mathbb{R}_+$. Then $T + S \circ \Theta$ is a stopping time with respect to $(\mathfrak{m}_t)$.

Proof. Let $R = T + S \circ \Theta$ and $t \in \mathbb{R}_+$. Then

$$[R < t] = \bigcup_{r \in \mathbb{Q}, 0<r<t} [S \circ \Theta < r] \cap [T + r \leq t].$$

But, for every $0 < r < t$, $[S \circ \Theta < r] = \Theta^{-1}([S < r]) \in \mathfrak{m}_{T+r}$ since $[S < r] \in \mathfrak{m}_r$, and hence $[S \circ \Theta < r] \cap [T + r \leq t] \in \mathfrak{m}_t$. Thus $[R < t] \in \mathfrak{m}_t$. ┘

5.11. EXERCISE. Let T be a stopping time with respect to $(\mathfrak{m}_t)$ such that $T(\Omega)$ is countable. Define $\varphi : \overline{\mathbb{R}}_+ \to \overline{\mathbb{R}}_+$ by

$$\varphi(t) = \inf \{s \geq 0 : [T = t] \in \mathfrak{m}_s\}.$$

Then $\varphi \circ T$ is a stopping time with respect to $(\mathfrak{m}_{t+})$.

6. Strong Markov Processes

Let $\mathfrak{X} = (\Omega, \mathfrak{m}, \mathfrak{m}_t, X_t, \Theta_t, P^x)$ be a Markov process with state space X. If $x \mapsto P^x(A)$ is universally measurable for every $A \in \mathfrak{m}$ we define probability measures P^μ, $\mu \in M_+^1(X_\Delta)$, on $(\Omega, \mathfrak{m})$ by

$$P^\mu(A) = \int P^x(A)\ \mu(dx)$$

and we shall say that $\mathfrak{X}$ is *complete* if

$$\mathfrak{m} = \bigcap_{\mu \in M_+^1(X_\Delta)} \mathfrak{m}^{P^\mu}, \quad \mathfrak{m}_t = \bigcap_{\mu \in M_+^1(X_\Delta)} \mathfrak{m}_t^{P^\mu} \qquad (t \in \overline{\mathbb{R}}_+).$$

For example, the modified process $(\Omega, \mathcal{F}, \mathcal{F}_t, X_t, \Theta_t, P^x)$ is complete.

Given a function $S : \Omega \to \overline{\mathbb{R}}_+$ we define $X_S : \Omega \to X_\Delta$ and $\Theta_S : \Omega \to \Omega$ by

$$X_S(\omega) := X_{S(\omega)}(\omega), \quad \Theta_S(\omega) = \Theta_{S(\omega)}(\omega).$$

If T is another positive numerical function on Ω then for every $\omega \in \Omega$

$$X_S \circ \Theta_T(\omega) = X_S(\Theta_{T(\omega)}(\omega)) = X_{S(\Theta_{T(\omega)}(\omega))}(\Theta_{T(\omega)}(\omega))$$
$$= X_{S \circ \Theta_T(\omega) + T(\omega)}(\omega) = X_{T + S \circ \Theta_T}(\omega),$$

i.e.

$$X_S \circ \Theta_T = X_{T + S \circ \Theta_T}.$$

In particular, $X_s \circ \Theta_T = X_{T+s}$ for every $s \in \overline{\mathbb{R}}_+$.

A stopping time with respect to $(\mathfrak{m}_t)$ is simply called a *stopping time* (of $\mathfrak{X}$). The process $\mathfrak{X}$ is called a *strong Markov process* if $\mathfrak{X}$ is complete and if for every stopping time T of $\mathfrak{X}$ the following *strong Markov property* holds:

(SM) $\Theta_T \in \mathfrak{m}/\mathfrak{m}$, $\Theta_T \in \mathfrak{m}_{T+s}/\mathfrak{m}_s$ for every $s \in \overline{\mathbb{R}}_+$, and for every $\mathfrak{m}$-measurable bounded real function Y on Ω and every $x \in X_\Delta$,

$$E^x[Y \circ \Theta_T | \mathfrak{m}_T] = E^{X_T}(Y).$$

6.1. EXAMPLE. Let $\mathfrak{X}^0$ be uniform motion to the left on $\mathbb{R}$. Then a function $T : \mathbb{R}_\Delta \to \overline{\mathbb{R}}_+$ is a stopping time with respect to $(\mathcal{F}_t)$ if and only if $T \in \mathcal{B}(\mathbb{R}_\Delta)^*/\mathcal{B}(\overline{\mathbb{R}}_+)$. Moreover, $\mathcal{F}_t = \mathcal{B}(\mathbb{R}_\Delta)^*$ and $X_T \in \mathcal{F}/\mathcal{B}(\mathbb{R}_\Delta)$ for every stopping time T, and (SM) reduces to the Markov property (M_3). Thus

$\mathfrak{X} = (\mathbb{R}_\Delta, \mathcal{B}(\mathbb{R}_\Delta)^*, \mathcal{B}(\mathbb{R}_\Delta)^*, X_t, \Theta_t, \varepsilon_x)$ is a strong Markov process called again uniform motion to the left on $\mathbb{R}$.

6.2. REMARKS. 1. The modified process $(\Omega, \mathcal{F}, \mathcal{F}_t, X_t, \Theta_t, P^x)$ satisfies (SM) for every constant time T by (4.4) and (4.5).

2. Let $\mathfrak{X}$ be a strong Markov process and T a stopping time of $\mathfrak{X}$. Then $X_T \in \mathfrak{m}_T/\mathcal{B}(X_\Delta)^*$ since $X_T = X_0 \circ \Theta_T$. Hence $E^{X_T}(Y)$ is $\mathfrak{m}_T$-measurable for every $\mathfrak{m}$-measurable bounded real function Y on Ω, and we obtain that for every $\mu \in M_+^1(X_\Delta)$

$$E^\mu[Y \circ \Theta_T \mid \mathfrak{m}_T] = E^{X_T}(Y).$$

6.3. PROPOSITION. If $\mathfrak{X}$ satisfies (SM) for every constant stopping time then $\mathfrak{X}$ satisfies (SM) for every stopping time T of $\mathfrak{X}$ such that $T(\Omega)$ is countable.

Proof. Fix $A \in \mathfrak{m}$. Then

$$\Theta_T^{-1}(A) = \bigcup_{t \in T(\Omega)} [T = t] \cap \Theta_t^{-1}(A) \in \mathfrak{m}.$$

If $s \in \overline{\mathbb{R}}_+$ and $A \in \mathfrak{m}_s$ then for every $r \in \mathbb{R}_+$

$$[T + s \leq r] \cap \Theta_T^{-1}(A) = \bigcup_{t \in T(\Omega), t+s \leq r} [T + s = t + s] \cap \Theta_t^{-1}(A) \in \mathfrak{m}_r,$$

hence $\Theta_T^{-1}(A) \in \mathfrak{m}_{T+s}$. Moreover, for every $x \in X_\Delta$ and every $B \in \mathfrak{m}_T$

$$P^x(\Theta_T^{-1}(A) \cap B) = \sum_{t \in T(\Omega)} P^x(\Theta_t^{-1}(A) \cap [T = t] \cap B)$$

$$= \sum_{t \in T(\Omega)} \int_{[T=t] \cap B} P^{X_t}(A)\, dP^x = \int_B P^{X_T}(A)\, dP^x,$$

i.e. $P^x[\Theta_T^{-1}(A) \mid \mathfrak{m}_T] = P^{X_T}(A)$. ∎

6.4. PROPOSITION. Suppose that $\mathfrak{X}$ satisfies (SM) for every constant stopping time and that the paths $t \mapsto X_t(\omega)$ are right continuous for every $\omega \in \Omega$. Then $X_{T \wedge t} \in \mathfrak{m}_t/\mathcal{B}(X_\Delta)$ for every stopping time T with respect to $(\mathfrak{m}_{t+})$ and every $t > 0$. In particular, $X_T \in \mathfrak{m}_T/\mathcal{B}(X_\Delta)$ for every stopping time T of $\mathfrak{X}$.

Proof. Let (T_n) be as in (5.9). By (6.3),

$X_{T_n\wedge t} = X_0 \circ \Theta_{T_n\wedge t} \in \mathfrak{m}_{T_n\wedge t}/\mathcal{B}(X_\Delta)$, hence $X_{T_n\wedge t} \in \mathfrak{m}_t/\mathcal{B}(X_\Delta)$ by (5.2.2). Since $T_n \wedge t \downarrow T \wedge t$ we have $\lim_{n\to\infty} X_{T_n\wedge t} = X_{T\wedge t}$.

Let U be an open subset of X_Δ and let (K_m) be a sequence of compact subsets of U such that $K_m \subset \overset{\circ}{K}_{m+1}$ and $\bigcup_{m=1}^{\infty} K_m = U$. Then

$$[X_{T\wedge t} \in U] = \bigcup_{m=1}^{\infty} \bigcap_{n=m}^{\infty} [X_{T_n\wedge t} \in K_m]$$

and hence $[X_{T\wedge t} \in U] \in \mathfrak{m}_t$. Thus $X_{T\wedge t} \in \mathfrak{m}_t / \mathcal{B}(X_\Delta)$.

Suppose now that T is a stopping time of $\mathfrak{X}$. Then

$$[X_T \in U] \cap [T \leq t] = [X_{T\wedge t} \in U] \cap [T \leq t] \in \mathfrak{m}_t$$

and hence $X_T \in \mathfrak{m}_T / \mathcal{B}(X_\Delta)$. $\square$

<u>6.5. PROPOSITION.</u> For every Markov process $\mathfrak{X} = (\Omega, \mathfrak{m}, \mathfrak{m}_t, X_t, \Theta_t, P^x)$ the following properties are equivalent:

(1) $\tilde{\mathfrak{X}} = (\Omega, \mathcal{F}, \mathcal{F}_t, X_t, \Theta_t, P^x|_{\mathcal{F}})$ is a strong Markov process.

(2) For every stopping time T of $\tilde{\mathfrak{X}}$, $X_T \in \mathcal{F}_T/\mathcal{B}(X_\Delta)$, and for every $s \in \mathbb{R}_+$, $B \in \mathcal{B}(X)$ and $x \in X$,

$$\text{(SM')} \quad P^x[X_{T+s} \in B] = E^x(P^{X_T}[X_s \in B]) .$$

Proof. (1) $\Rightarrow$ (2) : Let T be a stopping time of $\tilde{\mathfrak{X}}$, $s \in \mathbb{R}_+$, $B \in \mathcal{B}(X)$ and $x \in X$. Then $X_T \in \mathcal{F}_T/\mathcal{B}(X_\Delta)$ by (6.2.2). Choosing $Y = 1_{[X_s \in B]}$ we have $Y \circ \Theta_T = 1_{[X_{T+s} \in B]}$ and hence

$$P^x[X_{T+s} \in B] = E^x(Y \circ \Theta_T) = E^x(E^x[Y \circ \Theta_T \mid \mathfrak{m}_T] = E^x(E^{X_T}(Y)) = E^x(P^{X_T}[X_s \in B]) .$$

(2) $\Rightarrow$ (1) : The proof will be given in several steps.

a) For every stopping time T of $\tilde{\mathfrak{X}}$, every $s \in \overline{\mathbb{R}}_+$, $B \in \mathcal{B}(X_\Delta)$ and $x \in X_\Delta$,

$$P^x[X_{T+s} \in B \mid \mathcal{F}_T] = P^{X_T}[X_s \in B] .$$

Indeed, let $A \in \mathcal{F}_T$ and define $T' : \Omega \to \overline{\mathbb{R}}_+$ by

$$T' := \begin{cases} T & \text{on } A , \\ \infty & \text{on } \complement A . \end{cases}$$

Then T' is a stopping time of $\tilde{\mathfrak{X}}$ since, for every $t \in \mathbb{R}_+$,

$$[T' \leq t] = A \cap [T \leq t] \in \mathcal{F}_t .$$

Hence for every $s \in \mathbb{R}_+$, $B \in \mathcal{B}(X)$ and $x \in X$,

$$P^x([X_{T+s} \in B] \cap A) = P^x[X_{T'+s} \in B] = E^x(P^{X_{T'}}[X_s \in B]) = \int_A P^{X_T}[X_s \in B]\, dP^x$$

since $X_\infty = \Delta$ and $P^\Delta[X_s \in B] = 0$. This proves the assertion since the cases $s = \infty$, $B = \{\Delta\}$ or $x = \Delta$ are trivial.

b) $\Theta_T \in \mathcal{F}/\mathcal{F}^o$ and $\Theta_{T+s} \in \mathcal{F}_{T+s}/\mathcal{F}^o_s$ for every $s \in \mathbb{R}_+$ since

$$\Theta_T^{-1}([X_{s_1} \in B_1,\ldots,X_{s_n} \in B_n]) = [X_{T+s_1} \in B_1,\ldots,X_{T+s_n} \in B_n].$$

c) For every $\mathcal{F}^o$-measurable bounded real function Y on Ω, every stopping time T of $\tilde{\mathbf{X}}$ and every $x \in X_\Delta$,

$$E^x[Y \circ \Theta_T | \mathcal{F}_T] = E^{X_T}(Y).$$

It suffices to consider the case $Y = \prod_{j=1}^n f_j \circ X_{s_j}$ where $0 \le s_1 < \ldots < s_n \le \infty$ and $f_j \in \mathcal{B}_b(X_\Delta)$, $1 \le j \le n$. We argue by induction on n. For $n = 1$ the required equality follows from (a). Suppose that the statement holds for some n, fix a stopping time T of $\tilde{\mathbf{X}}$, and let $S = T + s_n$. By (5.5), $\mathcal{F}_T \subset \mathcal{F}_S$ and $f_j \circ X_{s_j} \circ \Theta_T = f_j \circ \Theta_{T+s_j}$ is $\mathcal{F}_S$-measurable for every $1 \le j \le n$. Fix $A \in \mathcal{F}_T$. Then

$$E^x(1_A \prod_{j=1}^{n+1} f_j \circ X_{s_j} \circ \Theta_T) = E^x(1_A \prod_{j=1}^{n} f_j \circ X_{T+s_j} \cdot f_{n+1} \circ X_{s_{n+1}-s_n} \circ \Theta_S)$$

$$= E^x(1_A \prod_{j=1}^{n} f_j \circ X_{T+s_j} \cdot E^{X_S}(f_{n+1} \circ X_{s_{n+1}-s_n})).$$

The last term has the form $E^x(1_A \prod_{j=1}^{n} g_j \circ X_{s_j} \circ \Theta_T)$ with $g_j = f_j$ for every $1 \le j \le n-1$ and $g_n(y) = f_n(y) E^y(f_{n+1} \circ X_{s_{n+1}-s_n})$. Therefore by induction hypothesis

$$E^x(1_A \prod_{j=1}^{n+1} f_j \circ X_{s_j} \circ \Theta_T) = E^x(1_A E^{X_T}\{\prod_{j=1}^{n} f_j \circ X_{s_j} E^{X_{s_n}}(f_{n+1} \circ X_{s_{n+1}-s_n})\})$$

$$= E^x(1_A E^{X_T}\{\prod_{j=1}^{n+1} f_j \circ X_{s_j}\})$$

since

$$E^{X_{s_n}}(f_{n+1} \circ X_{s_{n+1}-s_n}) = E^y[f_{n+1} \circ X_{s_{n+1}} | \mathcal{F}_{s_n}]$$

for every $y \in X_\Delta$. Since $y \mapsto E^y(\prod_{j=1}^{n+1} f_j \circ X_{s_j})$ is Borel measurable by (4.1) we obtain the required equality

$$E^x[\prod_{j=1}^{n+1} f_j \circ X_{s_j} \circ \Theta_T \mid \mathcal{F}_T] = E^{X_T}(\prod_{j=1}^{n+1} f_j \circ X_{s_j}) .$$

d) Fix a stopping time T of $\tilde{\mathbf{X}}$. It is easily checked that

$$\mathcal{F}_T = \bigcap_{\mu \in M_+^1(X_\Delta)} \mathcal{F}_T^{P^\mu} .$$

Therefore $X_T \in \mathcal{F}_T/\mathcal{B}(X_\Delta)$ implies by (4.2) that $X_T \in \mathcal{F}_T/\mathcal{B}(X_\Delta)^*$. Let $\mu \in M_+^1(X_\Delta)$ and $A \in \mathcal{F}^0$. Then by (c)

$$P^\mu[\Theta_T^{-1}(A) \mid \mathcal{F}_T] = P^{X_T}(A)$$

and hence

$$\Theta_T(P^\mu)(A) = P^\mu(\Theta_T^{-1}(A)) = E^\mu(P^{X_T}(A)) = \int P^y(A) \;\; X_T(P^\mu)(dy) = P^{X_T(P^\mu)}(A) .$$

Thus by (4.2), $\Theta_T \in \mathcal{F}/\mathcal{F}$ and $\Theta_T \in \mathcal{F}_{T+s}/\mathcal{F}_s$ for every $s \in \mathbb{R}_+$. Finally, let Y be an $\mathcal{F}$-measurable bounded real function on Ω. Then $Y \circ \Theta_T$ is $\mathcal{F}$-measurable and $E^{X_T}(Y)$ is $\mathcal{F}_T$-measurable. We now proceed as in the proof of (4.6). ⌟

<u>6.6. COROLLARY.</u> If $\mathbf{X} = (\Omega, \mathfrak{m}, \mathfrak{m}_t, X_t, \Theta_t, P^x)$ is a strong Markov process and the paths $t \mapsto X_t$ are right continuous then $(\Omega, \mathcal{F}, \mathcal{F}_t, X_t, \Theta_t, P^x|_{\mathcal{F}})$ is a strong Markov process.

If in addition $\mathfrak{m}_{t+} = \mathfrak{m}_t$ for every $t \in \mathbb{R}_+$, then $\mathcal{F}_{t+} = \mathcal{F}_t$ for every $t \in \mathbb{R}_+$.

Proof. The first part follows from (6.2.1), (6.4) and (6.5), and the second part from (4.6) using $\mathcal{F}_{t+}^0 \subset \mathfrak{m}_{t+}$. ⌟

<u>6.7. LEMMA.</u> Let T be a stopping time with respect to $(\mathcal{F}_{t+})$ and $\mu \in M_+^1(X_\Delta)$. Then there exists a stopping time T^μ with respect to $(\mathcal{F}_{t+}^0)$ such that $P^\mu[T^\mu \neq T] = 0$.

Proof. By (5.9) there exists a decreasing sequence (T_n) of stopping times with respect to $(\mathcal{F}_t)$ such that $T_n(\Omega) \subset \{\frac{k}{2^n} : k \in \mathbb{N}\} \cup \{\infty\}$ and $\inf T_n = T$. Fix $n \in \mathbb{N}$. Then for each $k \in \mathbb{N}$ there exists a set $A_{n,k} \in \mathcal{F}_{k2^{-n}}^0$ such that

$$P^\mu([T_n = \frac{k}{2^n}] \,\Delta\, A_{n,k}) = 0 .$$

Let $B_{n,1} = A_{n,1}$, $B_{n,k} = A_{n,k} \setminus \bigcup_{j=1}^{k-1} A_{n,j}$ for $k \geq 2$ and define $T_n^\mu : \Omega \to \overline{\mathbb{R}}_+$ by

$$T_n^\mu = \begin{cases} \frac{k}{2^n} & \text{on } B_{n,k}, \\ \infty & \text{on } \Omega \setminus \bigcup_{k=1}^{\infty} B_{n,k}. \end{cases}$$

Then T_n^μ is a stopping time with respect to $(\mathcal{F}_t^0)$ and $P^\mu[T_n^\mu \neq T_n] = 0$. Defining $T^\mu := \inf T_n^\mu$ we thus obtain a stopping time T^μ with respect to $(\mathcal{F}_{t+}^0)$ such that $P^\mu[T^\mu \neq T] = 0$. ┘

6.8. PROPOSITION. Let $\mathfrak{X} = (\Omega, \mathfrak{M}, \mathfrak{M}_t, X_t, \Theta_t, P^x)$ be a Markov process. Suppose that the paths $t \mapsto X_t(\omega)$ are right continuous for every $\omega \in \Omega$ and that for every stopping time T with respect to $(\mathcal{F}_{t+}^0)$, every $s \in \mathbb{R}_+$, $B \in \mathcal{B}(X)$, and $x \in X$,

$$P^x[X_{T+s} \in B] = E^x(P^{X_T}[X_s \in B]) .$$

Then $\tilde{\mathfrak{X}} = (\Omega, \mathcal{F}, \mathcal{F}_t, X_t, \Theta_t, P^x)$ is a strong Markov process and $\mathcal{F}_{t+} = \mathcal{F}_t$ for every $t \in \mathbb{R}_+$.

Proof. Let T be a stopping time of the Markov process $\tilde{\mathfrak{X}}$, $s \in \mathbb{R}_+$, $B \in \mathcal{B}(X)$, and $x \in X$. By (6.4), $X_T \in \mathcal{F}_T/\mathcal{B}(X_\Delta)$ and $X_{T+s} \in \mathcal{F}_{T+s}/\mathcal{B}(X_\Delta)$. By (6.7), there exists a stopping time T^x with respect to $(\mathcal{F}_{t+}^0)$ such that $P^x[T^x \neq T] = 0$. Then by assumption

$$P^x[X_{T+s} \in B] = P^x[X_{T^x+s} \in B] = E^x(P^{X_{T^x}}[X_s \in B]) = E^x(P^{X_T}[X_s \in B]) .$$

Therefore $\tilde{\mathfrak{X}}$ is a strong Markov process by (6.5).

Finally, let $t \in \mathbb{R}_+$, $A \in \mathcal{F}_{t+}^0$, and define $T : \Omega \to \overline{\mathbb{R}}_+$ by

$$T = \begin{cases} t & \text{on } A, \\ \infty & \text{on } \complement A. \end{cases}$$

Then T is a stopping time with respect to $(\mathcal{F}_{t+}^0)$, hence

$$P^x([X_{t+s} \in B] \cap A) = P^x[X_{T+s} \in B] = E^x(P^{X_T}[X_s \in B]) = \int_A P^{X_t}[X_s \in B]\, dP^x .$$

This shows that $(\Omega, \mathcal{F}^0, \mathcal{F}_{t+}^0, X_t, \Theta_t, P^x)$ is a Markov process, i.e. $\mathcal{F}_{t+} = \mathcal{F}_t$ by (4.6). ┘

6.9. REMARK. The proof of (6.8) shows that the assumption that $\mathfrak{X}$ is a Markov process may be weakened to the assumption that $\mathfrak{X}$ satisfies (M_0), (M_1)

and (M_2).

6.10. EXERCISES. 1. There exists a strong Markov process on $\mathbb{R}$ having the sub-Markov semigroup $\mathbb{P}$ of (3.7.1) as transition function.

2. There is no strong Markov process on $]-\infty,0]$ having the sub-Markov semigroup $\mathbb{P}$ of (3.7.2) as transition function. (Hint: Consider $T(\omega) = \inf\{t \geq 0 : X_t(\omega) < 0\}$ and show that $P^0(\theta_T^{-1}([T>0])) \neq E^0(P^{X_T}[T>0])$.)

7. Hunt Processes

A Markov process $\mathfrak{X} = (\Omega, \mathfrak{M}, \mathfrak{M}_t, X_t, \Theta_t, P^x)$ will be called a *Hunt process* if the following conditions are satisfied:

(HP_0) The path functions $t \to X_t(\omega)$ are right continuous on $[0,\infty[$ and have left-hand limits on $[0,\zeta[$.

(HP_1) $(\mathfrak{M}_t)$ is right continuous, i.e. $\mathfrak{M}_{t+} = \mathfrak{M}_t$ for every $t \in \mathbb{R}_+$.

(HP_2) $\mathfrak{X}$ is a strong Markov process.

(HP_3) $\mathfrak{X}$ is *quasi-left continuous on* $[0,\infty[$, i.e. whenever (T_n) is an increasing sequence of stopping times with limit T, then $\lim_{n\to\infty} X_{T_n} = X_T$ a.s. on $[T<\infty]$.

If $\mathfrak{X}$ satisfies (HP_0) - (HP_2) and is quasi-left continuous on $[0,\zeta[$ then $\mathfrak{X}$ is called a *standard process*.

7.1. EXAMPLE. The uniform motion $\mathfrak{X}$ to the left on $\mathbb{R}$ is a Hunt process. This follows immediately using (6.1).

7.2. REMARKS. Let $\mathfrak{X} = (\Omega, \mathfrak{M}, \mathfrak{M}_t, X_t, \Theta_t, P^x)$ be a strong Markov process with state space X.

1. If the transition function $(P_t)_{t>0}$ of $\mathfrak{X}$ is Markov then $\zeta=\infty$ a.s. on $[X_0 \in X]$ since for all $x \in X$

$$P^x[\zeta=\infty] = \inf_m P^x[\zeta \geq m] \geq \inf_m P^x[X_m \in X] = \inf_m P_m(x,X) = 1\ ,$$

and hence the process $\mathfrak{X}$ is already a Hunt process if it is a standard process.

2. If (T_n) is an increasing sequence of stopping times and $T = \lim_{n\to\infty} T_n$ then $[\lim_{n\to\infty} X_{T_n} = X_T] \in \mathfrak{m}_T$.

3. If $\mathfrak{X}$ is a Hunt process (standard process resp.) then $\tilde{\mathfrak{X}} = (\Omega, \mathcal{F}, \mathcal{F}_t, X_t, \Theta_t, P^x|_{\mathcal{F}})$ is a Hunt process (standard process resp.). Indeed, by (6.6) the process $\tilde{\mathfrak{X}}$ satisfies (HP_0) - (HP_2) and the quasi-left continuity follows from the preceding remark.

4. Let $\Omega' \in \mathfrak{m}$ such that $P^x(\Omega') = 1$ for every $x \in X_\Delta$ and $\Theta_t(\Omega') \subset \Omega'$ for every $t>0$. Fix $\omega_0 \in \Omega'$ such that $X_0(\omega_0) = \Delta$ and let for every $t > 0$

$$X_t'(\omega) = \begin{cases} X_t(\omega), & \omega \in \Omega' \\ \Delta, & \omega \in \complement\Omega', \end{cases} \qquad \Theta_t'(\omega) = \begin{cases} \Theta_t(\omega), & \omega \in \Omega', \\ \omega_0, & \omega \in \complement\Omega'. \end{cases}$$

Then $\mathfrak{X}' = (\Omega, \mathfrak{m}, \mathfrak{m}_t, X_t', \Theta_t, P^x)$ is a strong Markov process having the same transition function. $\mathfrak{X}'$ is a Hunt process (standard process resp.) if $\mathfrak{X}$ is a Hunt process (standard process resp.).

In the following let $\mathfrak{X} = (\Omega, \mathfrak{m}, \mathfrak{m}_t, X_t, \Theta_t, P^x)$ be a Markov process with state space X and let $\mathbb{P} = (P_t)_{t>0}$ denote the transition function of $\mathfrak{X}$. We define the set $E_{\mathfrak{X}}$ of excessive functions and the set $S_{\mathfrak{X}}$ of supermedian functions of $\mathfrak{X}$ by

$$E_{\mathfrak{X}} = E_{\mathbb{P}}, \quad S_{\mathfrak{X}} = S_{\mathbb{P}}.$$

We shall see that the existence of many continuous functions in $E_{\mathfrak{X}}$ implies the existence of a Hunt process $\tilde{\mathfrak{X}}$ which is equivalent to $\mathfrak{X}$.

Let us make the following convention which will be useful in the sequel: Every numerical function f on X will be extended to X_Δ by setting $f(\Delta) = 0$.

<u>7.3. LEMMA.</u> For every real $v \in S_{\mathfrak{X}}$ and $x \in X_\Delta$, the family $(v(X_t))_{t\geq 0}$ is an $(\mathfrak{m}_t)$-supermartingale on $(\Omega, \mathfrak{m}, P^x)$.

Proof. For every $s,t \geq 0$ such that $s > t$

$$E^x[v(X_s) \mid \mathfrak{m}_t] = E^{X_t}(v(X_{s-t})) = P_{s-t}v(X_t) \leq v(X_t).$$ ┘

<u>7.4. LEMMA.</u> Assume that there exists a min-stable function cone $P \subset C^+(X)$ such that $E_{\mathfrak{X}} = S(P)$. Then there exists a set $\Omega_1 \in \mathfrak{m}$ such that $P^x(\Omega_1) = 1$ for every $x \in X_\Delta$ and such that for every $\omega \in \Omega_1$ the restriction of $s \to X_s(\omega)$

to $\mathbb{Q}_+$ has right- and left-hand limits at all $t \in \mathbb{R}_+$ where the limits are in X for every $t < \zeta(\omega)$. If for every $t \in \mathbb{R}_+$

$$X'_t(\omega) := \begin{cases} \lim\limits_{s\downarrow t, s\in\mathbb{Q}} X_s(\omega) & , \ \omega \in \Omega_1 \\ \Delta & , \ \omega \in \Omega \setminus \Omega_1 \end{cases}$$

then $X'_t = X_t$ a.s.

Proof. Let (p_n) be a sequence in P separating the points of X and such that p_1, p_2 are strictly positive and $\frac{p_1}{p_2}$ tends to zero at Δ. Let Ω_1 be the set of all $\omega \in \Omega$ having the following three properties:

i) For every $n \in \mathbb{N}$, the restriction of $s \mapsto p_n(X_s(\omega))$ to $\mathbb{Q}_+$ has left- and right-hand limits at every $t \in [0,\infty[$.

ii) $\inf \{p_1(X_s(\omega)) : s \in \mathbb{Q}_+, s \leq t\} > 0$ for every $t \in \mathbb{Q}_+$ such that $p_1(X_t(\omega)) > 0$.

iii) $\sup \{p_2(X_s(\omega)) : s \in \mathbb{Q}_+, s \leq t\} < \infty$ for every $t \in \mathbb{Q}_+$.

Then $\Omega_1 \in \mathfrak{m}$ and $P^x(\Omega_1) = 1$ for every $x \in X_\Delta$ by (1.3) since by (7.3), for every $n \in \mathbb{N}$, the family $(p_n(X_t))_{t\geq 0}$ is an $(\mathfrak{m}_t)$-supermartingale on $(\Omega, \mathfrak{m}, P^x)$. We recall that the lifetime $\zeta(\omega)$ of $\omega \in \Omega$ is characterized by $X_t(\omega) \in X$ for every $0 \leq t < \zeta(\omega)$ and $X_t(\omega) = \Delta$ for every $t > \zeta(\omega)$. Let $\omega \in \Omega_1$, $0 \leq t < \zeta(\omega)$. Let $t_0 \in \mathbb{Q}_+$ such that $t < t_0 < \zeta(\omega)$ and $A := \{X_s(\omega) : s \in \mathbb{Q}_+, s \leq t_0\}$. Then $\alpha := \inf p_1(A) > 0$ by (ii) and $\beta = \sup p_2(A) < \infty$ by (iii), hence A is contained in the compact subset $K := \{\frac{p_1}{p_2} \geq \frac{\alpha}{\beta}\}$ of X. Let (s_m) be a sequence in $\mathbb{Q}_+$ such that $s_m > t$ and $\lim s_m = t$ and let z and z' be two limit points of the sequence $(X_{s_m}(\omega))$. Then $z, z' \in K$ and for every $n \in \mathbb{N}$ by the continuity of p_n,

$$p_n(z) = \lim_{s\downarrow t, s\in\mathbb{Q}_+} p_n(X_s(\omega)) = p_n(z') \quad ,$$

hence $z = z'$. Thus $\lim\limits_{s\downarrow t, s\in\mathbb{Q}_+} X_s(\omega)$ exists and is a point of K. Similarly, for every $0 < t < \zeta(\omega)$, $\lim\limits_{s\uparrow t, s\in\mathbb{Q}_+} X_s(\omega)$ exists and is a point of X .

It remains to consider the left-hand limit at $\zeta(\omega)$. So let $t = \zeta(\omega) < \infty$. If we do not have $\lim\limits_{s\uparrow t, s\in\mathbb{Q}_+} X_s(\omega) = \Delta$, there exists a sequence (s_m) in $\mathbb{Q}_+$ such that $s_m < t$, $\lim s_m = t$ and $(X_{s_m}(\omega))$ is converging to a point $z \in X$. Let

(s'_m) be another sequence in $\mathbb{Q}_+$ such that $s'_m < t$, $\lim s'_m = t$ and such that the sequence $(z'_m) = (X_{s'_m}(\omega))$ is converging to a point $z' \in X_\Delta$. Then for every $n \in \mathbb{N}$

$$\lim_{m\to\infty} p_n(z'_m) = \lim_{s\uparrow t, s\in\mathbb{Q}_+} p_n(X_s(\omega)) = p_n(z) .$$

In particular, $\lim_{m\to\infty} \frac{p_1}{p_2}(z'_m) = \frac{p_1}{p_2}(z) \neq 0$ and hence $z' \in X$ since $\lim_{y\to\Delta} \frac{p_1}{p_2}(y) = 0$. Then furthermore $p_n(z') = \lim_{m\to\infty} p_n(z'_m) = p_n(z)$ for every $n \in \mathbb{N}$, hence $z' = z$. Thus $\lim_{s\uparrow t, s\in\mathbb{Q}_+} X_s(\omega) = z$.

Let $t \in \mathbb{R}_+$, $x \in X$ and let (s_n) be a sequence of rationals strictly decreasing to t. Let $f \in K(X)$ and $g \in C(X_\Delta)$. Using the Markov property for $\mathfrak{X}$ we have

$$E^x\{f(X_{s_n})g(X_t)\} = E^x\{P_{s_n - t} f(X_t)g(X_t)\}$$

for each n. Letting n tend to infinity we conclude by (II.3.16) that

$$E^x\{f(X'_t)g(X_t)\} = E^x\{f(X_t)g(X_t)\} .$$

It now follows from monotone class theorems that $X'_t = X_t$ P^x-a.s. ┘

The mappings $t \mapsto X'_t(\omega)$ constructed in (7.4) are right-continuous and have left-hand limits on $[0,\infty[$. But we cannot expect that $X'_{t+s} = X'_t \circ \Theta_s$. Hence we proceed in a similar way as we did in the proof of (3.4).

Let $\widetilde{W}$ be the set of all maps $w \in W$ which are right continuous and have left-hand limits on $[0,\infty[$ such that $\lim_{s\uparrow t} w(s) \in X$ if $w(t) \in X$. Let $\widetilde{\pi}_t : \widetilde{W} \to X_\Delta$ denote the coordinate map $w \mapsto w(t)$ and define in $\widetilde{W}$ the σ-algebras $\widetilde{M} = \sigma(\widetilde{\pi}_t : t \in \overline{\mathbb{R}}_+)$, $\widetilde{M}_t = \sigma(\widetilde{\pi}_s : s \in \overline{\mathbb{R}}_+, s \le t)$, $t \in \overline{\mathbb{R}}_+$. Finally, let $\widetilde{\varphi}_t : \widetilde{W} \to \widetilde{W}$ be the shift operator defined by $(\widetilde{\varphi}_t(w))(s) := w(t+s)$.

<u>7.5. PROPOSITION.</u> Assume that there exists a min-stable function cone $P \subset C^+(X)$ such that $E_{\mathfrak{X}} = S(P)$. Then there exists a unique family $(\widetilde{P}^x)_{x\in X_\Delta}$ of probability measures $\widetilde{P}^x$ on $(\widetilde{W}, \widetilde{M})$ such that $\widetilde{\mathfrak{X}} = (\widetilde{W}, \widetilde{M}, \widetilde{M}_t, \widetilde{\pi}_t, \widetilde{\varphi}_t, \widetilde{P}^x)$ is a Markov process which is equivalent to $\mathfrak{X}$.

Proof. By (4.3), we may assume that $\mathfrak{M} = \mathcal{F}$ and $\mathfrak{M}_t = \mathcal{F}_t$ for every $t \in \overline{\mathbb{R}}_+$.

Defining X'_t as in the preceding lemma we then have $X'_t \in \mathfrak{m}_t / \mathcal{B}(X_\Delta)$. We now consider the mapping $\phi : \Omega \to \widetilde{W}$ defined by $\phi(\omega)(t) = X'_t(\omega)$. Then $\widetilde{\pi}_t \circ \phi = X'_t$ for every $t \in \overline{\mathbb{R}}_+$. Proceeding as in the proof of (3.4) it is easily verified that the desired probability measures $\widetilde{P}^x$ are obtained by $\widetilde{P}^x := \phi(P^x)$. ┘

7.6. PROPOSITION. Assume that there exists a min-stable function cone $P \subset C^+(X)$ such that $E_{\mathfrak{X}} = S(P)$. Suppose furthermore that the path functions $t \mapsto X_t(\omega)$ are contained in $\widetilde{W}$. Then $\widetilde{\mathfrak{X}} = (\Omega, \mathcal{F}, \mathcal{F}_t, X_t, \Theta_t, P^x)$ is a Hunt process.

Proof. By (II.4.7), $(X, E_{\mathfrak{X}})$ is a balayage space. In particular, we may assume that $\inf(p, a) \in P$ for every $p \in P$ and $a \in \mathbb{R}_+$.

By (4.3), $\widetilde{\mathfrak{X}}$ is a Markov process. Obviously, $\widetilde{\mathfrak{X}}$ satisfies (HP_0). Let T be a stopping time with respect to $(\mathcal{F}_{t+})$. By (5.9), there exists a sequence (T_n) of stopping times with respect to $(\mathcal{F}_t)$ such that $T_n(\Omega) \subset \mathbb{Q}_+ \cup \{\infty\}$, $T_n \downarrow T$ and $T_n > T$ on $[T < \infty]$. By (7.3), for every $v \in E_{\mathfrak{X}}$, the family $(v(X_t))_{t \geq 0}$ is an $(\mathcal{F}_t)$-supermartingale and hence the sequence $(E^x\{v(X_{T_n})\})$ is increasing for every $x \in X_\Delta$.

Let $x \in X_\Delta$, $s \in \mathbb{R}_+$ and $q \in P_b$. By (6.3), we obtain for every $n \in \mathbb{N}$

$$E^x\{q(X_{T_n+s})\} = E^x\{E^{X_{T_n}}(q(X_s))\} = E^x\{P_s q(X_{T_n})\} .$$

Since $P_s q \in E_{\mathfrak{X}}$ there exists an increasing sequence (q_m) in P such that $\sup q_m = P_s q$. By the right continuity of $t \mapsto X_t(\omega)$ and the continuity of q and q_m we have

$$\lim_{n\to\infty} q(X_{T_n+s}) = q(X_{T+s}) \quad \text{and} \quad \lim q_m(X_{T_n}) = q_m(X_T)$$

for every $m \in \mathbb{N}$. Therefore

$$\begin{aligned} E^x\{q(X_{T+s})\} &= \lim_{n\to\infty} E^x\{q(X_{T_n+s})\} = \lim_{n\to\infty} E^x\{P_s q(X_{T_n})\} \\ &= \sup_n E^x\{P_s q(X_{T_n})\} = \sup_n \sup_m E^x\{q_m(X_{T_n})\} \\ &= \sup_m \sup_n E^x\{q_m(X_{T_n})\} = \sup_m \lim_{n\to\infty} E^x\{q_m(X_{T_n})\} \\ &= \sup_m E^x\{q_m(X_T)\} = E^x\{P_s q(X_T)\} = E^x\{E^{X_T}(q(X_s))\} . \end{aligned}$$

Since any $f \in K(X)$ is the uniform limit of a sequence in $P_b - P_b$ we obtain that for every $B \in B(X)$

$$P^x[X_{T+s} \in B] = E^x(P^{X_T}[X_s \in B]) .$$

This yields (HP_1) and (HP_2) by (6.8).

It remains to show that $\tilde{\mathfrak{X}}$ is quasi-left continuous on $[0,\infty[$. To this end let (T_n) be an increasing sequence of stopping times of $\tilde{\mathfrak{X}}$ with limit T. In trying to prove that $X_{T_n} \to X_T$ P^x-a.s. on $[T<\infty]$ there is no loss of generality in assuming that T is bounded, as we do now. Let $L = \lim_{n\to\infty} X_{T_n}$. Similarly for $s > 0$ let $L_s = \lim_{n\to\infty} X_{T_n+s}$. Since $T_n + s \in [T, T+s]$ for all large n and $t \mapsto X_t(\omega)$ is right continuous, it follows that $\lim_{s\to 0} L_s = X_T$.

Let $p,q \in P_b$ and suppose that $C(q)$ is compact. Given $\varepsilon > 0$ there exists an $s > 0$ such that $P_s q + \varepsilon > q$ on $C(q)$ and hence $P_s q + \varepsilon \geq q$ on X. Therefore $P_s q$ converges uniformly to q as s tends to zero. Applying the strong Markov property we thus obtain for every $x \in X_\Delta$

$$E^x\{p(L)q(X_T)\} = \lim_{s\to 0} E^x\{p(L)q(L_s)\} = \lim_{s\to 0}\lim_{n\to\infty} E^x\{p(X_{T_n})q(X_{T_n+s})\}$$

$$= \lim_{s\to 0}\lim_{n\to\infty} E^x\{p(X_{T_n})P_s q(X_{T_n})\} = \lim_{n\to\infty} E^x\{p(X_{T_n})q(X_{T_n})\} = E^x\{p(L)q(L)\} .$$

By (II.6.19), we obtain the equality $E^x\{p(L)q(X_T)\} = E^x\{p(L)q(L)\}$ for all functions $p, q \in P_b$ and hence

$$E^x\{f(L)g(X_T)\} = E^x\{f(L)g(L)\}$$

for all $f, g \in K(X)$. Since the path functions $t \to X_t(\omega)$ are contained in $\tilde{W}$ we have $[L = \Delta] \subset [X_T = \Delta]$ and hence in addition

$$E^x\{1_{\{\Delta\}}(L)g(X_T)\} = 0 = E^x\{1_{\{\Delta\}}(L)g(L)\}$$

for every $g \in K(X)$. Therefore an application of monotone class theorems yields $X_T = L$ P^x-a.s. ┘

The following proposition is useful for the construction of examples (see VI.3.19)).

<u>7.7. PROPOSITION.</u> Let $\mathfrak{X} = (\Omega, \mathfrak{M}, \mathfrak{M}_t, X_t, \Theta_t, P^x)$ be a standard process with

state space X . Let $\varphi : X \to \tilde{X}$ be a continuous surjection of X onto a locally compact space $\tilde{X}$ with a countable base and ψ a Borel measurable mapping of $\tilde{X}$ into X such that $\varphi \circ \psi$ is the identity on $\tilde{X}$ and $P^x_{\varphi \circ X_t} = P^y_{\varphi \circ X_t}$ for all $t > 0$ and $x, y \in X$ satisfying $\varphi(x) = \varphi(y)$. Extend φ and ψ by $\varphi(\Delta) = \tilde{\Delta}$, $\psi(\tilde{\Delta}) = \Delta$, and define $\tilde{X}_t = \varphi \circ X_t$, $\tilde{\mathcal{F}}^o_t = \sigma(\tilde{X}_s : s \leq t)$, $\tilde{\mathcal{F}}^o = \tilde{\mathcal{F}}^o_\infty$, $\tilde{P}^{\tilde{x}} = P^{\psi(\tilde{x})}|_{\tilde{\mathcal{F}}^o}$.

Then the completion $\tilde{\mathfrak{X}} = (\Omega, \tilde{\mathcal{F}}, \tilde{\mathcal{F}}_t, \tilde{X}_t, \Theta_t, \tilde{P}^{\tilde{x}})$ is a standard process with state space $\tilde{X}$. $\tilde{\mathfrak{X}}$ is called the *image of* $\mathfrak{X}$ *under* φ , $\tilde{\mathfrak{X}} = \varphi(\mathfrak{X})$.

Proof. Since φ is continuous, $\varphi(X) = \tilde{X}, \varphi(\Delta) = \tilde{\Delta}$, the paths $t \to \tilde{X}_t$ are right continuous on $[0,\infty[$ and have left-hand limits on $[0,\zeta[= [0,\tilde{\zeta}[$. Moreover, $\tilde{\mathcal{F}}^o_t \subset \mathfrak{M}_t$ for every $t \in \bar{\mathbb{R}}_+$.

Let T be a stopping time with respect to $(\tilde{\mathcal{F}}^o_{t+})$, $s \in \mathbb{R}_+$, $\tilde{B} \in \mathcal{B}(\tilde{X})$ and $\tilde{x} \in \tilde{X}$. Then T is a stopping time with respect to $(\mathfrak{M}_{t+}) = (\mathfrak{M}_t)$ and hence

$$\tilde{P}^{\tilde{x}}[\tilde{X}_{T+s} \in \tilde{B}] = P^{\psi(\tilde{x})}[X_{T+s} \in \varphi^{-1}(\tilde{B})] = E^{\psi(\tilde{x})}(P^{X_T}[X_s \in \varphi^{-1}(\tilde{B})]) = E^{\psi(\tilde{x})}(P^{X_T}[\tilde{X}_s \in \tilde{B}]).$$

Since by assumption $P^{\psi \circ \varphi(y)}_{\tilde{X}_s} = P^y_{\tilde{X}_s}$ for every $y \in X$ we have

$$P^{X_T}[\tilde{X}_s \in \tilde{B}] = P^{\psi(\varphi \circ X_T)}[\tilde{X}_s \in \tilde{B}] = \tilde{P}^{\tilde{X}_T}[\tilde{X}_s \in \tilde{B}]$$

and therefore

$$\tilde{P}^{\tilde{x}}[\tilde{X}_{T+s} \in \tilde{B}] = E^{\psi(\tilde{x})}(\tilde{P}^{\tilde{X}_T}[\tilde{X}_s \in \tilde{B}]) = \tilde{E}^{\tilde{x}}(\tilde{P}^{\tilde{X}_T}[\tilde{X}_s \in \tilde{B}]) .$$

Applying (6.8) and (6.9) we conclude that $\tilde{\mathfrak{X}}$ is a strong Markov process and $\tilde{\mathcal{F}}_{t+} = \tilde{\mathcal{F}}_t$ for every $t \in \mathbb{R}_+$.

Hence it remains to show that $\tilde{\mathfrak{X}}$ is quasi-left continuous on $[0,\tilde{\zeta}[$. So let (T_n) be an increasing sequence of stopping times of $\tilde{\mathfrak{X}}$ and $T = \lim_{n\to\infty} T_n$. Fix $\tilde{\mu} \in M^1_+(\tilde{X}_{\tilde{\Delta}})$ and let $\mu = \psi(\tilde{\mu})$. Then $\mu \in M^1_+(X_\Delta)$ and for every $A \in \tilde{\mathcal{F}}^o \subset \mathfrak{M}$

$$\tilde{P}^{\tilde{\mu}}(A) = \int \tilde{P}^{\tilde{x}}(A)\, \tilde{\mu}(d\tilde{x}) = \int P^{\psi(\tilde{x})}(A)\, \tilde{\mu}(d\tilde{x}) = \int P^x(A)\, \mu(dx) = P^\mu(A) .$$

(We have to be a bit careful since $\tilde{\mathcal{F}}$ may not be contained in $\mathfrak{M}$.) Using (6.7) and (5.2) we obtain an increasing sequence $(\tilde{T}_n)$ of stopping times with respect to $(\tilde{\mathcal{F}}^o_{t+})$ such that $\tilde{T}_n = T_n$ $\tilde{P}^{\tilde{\mu}}$-a.s. for every $n \in \mathbb{N}$. Let $\tilde{T} = \sup \tilde{T}_n$. Since $\mathfrak{X}$ is quasi-left continuous on $[0,\zeta[$ we have $\lim_{n\to\infty} X_{\tilde{T}_n} = X_{\tilde{T}}$ P^μ-a.s. on $[\tilde{T} < \zeta]$, hence $\lim_{n\to\infty} \tilde{X}_{\tilde{T}_n} = \tilde{X}_{\tilde{T}}$ P^μ-a.s. on $[\tilde{T} < \zeta]$.

By (6.4) we know that

$$A := [\tilde{T} < \zeta] \cap [\lim_{n\to\infty} \tilde{X}_{\tilde{T}_n} \neq \tilde{X}_{\tilde{T}}] \in \tilde{\mathcal{F}}^o$$

and hence $\widetilde{\tilde{P}^\mu}(A) = P^\mu(A) = 0$. So $\lim_{n\to\infty} \tilde{X}_{\tilde{T}_n} = \tilde{X}_{\tilde{T}}$ $\widetilde{\tilde{P}^\mu}$-a.s. on $[\tilde{T}<\tilde{\zeta}]$ and therefore

$$\lim_{n\to\infty} \tilde{X}_{T_n} = \tilde{X}_T \quad \widetilde{\tilde{P}^\mu}\text{-a.s. on } [\tilde{T}<\tilde{\zeta}] .$$

┘

7.8. EXERCISES. 1. There exists a standard process on $\mathbb{R}$ having the sub-Markov semigroup of (3.7.1) as transition function. Any such process is not a Hunt process.

2. Let $(X, \mathcal{W})$ be a balayage space, let $\mathfrak{X}$ be a Hunt process such that $E_{\mathfrak{X}} = \mathcal{W}$ and the potential kernel V of $\mathfrak{X}$ satisfies $V1 \in \mathcal{P}_b$, and let $g \in \mathcal{B}_b^+(X)$. For every $t > 0$, $f \in \mathcal{B}^+(X)$ and $x \in X$ define

$$\tilde{P}_t f(x) = E^x\left(f(X_t) e^{-\int_0^t g(X_s)ds}\right).$$

Then $\tilde{\mathbb{P}} = (\tilde{P}_t)_{t>0}$ is a sub-Markov semigroup on X.

8. Four Equivalent Views of Potential Theory

Summarizing the main results we obtain that our four descriptions of potential theory, namely by

(1) balayage spaces,

(2) families of harmonic kernels,

(3) sub-Markov semigroups,

(4) Hunt processes,

are equivalent in the following sense:

8.1. THEOREM. Let $\mathcal{P} \subset C^+(X)$ be a function cone and $\mathcal{W} = S(\mathcal{P})$ such that $1 \in \mathcal{W}$. Then the following statements are equivalent:

(1) $(X, \mathcal{W})$ is a balayage space.

(2) There exists a family $(H_U)_{U\in\mathcal{U}}$ of harmonic kernels on X such that ${}^*\mathcal{H}^+(X) = \mathcal{W}$.

(3) $\mathcal{W}$ is min-stable, and there exists a sub-Markov semigroup $\mathbb{P} = (P_t)_{t>0}$ on X such that $E_{\mathbb{P}} = \mathcal{W}$.

(4) There exists a Hunt process $\mathfrak{X}$ with state space X such that $E_{\mathfrak{X}} = \mathcal{W}$.

Proof. (1) $\Leftrightarrow$ (2): (III.6.11) and (III.2.8).

(1) $\Leftrightarrow$ (3): (II.8.6) and (II.4.7).

(3) $\Rightarrow$ (4): By (3.4), there exists a Markov process $\mathfrak{Y}$ on X having $\mathbb{P}$ as transition function. We may assume that P is min-stable. Hence by (7.5) there exists a Markov process $\mathfrak{X} = (\Omega, \mathfrak{M}, \mathfrak{M}_t, X_t, \Theta_t, P^x)$ such that $\mathfrak{X}$ is equivalent to $\mathfrak{Y}$ and the path functions $t \mapsto X_t(\omega)$ of $\mathfrak{X}$ are contained in $\tilde{W}$. Then $\tilde{\mathfrak{X}} = (\Omega, \mathcal{F}, \mathcal{F}_t, X_t, \Theta_t, P^x)$ is a Hunt process by (7.6), and $E_{\tilde{\mathfrak{X}}} = E_{\mathbb{P}} = \mathcal{W}$.

(4) $\Rightarrow$ (3): By definition of the transition function $\mathbb{P} = (P_t)_{t>0}$ of $\mathfrak{X}$, we have $E_{\mathbb{P}} = E_{\mathfrak{X}} = \mathcal{W}$. Since $\mathfrak{X}$ is right continuous we have $\lim_{t\to 0} P_t f(x) = f(x)$ for every $f \in K(X)$ and $x \in X$, hence any l.s.c. $\mathbb{P}$-supermedian function is $\mathbb{P}$-excessive. In particular, $E_{\mathbb{P}} = \mathcal{W}$ is min-stable. ⌋

<u>8.2. REMARK.</u> By the theorem we have Hunt processes associated with classical potential theory and discrete potential theory. In fact, there exist corresponding processes (Brownian motion, Poisson process) having additional nice properties. However, some more knowledge of potential theory will facilitate their treatment. Therefore these processes will be considered later (see (VI.3.19)).

V. Examples

In this chapter we study permanence properties of balayage spaces. In particular, it will be possible to complete our list of standard examples by the two balayage spaces corresponding to Riesz potentials and the heat equation.

First of all we show that restriction of a balayage space to open or absorbing subsets gives rise to balayage spaces (section 1). Next we investigate which strong Feller resolvents and semigroups generate balayage spaces. These results are used to obtain new balayage spaces by perturbations (section 2), subordination by convolution semigroups (section 3), and by products with translation on $\mathbb{R}$ (section 5). The symmetric stable semigroup which is subordinated to Brownian semigroup by means of the one-sided stable semigroup generates the balayage space of the Riesz potentials studied in section 4 in great detail. We note that this contains as well further theorems of classical potential theory (e.g. representation of potentials, continuity principle, domination principle, Wiener's criterion). The product of Brownian semigroup on $\mathbb{R}^n$ and translation on $\mathbb{R}$, the heat semigroup, generates a harmonic space. Moreover, in section 6 we prove that the solutions of the heat equation on $\mathbb{R}^{n+1}$ also generate a harmonic space. A further study of this space yields among other things the expected identity of these two harmonic spaces. Brownian semigroups on the infinite dimensional torus are treated in section 7, and section 8 discusses images of balayage spaces. The final section 9 contains an example which will occasionally serve as a counterexample in the next chapter.

1. Subspaces

Let $(X,\mathcal{W})$ be a balayage space. We shall show that all open and certain closed subsets of X give rise to subspaces of $(X,\mathcal{W})$.

1.1. PROPOSITION. Let U be an open subset of X and $\mathcal{W}^U = {}^*\mathcal{H}^+(U)|_U$. Then $(U,\mathcal{W}^U)$ is a balayage space having the following properties:

1. $(H_{V|U})_{V\in\mathcal{U}(U)}$ is its associated family of harmonic kernels.
2. The $\mathcal{W}^U$-fine topology is the relative $\mathcal{W}$-fine topology of U .
3. For every $p\in P$, the function $p^U := (p - R_p^{\complement U})|_U$ is a continuous potential on the open subspace U .

$(U,\mathcal{W}^U)$ is called the *restriction of* $(X,\mathcal{W})$ *on* U .

Proof. Define for every $V\in\mathcal{U}(U)$

$$^U H_V := H_{V|U} \quad .$$

Then $(^U H_V)_{V\in\mathcal{U}(U)}$ satisfies the axioms (H_1) - (H_5) . To prove this, we shall make the usual convention that every numerical function on U will be extended by the value 0 to a numerical function on X .

(H_1): Let $x\in U$. If x is not finely isolated then by (III.2.7)

$$\varepsilon_x = \lim_{V,\mathcal{U}_x} H_V(x,\cdot) = \lim_{V,\mathcal{U}_x(U)} {}^U H_V(x,\cdot) \quad .$$

If x is finely isolated then $R_c^{\{x\}}$ is l.s.c. at x for every $c\in\mathbb{R}_+$ by (B_2) . Using (III.2.2) there exist $q\in P$ and $V\in\mathcal{U}_x(U)$ such that

$$R_{1+q(x)}^{\{x\}} \le {}^U R_1^{\{x\}} + q \quad \text{on} \quad V$$

where ${}^U R_f^A$ denotes the reduit of f on $A\subset U$ with respect to $\mathcal{W}^U$. Hence in this case

$$\liminf_{y\to x} {}^U R_1^{\{x\}}(y) \ge \liminf_{y\to x,\, y\in V} (R_{1+q(x)}^{\{x\}}(y) - q(y)) = 1 = {}^U R_1^{\{x\}}(x) \quad .$$

(H_2): If $V,W\in\mathcal{U}(U)$ satisfy $\overline{V}\subset W$ then for every $f\in K(U)$

$$^U H_V {}^U H_W f = {}^U H_V H_W f = H_V H_W f = H_W f = {}^U H_W f \quad .$$

(H_3)-(H_5) follow immediately from the definitions, whereas property (2) is a consequence of (III.2.2).

Finally, let $p \in P$. Then $p^U = p - R_p^{\complement U} \in S^+(U) \cap C(U)$ by (III.2.6). Let h be a harmonic function on the subspace U such that $0 \leq h \leq p^U$ and let $w \in \mathcal{W}$ such that $w \geq p$ on $\complement U$. Then $v = w - h - R_p^{\complement U} \in {}^*H(U)$, $v \geq 0$ on $\complement U$ and $\liminf_{x\to z} v(x) \geq \liminf_{x \to z} (w(x) - p(x)) = w(z) - p(z) \geq 0$ for every $z \in U^*$. Hence $v \geq 0$ by (III.4.1), i.e. $h + R_p^{\complement U} \leq w$. Then $h + R_p^{\complement U} \leq R_p^{\complement U}$, $h \leq 0$, and p^U is a continuous potential on U. ┘

A closed subset A of X is called an *absorbing set* if for every $x \in A$,

$$\varepsilon_x^{\circ\complement A} = 0 \quad .$$

This name is motivated by property (8) of the following proposition.

<u>1.2. PROPOSITION.</u> Let A be a subset of X. Then the following properties are equivalent:

(1) A is an absorbing set.

(2) For every $u \in \mathcal{W}$, $u1_{\complement A} \in \mathcal{W}$.

(3) There exists a function $u \in \mathcal{W}$ such that $A = \{u = 0\}$.

(4) There exists a function $u \in \mathcal{W}$ such that $A = \overline{\{u < +\infty\}}$.

(5) A is closed and for every $x \in A$ and all $U \in \mathcal{U}_x$, $H_U(x,\complement A) = 0$.

(6) A is closed and for every $x \in A$ and every neighborhood V of x there exists $U \in \mathcal{U}_x(V)$ such that $H_U(x,\complement A) = 0$.

If $1 \in \mathcal{W}$ then there are two more equivalent conditions:

(7) A is closed and for every sub-Markov semigroup $\mathbb{P} = (P_t)_{t>0}$ such that $\mathcal{W} = E_{\mathbb{P}}$, for every $x \in A$, and for every $t > 0$, $P_t(x, X \setminus A) = 0$.

(8) A is closed and for every Markov process $\mathfrak{X}$ such that $\mathcal{W} = E_{\mathfrak{X}}$, for every $x \in A$ and every $t > 0$,

$$P^x[X_t \in A \cup \{\Delta\}] = 1 \quad .$$

Proof. (1) ⇒ (2): Let $p \in P$. Then $R_p^{\complement A} \in \mathcal{W}$ since $\complement A$ is open. Moreover, $R_p^{\complement A} = p$ on $\complement A$ whereas $R_p^{\complement A}(x) = \varepsilon_x^{\circ\complement A}(p) = 0$ for every $x \in A$. Therefore $p1_{\complement A} = R_p^{\complement A} \in \mathcal{W}$. Given $u \in \mathcal{W}$ we choose an increasing sequence (p_n) in P

such that $\sup_n p_n = u$ and obtain $u1_{\complement A} = \sup_n p_n 1_{\complement A} \in \mathcal{W}$.

(2) ⇒ (3): Obvious.

(3) ⇒ (4): Let $u \in \mathcal{W}$ such that $A = \{u = 0\}$. Then $v := \sup_n (nu) \in \mathcal{W}$, $v = \infty$ on $\complement A$ and $v = 0$ on the closed set $A = \{u \leq 0\}$.

(4) ⇒ (3): Let $u \in \mathcal{W}$ such that $A = \overline{\{u < +\infty\}}$. Then $v := \widehat{\inf_n \frac{u}{n}} \in \mathcal{W}$, $v = 0$ on A and $v = +\infty$ on $\complement A$.

(3) ⇒ (2): Let $u \in \mathcal{W}$ such that $A = \{u = 0\}$ and let $v \in \mathcal{W}$. Then $v1_{\complement A} = \sup_n \inf (v, nu) \in \mathcal{W}$.

(2) ⇒ (1): Let $x \in A$ and $p \in \mathcal{P}$. Then $R_p^{\complement A} = p1_{\complement A}$, in particular $\overset{\circ}{\varepsilon}{}_x^{\complement A}(p) = 0$. Hence $\overset{\circ}{\varepsilon}{}_x^{\complement A} = 0$.

(2) ⇒ (5): Let $p \in \mathcal{P}$, $p > 0$. Then $v := p1_{\complement A} \in \mathcal{W}$, $A = \{v \leq 0\}$ is closed and for every $x \in A$, $U \in \mathcal{U}_x$ the inequality $H_U v(x) \leq v(x) \leq 0$ implies that $H_U(x, X \setminus A) = 0$.

(5) ⇒ (6): Obvious.

(6) ⇒ (2): (II.5.5) or (III.4.4).

(3) ⇒ (7): Immediate since $\sup_{t>0} P_t u = u$ for every $u \in \mathcal{W}$.

(7) ⇒ (2): If $u \in \mathcal{W}$ then $u1_{\complement A}$ is l.s.c. and supermedian, hence $u1_{\complement A} \in \mathcal{W}$ by (II.4.9).

(7) ⟺ (8): (IV.3.4). ┘

The following properties follow from the definition and (1.2).

<u>1.3. REMARKS.</u> 1. $\emptyset$ and X are always absorbing sets.

2. Finite unions and arbitrary intersections of absorbing sets are absorbing sets.

3. Every absorbing set is finely open.

4. If $u \in \mathcal{W}$ and if A is an absorbing set then $u_A \in \mathcal{W}$ where

$$u_A := \begin{cases} +\infty & \text{on } \complement A, \\ u & \text{on } A. \end{cases}$$

5. If $x \in X$ then x is an absorbing point if and only if $\{x\}$ is an absorbing set.

6. If X is connected and locally connected and if $V^* \subset \operatorname{supp} H_V(x,\cdot)$ for

all $x \in V \in \mathcal{U}$, then for each $u \in \mathcal{W}$ the following holds:

(a) $u = 0$ or $u > 0$,

(b) $u = +\infty$ or $\overline{\{u < +\infty\}} = X$.

Proof. By (1.2), the assertion is equivalent to the condition that $\emptyset$ and X are the only absorbing sets of X. Let $A \neq X$ be an absorbing set and let Z be a connected component of $\complement A$. Z is open since X is locally connected. If we had a point $x \in \overline{Z} \setminus Z$ then choosing $V \in \mathcal{U}_x$ such that $Z \not\subset V$ we would have $Z \cap V \neq \emptyset$, $Z \setminus \overline{V} \neq \emptyset$ and $Z = (Z \cap V) \cup (Z \setminus \overline{V})$ since $V^* \subset \operatorname{supp} H_V(x,\cdot) \subset A$. This is impossible since Z is connected. Therefore Z is closed, hence $Z = X$, $A = \emptyset$. ┘

1.4. EXAMPLES. 1. Classical potential theory in $\mathbb{R}^n$. The Poisson formula and (1.3.6) imply that $\emptyset$ and $\mathbb{R}^n$ are the only absorbing sets in $\mathbb{R}^n$. More generally, for every open connected subspace U of $\mathbb{R}^n$, there are only the trivial absorbing subsets $\emptyset$ and U.

2. Translation on $\mathbb{R}$. Since $\mathcal{W}$ consists of all positive numerical l.s.c. increasing functions on $\mathbb{R}$, (1.2.3) implies that all absorbing sets are given by $A_t := \{s \in \mathbb{R} : s \leq t\}$, $t \in \overline{\mathbb{R}}$.

3. Discrete potential theory. A subset A of X is absorbing if and only if $P(x,\complement A) = 0$ for every $x \in A$. This follows immediately from (III.1.1.3) and (1.2).

1.5. PROPOSITION. Let $A \subset X$ be an absorbing set. Then $(A,\mathcal{W}_{|A})$ is a balayage space having the following properties:

1. If $U \in \mathcal{U}$ such that $A \cap U \neq \emptyset$ and

$$ {}^{A}H_U(x,\cdot) := \begin{cases} H_U(x,\cdot) & \text{if } x \in A \cap U, \\ \varepsilon_x & \text{if } x \in A \setminus U, \end{cases} $$

then $({}^{A}H_U)_{U \in \mathcal{U}, A \cap U \neq \emptyset}$ is the family of harmonic kernels associated with $(A,\mathcal{W}_{|A})$.

2. The $\mathcal{W}|_A$-fine topology on A is the relative $\mathcal{W}$-fine topology on A.

3. For every $p \in \mathcal{P}$, the function $p|_A$ is a potential on the subspace A.

Proof. Since by (1.3.3) A is finely open, we have (2). If $u \in \mathcal{W}|_A$ then

$u_A \in \mathcal{W}$ by (1.3.4). This remark immediately implies the validity of the axioms (B_1) - (B_3) for $(A, \mathcal{W}|_A)$. Moreover, by (I.1.1.1) $\mathcal{P}|_A$ is a function cone on A such that $\mathcal{S}(\mathcal{P}|_A) = \mathcal{W}|_A$. Thus $(A, \mathcal{W}|_A)$ is a balayage space, and (3) holds by (II.5.2). Using (1.3.4) again we obtain that for every $U \in \mathcal{U}$, $x \in A \cap U$, and $p \in \mathcal{P}$,

$$\inf \{u(x) : u \in \mathcal{W}|_A , u \geq p \text{ on } A \setminus U\} = R_p^{\complement U}(x) .$$

This proves (1). ∎

Later we shall be able to consider restrictions on arbitrary closed basic subsets (see (VI.6.16)).

1.6. EXERCISES. 1. Let $(X, \mathcal{W})$ be a balayage space and let V be the potential kernel of $p \in \mathcal{P}$. Then, for every open subset U of X, $V - H_U V$ is the potential kernel of p^U.

2. Let A be an absorbing subset of a balayage space $(X, \mathcal{W})$. Then $p - R_p^A \in \mathcal{P}$ for every $p \in \mathcal{P}$.

3. Let $(A_i)_{i \in I}$ be a family of absorbing subsets of a balayage space $(X, \mathcal{W})$ and let $A = \bigcup_{i \in I} A_i$. Then $\bar{A} = \bar{A}^f$ is an absorbing set.

4. Continuation of exercise (III.2.11). If $\mathcal{W} = {}^*\mathcal{H}^+(X)$ then $1 \in \mathcal{W}$ and $\mathcal{W}^{X_n} = \mathcal{W}_n$ for every $n \in \mathbb{N}$. If $\tau_n \neq 0$ for every $n \in \mathbb{N}$ then $(X, \mathcal{W})$ has no absorbing points.

2. Strong Feller Kernels

We recall that a kernel V on X is called a *strong Feller kernel* if $V(\mathcal{B}_b(X)) \subset \mathcal{C}_b(X)$. A sub-Markov resolvent $\mathbb{V} = (V_\lambda)_{\lambda > 0}$ is called a *strong Feller resolvent* if every V_λ, $\lambda > 0$, is a strong Feller kernel. If the potential kernel V of $\mathbb{V}$ is bounded then the resolvent equation $V = V_\lambda + \lambda V V_\lambda = V_\lambda + \lambda V_\lambda V$ shows that $\mathbb{V}$ is a strong Feller resolvent if and only if V is a strong Feller kernel. A sub-Markov semigroup $\mathbb{P} = (P_t)_{t > 0}$ is called a *strong Feller semigroup* if every P_t, $t > 0$, is a strong Feller kernel. If $\mathbb{V}$ is the resolvent of a strong Feller semigroup $\mathbb{P}$ then obviously $\mathbb{V}$ is a strong Feller resolvent by Lebesgue's convergence theorem.

2.1. EXAMPLES. 1. If $(X,\mathcal{W})$ is a balayage space and $p \in \mathcal{P}_b$ then the potential kernel $(x,B) \mapsto p_B(x)$ associated with p is a strong Feller kernel.

2. The Brownian semigroup on $\mathbb{R}^n$ is a strong Feller semigroup by (0.2.2).

3. The translation semigroup on $\mathbb{R}$ is obviously not a strong Feller semigroup. However, its resolvent is a strong Feller resolvent.

2.2. PROPOSITION. Let $\mathbb{V} = (V_\lambda)_{\lambda > 0}$ be a strong Feller sub-Markov resolvent on X such that $\lim_{\lambda \to \infty} \lambda V_\lambda f = f$ for every $f \in \mathcal{K}(X)$ and such that there exist strictly positive functions $u,v \in E_{\mathbb{V}} \cap \mathcal{C}(X)$ satisfying $\frac{u}{v} \in \mathcal{C}_0(X)$. Then $(X,E_{\mathbb{V}})$ is a balayage space if $E_{\mathbb{V}}$ is linearly separating or the potential kernel V of $\mathbb{V}$ is proper.

Proof. By (II.4.7.) and (II.4.8) it suffices to show that $E_{\mathbb{V}} = S(E_{\mathbb{V}} \cap \mathcal{C}(X))$. So fix $s \in E_{\mathbb{V}}$ and define $s_n = \inf(s,n)$, $n \in \mathbb{N}$. Then by (II.3.10) the sequence $(nV_n s_n)$ is contained in $E_{\mathbb{V}} \cap \mathcal{C}(X)$. As in the proof of (II.3.11) we conclude that $(nV_n s_n)$ is increasing to s. Thus $s \in S(E_{\mathbb{V}} \cap \mathcal{C}(X))$. ┘

2.3. COROLLARY. Let $\mathbb{V}$ be a strong Feller sub-Markov resolvent on X such that $(X,E_{\mathbb{V}})$ is a balayage space. Then $(X,E_{\mathbb{V}^\alpha})$ is a balayage space for every $\alpha > 0$. If the potential kernel V of $\mathbb{V}$ is bounded then for every $\alpha > 0$

$$E_{\mathbb{V}^\alpha} \cap \mathcal{B}_b(X) = \{u \in \mathcal{B}_b(X) : u + \alpha Vu \in E_{\mathbb{V}}\} = (I + \alpha V)^{-1}(E_{\mathbb{V}} \cap \mathcal{B}_b(X)),$$

in particular the $E_{\mathbb{V}^\alpha}$-fine topology is the $E_{\mathbb{V}}$-fine topology.

Proof. Let $\alpha > 0$. Then $\mathbb{V}^\alpha = (V_{\alpha+\lambda})_{\lambda > 0}$ is a strong Feller resolvent with bounded potential kernel V_α. Moreover, $E_{\mathbb{V}} \subset E_{\mathbb{V}^\alpha}$ by (II.3.12) and clearly $\lim_{\lambda \to \infty} \lambda V_{\lambda+\alpha} f = f$ for every $f \in \mathcal{K}(X)$. Hence $(X,E_{\mathbb{V}^\alpha})$ is a balayage space by (2.2).

Suppose finally that V is bounded and let $u \in \mathcal{B}_b(X)$. Then $Vu \in \mathcal{C}_b(X)$ and for every $\lambda > 0$ by the resolvent equation

$$\lambda V_\lambda(u + \alpha Vu) = (\lambda - \alpha)V_\lambda u + \alpha Vu \quad ,$$

hence $(\lambda-\alpha)V_\lambda u \leq u$ if and only if $\lambda V_\lambda(u+\alpha Vu) \leq u + \alpha Vu$. Using (II.4.7) we thus conclude that $u \in E_{\mathbb{V}\alpha}$ if and only if $u + \alpha Vu \in E_{\mathbb{V}}$. ┘

2.4. COROLLARY. Let $\mathbb{P} = (P_t)_{t>0}$ be a sub-Markov semigroup on X such that $\lim_{t\to 0} P_t f = f$ for every $f \in K(X)$ and the resolvent of $\mathbb{P}$ (or even $\mathbb{P}$ itself) is strong Feller. Suppose furthermore that there exist strictly positive functions $u,v \in E_{\mathbb{P}} \cap C(X)$ satisfying $\frac{u}{v} \in C_0(X)$ and that $E_{\mathbb{P}}$ is linearly separating or the potential kernel of $\mathbb{P}$ is proper. Then $(X, E_{\mathbb{P}^\alpha})$ is a balayage space for every $\alpha \geq 0$.

2.5. EXAMPLE. Let $\mathbb{P}$ be the Brownian semigroup on $\mathbb{R}^n$, $n \geq 1$, and $\alpha > 0$. Then $(\mathbb{R}^n, E_{\mathbb{P}^\alpha})$ is a balayage space. This is true for $n \geq 3$ by (II.4.10.1) and (2.4). The general case follows easily from (2.4). It suffices to find a strictly positive function $u \in E_{\mathbb{P}^\alpha} \cap C_0(\mathbb{R}^n)$. To that end let $1 \leq i \leq n$ and define $u_i^\pm \in C(\mathbb{R}^n)$ by

$$u_i^\pm(x) = e^{\pm\sqrt{2\alpha}\, x_i} \quad .$$

Then, for every $t > 0$ and every $x \in \mathbb{R}^n$

$$P_t^\alpha u_i^\pm(x) = \left(\frac{1}{2\pi t}\right)^{n/2} e^{-\alpha t} \int e^{-\frac{\| y-x \|^2}{2t}} e^{\pm\sqrt{2\alpha}\, y_i} \lambda^n(dy)$$

$$= \frac{e^{-\alpha t}}{\sqrt{2\pi t}} \int_{-\infty}^{+\infty} e^{-\frac{(y_i - x_i)^2}{2t}} e^{\pm\sqrt{2\alpha}\, y_i} dy_i$$

$$= \frac{e^{\pm\sqrt{2\alpha}\, x_i}}{\sqrt{2\pi t}} \int_{-\infty}^{+\infty} e^{-\frac{(y_i - x_i \mp \sqrt{2\alpha}\, t)^2}{2t}} dy_i = e^{\pm\sqrt{2\alpha}\, x_i} = u_i^\pm(x) \quad ,$$

hence $u_i^\pm \in E_{\mathbb{P}^\alpha}$. Therefore

$$u := \inf(u_1^+, \ldots, u_n^+, u_1^-, \ldots, u_n^-) \in E_{\mathbb{P}^\alpha} \ .$$

Clearly, $u > 0$ and $u \in C_0(\mathbb{R}^n)$. Thus $(\mathbb{R}^n, E_{\mathbb{P}^\alpha})$ is a balayage space. In fact, $(\mathbb{R}^n, E_{\mathbb{P}^\alpha})$ is a harmonic space by (III.8.2) and (III.7.3).

Suppose now that $n = 1$, let $U =]a,b[\subset \mathbb{R}$ and let H_U be the corresponding harmonic kernel. Then for every $f \in C_b^+(\mathbb{R})$ the function $H_U f$ is the unique function $u \in C(\mathbb{R})$ such that $u = f$ on $\complement U$ and u is a linear combination of

$e^{\sqrt{2\alpha}\,x}$ and $e^{-\sqrt{2\alpha}\,x}$ on U , i.e. $u'' = 2\alpha u$ on U . Indeed, by (III.7.4), the functions $e^{\sqrt{2\alpha}\,x}$ and $e^{-\sqrt{2\alpha}\,x}$ are harmonic on X . Hence

$$v = \frac{\sinh\sqrt{2\alpha}\,(b-x)}{\sinh\sqrt{2\alpha}\,(b-a)} f(a) + \frac{\sinh\sqrt{2\alpha}\,(x-a)}{\sinh\sqrt{2\alpha}\,(b-a)} f(b)$$

is a harmonic function on $\mathbb{R}$ such that $v(a) = f(a)$ and $v(b) = f(b)$. Define $u \in C(\mathbb{R})$ by

$$u = \begin{cases} v & \text{on } U \\ f & \text{on } \complement U \end{cases} .$$

Then v is harmonic on U by (III.8.1). Using (III.4.3) we conclude that $H_U f = u$.

The constant 1 is a potential for the balayage space $(\mathbb{R}, E_{\mathbb{P}^\alpha})$ since $\lim_{n\to\infty} H_{]-n,n[}\, 1 = 0$. The translation invariance of $\mathbb{P}^\alpha$ shows that the fine support of 1 is the entire space $\mathbb{R}$, i.e. 1 is a strict potential by (II.7.8).

Let us finally note that every function $u \in E_{\mathbb{P}^\alpha}$ is continuous on $\mathbb{R}$. Indeed, let $x \in \mathbb{R}$ and (x_n) a sequence in $]x,\infty[$ converging to x. Choosing $x_0 < x$ we obtain that for every $n \in \mathbb{N}$

$$u(x) \geq H_{]x_0,x_n[}u(x) \geq \frac{\sinh\sqrt{2\alpha}\,(x-x_0)}{\sinh\sqrt{2\alpha}\,(x_n-x_0)}\, u(x_n) \quad ,$$

hence

$$u(x) \geq \limsup_{n\to\infty} u(x_n) \ .$$

Similarly for sequences (x_n) increasing to x .

The formula in (2.3) suggests a similar "perturbation" of balayage spaces $(X,\mathcal{W})$ replacing the kernel V by the potential kernel associated with $p \in \mathcal{P}_b$. Using (II.7.8) this can be achieved by an approximation procedure since we have the following general result.

2.6. LEMMA. Let $(\mathcal{W}_n)$ be a decreasing sequence of convex cones of positive l.s.c. functions on X such that (B_1) - (B_3) are satisfied for each $n \in \mathbb{N}$. Then $\mathcal{W}_\infty = \bigcap_{n=1}^{\infty} \mathcal{W}_n$ satisfies (B_1) - (B_3), too.

Proof. $\mathcal{W}_\infty$ clearly satisfies (B_1). If $\mathcal{V} \subset \mathcal{W}_\infty$ and $v = \inf \mathcal{V}$ then $\mathcal{V} \subset \mathcal{W}_n$ and $\hat{v} \in \mathcal{W}_n$ for every $n \in \mathbb{N}$, hence $\hat{v} \in \mathcal{W}_\infty$. Moreover, for every $x \in X$,

$$\hat{v}(x) \le \mathcal{W}_\infty\text{-}f\text{-}\liminf_{y\to x} v(y) \le \mathcal{W}_1\text{-}f\text{-}\liminf_{y\to x} v(y) = \hat{v}(x) \quad .$$

This shows that $\mathcal{W}_\infty$ satisfies (B_2).

Finally let $u,v \in \mathcal{W}_\infty$ such that $v \le u$ and $f \in \mathcal{B}^+(X)$ such that $v + f = u$ and $f = 0$ on $\{v = \infty\}$. For every $n \in \mathbb{N}$, let

$$w_n = \inf \{w \in \mathcal{W}_n : w \ge f\} \quad .$$

By (II.4.4), $w_n \in \mathcal{W}_n$ and there exists $v_n \in \mathcal{W}_n$ such that

$$w_n + v_n = u \quad .$$

Obviously, the sequence (w_n) is increasing and hence $w_\infty := \sup w_n \in \mathcal{W}_\infty$. Moreover,

$$w_\infty \ge \inf \{w \in \mathcal{W}_\infty : w \ge f\} \ge w_n$$

for every $n \in \mathbb{N}$ and hence

$$\inf \{w \in \mathcal{W}_\infty : w \ge f\} = w_\infty \in \mathcal{W}_\infty \quad .$$

The sequence v_n is decreasing on $\{u < \infty\}$, hence the sequence $(v_n + \frac{1}{n}u)$ is decreasing on X . Let $v_\infty = \inf (v_n + \frac{1}{n}u)$. Then $\hat{v}_\infty \in \mathcal{W}_\infty$ since, for every $m \in \mathbb{N}$, $\hat{v}_\infty = \widehat{\inf_{n \ge m} (v_n + \frac{1}{n}u)} \in \mathcal{W}_m$. Moreover, for every $x \in X$,

$$\hat{v}_\infty(x) = \mathcal{W}_\infty\text{-}f\text{-}\liminf_{y\to x} v_\infty(y) \quad .$$

Since $v_\infty + w_\infty = u$ and w_∞, u are $\mathcal{W}_\infty$-finely continuous we thus conclude that

$$\hat{v}_\infty + w_\infty = u \quad .$$

□

<u>2.7. PROPOSITION.</u> Let $(X,\mathcal{W})$ be a balayage space such that $1 \in \mathcal{W}$. Let $p \in \mathcal{P}_b$ and let K_p denote the potential kernel associated with p . Then there exists a unique balayage space $(X,\mathcal{W}^p)$ such that

$$\mathcal{W}_b^p = \{u \in \mathcal{B}_b^+(X) : u + K_p u \in \mathcal{W}\} = (I + K_p)^{-1} \mathcal{W}_b \quad .$$

$\mathcal{W}^p$ is the set of all functions $u \in \mathcal{B}^+(X)$ such that $\inf (u,n) + K_p(\inf (u,n)) \in \mathcal{W}$ for every $n \in \mathbb{N}$. In particular, $\mathcal{W} \subset \mathcal{W}^p$ and the corresponding fine topologies coincide.

Proof. If p is strict then by (II.7.8) there exists a strong Feller resolvent

$\mathbb{V}$ such that K_p is the potential kernel of $\mathbb{V}$ and $E_{\mathbb{V}} = \mathcal{W}$, hence the existence of $\mathcal{W}^p$ follows immediately from (2.3) : $\mathcal{W}^p = E_{\mathbb{V}^1}$.

Let us now consider the general case. We note first that $I + K_p : \mathcal{B}_b(X) \to \mathcal{B}_b(X)$ is an isomorphism by (II.7.8) and (II.7.4). The uniqueness of $\mathcal{W}^p$ follows easily. Indeed, let $(X,\mathcal{W}^p)$ be a balayage space such that $\mathcal{W}^p_b = (I+K_p)^{-1}\mathcal{W}_b$. Then $\mathcal{W}_b \subset \mathcal{W}^p_b$ since $(I+K_p)\mathcal{W}_b \subset \mathcal{W}_b$. In particular, $1 \in \mathcal{W}^p$. Given $u \in \mathcal{W}^p$ and $n \in \mathbb{N}$ we hence have $\inf(u,n) \in \mathcal{W}^p$ and therefore $\inf(u,n) + K_p(\inf(u,n)) \in \mathcal{W}$. If conversely $u \in \mathcal{B}^+(X)$ such that $\inf(u,n) + K_p(\inf(u,n)) \in \mathcal{W}$ for every $n \in \mathbb{N}$, then $\inf(u,n) \in \mathcal{W}^p$ for every $n \in \mathbb{N}$, hence $u = \sup_n \inf(u,n) \in \mathcal{W}^p$.

The existence of $\mathcal{W}^p$ will be established by approximation. Let $q \in \mathcal{P}_b$ be a strict potential. Then for every $\eta > 0$ the potential $p + \eta q$ is strict and we have a corresponding balayage space $(X,\mathcal{W}^{p+\eta q})$. If $0 < \eta < \eta'$ and $u \in \mathcal{W}^{p+\eta q}_b$ then

$$u + K_{p+\eta' q}u = u + K_{p+\eta q}u + (\eta' - \eta)K_q u \in \mathcal{W} \quad .$$

hence $u \in \mathcal{W}^{p+\eta' q}_b$. Therefore the family $(\mathcal{W}^{p+\eta q})_{\eta > 0}$ is increasing. Let

$$\widetilde{\mathcal{W}} = \bigcap_{\eta > 0} \mathcal{W}^{p+\eta q} \quad .$$

Then $\widetilde{\mathcal{W}}$ is a convex cone of positive l.s.c. numerical functions on X satisfying (B_1) - (B_3) by (2.6). Since $\mathcal{W} \subset \widetilde{\mathcal{W}}$, the set $\widetilde{\mathcal{W}} \cap \mathcal{C}_{\mathcal{P}}(X)$ is a function cone by (I.1.1.1).

If $u \in \mathcal{B}^+_b(X)$ such that $u + K_p u \in \mathcal{W}$, then $u + K_{p+\eta q}u = u + K_p u + \eta K_q u \in \mathcal{W}$, i.e. $u \in \mathcal{W}^{p+\eta q}$ for every $\eta > 0$, and hence $u \in \widetilde{\mathcal{W}}$. If conversely $u \in \widetilde{\mathcal{W}}_b$ then $u + K_p u = \inf_{\eta>0}(u + K_{p+\eta q}u) = \widehat{\inf_{\eta>0}(u + K_{p+\eta q}u)} \in \mathcal{W}$. Thus $\widetilde{\mathcal{W}}_b = (I + K_p)^{-1}\mathcal{W}_b$.

Finally, let $\varphi \in \mathcal{K}^+(X)$ and

$$u = \inf\{w \in \widetilde{\mathcal{W}} \ : w \geq \varphi\} \ .$$

Then $u \in \widetilde{\mathcal{W}}_b$ by (B_2). Let $\eta > 0$ and

$$u_\eta = \inf\{w \in \mathcal{W}^{p+\eta q} \ : w \geq \varphi\} \quad .$$

Then $u_\eta \in \mathcal{W}^{p+\eta q}_b \cap \mathcal{C}(X)$ since $(X,\mathcal{W}^{p+\eta q})$ is a balayage space. Obviously, $u \geq u_\eta$. On the other hand

$$u_\eta + \eta K_q u_\eta + K_p(u_\eta + \eta K_q u_\eta) = u_\eta + K_{p+\eta q} u_\eta + \eta K_p K_q u_\eta \in \mathcal{W}$$

hence $u_\eta + \eta K_q u_\eta \in \tilde{\mathcal{W}}$ and therefore $u_\eta + \eta K_q u_\eta \geq u$. Since $K_q u_\eta \leq K_q \|\varphi\| = \|\varphi\| q$, we conclude that (u_η) converges uniformly to u as η tends to zero. Therefore $u \in \tilde{\mathcal{W}} \cap \mathcal{C}_p(X)$. Proceeding as usual we conclude that $\tilde{\mathcal{W}} = S(\tilde{\mathcal{W}} \cap \mathcal{C}_p(X))$. Thus $(X,\tilde{\mathcal{W}})$ is a balayage space. ┘

For applications in section 5 it will be convenient to have an equivalent description for right continuous strong Feller semigroups. We first prove a simple lemma.

2.8. LEMMA. Let $\mathbb{P} = (P_t)_{t>0}$ be a sub-Markov semigroup such that, for every $f \in \mathcal{K}(X)$, $P_t f$ converges locally uniformly to f as t tends zero. Then, for every $f \in \mathcal{C}_b(X)$, $P_t f$ converges locally uniformly to f as t tends to zero.

Proof. Let $f \in \mathcal{C}_b(X)$, $0 \leq f \leq 1$. Let K be a compact subset of X and $\varepsilon > 0$. We choose $\varphi, \psi \in \mathcal{K}^+(X)$ such that $\varphi \leq f$ and $\varphi = f$ on K , $\psi \leq 1 - f$ and $\psi = 1 - f$ on K . Then there exists $t_0 > 0$ such that for every $0 < t < t_0$

$$P_t\varphi \geq \varphi - \varepsilon \text{ on } K , \quad P_t\psi \geq \psi - \varepsilon \text{ on } K .$$

Let $0 < t < t_0$. Then

$$P_t f \geq P_t\varphi \geq \varphi - \varepsilon = f - \varepsilon \text{ on } K .$$

Similarly $P_t(1-f) \geq 1 - f - \varepsilon$ on K and hence

$$P_t f \leq P_t 1 - 1 + f + \varepsilon \leq f + \varepsilon \text{ on } K .$$ ┘

2.9. REMARK. The proof shows that the same convergence holds for every $f \in \mathcal{C}(X)$ which is bounded by a function in $S_{\mathbb{P}} \cap \mathcal{C}(X)$.

2.10 PROPOSITION. Let $\mathbb{P} = (P_t)_{t>0}$ be a sub-Markov semigroup on X . Then the following properties are equivalent:

(1) $\mathbb{P}$ is a strong Feller semigroup and, for every $f \in \mathcal{K}(X)$ and $t > 0$, $\lim_{s \downarrow t} P_s f = P_t f$ locally uniformly.

(2) For every $f \in \mathcal{B}_b(X)$, the function $(x,t) \mapsto P_t f(x)$ is continuous on $X \times \mathbb{R}_+^*$.

Proof. (1) ⇒ (2): Let $f \in \mathcal{B}_b(X)$, $0 \leq f \leq 1$, $x \in X$, $0 < t < \infty$ and $\varepsilon > 0$. Let K be a compact neighborhood of x and $\varphi \in \mathcal{K}(X)$ such that $0 \leq \varphi \leq 1$ and $\varphi = 1$ on K . By (2.8) there exists $0 < \delta_1 < t$ such that for every $0 < r \leq \delta_1$

$$P_r(1-\varphi) < \varepsilon \quad \text{on} \quad K .$$

Let $t_0 = t - \delta_1$ and $g = P_{t_0} f$. Then $g \in \mathcal{C}_b(X)$, hence using (2.8) again we find $0 < \delta < \delta_1$ such that for every $0 < r < \delta$

$$|P_r g - g| < \varepsilon \quad \text{on} \quad S(\varphi) .$$

Consider now $t < s < t + \delta$. Since $1 \leq 1 - \varphi + 1_{S(\varphi)}$ and $|P_{s-t} g - g| \leq \max\,(P_{s-t} g, g) \leq 1$ we have

$$\begin{aligned} |P_s f - P_t f| &= |P_{t-t_0}(P_{s-t} g - g)| \\ &\leq P_{t-t_0}(1-\varphi) + P_{t-t_0}(1_{S(\varphi)} |P_{s-t} g - g|) \\ &< \varepsilon + P_{t-t_0}\, \varepsilon \leq 2\varepsilon \quad \text{on} \quad K . \end{aligned}$$

Similarly, if $t - \delta < s < t$ then

$$|P_s f - P_t f| = |P_{s-t_0}(g - P_{t-s} g)| < 2\varepsilon \quad \text{on} \quad K .$$

Thus $P_s f$ converges to $P_t f$ uniformly on K as s tends to t , and the statement follows easily since $P_t f \in \mathcal{C}(X)$.

(2) ⇒ (1): Trivial. ┘

2.11. REMARK. If $\mathbb{P}$ is a strong Feller semigroup on X and $\lim_{t \downarrow 0} P_t f = f$ locally uniformly for every $f \in \mathcal{K}(X)$ then $\lim_{s \downarrow t} P_s f = P_t f$ locally uniformly for every $f \in \mathcal{K}(X)$ and $t > 0$. This follows immediately from (2.8) using $P_s f = P_{s-t} P_t f$.

The following two results will be useful in section 5 and section 3.

2.12. LEMMA. Let $\mathbb{P} = (P_t)_{t>0}$ be a sub-Markov semigroup on X having a strong Feller resolvent $(V_\lambda)_{\lambda>0}$. Let $f \in \mathcal{B}_b(X \times \mathbb{R}_+^*)$ such that, for every $x \in X$, the function $t \mapsto f(x,t)$ is continuous on $\mathbb{R}_+^*$. Then, for every $\lambda > 0$, the function $x \mapsto \int_0^\infty e^{-\lambda t} P_t f_t(x)\,dt$ is continuous on X (where $f_t(y) = f(y,t)$) .

Proof. Fix $\lambda > 0$. We note first that for every $g \in \mathcal{B}_b(X)$ and $0<r<s<\infty$ the functions

$$\int_s^\infty e^{-\lambda t}P_t g\,dt = e^{-\lambda s}V_\lambda P_s g \quad\text{and}\quad \int_r^s e^{-\lambda t}P_t g\,dt = e^{-\lambda r}V_\lambda P_r g - e^{-\lambda s}V_\lambda P_s g$$

are continuous.

For every $n \in \mathbb{N}$ and $1 \le i \le n2^n + 1$ let

$$I_{ni} = \begin{cases}](i-1)2^{-n}, i\,2^{-n}] & \text{if } i \le n2^n, \\ ]n,\infty[& \text{if } i = n2^n+1, \end{cases}$$

define $g_{ni} \in \mathcal{B}_b(X)$ by

$$g_{ni}(x) = \sup\{f(x,t) : t \in I_{ni}\}$$

and $g_n \in \mathcal{B}_b(X \times \mathbb{R}_+^*)$ by

$$g_n(x,t) = \sum_{i=1}^{n2^n+1} 1_{I_{ni}}(t) g_{ni}(x) \quad .$$

Then, for every $n \in \mathbb{N}$, the function

$$\int_0^\infty e^{-\lambda t}P_t(g_n)_t\,dt = \sum_{i=1}^{n2^n+1} \int_{I_{ni}} e^{-\lambda t}P_t g_{ni}\,dt$$

is continuous. The sequence (g_n) is decreasing to f. Therefore the function

$$G := \int_0^\infty e^{-\lambda t}P_t f_t\,dt = \inf \int_0^\infty e^{-\lambda t}P_t(g_n)_t\,dt$$

is u.s.c. on X. Replacing f by $(-f)$ we conclude that G is l.s.c. as well. Thus G is continuous on X. ┘

2.13. COROLLARY. Let $\mathbb{P} = (P_t)_{t>0}$ be a sub-Markov semigroup on X having a strong Feller resolvent. Let $f \in \mathcal{B}_b(X)$ and let g be a numerical function on $\mathbb{R}_+^*$ which is integrable with respect to $\lambda_{\mathbb{R}_+}$. Then the function $x \mapsto \int_0^\infty P_t f(x) g(t)\,dt$ is continuous.

Proof. Fix $\varepsilon > 0$. Then there exists $h \in \mathcal{K}(\mathbb{R}_+^*)$ such that $\int_0^\infty |g-h|\,dt < \varepsilon$ and hence for every $x \in X$

$$\left|\int_0^\infty P_t f(x)g(t)\,dt - \int_0^\infty P_t f(x)h(t)\,dt\right| \le \|f\| \int_0^\infty |g-h|\,dt < \varepsilon\|f\| \quad .$$

Define $H \in \mathcal{B}_b(X \times \mathbb{R}_+^*)$ by $H(x,t) = f(x)e^t h(t)$. Then by (2.12) the function

$$x \mapsto \int_0^\infty e^{-t}P_t H_t(x)\,dt = \int_0^\infty P_t f(x)h(t)\,dt$$

is continuous on X. Hence the assertion follows. ┘

2.14. EXERCISES. 1. Let $(\mathbb{R}, \mathcal{W})$ be the harmonic space given by translation on $\mathbb{R}$, i.e. let $\mathcal{W}$ be the set of all l.s.c. increasing functions $u : \mathbb{R} \to \overline{\mathbb{R}}_+$. Let $\varphi \in C^+(\mathbb{R})$ such that $\int_{-\infty}^{\infty} \varphi(t)dt < \infty$ and define $p \in P_b$ by $p(x) = \int_{-\infty}^{x} \varphi(t)dt$. Then $\mathcal{W}^p$ is the set of all l.s.c. functions $u : \mathbb{R} \to \overline{\mathbb{R}}_+$ such that the function $e^p u$ is increasing. Moreover, $e^{-p} \in \mathcal{H}^p(\mathbb{R})$.

2. Continuation of exercise (IV.7.8.2). $E_{\tilde{\mathbb{P}}} = \mathcal{W}^{Vg}$.

3. Subordination by Convolution Semigroups

A family $(\mu_t)_{t>0}$ of measures on $\mathbb{R}_+^*$ is called a (vaguely continuous) *convolution semigroup* on $\mathbb{R}_+^*$ if the following conditions are satisfied:

(1) $\mu_t(\mathbb{R}_+^*) \leq 1$ for all $t > 0$,

(2) $\mu_s * \mu_t = \mu_{s+t}$ for all $s,t > 0$,

(3) $\lim_{t \to 0} \mu_t = \varepsilon_0$ (vaguely).

Note that (1) and (3) imply that $\lim_{t\to 0} \mu_t(f) = f(0)$ for every $f \in C_b(\mathbb{R}_+)$ (see the proof of (2.8)).

3.1. EXAMPLES. 1. For $\alpha \geq 0$ and $t > 0$ let $\mu_t = e^{-\alpha t}\varepsilon_t$. Then $(\mu_t)_{t>0}$ is obviously a convolution semigroup on $\mathbb{R}_+^*$.

2. For $\alpha \in]0,2]$ there exists a unique convolution semigroup $(\eta_t^\alpha)_{t>0}$ of probability measures on $\mathbb{R}_+^*$ such that for every $t,s > 0$

$$L\eta_t^\alpha(s) = \exp(-ts^{\frac{\alpha}{2}}) \quad .$$

$(\eta_t^\alpha)_{t>0}$ is called *one-sided stable semigroup*. Obviously, $\eta_t^2 = \varepsilon_t$ for every $t > 0$.

To prove the existence of $(\eta_t^\alpha)_{t>0}$, let $t > 0$ and define $f : \mathbb{R}_+^* \to \mathbb{R}_+$ by

$$f(s) = \exp(-ts^{\frac{\alpha}{2}}) \quad .$$

Using $\frac{\alpha}{2} \leq 1$ it is easily verified that for every $m \in \mathbb{N}$ there are $a_{1m}, \ldots, a_{mm} \in \mathbb{R}_+$ such that

$$f^{(m)} = (-1)^m \sum_{l=1}^{m} a_{lm} s^{l\frac{\alpha}{2}-m} f$$

and hence

$$(-1)^m f^{(m)} \geq 0 \quad .$$

By the theorem of Bernstein (I.4.2) there exists a unique measure η_t^α on $\mathbb{R}_+$ such that $L\eta_t^\alpha = f$. Moreover $\eta_t^\alpha(\mathbb{R}_+) = \lim_{s\to 0} f(s) = 1$, $\eta_t^\alpha(\{o\}) = \lim_{s\to\infty} f(s) = 0$, i.e. η_t^α is a probability measure on $]0,\infty[$. Since obviously

$$L\eta_{t_1}^\alpha \cdot L\eta_{t_2}^\alpha = L\eta_{t_1+t_2}^\alpha$$

we obtain that

$$\eta_{t_1}^\alpha * \eta_{t_2}^\alpha = \eta_{t_1+t_2}^\alpha$$

for all $t_1,t_2 > 0$. Finally, $\lim_{t\to 0} L\eta_t^\alpha = 1$ and hence $\lim_{t\to 0} \eta_t^\alpha = \varepsilon_0$, i.e. $(\eta_t^\alpha)_{t>0}$ is a convolution semigroup.

If $\mathbb{P} = (P_t)_{t>0}$ is a measurable sub-Markov semigroup on X and $(\mu_t)_{t>0}$ a convolution semigroup on $\mathbb{R}_+^*$ then for every $t>0$, a kernel P_t^μ on X is defined by

$$P_t^\mu f = \int P_s f \mu_t(ds) \qquad (f \in \mathcal{B}^+(X)) \quad .$$

$\mathbb{P}^\mu = (P_t^\mu)_{t>0}$ is called the *sub-Markov semigroup subordinated to* $\mathbb{P}$ *by means of* $(\mu_t)_{t>0}$. Since for all $t_1,t_2 > 0$ and $f \in \mathcal{B}^+(\mathbb{R}^n)$

$$\begin{aligned} P_{t_1}^\mu P_{t_2}^\mu f &= \int_0^\infty P_{s_1}(P_{t_2}^\mu f)\mu_{t_1}(ds_1) \\ &= \int_0^\infty (\int_0^\infty P_{s_1} P_{s_2} f \, \mu_{t_2}(ds_2))\mu_{t_1}(ds_1) = \int_0^\infty\int_0^\infty P_{s_1+s_2} f \, \mu_{t_2}(ds_2)\mu_{t_1}(ds_1) \\ &= \int_0^\infty P_s f(\mu_{t_1} * \mu_{t_2})(ds) = \int_0^\infty P_s f \, \mu_{t_1+t_2}(ds) = P_{t_1+t_2}^\mu f \quad , \end{aligned}$$

$\mathbb{P}^\mu$ is indeed a semigroup which is certainly sub-Markov. Moreover,

$$E_{\mathbb{P}} \subset E_{\mathbb{P}^\mu} :$$

Fix $u \in E_{\mathbb{P}}$. Then obviously $P_t^\mu u \leq u$ for every $t > 0$. Let $x \in X$, $a < u(x)$ and $0 < b < 1$. Then there exist $s_0 > 0$ and $t_0 > 0$ such that $P_s u(x) > a$ for every $0 < s < s_0$ and $\mu_{t_0}(]0,s_0[) > b$, hence

$$P^{\mu}_{t_0} u(x) \geq \int_0^{s_0} P_s u(x) \mu_{t_0}(ds) \geq ab \quad .$$

This implies that $u \in E_{\mathbb{P}^{\mu}}$.

3.2. EXAMPLES. 1. Let $\mathbb{P} = (P_t)_{t>0}$ be a measurable sub-Markov semigroup on X . If $\mu_t = e^{-\alpha t} \varepsilon_t$, $t > 0$, for some $\alpha \geq 0$, then $\mathbb{P}^{\mu} = (P^{\alpha}_t)_{t>0}$, where $P^{\alpha}_t = e^{-\alpha t} P_t$, $t > 0$ (see (II.3.5.4)).

2. Let $\mathbb{P} = (P_t)_{t>0}$ be the Brownian semigroup on $\mathbb{R}^n$ and $(\eta^{\alpha}_t)_{t>0}$ be the one-sided stable semigroup for some $\alpha \in]0,2]$. Then $\mathbb{P}^{\alpha} = (P^{\alpha}_t)_{t>0}$, where

$$P^{\alpha}_t f = \int P_s f \, \eta^{\alpha}_t(ds) \qquad (f \in \mathcal{B}^+(\mathbb{R}^n))$$

is defined in a manner not to be confused with the notation of (1).

$\mathbb{P}^{\alpha}$ is called the *symmetric stable semigroup of index* α on $\mathbb{R}^n$. For every $t > 0$, $P^{\alpha}_t 1 = \int 1 \, d\eta^{\alpha}_t = 1$, hence $\mathbb{P}^{\alpha}$ is a Markov semigroup. Note that $\mathbb{P}^2 = \mathbb{P}$.

3.3. PROPOSITION. Let $\mathbb{P} = (P_t)_{t>0}$ be a strong Feller sub-Markov semigroup on X and let $(\mu_t)_{t>0}$ be a convolution semigroup on $\mathbb{R}^*_+$. Then:

1. $\mathbb{P}^{\mu}$ is a strong Feller semigroup.

2. If $\lim_{t\to 0} P_t f = f$ for every $f \in K(X)$, if there are strictly positive $u,v \in E_{\mathbb{P}^{\mu}} \cap C(X)$ such that $\frac{u}{v} \in C_0(X)$ and if $E_{\mathbb{P}^{\mu}}$ is linearly separating or the potential kernel of $\mathbb{P}^{\mu}$ is proper, then $(X, E_{\mathbb{P}^{\mu}})$ is a balayage space.

3. If $(X, E_{\mathbb{P}})$ is a balayage space, then $(X, E_{\mathbb{P}^{\mu}})$ is a balayage space.

Proof. 1. Since $P_s(\mathcal{B}_b(X)) \subset C_b(X)$ for every $s > 0$ we obtain by Lebesgue's convergence theorem that $P^{\mu}_t(\mathcal{B}_b(X)) \subset C_b(X)$ for every $t > 0$.

2. It suffices to note that $\lim_{t\to 0} P^{\mu}_t f = f$ for every $f \in K(X)$ and to apply (2.4).

3. Consequence of (1) and (2). ┘

3.4. EXAMPLE. Let $\mathbb{P}^{\alpha}$ be the symmetric stable semigroup of index $\alpha \in]0,2]$ on $\mathbb{R}^n$, $0 < \alpha < n$. Then $(\mathbb{R}^n, E_{\mathbb{P}^{\alpha}})$ is a balayage space as we shall prove in a moment.

In the next section we shall study the potential theory of $(\mathbb{R}^n, E_{\mathbb{P}^{\alpha}})$ in more detail. In particular, we shall see that its potentials are just the Riesz potentials of order α .

In order to establish that $(\mathbb{R}^n, E_{\mathbb{P}^\alpha})$ is a balayage space we apply (2.1.2) and (3.3.3) if $n \geq 3$. The general case will follow from (3.3.2). Let us first determine the potential kernel V_α of $\mathbb{P}^\alpha$.

We define a measure κ_α on $\mathbb{R}_+$ by

$$\kappa_\alpha = \int_0^\infty \eta_t^\alpha \, dt \quad .$$

Then

$$L\kappa_\alpha(s) = \int_0^\infty L\eta_t^\alpha(s)dt = \int_0^\infty \exp(-t\, s^{\frac{\alpha}{2}})dt = s^{-\frac{\alpha}{2}} \quad .$$

On the other hand for every $s > 0$

$$\int_0^\infty e^{-ts}\, t^{\frac{\alpha}{2}-1} dt = s^{-\frac{\alpha}{2}} \int_0^\infty e^{-t}\, t^{\frac{\alpha}{2}-1} dt = \Gamma(\tfrac{\alpha}{2}) s^{-\frac{\alpha}{2}} \quad .$$

Therefore κ_α has the density

$$t \mapsto \frac{1}{\Gamma(\frac{\alpha}{2})} t^{\frac{\alpha}{2}-1}$$

with respect to the Lebesgue measure $\lambda_{\mathbb{R}_+^*}$. Consequently, for every $f \in \mathcal{B}^+(\mathbb{R}^n)$,

$$V_\alpha f = \int_0^\infty P_t^\alpha f\, dt = \int_0^\infty (\int_0^\infty P_s f\, \eta_t^\alpha(ds))dt = \int_0^\infty g_s * f\, \kappa_\alpha(ds) = k_\alpha * f$$

where

$$k_\alpha(x) = \int_0^\infty g_s(x)\kappa_\alpha(ds) = \frac{1}{\Gamma(\frac{\alpha}{2})} \int_0^\infty (2\pi s)^{-\frac{n}{2}} e^{-\frac{\|x\|^2}{2s}} s^{\frac{\alpha}{2}-1} ds$$

$$= \frac{\|x\|^{\alpha-n}}{\Gamma(\frac{\alpha}{2}) 2^{\alpha/2} \pi^{n/2}} \int_0^\infty s^{\frac{n-\alpha}{2}-1} e^{-s} ds = c_n^\alpha \|x\|^{\alpha-n}$$

with

$$c_n^\alpha = \frac{\Gamma(\frac{n-\alpha}{2})}{\Gamma(\frac{\alpha}{2}) 2^{\alpha/2} \pi^{n/2}} \quad .$$

V_α is a proper kernel since for every $r > 0$

$$\int_{B_r} \|y\|^{\alpha-n} \lambda^n(dy) = \frac{\pi^{n/2}}{\Gamma(\frac{n}{2}+1)} \int_0^r n\rho^{n-1} \rho^{\alpha-n} d\rho = \frac{\pi^{n/2}}{\Gamma(\frac{n}{2}+1)} \cdot \frac{n}{\alpha} r^\alpha \quad .$$

For every $x \in \mathbb{R}^n$ we define $k_\alpha^x : \mathbb{R}^n \to \overline{\mathbb{R}}_+$ by

$$k_\alpha^x(y) = k_\alpha(x-y) \quad .$$

Then $k_\alpha^x \in E_{\mathbb{P}^\alpha}$. Indeed, there exists a sequence (φ_m) in $\mathcal{K}^+(\mathbb{R}^n)$ such that

$\lambda^n(\varphi_m) = 1$ and $S(\varphi_m) \subset B_{\frac{1}{m}}(x)$ for every $m \in \mathbb{N}$. Then

$$\lim_{m\to\infty} k_\alpha * \varphi_m = k_\alpha^x .$$

Since $k_\alpha * \varphi_m = V_\alpha \varphi_m \in E_{\mathbb{P}^\alpha}$, we obtain that for every $t > 0$

$$P_t^\alpha k_\alpha^x \leq \liminf_{m\to\infty} P_t^\alpha(k_\alpha * \varphi_m) \leq \liminf_{m\to\infty} k_\alpha * \varphi_m = k_\alpha^x ,$$

i.e. $k_\alpha^x \in S_{\mathbb{P}^\alpha}$. Furthermore, k_α^x is continuous and $\lim_{t\to 0} P_t^\alpha f = f$ for every $f \in K^+(X)$. Hence $k_\alpha^x \in E_{\mathbb{P}^\alpha}$ and $\inf(k_\alpha^x, 1) \in E_{\mathbb{P}^\alpha} \cap C_0(\mathbb{R}^n)$.

In particular, we conclude from (3.3.2) that $(\mathbb{R}^n, E_{\mathbb{P}^\alpha})$ is a balayage space.

3.5. REMARK. Let $\mathbb{P}^\alpha$ be the symmetric stable semigroup of index α on $\mathbb{R}^n$, $0 < \alpha < n$. The balayage space $(\mathbb{R}^n, E_{\mathbb{P}^\alpha})$ has the following properties:

1. $\lambda^n(U) > 0$ for every non-empty finely open $U \in \mathcal{B}(\mathbb{R}^n)$.
2. Every countable subset of $\mathbb{R}^n$ is finely closed.
3. $E_{\mathbb{P}^\alpha} = \bigcap_{0<\beta<\alpha} E_{\mathbb{P}^\beta}$.

Proof. (1) and (2) follow in the same way as the corresponding properties for the Brownian semigroup (see (II.4.10.1)).

In order to prove (3) we note first that for every $0 < \beta < \alpha$, $k_\alpha = k_\beta * k_{\alpha-\beta}$. Indeed, for every $s > 0$

$$L\kappa_\beta(s)\, L\kappa_{\alpha-\beta}(s) = s^{-\frac{\beta}{2}} s^{-\frac{\alpha-\beta}{2}} = s^{-\frac{\alpha}{2}} = L\kappa_\alpha(s) .$$

Hence $\kappa_\beta * \kappa_{\alpha-\beta} = \kappa_\alpha$. Since $k_\gamma = \int_0^\infty g_s \kappa_\gamma(ds)$ the equation $k_\beta * k_{\alpha-\beta} = k_\alpha$ follows easily.

Now let $v \in E_{\mathbb{P}^\alpha}$ and $0 < \beta < \alpha$. By (II.3.11), there exists a sequence (f_m) in $\mathcal{B}^+(\mathbb{R}^n)$ such that $k_\alpha * f_m \uparrow v$. Then for every $m \in \mathbb{N}$

$$k_\alpha * f_m = k_\beta * (k_{\alpha-\beta} * f_m) \in E_{\mathbb{P}^\beta}$$

and hence $v \in E_{\mathbb{P}^\beta}$.

Suppose now conversely that $v \in \mathcal{B}_b^+(X)$ such that $v \in E_{\mathbb{P}^\beta}$ for every $0 < \beta < \alpha$. Let $t > 0$. Then

$$\lim_{\beta\to\alpha} L\eta_t^\beta = L\eta_t^\alpha$$

and hence

$$\lim_{\beta\to\alpha} \eta_t^\beta = \eta_t^\alpha \ .$$

For every $x \in \mathbb{R}^n$, the function $s \mapsto P_s v(x)$ is continuous and bounded on $]0,\infty[$. Hence

$$P_t^\alpha v = \int_0^\infty P_s v\, \eta_t^\alpha(ds) = \lim_{\beta\to\alpha} \int_0^\infty P_s v\, \eta_t^\beta(ds) = \lim_{\beta\to\alpha} P_t^\beta v \leq v \ .$$

Therefore $v \in S_{\mathbb{P}^\alpha}$. Since v is l.s.c., $v \in E_{\mathbb{P}^\alpha}$.

If v is an arbitrary function in $\bigcap_{0<\beta<\alpha} E_{\mathbb{P}^\beta}$ then $v_m = \inf (v,m) \in E_{\mathbb{P}^\beta}$ for every $m \in \mathbb{N}$ and $0 < \beta < \alpha$, hence $v_m \in E_{\mathbb{P}^\alpha}$ for every $m \in \mathbb{N}$ and $v = \sup v_m \in E_{\mathbb{P}^\alpha}$. ┘

We recall from chapter II that for every balayage space $(X,\mathcal{W})$ satisfying $1 \in \mathcal{W}$ there exists a sub-Markov semigroup $\mathbb{P}$ such that $\mathbb{P}$ has a strong Feller resolvent and $E_{\mathbb{P}} = \mathcal{W}$. Thus the following propositions show that the construction of a new balayage space from a given balayage space by means of subordination by a convolution semigroup $(\mu_t)_{t>0}$ on $\mathbb{R}_+^*$ is possible if the measure $\kappa = \int_0^\infty \mu_t dt$ is absolutely continuous with respect to the Lebesgue measure $\lambda_{\mathbb{R}_+^*}$ on $\mathbb{R}_+^*$. These results prove again that $(\mathbb{R}^n, E_{\mathbb{P}^\alpha})$ is a balayage space for the symmetric stable semigroup of index α on $\mathbb{R}^n$.

3.6. PROPOSITION. Let $(\mu_t)_{t>0}$ be a convolution semigroup on $\mathbb{R}_+^*$ and let $\kappa = \int_0^\infty \mu_t dt$. Then the following statements are equivalent:

(1) For every space X and every sub-Markov semigroup $\mathbb{P}$ on X with strong Feller resolvent the semigroup $\mathbb{P}^\mu$ has a strong Feller resolvent.

(2) For every $A \in \mathcal{B}(\mathbb{R})$ the function $1_A * \kappa$ is l.s.c.

(3) The measure κ is absolutely continuous with respect to the Lebesgue measure $\lambda_{\mathbb{R}_+^*}$.

Proof. (1) ⇒ (2): Let $A \in \mathcal{B}(\mathbb{R})$. Then for every $x \in \mathbb{R}$

$$1_A * \kappa(x) = \int_0^\infty 1_A(x-s)\kappa(ds) = \int_0^\infty \int_0^\infty T_s 1_A(x)\mu_t(ds)dt$$
$$= \int_0^\infty T_t^\mu 1_A(x)dt = \sup_{\alpha>0} \int_0^\infty e^{-\alpha t} T_t^\mu 1_A(x)dt$$

where $(T_t)_{t>0}$ is the translation semigroup $\mathbb{T}$ defined in (II.3.1.2). Since $\mathbb{T}$ has a strong Feller resolvent, the function $x \mapsto \int_0^\infty e^{-\alpha t} T_t^\mu 1_A(x)dt$ is by assumption continuous for each $\alpha > 0$. Therefore $1_A * \kappa$ is l.s.c.

(2) ⇒ (3): Let $A \in \mathcal{B}(\mathbb{R}_+)$ be a null set with respect to $\lambda_{\mathbb{R}_+}$. Then

$$\int_{\mathbb{R}} 1_{-A} * \kappa(x)dx = \int_{\mathbb{R}} \int_0^\infty 1_{-A}(x-s)\kappa(ds)dx = \int_0^\infty \int_{\mathbb{R}} 1_{-A}(x-s)dx\,\kappa(ds) = 0 .$$

Since $1_{-A} * \kappa$ is l.s.c., we conclude that $1_{-A} * \kappa = 0$. In particular,

$$\kappa(A) = 1_{-A} * \kappa(0) = 0 .$$

Hence κ is absolutely continuous with respect to $\lambda_{\mathbb{R}_+^*}$.

(3) ⇒ (1): Let $\mathbb{P} = (P_t)_{t>0}$ be a sub-Markov semigroup on a space X with strong Feller resolvent. Fix $\alpha > 0$. Since the finite measure $\nu = \int_0^\infty e^{-\alpha t}\mu_t dt$ is majorized by κ, we obtain by the theorem of Radon-Nikodym that ν has an integrable density g with respect to $\lambda_{\mathbb{R}_+^*}$. Fix $f \in \mathcal{B}_b(X)$. Then by (2.13) the function $\int_0^\infty P_s f g(s)ds$ is continuous. Since

$$\int_0^\infty P_s fg(s)ds = \int_0^\infty e^{-\alpha t}\left(\int_0^\infty P_s f\,\mu_t(ds)\right)dt = \int_0^\infty e^{-\alpha t}\, P_t^\mu f\,dt \quad ,$$

we finally conclude that $\mathbb{P}^\mu$ has a strong Feller resolvent. ┘

<u>3.7. COROLLARY.</u> Let $(\mu_t)_{t>0}$ be a convolution semigroup on $\mathbb{R}_+^*$ and let $\kappa = \int_0^\infty \mu_t\,dt$. Then the following statements are equivalent:

(1) For every space X and every sub-Markov semigroup $\mathbb{P}$ on X such that $(X, E_{\mathbb{P}})$ is a balayage space and $\mathbb{P}$ has a strong Feller resolvent, $(X, E_{\mathbb{P}^\mu})$ is a balayage space.

(2) $(\mathbb{R}, E_{\mathbb{T}^\mu})$ is a balayage space.

(3) The measure κ is absolutely continuous with respect to Lebesgue measure on $\mathbb{R}_+^*$.

Proof. (1) ⇒ (2): Trivial.

(2) ⇒ (3): Let $A \in \mathcal{B}(\mathbb{R})$. Then

$$1_A * \kappa = \int_0^\infty T_t^\mu 1_A\,dt \in E_{\mathbb{T}^\mu} \quad ,$$

hence $1_A * \kappa$ is l.s.c. Therefore (3) follows from (3.6).

(3) $\Rightarrow$ (1): Let $(X, E_{\mathbb{P}})$ be a balayage space such that the sub-Markov semigroup $\mathbb{P}$ has a strong Feller resolvent. Then $\mathbb{P}^\mu$ has a strong Feller resolvent by (3.6). Using (2.4) and the fact that $E_{\mathbb{P}} \subset E_{\mathbb{P}^\mu}$ we obtain that $(X, E_{\mathbb{P}^\mu})$ is a balayage space. ┘

<u>3.8. EXAMPLE.</u> Let $\alpha \in]0,2[$ and let $(\eta_t^\alpha)_{t>0}$ be the corresponding one-sided stable semigroup defined in (3.1). By (3.4) the measure $\int_0^\infty \eta_t^\alpha dt$ has the density $t \mapsto (\Gamma(\frac{\alpha}{2}))^{-1} t^{\frac{\alpha}{2}-1}$ with respect to $\lambda_{\mathbb{R}_+^*}$. Hence we may apply the preceding results to $(\eta_t^\alpha)_{t>0}$. In particular, $(\mathbb{R}, E_{\mathbb{T}^{\eta^\alpha}})$ is a balayage space. We note that for every $f \in \mathcal{B}_b(\mathbb{R})$, $t > 0$ and $x \in \mathbb{R}$,

$$T_t^{\eta^\alpha} f(x) = \int_0^\infty f(x-s)\eta_t^\alpha(ds) = f * \eta_t^\alpha(x) \quad .$$

<u>3.9. EXERCISES.</u> 1. Let $\mathbb{P}$ be a strong Feller semigroup on X and $\alpha, \beta \in]0,2]$. Then $(\mathbb{P}^\alpha)^\beta = \mathbb{P}^{\frac{\alpha\beta}{2}}$.

2. Let $\mu_t = \gamma_t \lambda_{\mathbb{R}_+^*}$, $t > 0$, where $\gamma_t(s) = (\Gamma(t))^{-1} s^{t-1} e^{-s}$. Then $(\mu_t)_{t>0}$ is a convolution semigroup on $\mathbb{R}_+^*$ (Γ-*semigroup*).

<u>4. Riesz Potentials</u>

In this section we continue the study of the balayage space $(\mathbb{R}^n, E_{\mathbb{P}^\alpha})$ introduced in (3.4) where $\mathbb{P}^\alpha$ is the symmetric stable semigroup of index α on $\mathbb{R}^n$, $n \geq 1$, $0 < \alpha \leq 2$, $\alpha < n$. Notions with respect to $(\mathbb{R}^n, E_{\mathbb{P}^\alpha})$ will be distinguished by adding the letter "α" : ${}^\alpha\mathcal{S}^+(\mathbb{R}^n)$, ${}^\alpha\mathcal{H}^+(U)$, ${}^\alpha R_V^{\complement U}$, α-superharmonic, α-harmonic etc.

First of all we determine the associated harmonic kernels of open balls. If $\alpha = 2$ (and $n \geq 3$) then these kernels are given by the Poisson integral (see section 0.1) since $\mathbb{P}^2$ is the Brownian semigroup on $\mathbb{R}^n$. So assume for the moment that $\alpha < 2$.

Let $\mathcal{U}$ be the family of all open balls $B_r(a)$. We shall write B_r instead of $B_r(0)$. For every $U = B_r(a)$ we define a sweeping kernel H_U^α by

$$H_U^\alpha f(x) = \int_{\complement U} f\,{}^\alpha p_x^U \, d\lambda^n \quad (f \in \mathcal{B}^+(X)\,, x \in U)$$

where the density ${}^\alpha p_x^U$ is defined by

$${}^\alpha p_x^U(y) = a_\alpha \frac{(r^2 - \|x-a\|^2)^{\frac{\alpha}{2}}}{(\|y-a\|^2 - r^2)^{\frac{\alpha}{2}}} \| y-x \|^{-n} \quad , \ \|x-a\| < r \leq \|y-a\|$$

and

$$a_\alpha = \pi^{-(\frac{n}{2}+1)} \Gamma(\tfrac{n}{2}) \sin \tfrac{\alpha\pi}{2} \ .$$

We shall see that H_U^α is the sweeping kernel corresponding to U and $(\mathbb{R}^n, E_{\mathbb{P}^\alpha})$. Let us note first that the family $(H_U^\alpha)_{U\in\mathcal{U}}$ as well as $E_{\mathbb{P}^\alpha}$ is invariant under translations (and rotations) of $\mathbb{R}^n$.

<u>4.1. PROPOSITION.</u> For every $U \in \mathcal{U}$, $H_U^\alpha 1 = 1$.

Proof. Because of the preceding remark it suffices to consider $U = B_r$ for some $r > 0$. Let $x \in B_r$. We have to show that

$$I_r(x) := \int_{\{\|y\| \geq r\}} \frac{\lambda^n(dy)}{(\|y\|^2 - r^2)^{\frac{\alpha}{2}} \|y-x\|^n} = \frac{1}{a_\alpha (r^2 - \|x\|^2)^{\frac{\alpha}{2}}} \ .$$

Using the normed surface measures σ_ρ on B_ρ^* we have

$$I_r(x) = \int_r^\infty \frac{\pi^{\frac{n}{2}}}{\Gamma(\frac{n}{2}+1)} \cdot n\rho^{n-1} \left(\int \frac{\sigma_\rho(dy)}{(\|y\|^2 - r^2)^{\frac{\alpha}{2}} \|y-x\|^n}\right) d\rho$$

$$= \frac{2\pi^{\frac{n}{2}}}{\Gamma(\frac{n}{2})} \int_r^\infty \frac{\rho^{n-1}}{(\rho^2 - r^2)^{\frac{\alpha}{2}}} \left(\int \frac{\sigma_\rho(dy)}{\|y-x\|^n}\right) d\rho \ .$$

By (0.1.2)

$$\int \frac{\rho^{n-2}(\rho^2 - \|x\|^2)}{\|y-x\|^n} \sigma_\rho(dy) = 1$$

for every $\rho > \|x\|$. Hence

$$I_r(x) = \frac{\pi^{\frac{n}{2}}}{\Gamma(\frac{n}{2})} \int_r^\infty \frac{2\rho \, d\rho}{(\rho^2 - r^2)^{\frac{\alpha}{2}} (\rho^2 - \|x\|^2)} \ .$$

Transforming the integral by $s = \frac{\rho^2 - r^2}{r^2 - \|x\|^2}$ and then by $t = \frac{s}{s+1}$ we obtain

$$\int_r^\infty \frac{2\rho\, d\rho}{(\rho^2-r^2)^{\frac{\alpha}{2}}(\rho^2-\|x\|^2)} = \frac{1}{(r^2-\|x\|^2)^{\frac{\alpha}{2}}}\int_0^\infty \frac{ds}{s^{\frac{\alpha}{2}}(s+1)}$$

$$= \frac{1}{(r^2-\|x\|^2)^{\frac{\alpha}{2}}}\int_0^1 t^{\frac{\alpha}{2}-1}(1-t)^{-\frac{\alpha}{2}}\,dt \;.$$

The last integral is

$$B(\tfrac{\alpha}{2}, 1-\tfrac{\alpha}{2}) = \frac{\Gamma(\frac{\alpha}{2})\Gamma(1-\frac{\alpha}{2})}{\Gamma(1)} = \frac{\pi}{\sin\frac{\pi\alpha}{2}}$$

where B is the beta function. Thus we have proved that indeed

$$I_r(x) = \frac{\pi^{\frac{n}{2}+1}}{\Gamma(\frac{n}{2})\sin\frac{\pi\alpha}{2}}\cdot(r^2-\|x\|^2)^{-\frac{\alpha}{2}} \;.$$ ┘

In the following two proofs we shall use a special transformation. Given $x \in \mathbb{R}^n$ and $r_0 > 0$ the transformation at $B_{r_0}(x)$ is defined by

$$w(y) = x - \frac{r_0^2}{\|y-x\|^2}(y-x) \qquad (y \in \mathbb{R}^n \setminus \{x\}) \quad ,$$

i.e. $w(y) \in \mathbb{R}^n \setminus \{x\}$ such that x lies on the open segment from y to $w(y)$ and

$$\|w(y)-x\|\cdot\|y-x\| = r_0^2 \;.$$

In particular, $w = w^{-1}$.

Let us first calculate the Jacobi determinant $\det Dw$ of w . Since

$$D_i w_j(y) = \begin{cases} \dfrac{r_0^2}{\|y-x\|^4}(\|y-x\|^2 - 2(y_i-x_i)^2) \;, & i = j \;, \\[2ex] \dfrac{r_0^2}{\|y-x\|^4}\cdot(-2)(y_i-x_i)(y_j-x_j) \;, & i \neq j \end{cases}$$

we obtain

$$|\det Dw\,(y)| = \frac{r_0^{2n}}{\|y-x\|^{2n}}|\det(\alpha_{ij})|$$

where defining $\alpha_i = \dfrac{y_i - x_i}{\|y-x\|}$ we have $\sum_{i=1}^n \alpha_i^2 = 1$ and

$$\alpha_{ij} = \begin{cases} 1 - 2\alpha_i^2 & , \quad i = j \\ -2\alpha_i\alpha_j & , \quad i \neq j \;. \end{cases}$$

It is easily checked that the columns of (α_{ij}) are orthonormal and hence $|\det(\alpha_{ij})| = 1$. Thus

$$|\det Dw(y)| = \frac{r_0^{2n}}{\|y-x\|^{2n}} = \frac{\|w(y) - x\|^n}{\|y-x\|^n} \quad .$$

Let $y,z \in \mathbb{R}^n \setminus \{x\}$. Then

$$\|w(y) - w(z)\|^2 = \|(w(y) - x) - (w(z) - x)\|^2$$

$$= \frac{r_0^4}{\|y-x\|^2\|z-x\|^2} (\|z-x\|^2 - 2<y-x,z-x> + \|y-x\|^2)$$

$$= \frac{r_0^4}{\|y-x\|^2\|z-x\|^2} \cdot \|y-z\|^2$$

and hence

$$\|w(y) - w(z)\| = \frac{r_0^2}{\|y-x\|\,\|z-x\|}\|y-z\|$$

(which could be used to deduce $|\det Dw(y)|$) .

If $r > 0$ such that $r^2 = r_0^2 + \|x\|^2$ then for every $y \in \mathbb{R}^n \setminus \{x\}$,

$$\|w(y)\|^2 - r^2 = \frac{\|w(y) - x\|^2}{r_0^2}(r^2 - \|y\|^2) \quad .$$

Indeed, let e be the unit vector $\frac{y-x}{\|y-x\|}$, $\alpha = \|y-x\|$ and $\beta = \|w(y) - x\|$. Then $y = x + \alpha e$, $w(y) = x - \beta e$, $\alpha\beta = r_0^2$ and

$$\|w(y)\|^2 - r^2 = \|x - \beta e\|^2 - r^2 = \beta^2 - 2\beta<x,e> - r_0^2$$

$$= \frac{\beta^2}{r_0^2}(r_0^2 - 2\alpha<x,e> - \alpha^2) = \frac{\beta^2}{r_0^2}(r^2 - \|x + \alpha e\|^2) = \frac{\|w(y) - x\|^2}{r_0^2}(r^2 - \|y\|^2) \; .$$

4.2. LEMMA. For every $r > 0$ and $x \in \overline{B}_r$,

$$a_\alpha \int_{B_r} k_\alpha^x(y)(r^2 - \|y\|^2)^{-\frac{\alpha}{2}} \lambda^n(dy) = 1 \quad .$$

Proof. For every $x \in \overline{B}_r$ and $y \in B_r$ let

$$\varphi(x,y) = a_\alpha k_\alpha^x(y)\,(r^2 - \|y\|^2)^{-\frac{\alpha}{2}} \quad .$$

If $x \in B_r$ and w is the transformation at $B_{r_0}(x)$ where $r_0^2 = r^2 - \|x\|^2$ then by (4.1)

$$\int_{B_r} \varphi(x,y)\lambda^n(dy) = a_\alpha \int_{\complement B_r} \frac{|\det Dw(y)|\lambda^n(dy)}{(r^2 - \|w(y)\|^2)^{\frac{\alpha}{2}}\|w(y) - x\|^{n-\alpha}}$$

$$= a_\alpha \int_{\complement B_r} \frac{(r^2 - \|x\|^2)^{\frac{\alpha}{2}} \lambda^n(dy)}{(\|y\|^2 - r^2)^{\frac{\alpha}{2}} \|y - x\|^n} = 1 .$$

Now fix $x \in B_r^*$. For every $0 < \delta < \frac{1}{2}$ let $x_\delta = (1 - 2\delta)x$. Since $\lim_{\delta \to 0} \varphi(x_\delta, y) = \varphi(x,y)$ for every $y \in B_r$, we clearly obtain by Fatou's lemma that

$$\int_{B_r} \varphi(x,y) \lambda^n(dy) \leq 1 .$$

Moreover, $\|y - x\| \leq \|y - x_\delta\| + \|x_\delta - x\| = \|y - x_\delta\| + 2\delta r \leq 3\|y - x_\delta\|$ if $\|y - x_\delta\| \geq \delta r$ and hence

$$\varphi(x_\delta, y) \leq 3^{n-\alpha} \varphi(x,y)$$

for all $y \in B_r \setminus B_{\delta r}(x_\delta)$. Finally $r^2 - \|y\|^2 = (r + \|y\|)(r - \|y\|) \geq r \cdot \delta r$ if $\|y - x_\delta\| < \delta_r$ and therefore by (3.4)

$$\int_{B_{\delta r}(x_\delta)} \varphi(x,y)\lambda^n(dy) \leq a_\alpha(\delta r^2)^{-\frac{\alpha}{2}} \int_{B_{\delta r}(x_\delta)} \|y - x_\delta\|^{\alpha - n} \lambda^n(dy) = \frac{2}{\alpha\pi} \sin \frac{\alpha\pi}{2} \delta^{\frac{\alpha}{2}} .$$

Thus an application of Lebesgue's theorem finishes the proof. ┘

<u>4.3. PROPOSITION.</u> For every $U \in \mathcal{U}$ and $z \in \complement U$, $H_U^\alpha k_\alpha^z = k_\alpha^z$.

Proof. It suffices to consider $U = B_r$ for some $r > 0$. Fix $z \in \complement U$ and $x \in U$. As before let w be the transformation at $B_{r_0}(x)$, $r_0^2 = r^2 - \|x\|^2$. Using

$$\| w(y)-z\| = \| w(y)-w(w(z))\| = \frac{r^2 - \|x\|^2}{\| y-x\| \, \| w(z)-x\|} \| y-w(z)\|$$

$$= \frac{\| z-x\| \, \| y-w(z)\|}{\| y-x\|}$$

we obtain by (4.2) that

$$\int \frac{H_U^\alpha(x,dy)}{\|y-z\|^{n-\alpha}} = a_\alpha(r^2 - \|x\|^2)^{\frac{\alpha}{2}} \int_{\complement U} \frac{\lambda^n(dy)}{(\| y \|^2 - r^2)^{\frac{\alpha}{2}} \| y-x\|^n \| y-z\|^{n-\alpha}}$$

$$= a_\alpha(r^2 - \| x \|^2)^{\frac{\alpha}{2}} \int_U \frac{|\det Dw(y)| \lambda^n(dy)}{(\| w(y)\| - r^2)^{\frac{\alpha}{2}} \| w(y)-x\|^n \| w(y)-z\|^{n-\alpha}}$$

$$= \frac{a_\alpha}{\| x-z\|^{n-\alpha}} \int_U \frac{\lambda^n(dy)}{(r^2 - \| y\|^2)^{\frac{\alpha}{2}} \| y-w(z)\|^{n-\alpha}} = \frac{1}{\| x-z\|^{n-\alpha}} .$$ ┘

<u>4.4. THEOREM.</u> $(H_U^\alpha)_{U\in\mathcal{U}}$ is a family of harmonic kernels on $\mathbb{R}^n$ associated with $(\mathbb{R}^n, E_{\mathbb{P}\alpha})$. The constants are α-harmonic on $\mathbb{R}^n$ and for every $z\in\mathbb{R}^n$ the function k_α^z is an α-potential on $\mathbb{R}^n$ which is α-harmonic on $\complement\{z\}$.

Proof. Let $U\in\mathcal{U}$, $f\in\mathcal{B}^+(\mathbb{R}^n)$ and

$$v = V_\alpha f = \int k_\alpha^y f(y)\,\lambda^n(dy) .$$

We claim that $H_U^\alpha v = {}^\alpha R_v^{\complement U}$. Indeed, defining $\tilde f\in\mathcal{B}^+(\mathbb{R}^n)$ by

$$\tilde f(y) = \begin{cases} 0 & , \ y\in U \\ f(y)+\int_U {}^\alpha p_z^U(y) f(z)\lambda^n(dz) & , \ y\in\complement U \end{cases}$$

we know by (II.7.1) that ${}^\alpha R_{V_\alpha\tilde f}^{\complement U} = V_\alpha\tilde f$. If $x\in\complement U$ then by (4.3) for every $z\in\mathbb{R}^n$

$$\int_{\complement U} k_\alpha(x-y)\ {}^\alpha p_z^U(y)\ \lambda^n(dy) = H_U^\alpha k_\alpha^x(z) = k_\alpha^x(z) = k_\alpha(x-z) ,$$

and hence

$$V_\alpha\tilde f(x) = \int_{\complement U} k_\alpha(x-y)f(y)\ \lambda^n(dy) + \int_{\complement U} k_\alpha(x-y)\Big(\int_U {}^\alpha p_z^U(y)f(z)\lambda^n(dz)\Big)\lambda^n(dy)$$

$$= \int_{\complement U} k_\alpha(x-y)f(y)\lambda^n(dy) + \int_U k_\alpha(x-z)f(z)\lambda^n(dz) = V_\alpha f(x) .$$

Therefore

$$V_\alpha\tilde f = {}^\alpha R_{V_\alpha\tilde f}^{\complement U} = {}^\alpha R_v^{\complement U} ,$$

in particular ${}^\alpha R_v^{\complement U}\in E_{\mathbb{P}\alpha}$, and using (4.3) again

$$H_U^\alpha v = H_U^\alpha V_\alpha\tilde f = H_U^\alpha\Big(\int k_\alpha^y\,\tilde f(y)\lambda^n(dy)\Big)$$

$$= \int H_U^\alpha k_\alpha^y\,\tilde f(y)\lambda^n(dy) = \int k_\alpha^y\,\tilde f(y)\lambda^n(dy) = V_\alpha\tilde f = {}^\alpha R_v^{\complement U} .$$

Now let p be a continuous real α-potential on $\mathbb{R}^n$. By (II.3.11), there exists a sequence (f_n) in $\mathcal{B}_b^+(\mathbb{R}^n)$ such that $(V_\alpha f_n)$ is increasing to p. Then

$$H_U^\alpha p = \sup H_U^\alpha(V_\alpha f_n) = \sup {}^\alpha R_{V_\alpha f_n}^{\complement U} = {}^\alpha R_p^{\complement U} .$$

Thus $(H_U^\alpha)_{U\in\mathcal{U}}$ is a family of harmonic kernels on $\mathbb{R}^n$ which is associated with $(\mathbb{R}^n, E_{\mathbb{P}\alpha})$.

Fix $z\in\mathbb{R}^n$. By (4.3) and (III.4.4), the function k_α^z is α-hyperharmonic on $\mathbb{R}^n$ and α-harmonic on $\mathbb{R}^n\setminus\{z\}$. Let $U\in\mathcal{U}$. If $z\in U$ then $H_U^\alpha k_\alpha^z|_U\in C(U)$ since k_α^z is bounded on $\complement U$. If $z\in\complement U$ then $H_U^\alpha k_\alpha^z|_U = k_\alpha^z|_U\in C(U)$ by (4.3). Therefore

k_α^z is α-superharmonic. Moreover, $\lim_{r\to\infty} H_{B_r}^\alpha k_\alpha^z = 0$ since $H_{B_r}^\alpha 1 = 1$ and $\lim_{\|x\|\to\infty} k_\alpha^z(x) = 0$. Thus k_α^z is an α-potential. ┘

Suppose now that $n \geq 2$ and $1 \leq \alpha < 2$. Let $U = B_r$ and $x \in U$. Then for every $y \in \mathbb{R}^n$ such that $\|y\| \geq 2 \max(r,1)$

$$(\|y\|^2 - r^2)^{-\alpha/2} \|x-y\|^{-n} \leq \left(\frac{2}{\|y\|}\right)^{n+\alpha} \leq \left(\frac{2}{\|y\|}\right)^{n+1} .$$

Since $y \mapsto \|y\|^{-(n+1)}$ is integrable on $\complement B_r$ and ${}^\alpha p_x^U$ converges to zero locally uniformly on $\complement \overline{B}_r$ as α tends to 2 an application of Lebesgue's theorem yields that for every $R > r$

$$\lim_{\alpha\to 2} H_U^\alpha(x, \complement B_R) = 0 .$$

This implies that there is a constant $c_r \geq 0$ such that for every $f \in C_b(\mathbb{R}^n)$

$$\lim_{\alpha\to 2} H_U^\alpha f(x) = c_r(r^2 - \|x\|^2) \int \frac{f(y)}{\|y-x\|^n} \sigma_r(dy) .$$

In particular,

$$1 = \lim H_U^\alpha 1(0) = c_r r^2 \int \frac{\sigma_r(dy)}{r^n} ,$$

i.e. $c_r = r^{n-2}$.

So $\lim_{\alpha\to 2} H_U^\alpha$ is the harmonic kernel $H_U^2 = H_U$ corresponding to the Laplace equation. Of course this could already be expected from (3.5.3). Indeed, (3.5.3) can be used to give another proof for this fact at least if $n \geq 3$.

Let us return to the general situation where $n \geq 1$, $0 < \alpha \leq 2$, $\alpha < n$.

4.5. LEMMA. If $f \in \mathcal{B}^+(\mathbb{R}^n)$ and $U \in \mathcal{U}$ such that $H_U^\alpha f(x_0) < \infty$ for some $x_0 \in U$ then $H_U^\alpha f$ is α-harmonic on U.

Proof. Let $V \in \mathcal{U}$ such that $\overline{V} \subset U$. There exists $c \in \mathbb{R}_+$ such that $H_U^\alpha(x,\cdot) \leq c\, H_U^\alpha(x_0,\cdot)$ for every $x \in V$. Thus by Lebesgue's theorem $H_U^\alpha f$ is continuous and real-valued on V. Since trivially $H_V^\alpha H_U^\alpha f = H_U^\alpha f$ we conclude that $H_U^\alpha f$ is α-harmonic on U. ┘

4.6. PROPOSITION. For every domain D of $\mathbb{R}^n$ the following properties hold:

1. If $u \geq 0$ is α-hyperharmonic on D and $u(x_0) > 0$ for some $x_0 \in D$ then $u > 0$ on D. In particular, $\emptyset$ and D are the only absorbing subsets of D.

2. If $u \geq 0$ is α-hyperharmonic on D and $u(x_o) < \infty$ for some $x_o \in D$ then $u < \infty$ on a dense subset of D.

3. If (h_n) is an increasing sequence of positive α-harmonic functions on D and $\sup h_n(x_o) < \infty$ for some $x_o \in D$ then the function $\sup h_n$ is α-harmonic on D.

Proof. Since $U^* \subset \text{supp}\,(H_U(x,\cdot))$ for every $x \in U \in \mathcal{U}$ the first two statements follow immediately from (1.3.6). The last statement is an easy consequence of (4.5). ┘

We define the *Riesz kernel* (of order α) $N_\alpha : \mathbb{R}^n \times \mathbb{R}^n \to \overline{\mathbb{R}}$ by

$$N_\alpha(x,y) = \begin{cases} \infty & , x = y \\ \| x-y \|^{\alpha-n} & , x \neq y \end{cases} .$$

By (4.4), for every $x \in \mathbb{R}^n$, the function $N_\alpha^x : y \mapsto N_\alpha(x,y)$ is an α-potential and α-harmonic on $\complement\{x\}$ since

$$c_n^\alpha N_\alpha^x = k_\alpha^x .$$

For every measure μ on $\mathbb{R}^n$ we define the *Riesz potential* (of order α) of μ by

$$N_\alpha^\mu(x) = N_\alpha^o * \mu(x) = \int \frac{\mu(dy)}{\|x-y\|^{n-\alpha}} \qquad (x \in \mathbb{R}^n) .$$

We note that N_2 is just the Newtonian kernel N, the corresponding potentials $N^\mu = N_2^\mu$ are called *Newtonian potentials.*

4.7. PROPOSITION. Let μ be a measure on $\mathbb{R}^n$ such that N_α^μ is not identically $+\infty$. Then N_α^μ is an α-potential and $C_\alpha(N_\alpha^\mu) = \text{supp}\,(\mu)$.

Proof. By Fatou's lemma N_α^μ is l.s.c. For every $x \in U \in \mathcal{U}$,

$$H_U^\alpha N_\alpha^\mu(x) = \iint N_\alpha^z(y)\mu(dz)H_U^\alpha(x,dy) = \iint N_\alpha^z(y)H_U^\alpha(x,dy)\mu(dz)$$

$$= \int H_U^\alpha N_\alpha^z(x)\mu(dz) \leq \int N_\alpha^z(x)\mu(dz) = N_\alpha^\mu(x) .$$

Therefore N_α^μ is α-hyperharmonic.

By (4.6), N_α^μ is finite on a dense subset of $\mathbb{R}^n$ and hence for every $U \in \mathcal{U}$ the function $H_U^\alpha N_\alpha^\mu$ α-harmonic on U by (4.5). Furthermore,

$$\lim_{r\to\infty} H_{B_r}^\alpha N_\alpha^\mu = \lim_{r\to\infty} \int H_{B_r}^\alpha N_\alpha^z \mu(dz) = 0$$

since $r \mapsto H^{\alpha}_{B_r} N^z_\alpha$ is decreasing and $\inf_{r>0} H^{\alpha}_{B_r} N^z_\alpha = 0$ by (4.4). Thus N^μ_α is an α-potential.

If $\mu(U) > 0$ then $H^\alpha_U N^\mu_\alpha < N^\mu_\alpha$ on U since by (4.6) $H^\alpha_U N^z_\alpha < N^z_\alpha$ on U for every $z \in U$. If $\mu(\overline{U}) = 0$ then $H^\alpha_U N^\mu_\alpha = N^\mu_\alpha$ by (4.3) and (0.1.6). Therefore $C_\alpha(N^\mu_\alpha) = \operatorname{supp}(\mu)$. ┘

We intend to show next that every α-potential on $\mathbb{R}^n$ is the Riesz potential of a measure μ. To that end we shall first improve the statement of (4.3).

4.8. PROPOSITION. For every $r > 0$ and $x \in B_r$ the function $\varphi_r : z \mapsto H^\alpha_{B_r} N^z_\alpha(x)$ is real-valued and continuous on $\mathbb{R}^n$.

Proof. If $\|z\| > r$ then $\varphi_r(z) = N^z_\alpha(x)$ by (4.3) and (0.1.6). Moreover, φ_r is real-valued and continuous on B_r by Lebesgue's theorem. Hence it remains to study the behavior at B^*_r.

So let $z_0 \in B^*_r$. Let us note first that the function $\rho \mapsto \varphi_\rho(z_0)$ is decreasing since $N^{z_0}_\alpha$ is α-hyperharmonic. Indeed, if $0 < \rho < \sigma < \infty$ then

$$\varphi_\sigma(z_0) = H^\alpha_{B_\sigma} N^{z_0}_\alpha(x) = H^\alpha_{B_\rho} H^\alpha_{B_\sigma} N^{z_0}_\alpha(x) \leq H^\alpha_{B_\rho} N^{z_0}_\alpha(x) = \varphi_\rho(z_0) .$$

Choosing $\|x\| < r_0 < r$ we therefore obtain that $\varphi_r(z_0) \leq \varphi_{r_0}(z_0) < \infty$. Furthermore, $\lim_{\rho\to r} \varphi_\rho(z_0) = \varphi_r(z_0)$.

Indeed, if $\alpha < 2$ then $\varphi_\rho(z_0) = \varphi_r(z_0)$ for every $0 < \rho < r$ by (4.3). Moreover, $\varphi_\rho(z_0) \leq \varphi_r(z_0)$ for every $\rho > r$ and $\liminf_{\rho \to r} \varphi_\rho(z_0) \geq \varphi_r(z_0)$ by Fatou's lemma. Therefore $\lim_{\rho\to r} \varphi_\rho(z_0) = \varphi_r(z_0)$. If $\alpha = 2$ then

$$\varphi_\rho(z_0) = \int \frac{\rho^{n-2}(\rho^2 - \|x\|^2)}{\|z_0 - y\|^{n-2}\,\|x-y\|^n}\,\sigma_\rho(dy) = \int \frac{\rho^{n-2}(\rho^2 - \|x\|^2)}{\|z_0 - \rho y\|^{n-2}\,\|x - \rho y\|^n}\,\sigma_1(dy)$$

where $\|z_0 - ry\| \leq 2\|z_0 - \rho y\|$ provided $\|y\| = 1$. Hence $\lim_{\rho\to r} \varphi_\rho(z_0) = \varphi_r(z_0)$ by Lebesgue's theorem.

Finally let $a < \varphi_r(z_0) < b$. Then there exist $r_0 < r_1 < r < r_2$ such that $\varphi_{r_1}(z_0) < b$ and $a < \varphi_{r_2}(z_0)$. Hence

$$\limsup_{z\to z_0} \varphi_r(z) \le \limsup_{z\to z_0} \varphi_{r_1}(z) = \varphi_{r_1}(z_0) < b$$

and

$$\liminf_{z\to z_0} \varphi_r(z) \ge \liminf_{z\to z_0} \varphi_{r_2}(z) = \varphi_{r_2}(z_0) > a \quad .$$

This shows that $\lim_{z\to z_0} \varphi_r(z) = \varphi_r(z_0)$. ┘

4.9. PROPOSITION. Let $p \in {}^{\alpha}P(\mathbb{R}^n)$. Then there exists a unique measure μ on $\mathbb{R}^n$ such that $N_\alpha^\mu = p$. Moreover, $\operatorname{supp}(\mu) = C_\alpha(p)$.

Proof. We recall that the potential kernel V_α of $\mathbb{P}^\alpha$ is given by $V_\alpha f = c_n^\alpha N_\alpha^0 * f$. Hence by (II.3.11) there exists a sequence (f_m) in $B_b^+(\mathbb{R}^n)$ such that the sequence $(N_\alpha^0 * f_m)$ is increasing and $\sup N_\alpha^0 * f_m = p$. For every $m \in \mathbb{N}$ let $\mu_m = f_m\lambda^n$.

Let $x_0 \in \mathbb{R}^n$ such that $p(x_0) < \infty$, let K be a compact subset of $\mathbb{R}^n$ and $\gamma = \inf N_\alpha^{x_0}(K)$. Then $\gamma > 0$ and for every $m \in \mathbb{N}$

$$\gamma\mu_m(K) \le \mu_m(N_\alpha^{x_0}) = N_\alpha^0 * f_m(x_0) \le p(x_0) < \infty \quad .$$

Hence there exists a subsequence of (μ_m) which converges weakly to a measure μ on $\mathbb{R}^n$. We may assume without loss of generality that the sequence (μ_m) itself converges to μ .

Let $x \in \mathbb{R}^n$. If $\psi \in K^+(\mathbb{R}^n)$ such that $\psi \le N_\alpha^x$ then

$$\mu(\psi) = \lim_{m\to\infty} \mu_m(\psi) \le \lim_{m\to\infty} \mu_m(N_\alpha^x) = p(x) \ .$$

Therefore $N_\alpha^\mu(x) = \mu(N_\alpha^x) \le p(x)$.

Let $a < p(x)$. Since $\inf_{r>0} H^\alpha_{B_r(x)}p(x) = 0$ and $\sup_{r>0} H^\alpha_{B_r(x)}p(x) = p(x)$ there exist $0 < r < R < \infty$ such that choosing $V = B_r(x)$ and $U = B_R(x)$ we have $a < H_V^\alpha p(x) - H_U^\alpha p(x)$. For every $z \in \mathbb{R}^n$ let

$$\varphi(z) = H_V^\alpha N_\alpha^z(x) - H_U^\alpha N_\alpha^z(x) \quad .$$

Obviously, $\varphi = 0$ on $\complement U$ and hence $\varphi \in K(\mathbb{R}^n)$ by (4.8). Therefore

$$a < H_V^\alpha p(x) - H_U^\alpha p(x) = \lim_{m\to\infty} (H_V^\alpha N_\alpha^{\mu_m}(x) - H_U^\alpha N_\alpha^{\mu_m}(x))$$

$$= \lim_{m\to\infty} \mu_m(\varphi) = \mu(\varphi) = H_V^\alpha N_\alpha^\mu(x) - H_U^\alpha N_\alpha^\mu(x) \le N_\alpha^\mu(x) \quad .$$

Thus $N_\alpha^\mu = p$.

Finally let ν be a measure on $\mathbb{R}^n$ such that $N_\alpha^\nu = p$. Let U be a relatively compact open subset of $\mathbb{R}^n$ and $\mu_1 = 1_U\mu$, $\mu_2 = 1_{\complement U}\mu$, $\nu_1 = 1_U\nu$ and $\nu_2 = 1_{\complement U}\nu$. Then $N_\alpha^{\mu_1} - N_\alpha^{\nu_1} = N_\alpha^{\nu_2} - N_\alpha^{\mu_2}$ is α-harmonic on U by (4.7). Let K be a compact subset of U and $\nu_1' = 1_K\nu$, $\nu_1'' = 1_{U\setminus K}\nu$. Then $N_\alpha^{\mu_1} - N_\alpha^{\nu_1'} = N_\alpha^{\mu_1} - N_\alpha^{\nu_1} + N_\alpha^{\nu_1''}$ is α-superharmonic on U . Moreover $N_\alpha^{\mu_1} - N_\alpha^{\nu_1'}$ is α-superharmonic on $\complement K$ by (4.7). Since $N_\alpha^{\mu_1} - N_\alpha^{\nu_1'} \geq -N_\alpha^{\nu_1'}$ and $N_\alpha^{\nu'_1}$ is an α-potential we conclude by (III.6.4) that $N_\alpha^{\mu_1} - N_\alpha^{\nu_1'} \geq 0$. Therefore $N_\alpha^{\mu_1} \geq N_\alpha^{\nu_1}$. Analogously $N_\alpha^{\nu_1} \geq N_\alpha^{\mu_1}$.

Now let q be a continuous real α-potential on $\mathbb{R}^n$. By the first part of the proof there exists a measure σ on $\mathbb{R}^n$ such that $N_\alpha^\sigma = q$. Hence

$$\mu_1(q) = \mu_1(N_\alpha^\sigma) = \sigma(N_\alpha^{\mu_1}) = \sigma(N_\alpha^{\nu_1}) = \nu_1(N_\alpha^\sigma) = \nu_1(q) < \infty \quad .$$

Therefore $\mu_1 = \nu_1$ and hence $\mu = \nu$. ⌟

4.10. REMARK. If $p = N_\alpha^\mu \in {}^\alpha\mathcal{P}(\mathbb{R}^n) \cap C(\mathbb{R}^n)$ then it is easily seen from (4.8) and (4.9) that for every $B \in \mathcal{B}(\mathbb{R}^n)$ and $x \in \mathbb{R}^n$

$$p_B(x) = \int_B \frac{\mu(dy)}{\| x-y \|^{n-\alpha}} \quad .$$

The characterization of fine supports given in (VI.8.2) will show that the α-fine support $\delta_\alpha(p)$ of p is the α-fine support of μ , i.e. $\delta_\alpha(p)$ is the smallest α-finely closed subset F of $\mathbb{R}^n$ such that $\mu(\complement F) = 0$. Since a Lebesgue null set of $\mathbb{R}^n$ has no α-finely interior points by (3.5.1) we conclude in particular that p is strict if μ has a strictly positive density with respect to Lebesgue measure.

The next property is known in classical potential theory as *continuity principle of* EVANS-VASILESCO.

4.11. PROPOSITION. Let μ be a measure on $\mathbb{R}^n$, $A = \operatorname{supp}(\mu)$ and $x_0 \in A$ such that $N_\alpha^\mu|_A$ is continuous at x_0 . Then N_α^μ is continuous at x_0 .

Proof. If $N_\alpha^\mu(x_0) = +\infty$ then N_α^μ is continuous at x_0 . So let $N_\alpha^\mu(x_0) < \infty$. If K is a compact neighborhood of x_0 and $\mu' = 1_K\mu$, $\mu'' = 1_{\complement K}\mu$ then

$N_\alpha^\mu = N_\alpha^{\mu'} + N_\alpha^{\mu''}$ and $N_\alpha^{\mu''}$ is continuous at x_0 by (4.7). Hence we may assume without loss of generality that the support A of μ is compact. For every $x \in \mathbb{R}^n$ let $y_x \in A$ such that

$$\| x-y_x \| = \min \{ \| x-y \| : y \in A\} .$$

Then for every $y \in A$

$$\| y_x - y \| \leq \| y_x - x \| + \| x-y \| \leq 2 \| x-y \|$$

and hence for every measure ν on A

$$N_\alpha^\nu(x) = \int \frac{\nu(dy)}{\| x-y \|^{n-\alpha}} \leq 2^{n-\alpha} \int \frac{\nu(dy)}{\| y_x - y \|^{n-\alpha}} = 2^{n-\alpha} N_\alpha^\nu(y_x) .$$

Let $\varepsilon > 0$. Since $N_\alpha^\mu(x_0) < \infty$ we have $\mu(\{x_0\}) = 0$ and there exists $r > 0$ such that

$$\int_{B_r(x_0)} \frac{\mu(dy)}{\| x_0 - y \|^{n-\alpha}} < \varepsilon .$$

Let $\mu_1 = 1_{B_r(x_0)}\mu$ and $\mu_2 = 1_{\complement B_r(x_0)}\mu$. Then $N_\alpha^\mu = N_\alpha^{\mu_1} + N_\alpha^{\mu_2}$ and $N_\alpha^{\mu_2}$ is continuous at x_0 by (4.7). Since $N_\alpha^{\mu_1}(x_0) < \varepsilon$ and $N_\alpha^{\mu_1}|_A$ is continuous at x_0 there exists $\delta > 0$ such that $N_\alpha^{\mu_1}(y) < 2^{-n}\varepsilon$ for every $y \in B_{2\delta}(x_0) \cap A$. If $x \in B_\delta(x_0)$ then $\| x-y_x \| \leq \| x-x_0 \| < \delta$ and hence $y_x \in B_{2\delta}(x_0)$. Thus for every $x \in B_\delta(x_0)$

$$N_\alpha^{\mu_1}(x) \leq 2^{n-\alpha} N_\alpha^{\mu_1}(y_x) < \varepsilon .$$

This shows that N_α^μ is continuous at x_0. ┘

4.12. PROPOSITION. For every locally bounded $p \in {}^\alpha P(\mathbb{R}^n)$ there exists a sequence (p_m) in ${}^\alpha P(\mathbb{R}^n) \cap C(\mathbb{R}^n)$ such that $p = \sum_{m=1}^\infty p_m$ and the α-harmonic supports $C_\alpha(p_m)$, $m \in \mathbb{N}$, are pairwise disjoint compact sets.

Proof. Let $p \in {}^\alpha P(\mathbb{R}^n)$ be locally bounded. By (4.9) there exists a measure μ on $\mathbb{R}^n$ such that $N_\alpha^\mu = p$. By Lusin's theorem*) there exists a sequence (K_n) of pairwise disjoint compact subsets of $\mathbb{R}^n$ such that $\mu(\complement \bigcup_{m=1}^\infty K_m) = 0$ and $p|_{K_m}$ is continuous for every $m \in \mathbb{N}$. Let

---

*) See e.g. H. BAUER [11]

$$\mu_m = 1_{K_m}\mu \;,\; p_m = N_\alpha^{\mu_m} \quad .$$

Then $\sum_{m=1}^{\infty} p_m = N_\alpha^\mu = p$. For every $m \in \mathbb{N}$, $p_m|_{K_m}$ is continuous since $p|_{K_m}$ is continuous and p_m , $\sum_{k \neq m} p_k$ are l.s.c. Thus $p_m \in C(\mathbb{R}^n)$ by (4.11). ┘

The next property is known in classical potential theory as *domination principle*.

4.13. COROLLARY. Let v be a positive α-hyperharmonic function and $p = N_\alpha^\mu$ an α-potential on $\mathbb{R}^n$ such that $v \geq p$ on supp (μ) . Then $v \geq p$ on $\mathbb{R}^n$.

Proof. By (4.12), there exists a sequence (p_m) in ${}^\alpha P(\mathbb{R}^n) \cap C(\mathbb{R}^n)$ such that $p = \sum_{m=1}^{\infty} p_m$. Let $k \in \mathbb{N}$ and $q_k = \sum_{m=1}^{k} p_m$. Then $q_k \in {}^\alpha P(\mathbb{R}^n) \cap C(\mathbb{R}^n)$ and $C_\alpha(q_k) \subset C_\alpha(p)$, hence $v \geq p \geq q_k$ on $C_\alpha(a_k)$. By (II.6.3), $v \geq q_k$ on $\mathbb{R}^n$. Thus $v \geq \sup q_k = p$ on $\mathbb{R}^n$. ┘

4.14. REMARK. In fact, the following result holds: If v is a positive α-hyperharmonic function, $p = N_\alpha^\mu$ an α-potential on $\mathbb{R}^n$ and $A \in \mathcal{B}(X)$ such that $\mu(\complement A) = 0$ and $v \geq p$ on A , then $v \geq p$ on $\mathbb{R}^n$. Indeed, let (K_m) be an increasing sequence of compact subsets of A such that $\mu(A \setminus \bigcup_{m=1}^{\infty} K_m) = 0$ and define $\mu_m = \mu|_{K_m}$, $m \in \mathbb{N}$. Then $v \geq p \geq N_\alpha^{\mu_m}$ on supp $(\mu_m) \subset K_m$, hence $v \geq N_\alpha^{\mu_m}$ for every $m \in \mathbb{N}$, and therefore $v \geq \sup N_\alpha^{\mu_m} = N_\alpha^\mu$.

4.15. PROPOSITION. Let V be a subset of $\mathbb{R}^n$ and $x_0 \in V$. Let (s_m) be a sequence in $\mathbb{R}_+^*$ and $0 < \delta < 1$ such that $s_{m+1} \leq \delta s_m$ for every $m \in \mathbb{N}$ and define $A_m = \{x \in \complement V : s_{m+1} \leq \|x-x_0\| < s_m\}$, $m \in \mathbb{N}$. Then V is an α-fine neighborhood of x_0 if and only if $\sum_{m=1}^{\infty} {}^\alpha R_1^{A_m}(x_0) < \infty$.

Proof. Suppose first that $\sum_{m=1}^{\infty} {}^\alpha R_1^{A_m}(x_0) < \infty$. Then there exists $m_0 \in \mathbb{N}$ such that $\sum_{m=m_0}^{\infty} {}^\alpha R_1^{A_m}(x_0) < \frac{1}{2}$. Defining $B = B_{s_{m_0}}(x_0)$ and $A = B \setminus V$ we have $A \subset \bigcup_{m=m_0}^{\infty} A_m$ and hence

$${}^\alpha R_1^A(x_0) \leq \sum_{m=m_0}^{\infty} {}^\alpha R_1^{A_m}(x_0) < \frac{1}{2} \quad .$$

Therefore there exists an α-hyperharmonic function $u \geq 0$ on X such that

$u \geq 1$ on A and $u(x_0) < \frac{1}{2}$. Thus $x_0 \in \{u < 1\} \cap B \subset V$. This proves that V is an α-fine neighborhood of x_0.

Assume now conversely that V is an α-fine neighborhood of x_0. By (II.4.1), there exists an α-hyperharmonic function $u \geq 0$ on X and $m_1 \in \mathbb{N}$ such that $u(x_0) < 1$ and $\{x \in \mathbb{R}^n : \|x-x_0\| < s_{m_1}, u(x) \leq 1\} \subset V$. For every $m \geq m_1$ let

$$G_m = \{x \in \mathbb{R}^n : s_{m+2} < \|x-x_0\| < s_m , u(x) > 1\} .$$

and define $G = \bigcup_{m=m_1}^{\infty} G_m$. Then $A_m \subset G_m$ for every $m \geq m_1$, hence it suffices to show that $\sum_{m=m_1}^{\infty} {}^{\alpha}R_1^{G_m}(x_0) < \infty$. Since $u > 1$ on G we have

$${}^{\alpha}R_1^{G}(x_0) \leq u(x_0) < 1 .$$

Fix $k \in \mathbb{N}$, $k \geq 3$, such that

$$c := \frac{u(x_0)}{(1-\delta^{k-2})^{n-\alpha}} < 1 ,$$

and fix $1 \leq i \leq k$. For every $m \in \mathbb{N}$ let

$$U_m = G_{m_1+mk+i} .$$

Obviously we have only to show that $\sum_{m=1}^{\infty} {}^{\alpha}R_1^{U_m}(x_0) < \infty$.

Let $U = \bigcup_{m=1}^{\infty} U_m$. Then $U \subset G$ and hence ${}^{\alpha}R_1^{U}(x_0) \leq u(x_0)$. Since U is a relatively compact open set, ${}^{\alpha}R_1^{U}$ is an α-potential and hence by (4.9) there exists a measure μ on X such that ${}^{\alpha}R_1^{U} = N_\alpha^{\mu}$. By (III.2.5), ${}^{\alpha}R_1^{U} \in {}^{\alpha}\mathcal{H}(\complement \overline{U})$ and hence $\mu(\complement \overline{U}) = 0$. Defining $\mu_m = \mu|_{\overline{U}_m}$, $m \in \mathbb{N}$, we have $N_\alpha^{\mu} = \sum_{m=1}^{\infty} N_\alpha^{\mu_m}$.

Fix $m \in \mathbb{N}$ and consider $l \in \mathbb{N}$, $l \neq m$, $x \in U_m$ and $y \in \overline{U}_l$. If $l < m$, then $\|y-x_0\| \leq \delta^{k-2}\|x-x_0\|$. If $l > m$, then $\|x-x_0\| \leq \delta^{k-2}\|y-x_0\|$. In both cases $\|x-y\| \geq (1-\delta^{k-2})\|y-x_0\|$. Defining $\mu_m' = \mu - \mu_m$ we hence obtain that for every $x \in U_m$

$$N_\alpha^{\mu_m'}(x) = \int \frac{\mu_m'(dy)}{\|x-y\|^{n-\alpha}} \leq \frac{1}{(1-\delta^{k-2})^{n-\alpha}} \int \frac{\mu_m'(dy)}{\|y-x_0\|^{n-\alpha}}$$

$$= \frac{1}{(1-\delta^{k-2})^{n-\alpha}} N_\alpha^{\mu_m'}(x_0) \leq \frac{u(x_0)}{(1-\delta^{k-2})^{n-\alpha}} = c .$$

Since $N_\alpha^{\mu_m} + N_\alpha^{\mu_m'} = N_\alpha^{\mu} = {}^{\alpha}R_1^{U}$ and ${}^{\alpha}R_1^{U} = 1$ on U we therefore conclude that $N_\alpha^{\mu_m} \geq 1 - c$ on U_m and hence $N_\alpha^{\mu_m} \geq (1-c)\,{}^{\alpha}R_1^{U_m}$. Thus

$$\sum_{m=1}^{\infty} {}^{\alpha}R_1^{U_m}(x_0) \leq \frac{1}{1-c} \sum_{m=1}^{\infty} N_{\alpha}^{\mu_m}(x_0) = \frac{1}{1-c} N_{\alpha}^{\mu}(x_0) \leq \frac{1}{1-c} < \infty \ .$$ ┘

Let A be a relatively compact subset of X. Then ${}^{\alpha}\hat{R}_1^A := \widehat{{}^{\alpha}R_1^A}$ is an α-potential on $\mathbb{R}^n$, hence by (4.9) there exists a unique measure μ_A on X such that ${}^{\alpha}\hat{R}_1^A = N_{\alpha}^{\mu_A}$. Assuming that ${}^{\alpha}\hat{R}_1^A$ is α-harmonic on $\complement\overline{A}$ (which will turn out in (VI.5.16) to be no restriction at all) we have $\operatorname{supp}(\mu_A) \subset \overline{A}$. The measure μ_A is called *α-equilibrium measure of* A and its total mass $\|\mu_A\| = \mu_A(\mathbb{R}^n)$ is called the *α-capacity* $\operatorname{cap}_{\alpha}(A)$ *of* A. As usual we omit "α" if $\alpha = 2$. Characteristic properties of the α-capacity will be discussed in (VI.5.22).

4.16. EXAMPLES. 1. Classical potential theory ($\alpha = 2$). For every $r > 0$, the equilibrium measure of the open ball $B_r = B_r(0)$ has the constant density r^{n-2} with respect to the (normed) surface measure σ_r on S_r. In particular,

$$\operatorname{cap}(B_r) = \operatorname{cap}(S_r) = r^{n-2} .$$

Indeed, the continuous real potential $p_r := \inf(r^{n-2}N^0, 1)$ is equal to 1 on $\overline{B}_r$ and equal to $r^{n-2}N^0$ on $\complement B_r$, hence harmonic on $\complement S_r$. Thus $R_1^{B_r} = R_1^{S_r} = p_r$. Since p_r is invariant under rotations, the measure $\mu_{B_r} = \mu_{S_r}$ has to be a multiple of σ_r. Using $N^{\sigma_r}(0) = r^{2-n}$ we finally conclude that $\mu_{B_r} = \mu_{S_r} = r^{n-2}\sigma_r$, $\operatorname{cap}(B_r) = \operatorname{cap}(S_r) = r^{n-2}$.

2. Riesz potentials ($\alpha < 2$). For every $r > 0$, the α-equilibrium measure of B_r is the measure ν_r on B_r having density $y \mapsto a_{\alpha}(r^2 - \|y\|^2)^{-\alpha/2}$ with respect to Lebesgue measure. In particular

$$\operatorname{cap}_{\alpha}(B_r) = r^{n-\alpha} \operatorname{cap}_{\alpha}(B_1) \ .$$

Indeed, $N_{\alpha}^{\nu_r} = 1$ on B_r by (4.2). Since $\nu_r(\complement B_r) = 0$ we conclude immediately from (4.14) that $R_1^{B_r} = N_{\alpha}^{\nu_r}$. Thus $\mu_{B_r} = \nu_r$ and hence in particular

$$\operatorname{cap}_{\alpha}(B_r) = a_{\alpha} \int_{B_r} \frac{\lambda^n(dy)}{(r^2 - \|y\|^2)^{\alpha/2}} = r^{n-\alpha} a_{\alpha} \int_{B_1} \frac{\lambda^n(dy)}{(1 - \|y\|^2)^{\alpha/2}} = r^{n-\alpha} \operatorname{cap}_{\alpha}(B_1) \ .$$

Calculating the integral we obtain that

$$\operatorname{cap}_{\alpha}(B_1) = \frac{\Gamma(\frac{n}{2})}{\Gamma(\frac{\alpha}{2})\Gamma(\frac{n-\alpha}{2}) + 1)} = \frac{2}{(n-\alpha)B(\frac{\alpha}{2}, \frac{n-\alpha}{2})}$$

but this will not be used in the following. Note, however, that the formula

$cap_\alpha(B_r) = r^{n-\alpha} cap_\alpha(B_1)$ holds as well if $\alpha = 2$.

Suppose now that $\alpha \leq 1$ and $\alpha < n$. Let μ_r be the image of ν_r under the injection $x \mapsto (x,0)$ from $\mathbb{R}^n$ into $\mathbb{R}^{n+1}$ and let p_r be the $(\alpha+1)$-potential of μ_r on $\mathbb{R}^{n+1}$. Since $(n+1)-(\alpha+1) = n-\alpha$, $p_r(x,0) = N_\alpha^{\nu_r}(x)$ for every $x \in \mathbb{R}^n$. Using (4.14) we obtain that μ_r is the $(\alpha+1)$-equilibrium measure of $B_r \times \{0\}$ in $\mathbb{R}^{n+1}$ and hence

$$cap_{\alpha+1}(B_r \times \{0\}) = cap_\alpha(B_r) = r^{n-\alpha} cap_\alpha(B_1) \quad .$$

In particular,

$$cap(B_r \times \{0\}) = r^{n-1} cap_1(B_1) = \frac{\Gamma(\frac{n}{2})}{\sqrt{\pi}\,\Gamma(\frac{n+1}{2})} r^{(n+1)-2} \quad .$$

<u>4.17. COROLLARY</u> (WIENER criterion). Let V be a subset of $\mathbb{R}^n$, $x_0 \in V$, and $\lambda > 1$. For every $m \in \mathbb{N}$ define $A_m = \{x \in \complement V : \lambda^{-(m+1)} \leq \|x-x_0\| < \lambda^{-m}\}$ and assume that ${}^\alpha\hat{R}_1^{A_m}(x_0) = {}^\alpha R_1^{A_m}(x_0)$ and ${}^\alpha\hat{R}_1^{A_m}$ is α-harmonic on $\complement \overline{A}_m$ (which by (VI.2.3) and (VI.5.16) is no restriction). Then V is an α-fine neighborhood of x_0 if and only if $\sum_{m=1}^{\infty} \lambda^{m(n-\alpha)} cap_\alpha(A_m) < \infty$.

Proof. Fix $m \in \mathbb{N}$ and let μ_m denote the α-equilibrium measure of A_m . Since $supp(\mu_m) \subset \overline{A}_m$ we obtain that

$$cap_\alpha(A_m)\lambda^{m(n-\alpha)} \leq R_1^{A_m}(x_0) = \hat{R}_1^{A_m}(x_0) = \int \frac{\mu_m(dx)}{\|x-x_0\|^{n-\alpha}} \leq cap_\alpha(A_m)\lambda^{(m+1)(n-\alpha)} \quad .$$

Thus the statement follows immediately from (4.15). ∎

<u>4.18. EXAMPLE.</u> For every $m \in \mathbb{N}$ let $a_m = 2 \cdot 3^{-m}$, $r_m = (\frac{1}{m})^{n-\alpha} 3^{-m}$, and define $V = \complement \bigcup_{m=1}^{\infty} B_{r_m}(a_m)$. Then V is not an α-fine neighborhood of the origin, but it is a β-fine neighborhood of 0 for every $0 < \beta < \alpha$. In particular, every β-fine topology with $\beta < \alpha$ is strictly finer than the α-fine topology.

Indeed, for every $m \in \mathbb{N}$, $B_{r_m}(a_m)$ is the set of all $x \in \complement V$ such that $3^{-(m+1)} \leq \|x-x_0\| < 3^{-m}$ and for every $0 < \beta \leq \alpha$ by (4.16)

$$\sum_{m=1}^{\infty} 3^{m(n-\beta)} cap_\beta(B_{r_m}(a_m)) = cap_\beta(B_1) \sum_{m=1}^{\infty} (3^m r_m)^{n-\beta} = cap_\beta(B_1) \sum_{m=1}^{\infty} (\frac{1}{m})^{\frac{n-\beta}{n-\alpha}} \quad .$$

Thus the assertion follows from (4.17).

<u>4.19. EXERCISES.</u> 1. Let μ be a measure on $\mathbb{R}^n$. Then N_α^μ is an α-potential

on $\mathbb{R}^n$ if and only if $\int_{\complement B_1(0)} \|y\|^{\alpha-n}\mu(dy) < \infty$.

2. An α-potential N_α^μ on $\mathbb{R}^n$ is a real continuous function if and only if, for every $x \in \mathbb{R}^n$, $\lim_{\varepsilon\to 0} \sup_{z\in\mathbb{R}^n} \int_{B_\varepsilon(x)} \|z-y\|^{\alpha-n}\mu(dy) = 0$.

3. Let $f \in \mathcal{B}^+(\mathbb{R}^n)$ such that f is locally bounded and $\int f(y)\|y\|^{\alpha-n}\lambda^n(dy) < \infty$. Then $N_\alpha^{f\lambda^n}$ is a real continuous α-potential on $\mathbb{R}^n$. It is a strict α-potential if f is strictly positive.

5. Products

In this section let $\mathbb{P} = (P_t)_{t>0}$ be a sub-Markov semigroup on X and let $\widetilde{\mathbb{P}} = (\widetilde{P}_t)_{t>0}$ be a sub-Markov semigroup on $\widetilde{X}$, $\widetilde{X}$ locally compact with countable base. Let $\mathbb{P}\otimes\widetilde{\mathbb{P}} = (P_t\otimes\widetilde{P}_t)_{t>0}$ where $P_t\otimes\widetilde{P}_t$ is the product kernel on $X\times\widetilde{X}$ introduced in (II.1.1.6), i.e. for every $f \in \mathcal{B}^+(X\times\widetilde{X})$

$$(P_t\otimes\widetilde{P}_t)f(x,\widetilde{x}) = \iint f(y,\widetilde{y})P_t(x,dy)P_t(\widetilde{x},d\widetilde{y}) \quad .$$

A simple calculation shows that $\mathbb{P}\otimes\widetilde{\mathbb{P}}$ is a sub-Markov semigroup on $X\times\widetilde{X}$. $\mathbb{P}\otimes\widetilde{\mathbb{P}}$ is called *product of* $\mathbb{P}$ *and* $\widetilde{\mathbb{P}}$.

5.1. EXAMPLE. Classical potential theory. The product of Brownian semigroup on $\mathbb{R}^n$ and Brownian semigroup on $\mathbb{R}^m$ is Brownian semigroup on $\mathbb{R}^{n+m}$.

5.2. PROPOSITION. The product semigroup $\mathbb{P}\otimes\widetilde{\mathbb{P}}$ has the following properties:

1. $E_{\mathbb{P}}\otimes E_{\widetilde{\mathbb{P}}} \subset E_{\mathbb{P}\otimes\widetilde{\mathbb{P}}}$, i.e. if $u \in E_{\mathbb{P}}$ and $\widetilde{u} \in E_{\widetilde{\mathbb{P}}}$ then the function $u\otimes\widetilde{u} : (x,\widetilde{x}) \mapsto u(x)\widetilde{u}(\widetilde{x})$ is excessive with respect to $\mathbb{P}\otimes\widetilde{\mathbb{P}}$.

2. If $\lim_{t\downarrow 0} P_t f = f$ (locally uniformly) ($\lim_{t\downarrow 0} \frac{1}{t}P_t f = 0$ locally uniformly on $\complement S(f)$ resp.) for every $f \in \mathcal{K}(X)$ and $\lim_{t\downarrow 0} \widetilde{P}_t\widetilde{f} = \widetilde{f}$ (locally uniformly) ($\lim_{t\downarrow 0} \frac{1}{t}\widetilde{P}_t\widetilde{f} = 0$ locally uniformly on $\complement S(\widetilde{f})$ resp.) for every $\widetilde{f} \in \mathcal{K}(\widetilde{X})$ then $\lim_{t\downarrow 0}(P_t\otimes\widetilde{P}_t)g = g$ (locally uniformly) ($\lim_{t\downarrow 0} \frac{1}{t}P_t\otimes\widetilde{P}_t)g = 0$ locally uniformly on $\complement S(g)$ resp.) for every $g \in \mathcal{K}(X\otimes\widetilde{X})$.

3. If $\mathbb{P}$ and $\widetilde{\mathbb{P}}$ are measurable and if the potential kernel V of $\mathbb{P}$ or the potential kernel $\widetilde{V}$ of $\widetilde{\mathbb{P}}$ is proper then the potential kernel W of $\mathbb{P}\otimes\widetilde{\mathbb{P}}$

is proper.

4. $\mathbb{P} \otimes \widetilde{\mathbb{P}}$ is a strong Feller semigroup if and only if $\mathbb{P}$ and $\widetilde{\mathbb{P}}$ are strong Feller semigroups.

Proof. Since $(P_t \otimes \widetilde{P}_t)(u \otimes \widetilde{u}) = (P_t u) \otimes (\widetilde{P}_t \widetilde{u})$, (1) follows immediately. In order to prove (2) it suffices to consider functions $g \in K(X \times \widetilde{X})$ which are of the form $g = f \otimes \widetilde{f}$, $f \in K(X)$, $\widetilde{f} \in K(\widetilde{X})$. Then $S(g) = S(f) \times S(\widetilde{f})$ and for every $t > 0$

$$\begin{aligned}
|(P_t \otimes \widetilde{P}_t)g - g| &= |P_t f \otimes \widetilde{P}_t \widetilde{f} - f \otimes \widetilde{f}| \\
&\le |(P_t f - f) \otimes \widetilde{P}_t \widetilde{f}| + |f \otimes (\widetilde{P}_t \widetilde{f} - \widetilde{f})| \\
&\le |P_t f - f| \otimes \| \widetilde{f} \| + \| f \| \otimes |\widetilde{P}_t \widetilde{f} - \widetilde{f}| , \\
|(P_t \otimes \widetilde{P}_t)g| &\le \inf(|P_t f| \otimes \| \widetilde{f} \|, \| f \| \otimes |\widetilde{P}_t \widetilde{f}|) .
\end{aligned}$$

Now (2) follows easily.

Suppose next that $\mathbb{P}$ and $\widetilde{\mathbb{P}}$ are measurable and one of the potential kernels, say $\widetilde{V}$, is proper. Let L be a compact subset of $X \times \widetilde{X}$. Then there exists a compact subset $\widetilde{K}$ of $\widetilde{X}$ such that $L \subset X \times \widetilde{K}$. For all $(x,\widetilde{x}) \in X \times \widetilde{X}$,

$$\begin{aligned}
W 1_L(x,\widetilde{x}) &\le W 1_{X \times \widetilde{K}}(x,\widetilde{x}) = \int_0^\infty \int_{\widetilde{K}} \int_X P_t(x,dy)\widetilde{P}_t(\widetilde{x},d\widetilde{y})dt \\
&\le \int_0^\infty \int_{\widetilde{K}} \widetilde{P}_t(\widetilde{x},d\widetilde{y})dt = \widetilde{V}(\widetilde{x},\widetilde{K}) .
\end{aligned}$$

Thus $W 1_L$ is bounded, i.e. W is a proper kernel.

We finally note that (4) follows from (II.1.6). ∎

5.3. PROPOSITION. Let $(X,E_{\mathbb{P}})$ and $(\widetilde{X},E_{\widetilde{\mathbb{P}}})$ be balayage spaces. Then $(X \times \widetilde{X}, E_{\mathbb{P} \otimes \widetilde{\mathbb{P}}})$ is a balayage space if and only if $E_{\mathbb{P} \otimes \widetilde{\mathbb{P}}} = S(E_{\mathbb{P} \otimes \widetilde{\mathbb{P}}} \cap C(X \times \widetilde{X}))$. In particular, $(X \times \widetilde{X}, E_{\mathbb{P} \otimes \widetilde{\mathbb{P}}})$ is a balayage space if $\mathbb{P} \otimes \widetilde{\mathbb{P}}$ has a strong Feller resolvent.

Proof. Since $E_{\mathbb{P}}$ and $E_{\widetilde{\mathbb{P}}}$ are linearly separating, the points of $X \times \widetilde{X}$ are linearly separated by $\{u \otimes 1 : u \in E_{\mathbb{P}}\} \cup \{1 \otimes \widetilde{u} : \widetilde{u} \in E_{\widetilde{\mathbb{P}}}\}$. Therefore $E_{\mathbb{P} \otimes \widetilde{\mathbb{P}}}$ is linearly separating by (5.2.1).

There exist strictly positive functions $u,v \in E_{\mathbb{P}} \cap C(X)$ such that $\frac{u}{v} \in C_0(X)$. Since we could replace u by $\inf(u,1)$ and v by $v+1$, we may assume that $u \le 1 \le v$. Analogously there exist $\tilde{u},\tilde{v} \in E_{\tilde{\mathbb{P}}} \cap C(\tilde{X})$ such that $0 < \tilde{u} \le 1 \le \tilde{v}$ and $\frac{\tilde{u}}{\tilde{v}} \in C_0(\tilde{X})$. Then $u \otimes \tilde{u}, v \otimes \tilde{v}$ are strictly positive functions in $E_{\mathbb{P}\otimes\tilde{\mathbb{P}}} \cap C(X \times \tilde{X})$ such that $\frac{u \otimes \tilde{u}}{v \otimes \tilde{v}} = \frac{u}{v} \otimes \frac{\tilde{u}}{\tilde{v}} \in C_0(X \otimes \tilde{X})$.

Moreover, $\lim_{t \to 0} (P_t \otimes \tilde{P}_t)g = g$ for every $g \in K(X \times \tilde{X})$ by (5.2.2). Thus the proposition follows from (II.4.9) and (2.4). ┘

5.4. COROLLARY. Let $(X,E_{\mathbb{P}})$ and $(\tilde{X},E_{\tilde{\mathbb{P}}})$ be balayage spaces and suppose that $\mathbb{P}$ and $\tilde{\mathbb{P}}$ are strong Feller semigroups. Then $(X \times \tilde{X}, E_{\mathbb{P}\otimes\tilde{\mathbb{P}}})$ is a balayage space.

Proof. (5.3) and (5.2.4). ┘

We shall now investigate the case where $\tilde{X} = \mathbb{R}$ and $\tilde{\mathbb{P}}$ is the translation semigroup $\mathbb{T} = (T_t)_{t>0}$. For every $f \in B_b(X \times \mathbb{R})$ and $t \in \mathbb{R}$ we define $f_t \in B_b(X)$ by $f_t(x) = f(x,t)$. Then for all $(x,s) \in X \times \mathbb{R}$

$$(P_t \otimes T_t)f(x,s) = P_t f_{s-t}(x) \quad .$$

5.5. LEMMA. Suppose that $\mathbb{P}$ is a strong Feller semigroup such that $\lim_{s \downarrow t} P_s f = P_t f$ locally uniformly for every $f \in K(X)$ and $t > 0$. Then $\mathbb{P} \otimes \mathbb{T}$ has a strong Feller resolvent $W = (W_\lambda)_{\lambda>0}$.

Proof. Fix $\lambda > 0$ and $f \in B_b(X \times \mathbb{R})$. Then for every $(x,s) \in X \times \mathbb{R}$

$$W_\lambda f(x,s) = \int_0^\infty e^{-\lambda t} P_t f_{s-t}(x)dt \quad .$$

Given $\varepsilon > 0$ we choose $T \in \mathbb{R}_+$ such that $\|f\|e^{-\lambda T} < \lambda\varepsilon$. Then

$$|W_\lambda f(x,s) - \int_0^T e^{-\lambda t} P_t f_{s-t}(x)dt| < \varepsilon$$

and if $-\infty < a < s < b < \infty$ then

$$F_T(x,s) := \int_0^T e^{-\lambda t} P_t f_{s-t}(x)dt = \int_{s-T}^s e^{-\lambda(s-t)} P_{s-t} f_t(x)dt$$
$$= \int_{a-T}^b 1_{]s-T,s[}(t) e^{-\lambda(s-t)} P_{s-t} f_t(x)dt \quad .$$

By Lebesgue's convergence theorem and (2.10) the function F_T is continuous on $X \times]a,b[$. Being the uniform limit of continuous functions F_T the function $W_\lambda f$ is continuous. ┘

5.6. PROPOSITION. The following statements are equivalent:

(1) $(X \times \mathbb{R}, E_{\mathbb{P}\otimes\mathbb{T}})$ is a balayage space.

(2) $\mathbb{P}$ is a strong Feller semigroup, $\lim_{t\to 0} P_t f = f$ locally uniformly for every $f \in K(X)$, and there exist strictly positive functions $u,v \in E_{\mathbb{P}\otimes\mathbb{T}} \cap C(X\times\mathbb{R})$ such that $u \in o(v)$.

Proof. (1) ⇒ (2): We note first that for every $x \in X$,

$$\sup_{t>0} P_t 1(x) = \sup_{t>0} (P_t \otimes T_t)\ 1\ (x,0) = 1 \quad .$$

Moreover, given $f \in K(X)$ we consider the continuous function $g : (x,s) \mapsto f(x)$ on $X \times \mathbb{R}$ and obtain by (II.3.16) and (2.8) that for every $x \in X$ $\lim_{t\to 0} (P_t \otimes T_t) g = g$ locally uniformly and hence $\lim_{t\to 0} P_t f = f$ locally uniformly.

It remains to show that $\mathbb{P}$ is a strong Feller semigroup. So let $f \in \mathcal{B}_b(X)$, $0 \le f \le 1$, and define $F : X \times \mathbb{R} \to [0,1]$ by

$$F(x,s) = \begin{cases} P_s f(x) & , \ x \in X, s > 0 , \\ 0 & , \ x \in X, s \le 0 . \end{cases}$$

Fix $t > 0$ and $(x,s) \in X \times \mathbb{R}$. If $s > t$ then

$$(P_t \otimes T_t)F(x,s) = P_t F_{s-t}(x) = P_t P_{s-t} f(x) = P_s f(x) = F(x,s)$$

and

$$(P_t \otimes T_t)(1_{X\times\mathbb{R}_+^*} - F)(x,s) = P_t 1(x) - P_t F_{s-t}(x) = P_t 1(x) - F(x,s) \ .$$

If $s \le t$ then

$$(P_t \otimes T_t)F(x,s) = P_t F_{s-t}(x) = P_t 0(x) = 0$$

and

$$(P_t \otimes T_t)(1_{X\times\mathbb{R}_+^*} - F)(x,s) = 0 \quad .$$

Therefore $F, 1_{X\times\mathbb{R}_+^*} - F \in E_{\mathbb{P}\otimes\mathbb{T}}$. In particular, these functions are l.s.c. such that $F + (1_{X\times\mathbb{R}_+^*} - F) = 1_{X\times\mathbb{R}_+^*}$. So F is continuous on $X \times \mathbb{R}_+^*$.

(2) ⇒ (1): Immediate consequence of (2.4) using (II.3.16), (5.2.2), (5.2.3) and (5.5). ┘

5.7. COROLLARY. Suppose that $(X, E_{\mathbb{P}})$ is a balayage space. Then the following statements are equivalent:

(1) $(X \times \mathbb{R}, E_{\mathbb{P} \otimes \mathbb{T}})$ is a balayage space.

(2) $\mathbb{P}$ is a strong Feller semigroup.

Proof. (1) $\Rightarrow$ (2) : (5.6).

(2) $\Rightarrow$ (1) : (5.3) and (5.5). ┘

5.8. EXAMPLE. Let $\mathbb{P}$ be the Brownian semigroup on $\mathbb{R}^n$, $n \geq 1$. Then $\mathbb{P} \otimes \mathbb{T}$ is called the *heat semigroup* on $\mathbb{R}^n \times \mathbb{R}$. $(\mathbb{R}^{n+1}, E_{\mathbb{P} \otimes \mathbb{T}})$ is a harmonic space. In the next section we shall study this harmonic space in more detail. In particular, we shall establish the connection to the heat equation.

First of all $(\mathbb{R}^{n+1}, E_{\mathbb{P} \otimes \mathbb{T}})$ is a balayage space. This follows at once from (5.7) if $n \geq 3$. For the general case we shall go back to (5.6). It suffices to find strictly positive functions $u, v \in E_{\mathbb{P} \otimes \mathbb{T}} \cap C(\mathbb{R}^{n+1})$ such that $u \in o(v)$. By (5.2.2) $\lim_{t \to 0} (P_t \otimes T_t) f = f$ for every $f \in K(\mathbb{R}^{n+1})$, so $E_{\mathbb{P} \otimes \mathbb{T}}$ is min-stable.

Let $1 \leq i \leq n$ and define $f_{\pm}^i : \mathbb{R}^n \times \mathbb{R} \to \mathbb{R}_+^*$ by

$$f_{\pm}^i(x,s) = e^{\pm x_i + \frac{1}{2}s} .$$

Then for all $t > 0$ and $(x,s) \in \mathbb{R}^n \times \mathbb{R}$

$$(P_t \otimes T_t) f_{\pm}^i(x,s) = \left(\frac{1}{2\pi t}\right)^{n/2} \int e^{-\frac{\|y-x\|^2}{2t}} e^{\pm y_i + \frac{1}{2}(s-t)} dy$$

$$= \frac{1}{\sqrt{2\pi t}} \int e^{-\frac{(y_i - x_i)^2}{2t}} e^{\pm y_i + \frac{1}{2}(s-t)} dy_i$$

$$= e^{\pm x_i + \frac{1}{2}s} \frac{1}{\sqrt{2\pi t}} \int e^{-\frac{(y_i - (x_i \pm t))^2}{2t}} dy_i = f_{\pm}^i(x,s) ,$$

and hence $f_{\pm}^i \in E_{\mathbb{P} \otimes \mathbb{T}}$. Let

$$u = \inf (f_+^1, \ldots, f_+^n, f_-^1, \ldots, f_-^n) ,$$

i.e.

$$u(x,s) = \exp\left(-\max_{1 \leq i \leq n} |x_i| + \frac{1}{2}s\right)$$

for all $(x,s) \in \mathbb{R}^n \times \mathbb{R}$. Then $u \in E_{\mathbb{P} \otimes \mathbb{T}}$.

Define $v_1 : \mathbb{R} \to \mathbb{R}_+^*$ by $v_1(s) = \sup (e^s, 1)$. Then obviously $v_1 \in E_{\mathbb{T}}$,

$v = 1 \otimes v_1 \in E_{\mathbb{P} \otimes \mathbb{T}}$ and $u \in o(v)$. So $(\mathbb{R}^{n+1}, E_{\mathbb{P} \otimes \mathbb{T}})$ is a balayage space.

Moreover, $(\mathbb{R}^{n+1}, E_{\mathbb{P} \otimes \mathbb{T}})$ has no absorbing points by (III.7.3) and has the local truncation property by (III.8.2) since $\lim_{t \downarrow 0} \frac{1}{t}(P_t \otimes T_t)g = 0$ locally uniformly on $\complement S(g)$ for every $g \in K(\mathbb{R}^{n+1})$ by (5.2.2). Thus $(\mathbb{R}^{n+1}, E_{\mathbb{P} \otimes \mathbb{T}})$ is a harmonic space.

Let W denote the potential kernel of the heat semigroup. Then for every $f \in \mathcal{B}_b(\mathbb{R}^n \times \mathbb{R})$ and $(x,t) \in \mathbb{R}^n \times \mathbb{R}$,

$$Wf(x,t) = \int_0^\infty (P_s f(\cdot,t-s))(x)ds = \int_0^\infty \int_{\mathbb{R}^n} \left(\frac{1}{2\pi s}\right)^{n/2} e^{-\frac{\|x-y\|^2}{2s}} f(y,t-s)\, \lambda^n(dy)\, ds$$

$$= \int_{-\infty}^t \int_{\mathbb{R}^n} \left(\frac{1}{2\pi(t-s)}\right)^{n/2} e^{-\frac{\|x-y\|^2}{2(t-s)}} f(y,s)\, \lambda^n(dy)ds = \int_{\mathbb{R}^n \times \mathbb{R}} w^z(x,t) f(z) \lambda^{n+1}(dz)$$

where for every $z = (y,s) \in \mathbb{R}^n \times \mathbb{R}$

$$w^z(x,t) = \begin{cases} \left(\frac{1}{2\pi(t-s)}\right)^{n/2} e^{-\frac{\|x-y\|^2}{2(t-s)}} & , t > s \\ 0 & , t \le s \end{cases} .$$

Let us finally note that the translation semigroup on $\mathbb{R}$ can be considered as a (degenerate) heat semigroup on $\mathbb{R}^0 \times \mathbb{R}$.

<u>5.9. PROPOSITION.</u> Let $\mathbb{P} = (P_t)_{t>0}$ be a strong Feller sub-Markov semigroup on X such that, for every $f \in K(X)$ and $t > 0$, $\lim_{s \downarrow t} P_s f = P_t f$ locally uniformly. Let $\tilde{\mathbb{P}} = (\tilde{P}_t)_{t>0}$ be a measurable sub-Markov semigroup on $\tilde{X}$ having a strong Feller resolvent $(\tilde{V}_\lambda)_{\lambda>0}$. Then $\mathbb{P} \otimes \tilde{\mathbb{P}}$ has a strong Feller resolvent $(W_\lambda)_{\lambda>0}$.

Proof. Fix $\lambda > 0$, $g \in \mathcal{B}_b(X \times \tilde{X})$ and let $(x_n,\tilde{x}_n)$ be a sequence in $X \times \tilde{X}$ converging to $(x_0,\tilde{x}_0)$. For every $n \in \mathbb{N} \cup \{0\}$ define $f_n \in \mathcal{B}_b(\tilde{X} \times \mathbb{R}_+^*)$ by

$$f_n(\tilde{x},t) = P_t g_{\tilde{x}}(x_n)$$

(where of course $g_{\tilde{x}}(x) = g(x,\tilde{x})$ for every $x \in X$). Then for all $n,m \in \mathbb{N}$

$$W_\lambda g(x_n,\tilde{x}_m) = \int_0^\infty \int_{\tilde{X}} \int_X e^{-\lambda t} g(x,\tilde{x}) P_t(x_n,dx) \tilde{P}_t(\tilde{x}_m,d\tilde{x})dt$$

$$= \int_0^\infty \int_{\tilde{X}} e^{-\lambda t} P_t g_{\tilde{x}}(x_n) \tilde{P}_t(\tilde{x}_m,d\tilde{x})dt = \int_0^\infty e^{-\lambda t} \tilde{P}_t (f_n)_t(\tilde{x}_m)dt \quad .$$

By (2.10), the functions $(x,t) \mapsto P_t g_{\tilde{x}}(x)$ are continuous on $X \times \mathbb{R}_+^*$ for every

$\tilde{x} \in \tilde{X}$. Applying (2.12) to $\tilde{\mathbb{P}}$ and the function f_0 we obtain that

$$\lim_{m\to\infty} W_\lambda g(x_0,\tilde{x}_m) = W_\lambda g(x_0,\tilde{x}_0) \quad .$$

Fix $\varepsilon > 0$ and choose $0 < r < s < \infty$ such that $2\|g\|(1 - e^{-\lambda r} + e^{-\lambda s}) < \lambda\varepsilon$. For every $n \in \mathbb{N}$ define $g_n : \tilde{X} \to \mathbb{R}$ by

$$g_n(\tilde{x}) = \sup_{r\leq t\leq s} |f_n(\tilde{x},t) - f_0(\tilde{x},t)| \quad .$$

Using the continuity of $(x,t) \mapsto P_t g_{\tilde{x}}(x)$ we conclude that (g_n) is a sequence in $\mathcal{B}^+(\tilde{X})$ which converges pointwise to zero. Moreover, $g_n \leq 2\|g\|$ for every $n \in \mathbb{N}$. Since

$$|W_\lambda g(x_n,\tilde{x}_n) - W_\lambda g(x_0,\tilde{x}_n)| \leq \int_0^\infty e^{-\lambda t}\tilde{P}_t|f_n - f_0|_t(\tilde{x}_n)dt$$

$$\leq \varepsilon + \int_r^s e^{-\lambda t}\tilde{P}_t g_n(\tilde{x}_n)dt \leq \varepsilon + \tilde{V}_\lambda g_n(\tilde{x}_n)$$

and $\lim_{n\to\infty} \tilde{V}_\lambda g_n(\tilde{x}_n) = 0$ by (II.1.5), we have

$$\lim_{n\to\infty} (W_\lambda g(x_n,\tilde{x}_n) - W_\lambda g(x_0,\tilde{x}_n)) = 0 \quad .$$

Hence

$$\lim_{n\to\infty} W_\lambda g(x_n,\tilde{x}_n) = W_\lambda g(x_0,\tilde{x}_0) \quad ,$$

i.e. W_λ is a strong Feller kernel. ┘

5.10. COROLLARY. Let $\mathbb{P}$ be a sub-Markov semigroup on X such that $(X,E_{\mathbb{P}})$ is a balayage space. Then the following statements are equivalent:

(1) $\mathbb{P}$ is a strong Feller semigroup.

(2) $(X \times \mathbb{R}, E_{\mathbb{P}\otimes\mathbb{T}})$ is a balayage space.

(3) $(X \times \tilde{X}, E_{\mathbb{P}\otimes\tilde{\mathbb{P}}})$ is a balayage space for every balayage space $(\tilde{X},E_{\tilde{\mathbb{P}}})$ such that $\tilde{\mathbb{P}}$ is a sub-Markov semigroup having a strong Feller resolvent.

Proof. (1) ⟺ (2) : (5.7).

(1) ⟹ (3) : (5.3) and (5.9).

(3) ⟹ (2) : Trivial. ┘

The following relation will be used in section 8 in order to show that $(X,E_{\mathbb{P}})$ is a balayage space provided $E_{\mathbb{P}}$ is linearly separating and $(X \times \tilde{X}, E_{\mathbb{P}\otimes\tilde{\mathbb{P}}})$ is a balayage space for some Markov semigroup $\tilde{\mathbb{P}}$ on $\tilde{X}$.

5.11. PROPOSITION. Let $\widetilde{\mathbb{P}}$ be a Markov semigroup on $\widetilde{X}$ such that $(X \times \widetilde{X}, E_{\mathbb{P} \otimes \widetilde{\mathbb{P}}})$ is a balayage space. Then $R_{f \otimes 1} = R_f \otimes 1$ for every l.s.c. function $f : X \to \overline{\mathbb{R}}_+$.

Proof. If $v \in E_{\mathbb{P}}$ such that $v \geq f$ then $v \otimes 1 \in E_{\mathbb{P} \otimes \widetilde{\mathbb{P}}}$ and $v \otimes 1 \geq f \otimes 1$. Therefore $R_f \otimes 1 \geq R_{f \otimes 1}$.

To prove the converse inequality suppose first that $f \in C_b(X)$ and define

$$u(x) = \inf_{\widetilde{x} \in \widetilde{X}} R_{f \otimes 1}(x,\widetilde{x}) \qquad (x \in X) .$$

Then $f \leq u$, $u \otimes 1 \leq R_{f \otimes 1}$, and $u \in \mathcal{B}^+(X)$ since $R_{f \otimes 1} \in E_{\mathbb{P} \otimes \widetilde{\mathbb{P}}} \cap C_b(X \times \widetilde{X})$ by (III.2.8) and (III.5.6). For every $t > 0$, $x \in X$, and $\widetilde{x} \in \widetilde{X}$

$$P_t u(x) = (P_t \otimes \widetilde{P}_t)(u \otimes 1)(x,\widetilde{x}) \leq (P_t \otimes \widetilde{P}_t) R_{f \otimes 1}(x,\widetilde{x}) \leq R_{f \otimes 1}(x,\widetilde{x}) ,$$

hence $u \in S_{\mathbb{P}}$. Let $v = \lim_{t \to 0} P_t u$. Then $v \in E_{\mathbb{P}}$, $v \leq u$ and $v \geq f$ since $\lim_{t \to 0} (P_t \otimes \widetilde{P}_t)(f \otimes 1) = f \otimes 1$. Therefore $R_f \otimes 1 \leq v \otimes 1 \leq R_{f \otimes 1}$ and we conclude that $R_{f \otimes 1} = R_f \otimes 1$ where $R_f = v \in E_{\mathbb{P}}$.

In the general case, we choose an increasing sequence (f_n) in $C_b^+(X)$ such that $\sup f_n = f$. Then $w := \sup R_{f_n} \in E_{\mathbb{P}}$, $w \geq f$, and hence

$$R_f \otimes 1 \leq w \otimes 1 = \sup R_{f_n} \otimes 1 = \sup R_{f_n \otimes 1} \leq R_{f \otimes 1} \leq R_f \otimes 1 . \qquad \lrcorner$$

5.12. EXERCISES. 1. Let $\alpha, \beta \in \mathbb{R}$, $1 \leq i \leq n$, and define $f : \mathbb{R}^n \times \mathbb{R} \to \mathbb{R}$ by $f(x,s) = e^{\alpha x_i + \beta s}$. Then f is excessive with respect to the heat semigroup if and only if $\alpha^2 \leq 2\beta$.

2. Let $P_{s,t}$, $0 \leq s < t < \infty$, be sub-Markov kernels on X such that $P_{s,u} = P_{s,t} P_{t,u}$ for all $0 \leq s < t < u < \infty$. Suppose that $\lim_{t \downarrow s} P_{s,t} f = f$ for every $f \in K(X)$ and $s \geq 0$ and that for every $f \in \mathcal{B}_b(X)$ and $t > 0$ the function $(x,s) \mapsto P_{s,t} f(x)$ is continuous on $X \times [0,t[$. For every $t > 0$, $x \in X$, $s \geq 0$ and $B \in \mathcal{B}(X \times \mathbb{R}_+)$ define

$$Q_t((x,s),B) = P_{s,s+t}(x, B_{s+t})$$

where $B_{s+t} = \{y \in X : (y,s+t) \in B\}$.

Then $\mathbb{Q} = (Q_t)_{t>0}$ is a sub-Markov semigroup on $X \times \mathbb{R}_+$ such that $\lim_{t \to 0} Q_t g = g$ for every $g \in K(X \times \mathbb{R}_+)$ and the resolvent of $\mathbb{Q}$ is a strong Feller resolvent. Moreover, if there exist strictly positive functions $u,v \in E_{\mathbb{Q}} \cap C(X \times \mathbb{R}_+)$ such that $u \in o(v)$ then $(X \times \mathbb{R}_+, E_{\mathbb{Q}})$ is a balayage space.

3. Let $\mathbb{P} = (P_t)_{t>0}$ be a strong Feller semigroup on X such that $(X, E_{\mathbb{P}})$ is a balayage space. Let $a,b \in \mathcal{B}^+(\mathbb{R}_+)$ such that $b > 0$ and $\int_0^t (a(\tau) + b(\tau))d\tau < \infty$ for every $t \in \mathbb{R}_+$. Define sub-Markov kernels $P_{s,t}$, $0 \leq s < t < \infty$, by

$$P_{s,t} = e^{-\varphi(s,t)} P_{\psi(s,t)}$$

where

$$\varphi(s,t) = \int_s^t a(\tau)d\tau \quad , \quad \psi(s,t) = \int_0^t b(\tau)d\tau \quad .$$

Then all assumptions made in (5.12.2) are satisfied.

6. Heat Equation

Given $n \in \mathbb{N}$ we consider the differential operator

$$\triangle = \frac{1}{2} \sum_{i=1}^{n} \frac{\partial^2}{\partial x_i^2} - \frac{\partial}{\partial t} \quad .$$

The main purpose of this section is to show that the solutions of the *heat equation* $\triangle h = 0$ generate a harmonic space $(\mathbb{R}^{n+1}, \mathcal{W})$ and to prove that $\mathcal{W} = E_{\mathbb{P} \otimes \mathbb{T}}$ where $\mathbb{P} \otimes \mathbb{T}$ is the heat semigroup introduced in (5.8).

For every open subset U of X let

$$\mathcal{H}(U) = \{h \in C^2(U) : \triangle h = 0\} \quad .$$

The harmonic kernel associated with a set $\prod_{i=1}^{n+1}]\alpha_i, \beta_i[$ will be derived from kernels associated with half-spaces using symmetry properties of the heat equation. For each $1 \leq i \leq n$ and $c \in \mathbb{R}$ the reflection at $\{x_i = c\}$ will be denoted by T_i^c, i.e.

$$T_i^c(x_1, \ldots, x_n, t) = (x_1, \ldots, x_{i-1}, 2c - x_i, x_{i+1}, \ldots, x_n, t) \quad .$$

As in section 5 we define for all $x,y \in \mathbb{R}^n$ and $t,s \in \mathbb{R}$

$$w(x,t,y,s) = \begin{cases} \left(\frac{1}{2\pi(t-s)}\right)^{n/2} e^{-\frac{\|x-y\|^2}{2(t-s)}} \quad , & t > s \quad , \\ 0 \quad , & t \leq s \quad . \end{cases}$$

It is easily verified that $w(\cdot,\cdot,y,s) \in C^\infty(\mathbb{R}^n \setminus \{(y,s)\})$ and $\triangle w(\cdot,\cdot,y,s) = 0$

on $\mathbb{R}^n \setminus \{(y,s)\}$ (see (0.2.1)).

Let $\alpha \in \mathbb{R}$, $U = \mathbb{R}^n \times]\alpha,\infty[$, $Y = \mathbb{R}^n \times \{\alpha\}$. Given $f \in \mathcal{B}_b(Y)$ we define $H_U f : U \to \mathbb{R}$ by

$$H_U f(x,t) = \int w(x,t,y,\alpha) f(y,\alpha) \lambda^n(dy) \quad .$$

In other words, H_U is obtained from the Brownian semigroup $(P_t)_{t>0}$ by

$$H_U f(x,t) = (P_{t-\alpha} f(\cdot,\alpha))(x) \quad .$$

Hence the following proposition is immediate from (0.2.2), (0.2.3), and the symmetry properties of w^0 and Lebesgue measure on $\mathbb{R}^n$.

6.1. PROPOSITION. 1. Let $f \in \mathcal{B}_b(Y)$. Then $H_U f \in \mathcal{H}_b(U)$ and for every $z^* \in Y$

$$\varliminf_{z \to z^*} f(z) \le \varliminf_{z \to z^*} H_U f(z) \le \varlimsup_{z \to z^*} H_U f(z) \le \varlimsup_{z \to z^*} f(z) \quad .$$

2. If $1 \le i \le n$ and $c \in \mathbb{R}$ such that $f \circ T_i^c = -f$ on Y then $(H_U f) \circ T_i^c = -H_U f$ on U .

3. If (f_m) is a uniformly bounded sequence in $\mathcal{B}_b(Y)$ such that $\lim_{m\to\infty} f_m = f$ then $\lim_{m\to\infty} H_U f_m = H_U f$.

For every $x,y \in \mathbb{R}^n$ and $s,t \in \mathbb{R}$ let

$$w_1(x,t,y,s) = \frac{1}{2}\left| \frac{\partial w}{\partial x_1}(x,t,y,s) - \frac{\partial w}{\partial x_1}(T_1^{y_1}(x,t),y,s)\right| = \begin{cases} \dfrac{|x_1-y_1|}{t-s}\, w(x,t,y,s), & t>s \\ 0, & t \le s \end{cases} .$$

Using the previous remarks on w we obtain that $w_1(\cdot,\cdot,y,s) \in C^\infty((\mathbb{R}\setminus\{y_1\}) \times \mathbb{R}^n)$ and $\Delta w_1(\cdot,\cdot,y,s) = 0$ on $(\mathbb{R}\setminus\{y_1\}) \times \mathbb{R}^n$.

Now let $a \in \mathbb{R}$, $\alpha,\beta \in \mathbb{R}$ with $\alpha < \beta$,

$$U = (\mathbb{R}\setminus\{a\}) \times \mathbb{R}^{n-1} \times]\alpha,\beta[\quad , \quad Y = \{a\} \times \mathbb{R}^{n-1} \times [\alpha,\beta] \; .$$

Given $f \in \mathcal{B}_b(Y)$ we define $H^a f : U \to \mathbb{R}$ by

$$H^a f(x,t) = \int_\alpha^\beta \int_{\mathbb{R}^{n-1}} w_1(x,t,a,y_2,\dots,y_n,s)\, f(a,y_2,\dots,y_n,s)\, \lambda^{n-1}(d(y_2,\dots,y_n))ds \quad .$$

Taking

$$G(\xi) := \frac{1}{\sqrt{2\pi}} \int_\xi^\infty e^{-\frac{\eta^2}{2}} d\eta$$

we have the following result.

6.2. PROPOSITION. 1. Let $f \in \mathcal{B}_b(Y)$. Then $H^a f \in \mathcal{H}_b(U))$.

2. For every $x \in \mathbb{R}^n$ with $x_1 \neq a$ the limit $\lim_{z\to(x,\beta)} H^a f(z)$ exists and $\lim_{z\to(x,\alpha)} H^a f(z) = 0$.

3. For every $z^* = (x,t) \in Y$ with $t > \alpha$,

$$\varliminf_{z\to z^*} f(z) \leq \varliminf_{z\to z^*} H^a f(z) \leq \varlimsup_{z\to z^*} H^a f(z) \leq \varlimsup_{z\to z^*} f(z) \quad .$$

4. For every $z^* = (x,\alpha) \in Y$,

$$\varlimsup_{z\to z^*} |H^a f(z)| \leq \varlimsup_{z\to z^*} |f(z)| \ .$$

5. If $2 \leq i \leq n$ and $c \in \mathbb{R}$ such that $f \circ T_i^c = -f$ on Y then $(H^a f) \circ T_i^c = -H^a f$ on U .

6. If (f_m) is a uniformly bounded sequence in $\mathcal{B}_b(Y)$ such that $\lim_{m\to\infty} f_m = f$ then $\lim_{m\to\infty} H^a f_m = H^a f$.

7. The mapping $H^a : \mathcal{B}_b(Y) \to \mathcal{H}(U)$ is positive, linear and $H^a 1(x,t) = 2G(\frac{|x_1-a|}{\sqrt{t-\alpha}})$.

Proof. Using $w_1(\cdot,\cdot,y,s) \in \mathcal{H}((\mathbb{R}\setminus\{y_1\}) \times \mathbb{R}^n)$ and proceeding as in the proof of (0.2.2) an application of Lebesgue's theorem yields that $H^a f \in \mathcal{H}(U)$. Consider the transformation

$$\rho : (y_2,\dots,y_n,s) \mapsto (\frac{y_2-x_2}{\sqrt{t-s}} \ , \ \dots\dots, \ \frac{y_n-x_n}{\sqrt{t-s}} \ , \ \frac{|x_1-a|}{\sqrt{t-s}}) \quad .$$

Since

$$|\det D\rho| = \frac{1}{2}(t-s)^{-\frac{n+2}{2}} |x_1-a|$$

we obtain that

$$H^a f(x,t) = 2 \cdot (\frac{1}{2\pi})^{n/2} \int_{\{z_1 \geq \frac{|x_1-a|}{\sqrt{t-\alpha}}\}} e^{-\frac{\|z\|^2}{2}} f(a, |x_1-a|\frac{z_2}{z_1} + x_2, \dots\dots\dots\dots\dots$$

$$\dots\dots\dots\dots, |x_1 - a|\frac{z_n}{z_1} + x_n, t - \frac{|x_1 - a|^2}{z_1^2})\lambda^n(d(z_1, \dots, z_n)) \ .$$

In particular,

$$H^a1(x,t) = 2 \int_{\{z_1 \geq \frac{|x_1-a|}{\sqrt{t-\alpha}}\}} e^{-\frac{\|z\|^2}{2}} \lambda(dz) = 2 \int_{\frac{|x_1-a|}{\sqrt{t-\alpha}}}^{\infty} e^{-\frac{\eta^2}{2}} d\eta = 2G\left(\frac{|x_1-a|}{\sqrt{t-\alpha}}\right) .$$

Applying Lebesgue's theorem the proof of (6.2) is now easily completed. ┘

Consider $a,b,\alpha,\beta \in \mathbb{R}$ such that $a < b$, $\alpha < \beta$ and let

$$U = \{(x,t) \in \mathbb{R}^n \times \mathbb{R} : a < x_1 < b , \alpha < t < \beta\} ,$$

$$Y = \{(x,t) \in \mathbb{R}^n \times \mathbb{R} : x_1 \in \{a,b\} , \alpha \leq t \leq \beta\} ,$$

$$Y^a = Y \cap \{x_1 = a\} , Y^b = Y \cap \{x_1=b\} .$$

For every $f \in \mathcal{B}_b(Y)$ let

$$f^a = f|_{Y^a} , \quad f^b = f|_{Y^b} .$$

Then we define a positive linear mapping $V : \mathcal{B}_b(Y) \to \mathcal{B}_b(Y)$ by

$$Vf(x,t) = \begin{cases} H^b f^b(x,t) , & (x,t) \in Y^a , \\ H^a f^a(x,t) , & (x,t) \in Y^b . \end{cases}$$

By (6.2), $V1 \leq 2G(\frac{b-a}{\sqrt{\beta-\alpha}}) < 1$. Therefore the inverse $(I+V)^{-1} : \mathcal{B}_b(Y) \to \mathcal{B}_b(Y)$ exists,

$$(I + V)^{-1} = \sum_{m=0}^{\infty} (-V)^m .$$

Given $f \in \mathcal{B}_b(Y)$ we now define $H_U f : U \to \mathbb{R}$ by

$$H_U f(x,t) = H^a((I+V)^{-1}f)^a(x,t) + H^b((I+V)^{-1}f)^b(x,t) .$$

<u>6.3. PROPOSITION.</u> 1. Let $f \in \mathcal{B}_b(Y)$. Then $H_U f \in \mathcal{H}_b(U)$.

2. For every $x \in \mathbb{R}^n$ with $a < x_1 < b$ the limit $\lim_{z \to (x,\beta)} H_U f(z)$ exists and $\lim_{z \to (x,\alpha)} H_U f(z) = 0$.

3. For every $z^* = (x,t) \in Y$ with $t > \alpha$

$$\varliminf_{z \to z^*} f(z) \leq \varliminf_{z \to z^*} H_U f(z) \leq \varlimsup_{z \to z^*} H_U f(z) \leq \varlimsup_{z \to z^*} f(z) .$$

4. For every $z^* = (x,\alpha) \in Y$

$$\varlimsup_{z \to z^*} |H_U f(z)| \leq \varlimsup_{z \to z^*} |f(z)| .$$

5. If $2 \leq i \leq n$ and $c \in \mathbb{R}$ such that $f \circ T_i^c = -f$ on Y then

$(H_U f) \circ T_i^c = -H_U f$ on U .

6. The mapping $H_U : \mathcal{B}_b(Y) \to \mathcal{B}_b(Y)$ is linear, and if $(f_m) \subset \mathcal{B}_b(Y)$ is uniformly bounded, $\lim_{m\to\infty} f_m = f$, then $\lim_{m\to\infty} H_U f_m = H_U f$.

Proof. The first two statements follow immediately from (6.2) and the definition of H_U . In order to see that (3) holds let $g = (I+V)^{-1}f$, i.e. $g \in \mathcal{B}_b(Y)$, $g + Vg = f$. Let $z^* = (x,t) \in Y$ such that $t > \alpha$. If $x_1 = a$ then by (6.2)

$$\underline{\lim}_{z\to z^*} g(z) \leq \underline{\lim}_{z\to z^*} H^a g^a(z) \leq \overline{\lim}_{z\to z^*} H^a g^a(z) \leq \overline{\lim}_{z\to z^*} g(z)$$

and

$$\lim_{z\to z^*} Vg(z) = H^b g^b(z^*) = \lim_{z\to z^*} H^b g^b(z) \quad ,$$

and hence $H_U f = H^a g^a + H^b g^b$ satisfies

$$\underline{\lim}_{z\to z^*} f(z) \leq \underline{\lim}_{z\to z^*} H_U f(z) \leq \overline{\lim}_{z\to z^*} H_U f(z) \leq \overline{\lim}_{z\to z^*} f(z) \quad .$$

Analogously if $x_1 = b$. Similarly (4) follows.

Assume now that $2 \leq i \leq n$ and $c \in \mathbb{R}$ such that $f \circ T_i^c = -f$ on Y . Using (6.2.5) we obtain that $(V^k f) \circ T_i^c = -V^k f$ on Y for every $k \in \mathbb{N}$, hence $g \circ T_i^c = -g$ on Y and $(H_U f) \circ T_i^c = -H_U f$ on U .

The linearity of H_U is obvious from the definition. Finally consider a sequence (f_m) in $\mathcal{B}_b(Y)$ such that $\| f_m \| \leq 1$ for every $m \in \mathbb{N}$ and $\lim_{m\to\infty} f_m = f$. Defining $g_m = (I+V)^{-1} f_m$ we obtain a uniformly bounded sequence (g_m) in $\mathcal{B}_b(Y)$. We claim that $\lim_{m\to\infty} g_m = g$ and therefore $\lim_{m\to\infty} H_U f_m = H_U f$ by (6.2). Indeed, for every $k \in \mathbb{N}$ we have $|V^k f_m| \leq \| V \|^k$ for every $m \in \mathbb{N}$ and $\lim_{m\to\infty} V^k f_m = V^k f$, hence

$$\lim_{m\to\infty} g_m = \lim_{m\to\infty} \sum_{k=0}^{\infty} (-V)^k f_m = \sum_{k=0}^{\infty} (-V)^k f = g \quad .$$ ∎

We are now ready to consider an open standard rectangle U in $\mathbb{R}^{n+1}$, i.e. a set U of the form $U =]a_1,b_1[\times\ldots\times]a_n,b_n[\times]\alpha,\beta[$ where $a_i, b_i, \alpha, \beta \in \mathbb{R}$, $a_i < b_i$, $\alpha < \beta$. Let

$$Y_0 = \mathbb{R}^n \times \{\alpha\} \ , \ Y_i = \{(x,t) \in \mathbb{R}^n \times \mathbb{R} : x_i \in \{a_i,b_i\},\ \alpha \leq t \leq \beta\} \quad (1 \leq i \leq n) \quad .$$

For every $0 \leq i \leq n$ let

$$F_i = U^* \cap Y_i$$

and define

$$U^*_\triangle = \bigcup_{i=0}^{n} F_i \ , \qquad P = \bigcup_{0\le i<j\le n} F_i \cap F_j \ .$$

Let $f \in \mathcal{B}_b(U^*_\triangle)$. Then, for each $0 \le i \le n$, there exists a (unique) function $f_i \in \mathcal{B}_b(Y_i)$ such that $f_i = f$ on $F_i \setminus P$ and

$$f_i \circ T_j^{c_j} = -f_i \quad \text{on } Y_i$$

for every $1 \le j \le n$, $j \neq i$, and $c_j \in \{a_j, b_j\}$. If $z_i \in F_i \cap P$ and $\lim_{z\to z_i} f(z) = f(z_i) = 0$ then $\lim_{z\to z_i} f_i(z) = f_i(z_i) = 0$.

Let

$$U_0 = \mathbb{R}^n \times]\alpha,\infty[\ , \quad U_i = \{(x,t) \in \mathbb{R}^n \times \mathbb{R} : a_i < x_i < b_i \ , \ \alpha < t < \beta\} \quad (1\le i\le n) \ .$$

According to (6.1) we then have a function $H_{U_0} f_0 \in H(U_0)$ and according to (6.3) a function $H_{U_1} f_1 \in H(U_1)$ and analogously $H_{U_2} f_2 \in H(U_2)$, ..., $H_{U_n} f_n \in H(U_n)$. We now define $H_U f : U \to \mathbb{R}^n$ by

$$H_U f(x,t) = \sum_{i=0}^{n} H_{U_i} f_i(x,t) \ .$$

<u>6.4. PROPOSITION.</u> Let $f \in \mathcal{B}_b(U^*_\triangle)$. Then $H_U f \in H_b(U)$. For every $z^* \in U^* \setminus U^*_\triangle$ the limit $\lim_{z\to z^*} H_U f(z)$ exists and for every $z^* \in U^*_\triangle \setminus P$

$$(*) \qquad \varliminf_{z\to z^*} f(z) \le \varliminf_{z\to z^*} H_U f(z) \le \varlimsup_{z\to z^*} H_U f(z) \le \varlimsup_{z\to z^*} f(z) \ .$$

Moreover, if $z^* \in U^*_\triangle$ and $\lim_{z\to z^*} f(z) = 0$ then $\lim_{z\to z^*} H_U f(z) = 0$.

Proof. By (6.1) and (6.3), $H_U f \in H_b(U)$ and $\lim_{z\to z^*} H_U f(z)$ exists for every $z^* \in U^* \setminus U^*_\triangle$. Consider now $z^* \in U^* \setminus P$. Then there exists a unique $0 \le i \le n$ such that $z^* \in F_i$. Then

$$\varliminf_{z\to z^*} f(z) = \varliminf_{z\to z^*} f_i(z) \le \varliminf_{z\to z^*} H_{U_i} f_i(z) \le \varlimsup_{z\to z^*} H_{U_i} f_i(z) \le \varlimsup_{z\to z^*} f_i(z) = \varlimsup_{z\to z^*} f(z)$$

whereas for every $j \neq i$, $0 \le j \le n$,

$$\lim_{z\to z^*} H_{U_j} f_j(z) = 0 \ .$$

Thus (*) holds.

Finally let $z \in U^*_\Delta$ such that $\lim\limits_{z\to z^*} f(z) = 0$. Then $\lim\limits_{z\to z^*} f_i(z) = 0$ for every i such that $z^* \in F_i$. Thus $\lim\limits_{z\to z^*} H_{U_i} f_i(z) = 0$ for every $0 \le i \le n$ and hence $\lim\limits_{z\to z^*} H_U f(z) = 0$. ┘

In the following we shall prove a minimum principle which shows that in fact (*) holds for every $z \in U^*_\Delta$.

6.5. LEMMA. Let $g \in C^2(U)$ such that $\Delta g \le 0$ and $\liminf\limits_{z\to z^*} g(z) \ge 0$ for every $z \in U^*_\Delta$. Then $g \ge 0$.

Proof. Assume that there exists a point $z_1 \in U$ such that $g(z_1) < 0$. Let $\alpha < \tilde{\beta} < \beta$ such that

$$z_1 \in \tilde{U} := \{(x,t) \in U : t < \tilde{\beta}\} .$$

Define $\tilde{w} : \tilde{U} \to \mathbb{R}$ by $\tilde{w}(x,t) = (\tilde{\beta}-t)^{-1}$. Then $\Delta\tilde{w}(x,t) = -(\tilde{\beta}-t)^2 < 0$ and $\lim\limits_{z\to z^*} \tilde{w}(z) = \infty$ for every $z^* \in \tilde{U}^* \setminus \tilde{U}^*_\Delta$. Let $\varepsilon > 0$ such that $g(z_1) + \varepsilon\tilde{w}(z_1) < 0$ and define

$$\tilde{g} = g|_{\tilde{U}} + \varepsilon\tilde{w} .$$

Then $\tilde{g} \in C^2(\tilde{U})$, $\Delta\tilde{g} < 0$, $\tilde{g}(z_1) < 0$ and $\liminf\limits_{z\to z^*} \tilde{g}(z) \ge 0$ for every $z \in \tilde{U}^*$. In particular, there exists a point $z_0 \in U$ such that $\tilde{g}(z_0) = \inf \tilde{g}(U)$. Obviously $\frac{\partial\tilde{g}}{\partial t}(z_0) = 0$ and $\frac{\partial^2\tilde{g}}{\partial x_i^2}(z_0) \ge 0$ for every $1 \le i \le n$ and hence $\Delta\,\tilde{g}(z_0) = \frac{1}{2}\sum\limits_{i=1}^{n} \frac{\partial^2\tilde{g}}{\partial x_i^2}(z_0) - \frac{\partial\tilde{g}}{\partial t}(z_0) \ge 0$. This contradiction completes the proof. ┘

6.6. LEMMA. Let $z_0 = (x_0,t_0) \in U$ and $M \in \mathbb{R}_+$. Then there exists a function $h \in H^+(U)$ such that $h(z_0) \le 1$ and $\lim\limits_{z\to z^*} h(z) \ge M$ for every $z^* \in P$.

Proof. Let $c = (\sum\limits_{i=1}^{n}(b_i-a_i)^2)^{\frac{1}{2}}$ and define $u : U \to \mathbb{R}$ by

$$u(x,t) = \sum_{\substack{1\le i<j\le n\\ c_k\in\{a_k,b_k\}}} \log \frac{c}{\sqrt{(x_i-c_i)^2+(x_j-c_j)^2}} .$$

Then $u \in H^+(U)$ and $\lim\limits_{z\to z^*} u(z) = \infty$ for all $z^* \in \bigcup\limits_{1\le i<j\le n} F_i \cap F_j$.

Let $\gamma > 0$ such that $\gamma(u(z_0) + 2n(t_0-\alpha)^{-\frac{1}{2}}) \le 1$, set $\delta = (\frac{\gamma}{M})^2$ and define $v : U \to \mathbb{R}$ by

$$v(x,t) = \sum_{\substack{1\le i\le n\\ c_i\in\{a_i,b_i\}}} (t-\alpha+\delta)^{-\frac{1}{2}} \exp\left(-\frac{(x_i-c_i)^2}{2(t-\alpha+\delta)}\right) .$$

Then obviously $v \in H^+(U)$, $v(z_0,t_0) \le 2n(t_0-\alpha)^{-\frac{1}{2}}$ and $\lim_{z\to z^*} v(z) \ge \delta^{-\frac{1}{2}}$ for every $z^* \in \bigcup_{1\le i\le n} F_0 \cap F_i$. Therefore the function $h = \gamma(u+v)$ has the desired properties. ┘

6.7. PROPOSITION. Let $g \in C^2(U)$ be bounded such that $\triangle g \le 0$ and $\liminf_{z\to z^*} g(z) \ge 0$ for every $z \in U^*_\triangle \setminus P$. Then $g \ge 0$.

Proof. Let $K \in \mathbb{R}_+$ such that $|g| \le K$. Let $z_0 \in U$, $\varepsilon > 0$ and $M \in \mathbb{R}_+$ such that $\varepsilon M \ge K$. Using (6.6) we choose a function $h \in H^+(U)$ such that $h(z_0) \le 1$ and $\lim_{z\to z^*} h(z) \ge M$ for every $z^* \in P$. Then $\tilde{g} := g + \varepsilon h \in C^2(U)$, $\triangle\tilde{g} = \triangle g \le 0$ and $\liminf_{z\to z^*} \tilde{g}(z) \ge 0$ for every $z \in U^*_\triangle$. Hence $\tilde{g} \ge 0$ by (6.5), in particular $g(z_0) + \varepsilon \ge 0$. Therefore $g \ge 0$. ┘

6.8. COROLLARY. Let $h \in H_b(U)$ such that $\lim_{z\to z^*} h(z) = 0$ for every $z^* \in U^*_\triangle \setminus P$. Then $h = 0$.

6.9. PROPOSITION. H_U is a kernel, $H_U 1 = 1$. For every $f \in C(U^*_\triangle)$ and $z^* \in U^*_\triangle$, $\lim_{z\to z^*} H_U f(z) = f(z^*)$.

Proof. Let $f \in B_b^+(U^*)$. Then $H_U f \in H_b(U)$ by (6.4) and for every $z^* \in U^*_\triangle \setminus P$

$$\liminf_{z\to z^*} H_U f(z) \ge \liminf_{z\to z^*} f(z) \ge 0 \quad ;$$

hence $H_U f \ge 0$ by (6.7). The linearity and the σ-continuity of H_U follow easily from (6.1) and (6.3).

Suppose that $f \in C(U^*_\triangle)$ and let $h = H_U f$. Then $\lim_{z\to z^*} h(z) = f(z^*)$ for every $z^* \in U^*_\triangle \setminus P$ by (6.4). Now let $z_0 \in P$, $\tilde{f} = f - f(z_0)$ and $\tilde{h} = f(z_0) + H_U\tilde{f}$. Then $\tilde{h} \in H_b(U)$ and for every $z^* \in U^*_\triangle \setminus P$

$$\lim_{z\to z^*} \tilde{h}(z) = f(z_0) + \tilde{f}(z^*) = f(z^*) .$$

Therefore $h = \tilde{h}$ by the preceding corollary. Since $\tilde{f} \in C(U^*_\triangle)$ and $\tilde{f}(z_0) = 0$ we know by (6.4) that $\lim_{z\to z^*} H_U\tilde{f}(z) = 0$. Therefore

$$\lim_{z\to z_0} h(z) = \lim_{z\to z_0} \tilde{h}(z) = f(z_0) \quad .$$

Moreover, since $1 \in H(U)$ we obtain from (6.8) that $H_U 1 = 1$. ┘

6.10. REMARKS. 1. If W is an open subset of $\mathbb{R}^n \times \mathbb{R}$ such that $\overline{U} \subset W$ and $h \in H(W)$ then $H_U(h|_{U^*_\triangle}) = h|_U$ by (6.9) and (6.8). Going back to the definition of H_U and using Lebesgue's theorem it is hence easily shown that $H(W) \subset C^\infty(W)$.

2. For every $f \in B_b(U^*_\triangle)$ and $z^* \in U^*_\triangle$

$$\varliminf_{z\to z^*} f(z) \leq \varliminf_{z\to z^*} H_U f(z) \leq \varlimsup_{z\to z^*} H_U f(z) \leq \varlimsup_{z\to z^*} f(z) \quad .$$

The first inequality is obtained from (6.9) by choosing $\tilde{f} \in C(U^*_\triangle)$ such that $\tilde{f} \leq f$ and $\tilde{f}(z^*) = \varliminf_{z\to z^*} f(z)$ and the last inequality follows from the first one if we replace f by $-f$.

Defining

$$H_U(z,B) = \begin{cases} H_U(1_B|_{U^*_\triangle})(z) & , z \in U \\ \varepsilon_z(B) & , z \in \complement U \end{cases}$$

we obtain a sweeping kernel H_U on X .

6.11. THEOREM. Let $\mathcal{U}$ be the set of all open standard rectangles in $\mathbb{R}^{n+1}$. Then $(H_U)_{U\in\mathcal{U}}$ is a family of harmonic kernels such that $(\mathbb{R}^{n+1}, {}^*H^+(\mathbb{R}^{n+1}))$ is a harmonic space.

For every open subset W of $\mathbb{R}^{n+1}$ the set $\{h \in C^2(W) : \triangle h = 0\}$ is the set of all harmonic functions on W , and every function $s \in C^2(W)$ such that $\triangle s \leq 0$ is superharmonic on W .

Proof. It is clear from the preceding results that the family $(H_U)_{U\in\mathcal{U}}$ satisfies the axioms (H_1) - (H_3).

Let $s \in C^2(W)$ such that $\triangle s \leq 0$. Let $U \in \mathcal{U}$ such that $\overline{U} \subset W$ and $g = (s - H_U s)|_U$. Then $g \in C^2(U)$, $\triangle g = \triangle s \leq 0$ on U and $\lim_{z\to z^*} g(z) = 0$ for every $z^* \in U^*_\triangle$. Therefore $g \geq 0$ by (6.5), i.e. $H_U s \leq s$. Hence s is superharmonic (with respect to $(H_U)_{U\in\mathcal{U}}$) . In particular, every function $h \in C^2(W)$ satisfying $\triangle h = 0$ is harmonic. The converse is already established in (6.10.1).

Let us now prove that (H_4) holds. So let $U =]a_1,b_1[\times\dots\times]a_n,b_n[\times]\alpha,\beta[$, $a_i < b_i$, $\alpha < \beta$. If $z^* \in U^*_\Delta$ then $\lim_{z\to z^*} H_U\varphi(z) = \varphi(z^*)$ for every $\varphi \in K(X)$, i.e. every ultrafilter on U converging to z^* is regular. The positive function $\tilde{w} : (x,t) \mapsto \frac{1}{\beta-t}$ is superharmonic on U since $\Delta\tilde{w} < 0$, and $\lim_{z\to z^*} \tilde{w}(z) = \infty$ for every $z^* \in U^* \setminus U^*_\Delta$. This proves (H_4).

Moreover $1 \in H^+(\mathbb{R}^{n+1})$ and the family $\{(x,t)\mapsto \exp(\sum_{i=1}^n a_ix_i + \frac{1}{2}\|a\|^2 t) : a \in \mathbb{R}^n\}$ of positive harmonic functions on $\mathbb{R}^{n+1}$ separates the points of $\mathbb{R}^{n+1}$. Hence (H_5) holds.

Thus $(\mathbb{R}^{n+1}, {}^*H^+(\mathbb{R}^{n+1}))$ is a harmonic space by (III.7.1) and (III.8.1). ┘

6.12 LEMMA. Let $u \in C^2(\mathbb{R})$ such that u is increasing and $\lim_{t\to-\infty} u(t) = 0$. Then $v : (x,t) \mapsto u(t)$ is a potential on $\mathbb{R}^n \times \mathbb{R}$.

Proof. Since $\Delta v(x,t) = -\frac{du}{dt}(t) \leq 0$ we know by (6.11) that v is superharmonic. Let h be the greatest harmonic minorant of v . Given $y \in \mathbb{R}^n$ we define $h^y : \mathbb{R}^n \times \mathbb{R} \to \mathbb{R}$ by

$$h^y(x,t) = h(x-y,t) \quad .$$

Then $\Delta h^y(x,t) = \Delta h(x-y,t) = 0$ and $h^y(x,t) \leq v(x-y,t) = v(x,t)$, i.e. h^y is a harmonic minorant of v . Hence $h^y \leq h$. In particular,

$$h(0,t) = h^y(y,t) \leq h(y,t) = h^{-y}(0,t) \leq h(0,t) \quad ,$$

i.e. $h(y,t) = h(0,t)$ for all $y \in \mathbb{R}^n$, $t \in \mathbb{R}$. Let $g(t) = h(0,t)$. Then for every $t \in \mathbb{R}$

$$\frac{dg}{dt}(t) = \Delta h(0,t) = 0 ,$$

hence g is a constant. Since $g \leq u$ and $\lim_{t\to-\infty} u(t) = 0$ we conclude that $g = 0$ and therefore $h = 0$. Thus v is a potential. ┘

6.13. PROPOSITION. For every $t \in \mathbb{R}$ the halfspace $A_t = \{(y,s) \in \mathbb{R}^n \times \mathbb{R} : s \leq t\}$ is an absorbing set.

Proof. By (6.12) the function

$$v : (x,s) \mapsto \sup\,(0,(s-t)^3)$$

is a potential. Hence $A_t = \{v = 0\}$ is absorbing by (1.2). $\lrcorner$

As before let $\mathbb{P} = (P_t)_{t>0}$ be the Brownian semigroup on $\mathbb{R}^n$.

<u>6.14. PROPOSITION.</u> Let $\alpha \in \mathbb{R}$, $U = \mathbb{R}^n \times]\alpha,\infty[$ and $(x,t) \in U$. Then for every $B \in \mathcal{B}(\mathbb{R}^n)$

$$\overset{\circ}{\varepsilon}{}^{\complement U}_{(x,t)}(B \times \{\alpha\}) = P_{t-\alpha}(x,B) \quad .$$

Proof. Let $p \in \mathcal{P}_b(\mathbb{R}^n \times \mathbb{R})$. Then $R_p^{\complement U} \in \mathcal{H}(U)$ by (III.2.6). Moreover $\hat{R}_p^{\complement U} = R_p^{\complement U}$ since $\complement U$ is absorbing by (6.13) and hence finely open. Therefore $\lim_{z\to z^*} R_p^{\complement U}(z) = p(z^*)$ for every $z \in U^*$.

On the other hand, let

$$h(x,t) = \begin{cases} (P_{t-\alpha}p(\cdot,\alpha))(x) & , t > \alpha \quad , \\ p(x,t) & , t \leq \alpha \quad . \end{cases}$$

By (6.1), $h \in \mathcal{H}(U)$ and $\lim_{z\to z^*} h(z) = p(z^*)$ for every $z^* \in U^*$. Let $g = R_p^{\complement U} - h$ and $a \in \mathbb{R}_+$ such that $p \leq a$. Then $g \in \mathcal{H}(U)$, $g = 0$ on $\complement U$, $\lim_{z\to z^*} g(z) = g(z^*)$ for every $z^* \in U^*$, and $|g| \leq a$ since $0 \leq R_p^{\complement U} \leq p$ and $(P_t)_{t>0}$ is sub-Markov. By (6.11), the function q defined by $q(x,t) = a \sup(0,(t-\alpha+1)^3)$ is a potential on $\mathbb{R}^n \times \mathbb{R}$, and of course $q \geq a$ on U. Applying (III.6.6) to g and $-g$ we thus conclude that $h = R_p^{\complement U}$. This proves the statement. $\lrcorner$

<u>6.15. REMARK.</u> The preceding proposition states that for every open set $U = \mathbb{R}^n \times]\alpha,\infty[$ the operator H_U defined at the beginning of this section is in fact the harmonic kernel associated with U (if we neglect the restriction of functions on U^* and the trivial extension by the identity on $\complement U$).

<u>6.16. COROLLARY.</u> $(A_t)_{t\in\overline{\mathbb{R}}}$ is the family of all absorbing subsets of $\mathbb{R}^n \times \mathbb{R}$.

Proof. Let A be an absorbing subset of $\mathbb{R}^n \times \mathbb{R}$. Then $1_{\complement A}$ is hyperharmonic on $\mathbb{R}^n \times \mathbb{R}$. Suppose that $(x,t) \in A$, let $\alpha < t$ and $U = \mathbb{R}^n \times]\alpha,\infty[$. Let $B \subset \mathbb{R}^n$ such that

$$(\mathbb{R}^n \times \{\alpha\}) \cap \complement A = B \times \{\alpha\} \quad .$$

Then B is open and by (6.14)

$$P_{t-\alpha}(x,B) = R^{\complement U}_{1_{\complement A}}(x,t) \leq 1_{\complement A}(x,t) = 0 \quad .$$

Therefore $B = \emptyset$, i.e. $\mathbb{R}^n \times \{\alpha\} \subset A$. Since A is closed we obtain that $A = A_s$ for some $s \in \overline{\mathbb{R}}$ $(A_{-\infty} = \emptyset$, $A_\infty = \mathbb{R}^n \times \mathbb{R})$. ┘

Given $z = (y,s) \in \mathbb{R}^n \times \mathbb{R}$ let w^z denote the function $(x,t) \mapsto w(x,t,y,s)$ on $\mathbb{R}^n \times \mathbb{R}$.

6.17. COROLLARY. For every $z \in \mathbb{R}^n \times \mathbb{R}$, the function w^z is a potential which is harmonic on $\complement\{z\}$.

Proof. Since $w^z \in C^\infty(\complement\{z\})$ and $\Delta w^z = 0$ on $\complement\{z\}$ we know by (6.11) that w^z is harmonic on $\complement\{z\}$. Moreover, $\liminf_{z' \to z} w^z(z') = 0 = w^z(z)$ and $H_U w^z(z) = 0$ for every $U \in \mathcal{U}$ such that $z \in U$ by (6.13) and (1.2). Hence w^z is hyperharmonic by (III.4.4). Proceeding as in the proof of (4.4) we finally conclude that w^z is a potential. ┘

6.18. THEOREM. $^*\mathcal{H}^+(\mathbb{R}^n \times \mathbb{R}) = E_{\mathbb{P} \otimes \mathbb{T}}$.

Proof. Let $u \in {}^*\mathcal{H}^+(\mathbb{R}^n \times \mathbb{R})$, $s > 0$, $(x,t) \in \mathbb{R}^n \times \mathbb{R}$ and $U = \mathbb{R}^n \times]t-s,\infty[$. Then by (6.14)

$$(P_s \otimes T_s)u(x,t) = (P_s u(\cdot,t-s))(x) = R^{\complement U}_u(x,t) \leq u(x,t) \quad .$$

Therefore $u \in S_{\mathbb{P} \otimes \mathbb{T}}$. Since $\lim_{s \to 0} (P_s \otimes T_s)f = f$ for every $f \in \mathcal{K}(\mathbb{R}^n \times \mathbb{R})$ we conclude as usual that $u \in E_{\mathbb{P} \otimes \mathbb{T}}$.

Let W denote the potential kernel of $\mathbb{P} \otimes \mathbb{T}$. Then for every $f \in \mathcal{B}^+(\mathbb{R}^n \times \mathbb{R})$ and $(x,t) \in \mathbb{R}^n \times \mathbb{R}$, by (5.8) and (6.17)

$$Wf = \int w^z f(z)\, \lambda^{n+1}(dz) \in {}^*\mathcal{H}^+(\mathbb{R}^{n+1}) \quad .$$

Since W is a proper kernel we thus obtain by (II.3.11) that $E_{\mathbb{P} \otimes \mathbb{T}} \subset {}^*\mathcal{H}^+(\mathbb{R}^{n+1})$. ┘

6.19. COROLLARY. Let $u \geq 0$ be a function on $\mathbb{R}^n$ which is hyperharmonic with respect to classical potential theory and let $v \geq 0$ be a l.s.c. increasing function on $\mathbb{R}$. Then the function $(x,t) \mapsto u(x)v(t)$ is hyperharmonic with respect to the heat equation.

Proof. (6.18) and (5.2.1). ┘

6.20. EXERCISES. 1. Let $u : \mathbb{R} \to [0,\infty]$ and define $v : \mathbb{R}^n \times \mathbb{R} \to [0,\infty]$ by $v(x,t) = u(t)$.

a) Then v is hyperharmonic (superharmonic resp.) if and only if u is l.s.c. and increasing (l.s.c., increasing and real resp.).

b) Suppose that u is l.s.c., increasing, and real. Then v is a potential (harmonic resp.) if and only if $\lim_{t\to-\infty} u(t) = 0$ (u is constant resp.).

2. Every bounded harmonic function on $\mathbb{R}^n \times \mathbb{R}$ is constant.

3. Let $\alpha \in \mathbb{R}$ and $B = \{(x,t) \in \mathbb{R}^n \times \mathbb{R} : t \geq \alpha\}$. Then $\overset{\circ}{\varepsilon}{}_x^B = 0$ for every $x \in \complement B$.

7. Brownian Semigroups on the Infinite Dimensional Torus

In this section we shall first consider the Brownian semigroup $\overline{\mathbb{P}} = (\overline{P}_t)_{t>0}$ on the torus $T = \mathbb{R}/_{2\pi\mathbb{Z}}$. Finally we shall see that for every sequence $A = (a_n)$ of strictly positive real numbers satisfying $\sum_{n=1}^{\infty} e^{-ta_n} < \infty$ for every $t > 0$ the infinite products $P_t^A = \bigotimes_{n=1}^{\infty} \overline{P}_{a_n t}$ form a strong Feller semigroup $\mathbb{P}^A$ on the infinite dimensional torus T^∞. This will lead to harmonic spaces $(T^\infty, E_{(\mathbb{P}^A)\alpha})$, $\alpha > 0$, and $(T^\infty \times \mathbb{R}, E_{\mathbb{P}^A \otimes \mathbb{T}})$.

Let $\mathbb{P} = (P_t)_{t>0}$ be the Brownian semigroup on $\mathbb{R}$. The Brownian semigroup $\overline{\mathbb{P}}$ on T is obtained by considering $\mathbb{P}$ modulo 2π. To be more specific let $j : x \mapsto \overline{x}$ denote the quotient map from $\mathbb{R}$ on $\mathbb{R}/_{2\pi\mathbb{Z}}$. The mapping $f \mapsto f \circ j$ yields a one-to-one correspondence between numerical functions on T and numerical functions on $\mathbb{R}$ which are periodic with period 2π. If $f \in \mathcal{B}^+(T)$ and $t > 0$ then $P_t(f \circ j)$ is periodic with period 2π. Hence we may define kernels $\overline{P}_t$ on T, $t > 0$, by

$$(\overline{P}_t f) \circ j = P_t(f \circ j) \quad (f \in \mathcal{B}^+(T)) \quad .$$

Clearly, $P_t 1 = 1$ and $P_t(\mathcal{B}_b(\mathbb{R})) \subset C_b(\mathbb{R})$ imply that $\overline{P}_t 1 = 1$ and $\overline{P}_t(\mathcal{B}_b(T)) \subset C(T)$. Moreover, for all $s,t \in \mathbb{R}_+^*$ and $f \in \mathcal{B}^+(T)$,

$$(\overline{P}_s\overline{P}_t f)\circ j = P_s[(\overline{P}_t f)\circ j] = P_sP_t(f\circ j) = P_{s+t}(f\circ j) = (\overline{P}_{s+t}f)\circ j \quad .$$

Therefore, $\overline{\mathbb{P}} = (\overline{P}_t)_{t>0}$ is a strong Feller Markov semigroup on the compact space T. For obvious reasons it is called *Brownian semigroup on* T. If $f \in C(T)$ then $\lim_{t\to 0} \|P_t(f\circ j) - f\circ j\| = 0$ by (0.2.3), hence $\lim_{t\to 0} \|\overline{P}_t f - f\| = 0$. Moreover, $\lim_{t\to 0} \frac{1}{t}\overline{P}_t f = 0$ locally uniformly on $T\setminus S(f)$ by (0.2.4).

The *Brownian semigroup on* T^n, $n \geq 1$, is defined as the direct product of n copies of $\overline{\mathbb{P}}$. By reasons which will become clear when we treat the infinite dimensional case we want to consider a slight generalization of this semigroup. Given a finite sequence $A = (a_1,\dots,a_n)$ of strictly positive real numbers we define Markov kernels P_t^A on T^n by

$$P_t^A = \overline{P}_{a_1t} \otimes \dots\dots \otimes \overline{P}_{a_nt} \qquad (t>0) .$$

<u>7.1. PROPOSITION.</u> Let $A = (a_1,\dots,a_n)$ be a finite sequence in $\mathbb{R}_+^*$. Then $\mathbb{P}^A = (P_t^A)_{t>0}$ is a strong Feller Markov semigroup on T^n such that $\lim_{t\to 0} \|P_t^A f - f\| = 0$ and $\lim_{t\to 0} \frac{1}{t}P_t^A f = 0$ locally uniformly on $\complement S(f)$ for every $f \in C(T^n)$. In particular, $(T^n, E_{(\mathbb{P}^A)^\alpha})$ is a harmonic space for every $\alpha > 0$. Moreover, $(T^n \times \mathbb{R}, E_{\mathbb{P}^A \otimes \mathbb{T}})$ is a harmonic space.

Proof. The first part of the proposition follows from (5.2) and the preceding considerations.

Fix $\alpha > 0$. Then the potential kernel of the semigroup $(\mathbb{P}^A)^\alpha = (e^{-\alpha t}P_t^A)_{t>0}$ is proper. Since $1 \in E_{\mathbb{P}^A}$ and $1 \in C(T^n) = C_0(T^n)$ we thus conclude by (II.3.17) and (2.4) that $(T^n, E_{(\mathbb{P}^A)^\alpha})$ is a balayage space. By (III.8.2) and (III.7.3) it is even a harmonic space.

Define $u,v : T^n \times \mathbb{R} \to \mathbb{R}_+^*$ by

$$u(x,s) = \begin{cases} 1 & , s>0 , \\ \frac{1}{|s|+1} & , s \leq 0 , \end{cases} \qquad v(x,s) = \begin{cases} s+1 , & s \geq 0 , \\ 1 & , s<0 . \end{cases}$$

Then $u,v \in E_{\mathbb{P}^A \otimes \mathbb{T}} \cap C(T^n \times \mathbb{R})$ and $u \in o(v)$. Hence $(T^n \times \mathbb{R}, E_{\mathbb{P}^A \otimes \mathbb{T}})$ is a balayage space by (5.6). It is a harmonic space by (III.8.2) and (III.7.3). ∎

A definition of Brownian semigroups on the infinite dimensional torus T^∞

requires a closer look at the particular form of the kernels $\overline{P}_t$, $t > 0$. For a moment fix $t > 0$. Then for every $f \in \mathcal{B}^+(T)$ and $x \in \mathbb{R}$

$$P_t(f \circ j)(x) = \frac{1}{\sqrt{2\pi t}} \int_{-\infty}^{+\infty} e^{-\frac{(x-y)^2}{2t}} (f \circ j)(y)dy = \frac{1}{\sqrt{2\pi t}} \sum_{k \in \mathbb{Z}} \int_0^{2\pi} e^{-\frac{(x-y+2\pi k)^2}{2t}} f(y)dy \quad .$$

Defining a function $\overline{g}_t : \mathbb{R} \to \mathbb{R}_+$ by

$$\overline{g}_t(x) = \frac{1}{\sqrt{2\pi t}} \sum_{k \in \mathbb{Z}} e^{-\frac{(x+2\pi k)^2}{2t}}$$

we hence obtain that

$$\overline{P}_t f(\overline{x}) = P_t(f \circ j)(x) = \int_0^{2\pi} \overline{g}_t(x-y)(f \circ j)(y)dy \quad .$$

Clearly, $\overline{g}_t$ is periodic with period 2π . Moreover, $\overline{g}_t$ is continuously differentiable and for every $n \in \mathbb{N}_0$

$$\frac{1}{\pi} \int_0^{2\pi} e^{-inx}\, \overline{g}_t(x)dx = \frac{1}{\pi} \int_{-\infty}^{+\infty} e^{-inx} \frac{1}{\sqrt{2\pi t}} e^{-\frac{x^2}{2t}} dx = \frac{1}{\pi} e^{-\frac{t}{2}n^2}$$

by a well known formula for the Fourier transform of a normal distribution. Hence for every $x \in \mathbb{R}$

$$\overline{g}_t(x) = \frac{1}{2\pi} + \frac{1}{\pi} \sum_{n=1}^{\infty} e^{-\frac{t}{2}n^2} \cos nx \quad .$$

Let

$$\varphi(t) = \sum_{n=1}^{\infty} e^{-\frac{t}{2}n^2} \quad .$$

Then

$$|\overline{g}_t - \frac{1}{2\pi}| \leq \frac{1}{\pi}\varphi(t) \quad .$$

We note that clearly

$$\lim_{t \to \infty} \varphi(t) e^{\frac{t}{2}} = 1 \quad .$$

Given an infinite sequence $A = (a_k)$ of strictly positive real numbers we define probability measures $P_t^A(x,\cdot)$, $t > 0$, $x \in T^\infty$, on T^∞ by

$$P_t^A(x,\cdot) = \bigotimes_{k=1}^{\infty} \overline{P}_{a_k t}(x_k,\cdot) \quad .$$

For every $f \in \mathcal{B}^+(T)$ we obtain a function $P_t^A f : T^\infty \to [0,\infty]$ by

$$P_t^A f(x) = \int f(y) P_t^A(x,dy) \qquad (x \in T^\infty) \; .$$

For every $n \in \mathbb{N}$ let $A|n = (a_1, \ldots, a_n)$ and let π_n be the canonical projection from T^∞ on T^n, i.e.

$$\pi_n((x_k)) = (x_1, \ldots, x_n) \qquad ((x_k) \in T^\infty) .$$

It is an immediate consequence of the definitions that for every $g \in \mathcal{B}^+(T^n)$

$$P_t^A(g \circ \pi_n) = (P_t^{A|n} g) \circ \pi_n .$$

This implies that $\mathbb{P}^A = (P_t^A)_{t>0}$ is a Markov semigroup on T^∞.

<u>7.2. THEOREM.</u> Let $A = (a_k)$ be an infinite sequence in $\mathbb{R}_+^*$ such that $\sum_{k=1}^{\infty} e^{-ta_k} < \infty$ for every $t > 0$. Then $\mathbb{P}^A$ is a strong Feller Markov semigroup on T^∞ such that $\lim_{t\to 0} \frac{1}{t} P_t^A f = 0$ locally uniformly on $\complement S(f)$ for every $f \in C(T^\infty)$. In particular, $(T^\infty, \mathcal{E}_{(\mathbb{P}^A)_\alpha})$ is a harmonic space for every $\alpha > 0$. Moreover, $(T^\infty \times \mathbb{R}, \mathcal{E}_{\mathbb{P}^A \otimes \mathbb{T}})$ is a harmonic space.

Proof. Let $s_0 > 0$ such that $e^{-\frac{s_0}{2}} < \frac{1}{15}$ and consider $s \geq s_0$. Then

$$\varphi(s) = \sum_{n=1}^{\infty} e^{-\frac{s}{2}n^2} \leq \frac{e^{-\frac{s}{2}}}{1 - e^{-\frac{s}{2}}} \leq \frac{15}{14} e^{-\frac{s}{2}} \leq \frac{1}{14} .$$

Let us recall that $|\bar{g}_s - \frac{1}{2\pi}| \leq \frac{1}{\pi}\varphi(s)$. This inequality implies that for all $z_1, z_2 \in \mathbb{R}$

$$\bar{g}_s(z_1) \leq \bar{g}_s(z_2) + \frac{2}{\pi}\varphi(s) \leq \bar{g}_s(z_2) + \frac{2}{\pi}\varphi(s) \frac{\bar{g}_s(z_2)}{\frac{1}{2\pi} - \frac{1}{\pi}\varphi(s)}$$

$$= \left(1 + \frac{4\varphi(s)}{1-2\varphi(s)}\right)\bar{g}_s(z_2) \leq \left(1 + 5e^{-\frac{s}{2}}\right)\bar{g}_s(z_2) .$$

Hence for all $x, y \in T$

$$\bar{P}_s(x,\cdot) \leq \left(1 + 5e^{-\frac{s}{2}}\right)\bar{P}_s(y,\cdot) .$$

Fix $t > 0$ and $f \in \mathcal{B}(T^\infty)$ such that $0 \leq f \leq 1$. There exists $n \in \mathbb{N}$ such that $a_k t \geq s_0$ for every $k \geq n$ and

$$\prod_{k=n+1}^{\infty} \left(1 + 5e^{-\frac{a_k t}{2}}\right) \leq 1 + \varepsilon .$$

Let $A' = (a_{n+1}, a_{n+2}, \ldots)$. Then for all $x, y \in T^\infty$

$$P_t^{A'}(x,\cdot) = \bigotimes_{k=1}^{\infty} \overline{P}_{a_{n+k}t}(x_k,\cdot) \leq (1+\varepsilon) \bigotimes_{k=1}^{\infty} \overline{P}_{a_{n+k}t}(y_k,\cdot) = (1+\varepsilon)P_t^{A'}(y,\cdot) \quad .$$

For every $x \in T^\infty$ define $g_x : T^n \to [0,1]$ by

$$g_x(z) = \int f(z,z')P_t^{A'}(x,dz') \qquad (z \in T^n) \quad .$$

The inequality above yields that $g_x \leq (1+\varepsilon)g_y \leq g_y + \varepsilon$ and $g_y \leq (1+\varepsilon)g_x \leq g_x + \varepsilon$ for all $x,y \in T^\infty$, i.e.

$$|g_x - g_y| \leq \varepsilon \quad .$$

Given $x \in T^\infty$ let $x' = (x_{n+1}, x_{n+2}, \ldots\ldots)$. Obviously

$$P_t^A f(x) = P_t^{A|n} g_{x'}(\pi_n(x)) \quad .$$

Now fix $x \in T^\infty$. Since $P_t^{A|n}$ is a strong Feller kernel, there exists an open neighborhood U of $\pi_n(x)$ in T^n such that

$$|P_t^{A|n} g_{x'} - P_t^{A|n} g_{x'}(\pi_n(x))| < \varepsilon \quad \text{on} \quad U \quad .$$

Then for all $y \in \pi_n^{-1}(U)$

$$|P_t^A f(x) - P_t^A f(y)| = |P_t^{A|n} g_{x'}(\pi_n(x)) - P_t^{A|n} g_{y'}(\pi_n(y))|$$

$$\leq |P_t^{A|n} g_{x'}(\pi_n(x)) - P_t^{A|n} g_{x'}(\pi_n(y))| + |P_t^{A|n} g_{x'}(\pi_n(y)) - P_t^{A|n} g_{y'}(\pi_n(y))| \leq 2\varepsilon$$

since $|g_{x'} - g_{y'}| \leq \varepsilon$. So P_t^A is a strong Feller kernel.

Suppose now that $f \in C(T^\infty)$, $0 \leq f \leq 1$, and consider $x \in \complement S(f)$. Then there exists $n \in \mathbb{N}$ and an open subset V of T^n such that $x \in \pi_n^{-1}(V) \subset \complement S(f)$. We choose a function $g \in C(T^n)$ such that $0 \leq g \leq 1$, $\pi_n(x) \in \complement S(g)$ and $g = 1$ on $\complement V$. Then $f \leq g \circ \pi_n$. Let K be a compact neighborhood of $\pi_n(x)$ in $\complement S(g)$. Then by (7.1) $\lim\limits_{t\to 0} \frac{1}{t} P_t^{A|n} g = 0$ uniformly on K. Since

$$\frac{1}{t}P_t^A f \leq \frac{1}{t}P_t^A(g \circ \pi_n) = \frac{1}{t}(P_t^{A|n} g) \circ \pi_n \quad ,$$

we obtain that $\lim\limits_{t\to 0} \frac{1}{t}P_t^A f = 0$ uniformly on the compact neighborhood $\pi_n^{-1}(K)$ of x.

This finishes the proof of the first part of the theorem. The second part follows exactly as in the finite dimensional case. ┘

7.3. REMARK. Let us finally note that our condition on the sequence (a_k) in R_+^* expresses a fairly slow convergence of (a_k) to infinity. If for example

$\sum_{k=1}^{\infty} \frac{1}{a_k^m} < \infty$ for some $m \in \mathbb{N}$ then $\sum_{k=1}^{\infty} e^{-ta_k} < \infty$ for every $t > 0$ since $e^{-ta_k} \leq \frac{m!}{t^m} \frac{1}{a_k^m}$.

7.4. EXERCISE. Let $\overline{\mathbb{P}} = (\overline{P}_t)_{t>0}$ be the pseudo-Poisson semigroup on $\{0,1\}$ given by $P(0,\cdot) = \varepsilon_1$, $P(1,\cdot) = \varepsilon_0$ (see (II.3.1.3)). Let $A = (a_k)$ be an infinite sequence in $\mathbb{R}_+^*$ such that $\sum_{k=1}^{\infty} e^{-ta_k} < \infty$ for every $t > 0$. Defining $P_t^A(x,\cdot) = \bigotimes_{k=1}^{\infty} \overline{P}_{a_k t}(x_k,\cdot)$ we obtain a strong Feller semigroup $\mathbb{P}^A$ on $\{0,1\}^{\mathbb{N}}$ such that $\lim_{t\to 0} P_t^A f = f$ for every $f \in C(\{0,1\}^{\mathbb{N}})$.

8. Images

Let $(X,\mathcal{W})$ be a balayage space and let π be an open continuous mapping of X onto a locally compact space X_π with countable base. Let $\mathcal{W}_\pi$ denote the set of all positive numerical functions w on X_π such that $w \circ \pi \in \mathcal{W}$. In order to show that $(X_\pi, \mathcal{W}_\pi)$ is a balayage space we suppose that the following properties hold:

(I_1) If $f : X_\pi \to \overline{\mathbb{R}}_+$ is l.s.c. then $R_{f\circ\pi} = g \circ \pi$ for some $g : X_\pi \to \overline{\mathbb{R}}_+$.

(I_2) $\mathcal{W}_\pi$ is linearly separating.

(I_3) There exists a strictly positive function in $\mathcal{W}_\pi \cap C(X_\pi)$.

Clearly, (I_2) and (I_3) are necessary properties. Moreover, (I_3) is trivially satisfied if $1 \in \mathcal{W}$ and then (I_2) reduces to point separation by $\mathcal{W}_\pi$.

Before going into details let us look at some special cases.

8.1. EXAMPLES. 1. Product spaces. Let $\mathbb{P}$ be a sub-Markov semigroup on X and $\widetilde{\mathbb{P}}$ a Markov semigroup on $\widetilde{X}$ such that $(X \times \widetilde{X}, E_{\mathbb{P}\otimes\widetilde{\mathbb{P}}})$ is a balayage space. If $\pi : X \times \widetilde{X} \to X$ denotes the projection then $((X \times \widetilde{X})_\pi , (E_{\mathbb{P}\otimes\widetilde{\mathbb{P}}})_\pi) = (X, E_{\mathbb{P}})$. Moreover, (I_1) holds by (5.11), and (I_3) is satisfied since $1 \in E_{\mathbb{P}}$.

2. Orbit spaces. Let G be a group of *automorphisms* on a balayage space $(X,\mathcal{W})$, i.e. a group of homeomorphisms $\sigma : X \to X$ such that $w \circ \sigma \in \mathcal{W}$ for every $w \in \mathcal{W}$, let $\pi : X \to X/G = X_\pi$ denote the quotient mapping $x \mapsto Gx$, and suppose that points lying on different orbits are linearly separated by the set $\mathcal{W}^G$ of all G-invariant

functions in $\mathcal{W}$ and that there exists a strictly positive function in $\mathcal{W}^G \cap C(X)$. Then π is open and continuous, and the mapping $w \mapsto w \circ \pi$ is a one-to-one correspondence between $\mathcal{W}_\pi$ and $\mathcal{W}^G$. If $X/_G$ is a Hausdorff space (which is obviously true if $\mathcal{W}^G \cap C(X)$ linearly separates the orbits of X), then $X/_G$ is a locally compact space with countable base. Now fix a positive numerical function f on X_π and let $\sigma \in G$. If $w \in \mathcal{W}$ such that $w \geq f \circ \pi$ then $w \circ \sigma \geq f \circ \pi \circ \sigma = f \circ \pi$ and hence $w \circ \sigma \geq R_{f \circ \pi}$. Therefore $R_{f \circ \pi} \circ \sigma \geq R_{f \circ \pi}$. Similarly, $R_{f \circ \pi} \circ \sigma^{-1} \geq R_{f \circ \pi}$ and hence $R_{f \circ \pi} \geq R_{f \circ \pi} \circ \sigma$. So $R_{f \circ \pi}$ is G-invariant and there exists a function g on X_π such that $g \circ \pi = R_{f \circ \pi}$.

Let us return to the general situation. Obviously, $\mathcal{W}_\pi$ is a convex cone of l.s.c. positive functions on X_π and the mapping π is finely continuous, i.e. the inverse image $\pi^{-1}(V)$ of every $\mathcal{W}_\pi$-finely open subset V of X_π is a $\mathcal{W}$-finely open subset of X.

8.2. LEMMA. Let f be a l.s.c. function on X_π. Then $R_f \in \mathcal{W}_\pi$, $R_f \circ \pi = R_{f \circ \pi}$.

Proof. The function $f \circ \pi$ is l.s.c. on X. So $R_{f \circ \pi} \in \mathcal{W}$ and there exists a positive numerical function g on X_π such that $R_{f \circ \pi} = g \circ \pi$. Then $g \in \mathcal{W}_\pi$ and $g \geq f$, hence $g \geq R_f$. Therefore $R_{f \circ \pi} = g \circ \pi \geq R_f \circ \pi$. On the other hand, if $w \in \mathcal{W}_\pi$ such that $w \geq f$, then $w \circ \pi \in \mathcal{W}$ and $w \circ \pi \geq f \circ \pi$, hence $w \circ \pi \geq R_{f \circ \pi}$. Thus conversely $R_f \circ \pi \geq R_{f \circ \pi}$. ┘

8.3. LEMMA. $\mathcal{W}_\pi \cap C(X_\pi)$ is linearly separating.

Proof. If suffices to show that $\mathcal{W}_\pi = S(\mathcal{W}_\pi \cap C(X_\pi))$. So fix $w \in \mathcal{W}_\pi$. Let w_0 be a strictly positive function in $\mathcal{W}_\pi \cap C(X_\pi)$. Then there exists an increasing sequence (f_n) in $C(X_\pi)$ such that $\sup f_n = w$ and each function f_n is bounded by a multiple of w_0. For every $n \in \mathbb{N}$, $R_{f_n} \circ \pi = R_{f_n \circ \pi} \in \mathcal{W} \cap C(X)$ since the continuous function $f_n \circ \pi$ is bounded by a multiple of $w_0 \circ \pi$, hence $R_{f_n} \in \mathcal{W}_\pi \cap C(X_\pi)$. Clearly, the sequence (R_{f_n}) is increasing to w. ┘

For every $w \in \mathcal{W}_\pi \cap C(X_\pi)$ define

$$h_w = \inf \{R_w^{\complement L} : L \text{ compact} \subset X_\pi\}$$

(where of course $R_w^{\complement L} = R_{1_{\complement L} w}$) . By (8.2),

$$h_w \circ \pi = \inf \{R_{w \circ \pi}^{\complement \pi^{-1}(L)} : L \text{ compact} \subset X_\pi\} \quad .$$

If (K_n) is an exhaustion of X then $(\pi(K_n))$ is an exhaustion of X_π and hence

$$h_w = \inf R_w^{\complement \pi(K_n)} \quad , \quad h_w \circ \pi = \inf R_{w \circ \pi}^{\complement \pi^{-1}(\pi(K_n))} \quad .$$

By (III.2.5), for every $n \in \mathbb{N}$, the function $R_{w \circ \pi}^{\complement \pi^{-1}(\pi(K_n))}$ is harmonic on the open subset $\overset{\circ}{K}_n$ of $\pi^{-1}(\pi(K_n))$. Thus $h_w \circ \pi$ is harmonic on X by (III.2.3) and $h_w \in \mathcal{W}_\pi \cap C(X_\pi)$.

Let P_π denote the set of all $p \in \mathcal{W}_\pi \cap C(X_\pi)$ such that $h_p = 0$.

<u>8.4. LEMMA.</u> For every $z \in X_\pi$ there exists a function $p \in P_\pi$ such that $p(z) > 0$.

Proof. The proof will be given in two steps.

1. Fix $w \in \mathcal{W}_\pi \cap C(X_\pi)$ and define $p = w - h_w$. Since $h_w \circ \pi$ is a harmonic minorant of $w \circ \pi$, we have $w \circ \pi - h_w \circ \pi \in \mathcal{W} \cap C(X)$ and therefore $p \in \mathcal{W}_\pi \cap C(X_\pi)$. We claim that $h_p = 0$. Indeed, let (K_n) be a sequence of compact subsets of X such that $K_n \subset \overset{\circ}{K}_{n+1}$ for every $n \in \mathbb{N}$ and $\bigcup_{n=1}^{\infty} K_n = X$. For every $n \in \mathbb{N}$ let $U_n = \complement \pi^{-1}(\pi(K_n))$. Then, for every $x \in X$ and every $m \in \mathbb{N}$,

$$\overset{\circ}{\varepsilon}{}_x^{U_m}(h_w \circ \pi) = \inf_{n \geq m} \overset{\circ}{\varepsilon}{}_x^{U_m}(R_{w \circ \pi}^{U_n}) = \inf_{n \geq m} R_{R_{w \circ \pi}^{U_n}}^{U_m}(x) = \inf_{n \geq m} R_{w \circ \pi}^{U_n}(x) = h_w \circ \pi(x) \quad .$$

and therefore

$$h_p \circ \pi(x) = \lim_{m \to \infty} R_{p \circ \pi}^{U_m}(x) = \lim_{m \to \infty} (R_{w \circ \pi}^{U_m}(x) - \overset{\circ}{\varepsilon}{}_x^{U_m}(h_w \circ \pi)) = 0 \quad .$$

2. Now fix $z \in X_\pi$ and let $x \in X$ such that $\pi(x) = z$. By (1) it suffices to show that there exists a function $w \in \mathcal{W}_\pi \cap C(X_\pi)$ such that $h_w(z) < w(z)$. Let L be a compact neighborhood of z , $U = \complement \pi^{-1}(L)$, and $\mu = \overset{\circ}{\varepsilon}{}_x^U$. If $\mu = 0$ we choose $w \in \mathcal{W}_\pi \cap C(X_\pi)$ such that $w(z) > 0$ and obtain that $h_w(z) = h_w \circ \pi(x) \leq \mu(w \circ \pi) = 0 < w(z)$. So assume that $\mu \neq 0$ and choose a point y in the support of μ . Obviously, $\pi(y) \neq z$ since $y \in \overline{U} \subset \complement \pi^{-1}(\overset{\circ}{L})$. Hence by (8.3) there exist

$w_1, w_2 \in \mathcal{W}_\pi \cap C(X_\pi)$ such that $w_1(z)w_2(\pi(y)) < w_1(\pi(y))w_2(z)$. Defining

$$w = \inf\,(w_1(z)w_2\,,\; w_2(z)w_1) \quad \text{and} \quad w' = w_2(z)w_1$$

we have $w, w' \in \mathcal{W}_\pi \cap C(X_\pi)$, $w \leq w'$, $w(\pi(y)) < w'(\pi(y))$ and $w(z) = w'(z)$. So

$$h_w(z) = h_w \circ \pi(x) \leq \mu(w \circ \pi) < \mu(w' \circ \pi) \leq w' \circ \pi(x) = w(z) \quad .$$

8.5. THEOREM. $(X_\pi, \mathcal{W}_\pi)$ is a balayage space.

Proof. If (w_n) is an increasing sequence in $\mathcal{W}_\pi$ then $(w_n \circ \pi)$ is an increasing sequence in $\mathcal{W}$, hence $(\sup w_n) \circ \pi = \sup (w_n \circ \pi) \in \mathcal{W}$, i.e. $\sup w_n \in \mathcal{W}_\pi$.

Next consider a subset $\mathcal{V}$ of $\mathcal{W}_\pi$ and let $u = \inf \mathcal{V}$. Then $u \circ \pi$ is the infimum of the subset $\{v \circ \pi : v \in \mathcal{V}\}$ of $\mathcal{W}$, hence $\widehat{u \circ \pi} \in \mathcal{W}$. Since $\widehat{u \circ \pi} = \hat{u} \circ \pi$, we obtain that $\hat{u} \in \mathcal{W}_\pi$. Moreover, $\hat{u}^f \circ \pi \leq \widehat{u \circ \pi}^f$ since π is finely continuous. Therefore

$$\widehat{u \circ \pi} = \hat{u} \circ \pi \leq \hat{u}^f \circ \pi \leq \widehat{u \circ \pi}^f = \widehat{u \circ \pi} \quad .$$

So $\hat{u}^f = \hat{u} \in \mathcal{W}_\pi$.

Furthermore, $u - R_{u-v} \in \mathcal{W}_\pi \cap C(X_\pi)$ for all $u, v \in \mathcal{W}_\pi \cap C(X_\pi)$ Indeed, we have $u \circ \pi$, $v \circ \pi \in \mathcal{W} \cap C(X)$ and hence $(u - R_{u-v}) \circ \pi = u \circ \pi - R_{u \circ \pi - v \circ \pi} \in \mathcal{W} \cap C(X)$ by (8.2).

Let us recall that by definition

$$P_\pi = \{p \in \mathcal{W}_\pi \cap C(X_\pi) : \inf \{R_p^{\complement L} : L \text{ compact} \subset X_\pi\} = 0\} \quad .$$

It is easily verified that P_π is a convex cone and $(P_\pi)_\sigma = P_\pi$. We have to show that P_π is a function cone and $S(P_\pi) = \mathcal{W}_\pi$.

By (8.4), for every $z \in X_\pi$, there exists $p \in P_\pi$ such that $p(z) > 0$.

If $p \in P_\pi$ and $f \in C(X_\pi)$ such that $f \leq p$ then $R_f \in P_\pi$ since $R_f \leq p$ and $R_f \in \mathcal{W}_\pi \cap C(X_\pi)$ by (8.2).

Now fix $p \in P_\pi$, $x \in X_\pi$, and $\varepsilon > 0$. Choose a compact subset L of X_π such that $R_p^{\complement L}(x) < \varepsilon$ and a function $f \in C(X_\pi)$ such that $f \leq p$, $f = 0$ on L , and $\{f < p\}$ is relatively compact. Then $q := R_f \in P_\pi$, $\{q < p\} \subset \{f < p\}$, and $q(x) \leq R_p^{\complement L}(x) < \varepsilon$.

Given $w \in \mathcal{W}_\pi$, we choose an increasing sequence (φ_n) in $K^+(X_\pi)$ such that

$\sup \varphi_n = w$. Then (R_{φ_n}) is an increasing sequence in P_π such that $\sup R_{\varphi_n} = w$. So $\mathcal{W}_\pi = S(P_\pi)$. In particular, P_π is linearly separating. Thus $P_\pi = (P_\pi)_\sigma$ is a function cone by (I.1.1.3). An application of (II.4.5) finishes the proof. ┘

8.6. COROLLARY. Let $\mathbb{P}$ be a sub-Markov semigroup on X and $\widetilde{\mathbb{P}}$ a Markov semigroup on $\widetilde{X}$ such that $(X \times \widetilde{X}, E_{\mathbb{P} \otimes \widetilde{\mathbb{P}}})$ is a balayage space. Then $(X, E_{\mathbb{P}})$ is a balayage space provided $E_{\mathbb{P}}$ separates the points of X .

Proof. (8.1.1) and (8.5). ┘

Let us recall that a balayage space $(X,\mathcal{W})$ is a harmonic space if and only if $(X,\mathcal{W})$ has the local truncation property and X has no absorbing points. It is easily verified that the balayage space $(X_\pi, \mathcal{W}_\pi)$ has the local truncation property if $(X,\mathcal{W})$ has the local truncation property. Moreover, given $x \in X_\pi$, we have $(\infty \cdot 1_{\complement\{x\}}) \circ \pi = \infty \cdot 1_{\complement\pi^{-1}(x)}$, hence x is an absorbing point in the balayage space $(X_\pi, \mathcal{W}_\pi)$ if and only if $\pi^{-1}(x)$ is an absorbing set in the balayage space $(X,\mathcal{W})$. Thus (8.5) leads to the following result.

8.7. COROLLARY. Let $(X,\mathcal{W})$ be a harmonic space. Then $(X_\pi, \mathcal{W}_\pi)$ is a harmonic space if and only if none of the sets $\pi^{-1}(x)$, $x \in X_\pi$, is an absorbing subset of $(X,\mathcal{W})$.

Starting from our standard examples the construction of orbit spaces leads to further examples.

8.8. EXAMPLES

1. Reflection at a hyperplane. Let $X =]-a,a[\times \mathbb{R}^{n-1}$, $a \in]0,\infty]$, $n \in \mathbb{N}$, and let G be the group consisting of the identity on X and the reflection $\sigma : (x_1,x_2,\dots,x_n) \mapsto (-x_1,x_2,\dots,x_n)$ at the hyperplane $\{0\} \times \mathbb{R}^{n-1}$. Then $X/_G$ can be identified with the space $[0,a[\times \mathbb{R}^{n-1}$.

a) Classical potential theory ($a < \infty$ if $n \leq 2$). The quotient is again a harmonic space. Indeed, if $n \geq 3$ then for every $x \in X$ the function $w : z \mapsto \| z-x\|^{2-n} + \| z-\sigma(x)\|^{2-n}$ is a G-invariant positive hyperharmonic function on X such that $w(y) < w(x)$ $(= \infty)$ for every $y \in \complement\{x,\sigma(x)\}$. If $n = 1$, it suffices to consider the function $z \mapsto a - |z|$, and if $n = 2$, we may obtain the desired separation by the G-invariant positive harmonic functions $(x_1,x_2) \mapsto \cosh a + \cosh x_1 \cdot \cos(x_2+b)$, $b \in \mathbb{R}$. Moreover, clearly no orbit $\{x,\sigma\{x\}\}$ is absorbing.

b) Riesz potentials $(a = \infty)$. The quotient is a balayage space. The necessary point separation is obtained by the functions $z \mapsto \| z-x\|^{\alpha-n} + \| z-\sigma(x)\|^{\alpha-n}$, $x \in X$.

c) Heat equation. As in the classical case the quotient is a harmonic space. It suffices to consider the functions $w^x + w^{\sigma(x)}$, $x \in X$ (see (6.17)).

2. Discrete translation in one direction. Let $X = \mathbb{R}^n \times]-a,a[$, $n \in \mathbb{N}$, $a \in]0,\infty]$, and let G be the group of all translations

$$x \mapsto x + 2\pi k\,(1,0,\ldots\ldots,0) \qquad (k \in \mathbb{Z}) \quad .$$

Then X/G can be identified with the space $T \times \mathbb{R}^{n-1} \times]-a,+a[$ where T is the torus $\mathbb{R}/2\pi\mathbb{Z}$.

a) Classical potential theory. Suppose that $a < \infty$ if $n \le 2$. Then the quotient is a harmonic space. If $n \ge 3$, it suffices to note that $\sum_{\sigma\in G} \| z-\sigma(x)\|^{1-n} < \infty$ for every $z \in \complement Gx$. If $n = 1$ or $n = 2$, we obtain the necessary point separation by the G-invariant positive harmonic functions $x \mapsto e^a + \cos(x_i+b)e^{x_n}$, $b \in \mathbb{R}$, $1 \le i < n$.

b) Riesz potentials $(a = \infty)$. If $\alpha < n$, then the quotient is a balayage space since $\sum_{\sigma\in G} \| z-\sigma(x)\|^{\alpha-n-1} < \infty$ for every $z \in \complement Gx$.

c) Heat equation. The quotient is a harmonic space since $\sum_{\sigma\in G} w^{\sigma(x)} < \infty$ on $\complement Gx$.

We obtain another quotient if we consider translations along the time axis: Let $X =]-a,+a[\times \mathbb{R}^n$ and let G be the group of all translations $x \mapsto x + 2\pi k(0,\ldots,0,1)$, $k \in \mathbb{Z}$. Suppose that $a < \infty$ if $n \le 2$. Then the quotient is another harmonic space.

3. Translation in one direction. Let $X = \mathbb{R}^n \times]-a,a[$, $n \ge 1$, $a \in]0,\infty]$, and let G be the group of all translations

$$x \mapsto x + t(1,0,\ldots\ldots,0) \qquad (t \in \mathbb{R}) \quad .$$

Then X/G can be identified with the space $\mathbb{R}^{n-1} \times]-a,a[$.

a) Classical potential theory. Suppose that $a < \infty$ if $n \le 2$. Then the quotient yields the classical example on $\mathbb{R}^{n-1}\times]-a,a[$.

b) Heat equation. The quotient corresponds to the heat equation on $\mathbb{R}^{n-1}\times]-a,a[$. If $X = \mathbb{R}^{n-1}\times]-a,a[\times\mathbb{R}$, $a < \infty$ if $n \leq 2$, and if G is the group of all translations $x \to x + t(0,\ldots,0,1)$, $t \in \mathbb{R}$, then the quotient obtained from the heat equation example on X is the classical Laplace example on $\mathbb{R}^{n-1}\times]-a,a[$. The verification of the details is left to the reader.

4. Rotations. Let $0 < a \leq \infty$ and $X = \{x \in \mathbb{R}^n : \|x\| < a\}$, $n \geq 1$. Let G be the group $SO(n)$ acting on X . Then $X/_G$ can be identified with the interval $[0,a[$.

a) Classical potential theory ($a < \infty$ if $n \leq 2$) . The quotient is again a harmonic space. Indeed, define a G-invariant positive hyperharmonic function w on X by $w(z) = \|z\|^{2-n}$ if $n \geq 3$, $w(z) = \log a - \log\|z\|$ if $n = 2$, and $w(z) = a - |z|$ if $n = 1$. Then w separates any two points lying on different orbits.

b) Riesz potentials ($a = \infty$) . The quotient is a balayage space since the function $z \to \|z\|^{\alpha-n}$ separates any two points lying on different orbits.

8.9. EXERCISE. Let $X = \mathbb{R}^3 \times \{0,1\}$ and $\mathcal{W}$ the set of all numerical functions $w \geq 0$ on X such that the functions $z \mapsto w(z,0)$ and $z \mapsto w(z,1)$ are hyperharmonic functions on $\mathbb{R}^3$ (in the classical sense). Then $(X,\mathcal{W})$ is a harmonic space. If G is the group of all translations $(z,t) \mapsto (z+a,t)$, $a \in \mathbb{R}^3$, then the orbit space is certainly a balayage space, but not a harmonic space.

9. Further Examples

9.1. EXAMPLE. Let X be the topological sum of $\mathbb{R}$ and $\mathbb{R}^3$ and fix $z \in \mathbb{R}^3$. Connecting the harmonic structure on $\mathbb{R}$ given in (2.5) and the classical structure on $\mathbb{R}^3$ we shall construct a balayage space $(X,\mathcal{W})$ such that:

(1) $R_1^{\{z\}} = 1_{\mathbb{R}\cup\{z\}} \in \mathcal{H}(\complement\{z\})$, $\hat{R}_1^{\{z\}} = 1_{\mathbb{R}} \notin \mathcal{H}(\complement\{z\})$.

(2) Every locally bounded potential on X is a countable sum of continuous potentials.

Let $\mathcal{U}_1 = \{]a,b[: a,b \in \mathbb{R}, a < b\}$, $\mathcal{U}_3 = \{B_r(a) : a \in \mathbb{R}^3, r > 0\}$. Then $\mathcal{U} = \mathcal{U}_1 \cup \mathcal{U}_3$ is a base of relatively compact open subsets of X. Let $\mathbb{P}_i$ be the Brownian semigroup on $\mathbb{R}^i$, $\mathcal{W}_1 = E_{\mathbb{P}^{\frac{1}{2}}}$, $\mathcal{W}_3 = E_{\mathbb{P}_3}$, and let $(H_U)_{U \in \mathcal{U}_i}$ be the family of harmonic kernels associated with $(\mathbb{R}^i, \mathcal{W}_i)$, $i = 1,3$. We recall from (2.5) that for every $U =]a,b[\in \mathcal{U}_1$ and every $x \in U$

$$H_U(x,\cdot) = \frac{\sinh(b-x)}{\sinh(b-a)}\varepsilon_a + \frac{\sinh(x-a)}{\sinh(b-a)}\varepsilon_b ,$$

in particular $H_U 1(x) < 1$. For every $x \in U \in \mathcal{U}_1$ we define a probability measure $H_U^z(x,\cdot)$ on X by

$$H_U^z(x,\cdot) = H_U(x,\cdot) + (1-H_U 1(x))\varepsilon_z .$$

For every $x \in U \in \mathcal{U}_3$ let $H_U^z(x,\cdot) = H_U(x,\cdot)$. Extending as usual by $H_U^z(x,\cdot) = \varepsilon_x$ for every $U \in \mathcal{U}$ and $x \in X \setminus U$ we obtain a family $(H_U^z)_{U \in \mathcal{U}}$ of sweeping kernels on X.

We claim that $(H_U^z)_{U \in \mathcal{U}}$ is a family of harmonic kernels on X such that $1 \in \mathcal{H}(X)$ and ${}^*\mathcal{H}^+(X)$ is the set of all $u \in \mathcal{B}^+(X)$ such that $u|_{\mathbb{R}^3} \in \mathcal{W}_3$, $u(z) \leq u|_{\mathbb{R}}$, and $u|_{\mathbb{R}} - u(z) \in \mathcal{W}_1$ if $u(z) < \infty$.

Indeed, (H_1), (H_3) and (H_4) follow immediately. In order to prove (H_2) it suffices to consider $V,U \in \mathcal{U}_1$ such that $\overline{V} \subset U$. Then for every $f \in \mathcal{B}_b(X)$ and $x \in V$

$$\begin{aligned} H_V^z H_U^z f(x) &= H_V H_U^z f(x) + (1 - H_V 1(x))\, H_U^z f(z) \\ &= H_V(H_U f + (1 - H_U 1) f(z))(x) + (1 - H_V 1(x)) f(z) \\ &= H_U f(x) + (1 - H_U 1(x)) f(z) = H_U^z f(x) . \end{aligned}$$

Therefore $H_V^z H_U^z = H_U^z$.

Obviously $1 \in \mathcal{H}(X)$. Let $u \in {}^*\mathcal{H}^+(X)$. Since $\lim\limits_{n\to\infty} H_{]-n,n[}1 = 0$ on $\mathbb{R}$ we obtain that for every $x \in \mathbb{R}$

$$u(z) \leq \lim_{n\to\infty} H^z_{]-n,n[}u(x) \leq u(x) .$$

If $u(z) < \infty$ then for every $x \in U \in \mathcal{U}_1$

$$H_U(u-u(z))(x) = H_U^z(u-u(z))(x) = H_U^z u(x) - u(z) \le u(x) - u(z)$$

and hence $u|_{\mathbb{R}} - u(z) \in \mathcal{W}_1$. Moreover, obviously $u|_{\mathbb{R}^3} \in \mathcal{W}_3$. Suppose now conversely that $u \in \mathcal{B}^+(X)$ such that $u|_{\mathbb{R}^3} \in \mathcal{W}_3$, $u(z) \le u|_{\mathbb{R}}$ and $u|_{\mathbb{R}} - u(z) \in \mathcal{W}_1$ if $u(z) < \infty$. If $u(z) = \infty$ then obviously $u \in {}^*\mathcal{H}^+(X)$. So let $u(z) < \infty$. Then for every $x \in U \in \mathcal{U}_1$

$$H_U^z u(x) = H_U(u-u(z))(x) + u(z) \le u(x) - u(z) + u(z) = u(x) \quad .$$

Therefore $u \in {}^*\mathcal{H}^+(X)$. Using the fact that $\mathcal{W}_i$ separates the points of $\mathbb{R}^i$, $i = 1,3$, we now conclude easily that ${}^*\mathcal{H}^+(X)$ separates the points of X .

Thus $(H_U^z)_{U \in \mathcal{U}}$ is a family of harmonic kernels on X . We have $R_1^{\{z\}} = 1_{\mathbb{R} \cup \{z\}}$, $\hat{R}_1^{\{z\}} = 1_{\mathbb{R}}$. Note that of course $R_1^{\{z\}} \in \mathcal{H}(X \setminus \{z\})$ whereas for every $U \in \mathcal{U}_1$ and $x \in U$, $H_U^z \hat{R}_1^{\{z\}}(x) < 1 = \hat{R}_1^{\{z\}}(x)$. Finally, let p be a locally bounded potential on X . Then $p|_{\mathbb{R}} - p(z) \in \mathcal{W}_1 \cap \mathcal{C}(\mathbb{R})$ since the $\mathcal{W}_1$-fine topology is the euclidean topology on $\mathbb{R}$. Furthermore, $p|_{\mathbb{R}^3}$ is a locally bounded potential with respect to $(\mathbb{R}^3, \mathcal{W}_3)$. Hence by (4.12) there exists a sequence $(p_n)_{n \ge 1}$ in $\mathcal{W}_3 \cap \mathcal{C}(\mathbb{R}^3)$ such that $p|_{\mathbb{R}^3} = \sum_{n=1}^{\infty} p_n$. Defining

$$q_0 = \begin{cases} p - p(z) & \text{on } \mathbb{R} \\ 0 & \text{on } \mathbb{R}^3 \end{cases}$$

and

$$q_n = \begin{cases} p_n(z) & \text{on } \mathbb{R} \\ p_n & \text{on } \mathbb{R}^3 \end{cases} \quad (n \ge 1)$$

we obtain a sequence $(q_n)_{n \ge 0}$ in $\mathcal{W} \cap \mathcal{C}(X)$ such that $p = \sum_{n=0}^{\infty} q_n$.

<u>9.2. EXERCISES.</u> 1. Let $(X,\mathcal{W})$ be a discrete balayage space such that $1 \in \mathcal{W}$. Given $x,y \in X$ let $p_{xy} = \overset{\circ}{\varepsilon}{}_x^{\complement\{x\}}(\{y\})$ and $e_x = 1 - \sum_{y \in X} p_{xy}$. Let X_a be the discrete space obtained from X by adding an isolated point a and let $\mathcal{W}$ be the set of all positive numerical functions on X_a such that for every $x \in X$

$$\sum_{y \in X} p_{xy} f(y) + e_x f(a) \le f(x) \quad .$$

Then $(X_a, \mathcal{W}_a)$ is a balayage space such that $1 \in \mathcal{H}(X)$, a is an absorbing point and $\mathcal{W}_a^X = \mathcal{W}$.

2. Let $(\{1, \ldots, n\}, \mathcal{W})$ be a (finite) discrete balayage space. For all $i,j \in \{1, \ldots, n\}$ let

$$\alpha_{ij} = \overset{o}{\varepsilon}_i^{\{i\}}(\{j\})$$

and suppose that $\sum_{j=1}^{n} \alpha_{ij} \in \{0,1\}$ for all $1 \leq i \leq n$. For every $1 \leq i \leq n$ let $z_i = (\delta_{1i}, \ldots, \delta_{ni})$ and let $X \subset \mathbb{R}^n$ be the union of all closed line segments $[z_1, z_j]$, $1 \leq i < j \leq n$. For $i,j \in \{1, \ldots, n\}$ and $0 \leq \mu < \lambda \leq 1$ define

$$z_{ij}(\lambda) = (1-\lambda)z_i + \lambda z_j \quad ,$$

$$U_i(\lambda) = \bigcup_{j=1}^{n} [z_i, z_{ij}(\lambda)[\quad ,$$

$$U_{ij}(\mu,\lambda) =]z_{ij}(\mu) \, , \, z_{ij}(\lambda)[\quad ,$$

$$H_{U_i(\lambda)}(z_i,\cdot) = \sum_{j=1}^{n} \alpha_{ij}\varepsilon_{z_{ij}(\lambda)} \, ,$$

$$H_{U_i(\lambda)}(z_{ij}(\rho),\cdot) = \frac{\lambda-\rho}{\lambda} H_{U_i(\lambda)}(z_i,\cdot) + \frac{\rho}{\lambda}\varepsilon_{z_{ij}(\lambda)} \, , \text{ if } 0 < \rho < \lambda \quad ,$$

$$H_{U_{ij}(\mu,\lambda)}(z_{ij}(\rho),\cdot) = \frac{\lambda-\rho}{\lambda-\mu}\varepsilon_{z_{ij}} + \frac{\rho-\mu}{\lambda-\mu}\varepsilon_{z_{ij}(\lambda)} \quad .$$

a) These definitions lead to a family $(H_U)_{U \in \mathcal{U}}$ of harmonic kernels on X.

b) For all $a_1, \ldots, a_n \in \mathbb{R}_+$ there exists a unique function $f \in \mathcal{H}(X \setminus \{z_1, \ldots, z_n\}) \cap \mathcal{C}(X)$ such that $f(z_i) = a_i$ for all $1 \leq i \leq n$. Moreover $f \in {}^*\mathcal{H}^+(X)$ if and only if the function $i \mapsto a_i$ is contained in $\mathcal{W}$.

c) For all $i,j \in \{1, \ldots, n\}$ and every subset A of $\{1, \ldots, n\}$,

$$\overset{o}{\varepsilon}_{z_i}^{\{z_k : k \in A\}}(\{z_j\}) = \overset{o}{\varepsilon}_i^A(\{j\}) \quad .$$

3. For every finite discrete balayage space $(X,\mathcal{W})$ there exists a connected balayage space $(\tilde{X},\tilde{\mathcal{W}})$ having the local truncation property such that $X \subset \tilde{X}$ and $\tilde{\mathcal{W}}|_X = \mathcal{W}$. (Hint: Use (II.7.9), (9.2.1) and (9.2.2).)

VI. Balayage Theory

The content of this chapter is the main issue of the book. Almost all principal results in potential theory depend on properties of the reduit and its regularization, the balayage. Therefore the study of the functions

$$R : (u,A,x) \mapsto R_u^A(x) \quad \text{and} \quad \hat{R} : (u,A,x) \mapsto \hat{R}_u^A(x)$$

is the most important tool.

The functions R and $\hat{R}$ are additive in $u \in \mathcal{W}$, the function R is a capacity in $A \in \mathfrak{P}(X)$ (section 1). In particular, the additivity of $\hat{R}$ allows the introduction of balayage of measures, and the important equality $\hat{R}_u^A = R_u^A$ on $\complement A$ follows by capacitability arguments (section 2). The identification of balayage measures as hitting distributions of associated Hunt processes is the key to probabilistic interpretations of analytic notions and results (section 3). The important base operation on subsets of X yields a characterization of the fine topology and finely continuous functions (section 4). Moreover, it allows an elegant and manageable description of various types of exceptional sets arising by the regularization of reduits (section 5). Usually the probabilistic characterization of all notions is given immediately after their introduction, but the probabilistic description of semipolar sets is postponed to section 7 where an exhaustive treatment is possible using the new concepts of essential base and essential balayage (section 6). In section 8 the fine support of continuous potentials will be integrated in this context. Further properties of the balayage of measures (section 9) will be used to study convergence properties of balayage measures (sections 10 and 11). Finally, the set of extreme representing measures is determined (section 12).

1. Balayage of Functions

In the following let $(X,\mathcal{W})$ be a balayage space. We recall from section (II.5) that for every $A \subset X$ and $u \in \mathcal{W}$ the *reduit* of u on A is defined by

$$R_u^A = \inf \{v \in \mathcal{W} : v \geq u \text{ on } A\}$$
$$= \inf \{v \in \mathcal{W} : v \leq u , v = u \text{ on } A\} .$$

The function

$$\hat{R}_u^A := \widehat{R_u^A} \in \mathcal{W}$$

is called the *balayage* of u on A .

Obviously $\hat{R}_u^A \leq R_u^A \leq u$, $\{\hat{R}_u^A < R_u^A\}$ is finely meager by (II.4.3), and $\hat{R}_u^A = R_u^A$ if A is finely open. We already know from (II.5.3) that $R_{u+v}^A = R_u^A + R_v^A$ if the functions $u,v \in \mathcal{W}$ are finite on A . In fact, this additivity holds even without this restriction.

1.1. PROPOSITION. Let $A \subset X$ and $u,v \in \mathcal{W}$. Then

$$R_{u+v}^A = R_u^A + R_v^A \quad , \quad \hat{R}_{u+v}^A = \hat{R}_u^A + \hat{R}_v^A .$$

Proof. Let $B = A \cap \{u + v < \infty\}$ and $x \in X$ such that $R_{u+v}^A(x) < \infty$. Let $w_0 \in \mathcal{W}$ such that $w_0(x) < \infty$ and $w_0 \geq u + v$ on A . For every $\varepsilon > 0$ and $w \in \mathcal{W}$ such that $w \geq u$ on B we have $w + \varepsilon w_0 \geq u$ on A , hence $w + \varepsilon w_0 \geq R_u^A$. This implies that $R_u^B(x) \geq R_u^A(x)$. Analogously $R_v^B(x) \geq R_v^A(x)$. Using (II.5.3) we obtain

$$R_{u+v}^A(x) \geq R_{u+v}^B(x) = R_u^B(x) + R_v^B(x) \geq R_u^A(x) + R_v^A(x) .$$

Hence $R_{u+v}^A \geq R_u^A + R_v^A$, and therefore the first equality holds.

Thus the second equality holds outside a finely meager subset of X and hence on X since X is a Baire space in the fine topology ((II.4.2)). ┘

The following approximation property is almost trivial, but very useful.

1.2. PROPOSITION. Let $A \subset X$ and $u \in \mathcal{W}$ such that $u < \infty$ on A . Then

$$R_u^A = \inf \{R_u^V : V \text{ finely open, } V \supset A\} .$$

If u is continuous then

$$R_u^A = \inf\{R_u^U : U \text{ open}, \ U \supset A\} .$$

Proof. Let $u_0 \in \mathcal{W}$ such that $0 < u_0 < \infty$ and $\varepsilon > 0$. Let $v \in \mathcal{W}$ such that $v \geq u$ on A. Then $V = \{v + \varepsilon u_0 > u\}$ is a finely open set containing A. Evidently $v + \varepsilon u_0 \geq R_u^V$ and hence

$$R_u^A \geq \inf\{R_u^V : V \text{ finely open}, \ V \supset A\} .$$

The converse inequality is trivial.

In order to obtain the second statement it suffices to note that the set $\{v + \varepsilon u_0 > u\}$ is even open if u is continuous. ┘

Let $A,B \subset X$ and $u \in \mathcal{W}$ such that $u < \infty$ on $A \cup B$. Then trivially $R_u^{A\cup B} \leq R_u^A + R_u^B$ and hence $\hat{R}_u^{A\cup B} \leq \hat{R}_u^A + \hat{R}_u^B$. A first application of (1.2) yields that in fact $\hat{R}_u^{A\cup B}$ is a specific minorant of $\hat{R}_u^A + \hat{R}_u^B$, i.e. $\hat{R}_u^{A\cup B} + w = \hat{R}_u^A + \hat{R}_u^B$ for some function $w \in \mathcal{W}$. Define

$$u_{A,B} = \inf\{R^{V\cup W}_{R_u^V \wedge R_u^W} : V,W \text{ finely open}, \ V \supset A, W \supset B\} .$$

Then clearly $\hat{u}_{A,B} \in \mathcal{W}$ and we have the following result.

<u>1.3. PROPOSITION.</u> Let A and B be two subsets of X and let $u \in \mathcal{W}$ such that $u < \infty$ on $A \cup B$. Then

$$R_u^{A\cup B} + u_{A,B} = R_u^A + R_u^B \quad \text{and} \quad \hat{R}_u^{A\cup B} + \hat{u}_{A,B} = \hat{R}_u^A + \hat{R}_u^B .$$

In particular,

$$R_u^{A\cup B} + R_u^{A\cap B} \leq R_u^A + R_u^B \quad \text{and} \quad \hat{R}_u^{A\cup B} + \hat{R}_u^{A\cap B} \leq \hat{R}_u^A + \hat{R}_u^B .$$

Proof. Again it suffices to prove the statements for the reduits. Let V and W be finely open subsets of X such that $V \supset A$ and $W \supset B$ and $u < \infty$ on $V \cup W$. Then $R_u^V, R_u^W \in \mathcal{W}$ and

$$u + R_u^V \wedge R_u^W = R_u^V + R_u^W \quad \text{on} \quad V \cup W .$$

Hence by (1.1)

$$R_u^{V\cup W} + R^{V\cup W}_{R_u^V \wedge R_u^W} = R^{V\cup W}_{R_u^V} + R^{V\cup W}_{R_u^W} = R_u^V + R_u^W .$$

Thus by (1.2)

$$R_u^{A\cup B} + u_{A,B} = R_u^A + R_u^B .$$

Finally, $A \subset V$ and $B \subset W$ imply that $R_u^V \wedge R_u^W = u$ on $A \cap B \subset U \cup W$, hence

$$R_u^{A\cap B} \leq u_{A,B} .$$

Proceeding in five steps we shall now prove that $R_u^A = \sup R_{f_n}$ for every increasing sequence (f_n) such that $\sup f_n = u\, 1_A$.

<u>1.4. LEMMA.</u> Let (A_n) be an increasing sequence of subsets of X and let $u \in \mathcal{W}$ such that $u < \infty$ on $A := \bigcup_{n=1}^{\infty} A_n$. Then $\sup_n R_u^{A_n} = R_u^A$.

Proof. Evidently $\sup_n R_u^{A_n} \leq R_u^A$. So let $x \in X$ such that $\sup_n R_u^{A_n}(x) < \infty$. Let $\varepsilon > 0$. Then for every $n \in \mathbb{N}$ there exists a finely open set V_n containing A_n such that

$$R_u^{V_n}(x) < R_u^{A_n}(x) + \frac{\varepsilon}{2^n} .$$

Define an increasing sequence (W_n) of finely open sets by

$$W_n = V_1 \cup V_2 \cup \ldots \cup V_n .$$

Then for every $n \in \mathbb{N}$

$$R_u^{W_n}(x) \leq R_u^{A_n}(x) + (1 - \frac{1}{2^n})\, \varepsilon .$$

Indeed, the inequality is obvious if $n = 1$. So suppose that the inequality holds for some $n \in \mathbb{N}$. Then by (1.3)

$$R_u^{W_{n+1}}(x) + R_u^{W_n \cap V_{n+1}}(x) \leq R_u^{W_n}(x) + R_u^{V_{n+1}}(x)$$

$$\leq R_u^{A_n}(x) + (1 - \frac{1}{2^n})\, \varepsilon + R_u^{A_{n+1}}(x) + \frac{\varepsilon}{2^{n+1}}$$

where $A_n \subset V_n \subset W_n$ and $A_n \subset A_{n+1} \subset V_{n+1}$ and hence $R_u^{A_n}(x) \leq R_u^{W_n \cap V_{n+1}}(x)$. Therefore

$$R_u^{W_{n+1}}(x) \leq R_u^{A_{n+1}}(x) + (1 - \frac{1}{2^{n+1}})\, \varepsilon .$$

Let $v = \sup_n R_u^{W_n}$. Then $v \in \mathcal{W}$ and $v = u$ on $\bigcup_{n=1}^{\infty} W_n = \bigcup_{n=1}^{\infty} V_n \supset A$, i.e. $v \geq R_u^A$. In particular,

$$R_u^A(x) \leq v(x) \leq \sup_n R_u^{A_n}(x) + \varepsilon .$$

<u>1.5. LEMMA.</u> Let $A \subset X$ and $u \in \mathcal{W}$ such that $u > 0$ on A. Then $R_\infty^A = \infty \cdot R_u^A$.

Proof. Let $v \in \mathcal{W}$ such that $v = \infty$ on A. Then $v \geq m\,u$ on A for every $m \in \mathbb{N}$ and hence

$$v \geq \sup_m R^A_{m\,u} = \sup_m m\, R^A_u = \infty \cdot R^A_u .$$

Thus $R^A_\infty \geq \infty \cdot R^A_u$.

The converse inequality trivially holds on $\{R^A_u > 0\}$. So let $x \in X$ such that $R^A_u(x) = 0$ and let $\varepsilon > 0$. Then for every $n \in \mathbb{N}$ there exists a function $v_n \in \mathcal{W}$ such that $v_n \geq u$ on A and $v_n(x) < \frac{\varepsilon}{2^n}$. Then $v := \sum_{n=1}^{\infty} v_n \in \mathcal{W}$, $v = \infty$ on A and $v(x) < \varepsilon$, hence $R^A_\infty(x) < \varepsilon$. Thus $R^A_\infty(x) = 0 = \infty \cdot R^A_u(x)$. ┘

1.6. COROLLARY. Let (A_n) be an increasing sequence of subsets of X and let (u_n) be an increasing sequence in $\mathcal{W}$ such that $u := \sup u_n = \infty$ on $A := \bigcup_{n=1}^{\infty} A_n$. Then $\sup R^{A_n}_{u_n} = R^A_u$.

Proof. Let $u_0 \in \mathcal{W}$ such that $0 < u_0 < \infty$, let a be a strictly positive real number, and define for every $n \in \mathbb{N}$

$$B_n = A_n \cap \{u_n \geq a\, u_0\} .$$

Then (B_n) is increasing to A and $R^{A_n}_{u_n} \geq R^{B_n}_{a\,u_0}$ and hence

$$\sup_n R^{A_n}_{u_n} \geq a \sup_n R^{B_n}_{u_0} = a\, R^A_{u_0} .$$

Thus

$$\sup_n R^{A_n}_{u_n} \geq \infty \cdot R^A_{u_0} = R^A_\infty = R^A_u .$$ ┘

1.7. PROPOSITION. Let (A_n) be an increasing sequence of subsets of X and let (u_n) be an increasing sequence in $\mathcal{W}$, let $A = \bigcup_{n=1}^{\infty} A_n$ and $u = \sup u_n$. Then $\sup R^{A_n}_{u_n} = R^A_u$ and $\sup \hat{R}^{A_n}_{u_n} = \hat{R}^A_u$.

Proof. 1. Suppose first that $u < \infty$ on A. Let $\varepsilon > 0$ and define for every $n \in \mathbb{N}$

$$B_n = A_n \cap \{(1-\varepsilon)u \leq u_n\} .$$

Then (B_n) is increasing to A and $R^{A_n}_{u_n} \geq (1-\varepsilon)\, R^{B_n}_u$, hence by (1.4)

$$\sup R^{A_n}_{u_n} \geq (1-\varepsilon) \sup R^{B_n}_u = (1-\varepsilon)\, R^A_u .$$

Therefore $\sup R^{A_n}_{u_n} \geq R^A_u$.

2. Let us now consider the general case. Let

$$A' = A \cap \{u < \infty\} \quad , \quad A'' = A \cap \{u = \infty\} .$$

Then by (1)

$$R_u^{A'} = \sup R_{u_n}^{A' \cap A_n} \leq \sup R_{u_n}^{A_n} .$$

Let $x \in X$ such that $\sup R_{u_n}^{A_n}(x) < \infty$. Then by (1.6)

$$R_u^{A''}(x) = \sup R_{u_n}^{A'' \cap A_n}(x) \leq \sup R_{u_n}^{A_n}(x) < \infty$$

and hence $R_u^{A''}(x) = 0$ by (1.5). Therefore

$$R_u^A(x) \leq R_u^{A'}(x) + R_u^{A''}(x) = R_u^{A'}(x) \leq \sup R_{u_n}^{A_n}(x) .$$

Thus $\sup R_{u_n}^{A_n} = R_u^A$. Using the fact that the set $\{\hat{R}_u^A < R_u^A\} \cup \bigcup_{n=1}^{\infty} \{\hat{R}_{u_n}^{A_n} < R_{u_n}^{A_n}\}$ is finely meager we conclude that $\sup \hat{R}_{u_n}^{A_n} = \hat{R}_u^A$. ┘

<u>1.8. COROLLARY.</u> Let $A \subset X$, $u \in \mathcal{W}$ and let (f_n) be an increasing sequence of numerical functions on X such that $\sup f_n = u\, 1_A$. Then $\sup R_{f_n} = R_u^A$ and $\sup \hat{R}_{f_n} = \hat{R}_u^A$.

Proof. Let $u_0 \in \mathcal{W}$ such that $0 < u_0 < \infty$. For every $m \in \mathbb{N}$ let

$$u_m = \inf\,(m\, u_0 \,,\, (1 - \frac{1}{m})\, u) .$$

Then (u_m) is an increasing sequence in $\mathcal{W}$ such that $\sup u_m = u$, $u_m < \infty$ and $u_m < u$ on $\{u < \infty\}$. Defining

$$A_{nm} = A \cap \{f_n \geq u_m\}$$

the sequence $(A_{nm})_{n \in \mathbb{N}}$ is increasing to A for every $m \in \mathbb{N}$ and $R_{f_n} \geq R_{u_m}^{A_{nm}}$ for all $m,n \in \mathbb{N}$. Hence by (1.7)

$$\sup_n R_{f_n} \geq \sup_n \sup_m R_{u_m}^{A_{nm}} = \sup_m \sup_n R_{u_m}^{A_{nm}} = \sup_m R_{u_m}^A = R_u^A .$$

The converse inequality is trivial. Thus $\sup R_{f_n} = R_u^A$ and, by (II.4.3), $\sup \hat{R}_{f_n} = \hat{R}_u^A$. ┘

Combining the preceding results we obtain capacities associated with the mapping $A \mapsto R_u^A$ if $u \in \mathcal{W} \cap C(X)$.

<u>1.9. PROPOSITION.</u> Let A be a Borel subset of X, $v \in \mathcal{W} \cap C(X)$ and $\mu \in M_+(X)$

such that $\int v\, d\mu < \infty$. Then

$$R_v^A = \sup\{R_v^K : K \text{ compact} \subset A\}\ , \quad \hat{R}_v^A = \sup\{\hat{R}_v^K : K \text{ compact} \subset A\}\ ,$$

and

$$\int R_v^A\, d\mu = \sup\{\int R_v^K\, d\mu : K \text{ compact} \subset A\} = \inf\{\int R_v^U\, d\mu : U \text{ open} \supset A\}\ ,$$

Proof. Define $c : \mathfrak{P}(X) \to \mathbb{R}_+$ by

$$c(B) = \inf\{\int R_v^U\, d\mu : U \text{ open} \supset B\}\ .$$

Let (B_n) be an increasing sequence of subsets of X and $B = \bigcup_{n=1}^{\infty} B_n$. Given $\varepsilon > 0$ there exist open sets V_n such that $B_n \subset V_n$ and

$$\int R_v^{V_n}\, d\mu < c(B_n) + \frac{\varepsilon}{2^n}\ .$$

Defining $U_n = V_1 \cup \ldots \cup V_n$ and proceeding as in the proof of (1.4) we conclude that

$$\int R_v^{U_n}\, d\mu \leq c(B_n) + (1 - \frac{1}{2^n})\, \varepsilon$$

for every $n \in \mathbb{N}$. Hence $U := \bigcup_{n=1}^{\infty} U_n$ is an open set containing B such that

$$\int R_v^U\, d\mu = \sup \int R_v^{U_n}\, d\mu \leq \sup c(B_n) + \varepsilon\ .$$

Therefore $c(B) = \sup c(B_n)$.

Now let $x \in X$, $a < R_v^A(x)$, $b < c(A)$, $\eta > 0$ and $\varphi \in K(X)$ such that $\varphi \leq \hat{R}_u^A$. Let

$$C = \{B \in \mathfrak{P}(X) : R_v^B(x) > a\ , c(B) > b\ , \hat{R}_v^B > \varphi - \eta\}\ .$$

Then obviously $A \in C$, and if $B \in C$ and $B \subset B' \subset X$ then $B' \in C$. Moreover if (B_n) is an increasing sequence in $\mathfrak{P}(X)$ such that $\bigcup_{n=1}^{\infty} B_n \in C$ then $B_n \in C$ for some $n \in \mathbb{N}$ by (1.7) and the preceding considerations.

Thus by (I.3.4) and (I.3.8) there exists a decreasing sequence (A_n) of relatively compact subsets of A such that $A_n \in C$ for every $n \in \mathbb{N}$ and the compact set

$$K := \bigcap_{n=1}^{\infty} \bar{A}_n$$

is contained in A . If U is an open subset of X containing K then $\bar{A}_n \subset U$ for some $n \in \mathbb{N}$ and hence $\inf R_v^{\bar{A}_n} \leq R_v^U$. Using (1.2) we conclude that

$$R_v^K = \inf R_v^{\bar{A}_n} \geq \inf R_v^{A_n}\ .$$

In particular, $R_v^K(x) \geq a$ and $\hat{R}_v^K \geq \varphi - \eta$.

Furthermore, there exists a decreasing sequence (U_m) of relatively compact open subsets of X containing K such that $\bigcap_{m=1}^{\infty} \bar{U}_m = K$. Then $R_v^K = \inf R_v^{U_m}$ by (1.2) and hence

$$\int R_v^K \, d\mu = \inf_m \int R_v^{U_m} \, d\mu \geq \inf_n c(A_n) \geq b \; .$$

Thus

$$c(A) \leq \sup \{\int R_v^K \, d\mu : K \text{ compact} \subset A\} \leq \int_* R_v^A \, d\mu$$

$$\leq \int^* R_v^A \, d\mu \leq \inf \{\int R_v^U \, d\mu : U \text{ open} \supset A\} = c(A) \; .$$

┘

1.10. COROLLARY. Let $v \in \mathcal{W} \cap \mathcal{C}(X)$ and $\mu \in M_+(X)$ such that $\int v \, d\mu < \infty$. Then the mapping $A \mapsto \int^* R_v^A \, d\mu$ is a capacity.

Proof. (1.7) and (1.9). ┘

1.11. EXERCISES. 1. Let A be a subset of X , $u \in \mathcal{W}$, and $B = A \cap \{u < \infty\}$. Then $R_u^A = R_u^B + R_u^{A \setminus B}$.

2. Let $u, v \in \mathcal{W}$. Then $R_u^{\{v<\infty\}} \prec u$. (Hint: Apply (B_3) to $u \leq \hat{R}_u^{\{v<\infty\}} + \hat{R}_u^{\{v=\infty\}}$.)

3. Let A be a finely open subset of X . Then there exists a sequence (K_n) of compact subsets of A such that $\hat{R}_u^{\cup K_n} = R_u^A$ for every $u \in \mathcal{W}$. (Hint: Use (I.1.7) and (II.4.1) to prove the statement for a strict potential.)

4. Let $(Y, \mathcal{A}, \nu)$ be a measure space and let $(u_y)_{y \in Y}$ be a family in $\mathcal{W}$ such that $(x,y) \mapsto u_y(x)$ is $\mathcal{B}(X) \otimes \mathcal{A}$ - measurable. Then the function $u : x \mapsto \int u_y(x) \, \nu(dy)$ is contained in $\mathcal{W}$ and $\hat{R}_u^A(x) = \int \hat{R}_{u_y}^A(x) \, \nu(dy)$ for every $A \subset X$ and $x \in X$.

2. Balayage of Measures

As usual let P be the cone of continuous real potentials of the balayage space $(X, \mathcal{W})$. We recall from section (II.5) that given $A \subset X$ and $x \in X$ we have a measure ε_x^A on X characterized by $\varepsilon_x^A(p) = R_p^A(x)$ for every $p \in P$. By (1.7) we now know that in fact $\overset{\circ}{\varepsilon}{}_x^A(u) = R_u^A(x)$ for every $u \in \mathcal{W}$. In chapter III the description of a balayage space by a family $(H_U)_{U \in \mathcal{U}}$ of harmonic kernels where $H_U(x,\cdot) = \overset{\circ}{\varepsilon}{}_x^{\complement U}$ illustrated the importance of these measures. Using $\hat{R}_u^A$ instead of R_u^A we shall now define balayage of measures on A. The measures ε_x^A obtained by balayage will be even more useful than the measures $\overset{\circ}{\varepsilon}{}_x^A$. This is partly due to the lower semi-continuity and the fine continuity of the functions $\hat{R}_u^A$ and partly due to the fact that the exceptional sets studied in section 5 are related to the balayage of functions. Moreover, $\overset{\circ}{\varepsilon}{}_x^A = \varepsilon_x$ for every $x \in A$ and we shall see in (2.3) that $\overset{\circ}{\varepsilon}{}_x^A = \varepsilon_x^A$ for every $x \in \complement A$. So the study of the measures ε_x^A is more general than the investigation of the measures $\overset{\circ}{\varepsilon}{}_x^A$.

2.1. PROPOSITION. Let $x \in X$ and $A \subset X$. Then there exists a unique measure ε_x^A on X such that

$$\int u \, d\varepsilon_x^A = \hat{R}_u^A(x)$$

for every $u \in \mathcal{W}$. Furthermore $\varepsilon_x^A(\complement \bar{A}) = 0$.

The measure ε_x^A is called *balayage of* ε_x *on* A.

Proof. Consider the mapping $p \mapsto \hat{R}_p^A(x)$ from $\mathcal{P}$ into $\mathbb{R}_+$. It is evidently increasing, positively homogeneous and it is additive by (1.1). Thus the first statement follows from (I.1.4). For every $u \in \mathcal{W}$ there exists an increasing sequence (p_n) in $\mathcal{P}$ such that $\sup p_n = u$. Then by (1.7)

$$\int u \, d\varepsilon_x^A = \sup \int p_n \, d\varepsilon_x^A = \sup \hat{R}_{p_n}^A(x) = \hat{R}_u^A(x) .$$

Finally let $y \in \complement \bar{A}$. By (I.1.2), there exist $p,q \in \mathcal{P}$ such that $q \leq p$, $q(y) < p(y)$ and $q = p$ on A. Then $R_q^A = R_p^A$ and hence

$$\varepsilon_x^A(p-q) = \hat{R}_p^A(x) - \hat{R}_q^A(x) = 0 .$$

The following result allows all considerations involving balayage on arbitrary subsets of X to be restricted on Borel sets. This will be particularly helpful when dealing with exceptional sets.

2.2. PROPOSITION. For every $A \subset X$ there exists a G_δ-set $A' \supset A$ such that $\hat{R}_u^{A'} = \hat{R}_u^A$ for every $u \in \mathcal{W}$.

Proof. Let $p \in P$ be a strict potential. By the lemma of Choquet (I.1.8) there exists a decreasing sequence (u_n) in W such that $u_n \geq p$ on A and

$$\widehat{\inf u_n} = \hat{R}_p^A .$$

Taking $q_0 \in P$ with $q_0 > 0$ we may replace u_n by $u_n + \frac{1}{n} q_0$, i.e. we may suppose that $u_n > p$ on A . Then

$$A' := \bigcap_{n=1}^{\infty} \{u_n > p\}$$

is a G_δ-set, $A' \supset A$, and for every $n \in \mathbb{N}$ $u_n \geq R_p^{A'} \geq R_p^A$. Hence

$$\hat{R}_p^A = \widehat{\inf u_n} \geq \hat{R}_p^{A'} \geq \hat{R}_p^A ,$$

i.e. $\hat{R}_p^{A'} = \hat{R}_p^A$. Furthermore for every $x \in X$ and every $u \in W$

$$\varepsilon_x^A(u) = \hat{R}_u^A(x) \leq \hat{R}_u^{A'}(x) = \varepsilon_x^{A'}(x) .$$

Since p is strict we conclude that $\varepsilon_x^A = \varepsilon_x^{A'}$ for every $x \in X$ and hence $\hat{R}_u^A = \hat{R}_u^{A'}$ for every $u \in W$. ┘

<u>2.3. PROPOSITION.</u> Let $A \subset X$ and $p \in P$. Then R_p^A is continuous on $\complement \bar{A}$ and $\hat{R}_p^A = R_p^A$ on $\complement A$. In particular, $\varepsilon_x^A = \overset{\circ}{\varepsilon}{}_x^A$ for every $x \in \complement A$.

Proof. By (2.2), there exists a G_δ-set $A_1 \supset A$ such that $\hat{R}_p^A = \hat{R}_p^{A_1}$. Let $x \in \complement A$. It follows immediately from (1.2) that there exists a G_δ-set $A_2 \supset A$ satisfying $x \notin A_2$ and $R_p^A(x) = R_p^{A_2}(x)$. Then $A' := A_1 \cap A_2$ is a G_δ-set such that $x \notin A$,

$$\hat{R}_p^A(x) = \hat{R}_p^{A'}(x) \quad \text{and} \quad R_p^A(x) = R_p^{A'}(x) .$$

For every compact subset K of A' the function R_p^K is continuous on $\complement K$ by (III.2.6), hence in particular $\hat{R}_p^K(x) = R_p^K(x)$. Thus by (1.9)

$$\begin{aligned} \hat{R}_p^A(x) = \hat{R}_p^{A'}(x) &= \sup \{\hat{R}_p^K(x) : K \text{ compact } \subset A'\} \\ &= \sup \{R_p^K(x) : K \text{ compact } \subset A'\} = R_p^{A'}(x) = R_p^A(x) , \end{aligned}$$

i.e. the function R_p^A is l.s.c. at x .

Furthermore, let $V \subset X$ be open such that $\bar{V} \cap \bar{A} = \emptyset$. By (1.2),

$$R_p^A = \inf \{R_p^U : U \text{ open} \supset A , \bar{U} \cap V = \emptyset\} .$$

By (III.2.5), every function R_p^U is continuous on $\complement\bar{U}$. Therefore R_p^A is u.s.c. on V. This shows that R_p^A is continuous on $\complement\bar{A}$. ┘

2.4. COROLLARY. For every $A \subset X$ and $u \in \mathcal{W}$, $\hat{R}_u^A = R_u^A$ on $\complement A$.

Proof. (1.7) and (2.3). ┘

The following lemma will be the key for a proof of the important fact that measures obtained by balayage on A are supported by the fine closure of A (see 4.6). Here it is used to prove the harmonicity of functions R_v^A on $\complement\bar{A}$.

2.5. LEMMA. Let $A \subset X$ and $p \in \mathcal{P}$. Then for every compact subset K of $\complement A$ and every $\varepsilon > 0$ there exists an open set $U \subset X$ such that $A \subset U$ and $R_p^U \leq R_p^A + \varepsilon$ on K.

Proof. Let us first suppose that $\bar{A} \cap K = \emptyset$. Then

$$R_p^A = \inf \{R_p^U : U \text{ open} \supset A, \ \bar{U} \cap K = \emptyset\}$$

where R_p^A and every R_p^U is continuous on K by (2.3). Hence the statement follows from Dini's theorem.

In the general case we choose an increasing sequence (A_n) of subsets of X such that $\bigcup_{n=1}^{\infty} A_n = A$ and $\bar{A}_n \cap K = \emptyset$ for every $n \in \mathbb{N}$. For every $n \in \mathbb{N}$ there exists an open set V_n containing A_n such that

$$R_p^{V_n} \leq R_p^{A_n} + \frac{\varepsilon}{2^n} \quad \text{on } K .$$

Defining $U_n = V_1 \cup \ldots\ldots \cup V_n$ and proceeding as in the proof of (1.4) we conclude that for every $n \in \mathbb{N}$

$$R_p^{U_n} \leq R_p^{A_n} + \left(1-\frac{1}{2^n}\right) \varepsilon \quad \text{on } K .$$

The sequence (U_n) is increasing to an open set U containing A. Furthermore,

$$R_p^U = \sup R_p^{U_n} \leq \sup R_p^{A_n} + \varepsilon = R_p^A + \varepsilon \quad \text{on } K .$$ ┘

2.6. COROLLARY. Let $A \in \mathcal{B}(X)$, $w \in \mathcal{W} \cap C(X)$ and $v \in \mathcal{W}$ such that $v \leq w$. Then $R_v^A \in \mathcal{H}^+(\complement\bar{A})$.

Proof. Let $p \in P$ and $V \in \mathcal{U}$ such that $\bar{V} \cap \bar{A} = \emptyset$. Let $x \in V$ and $\eta > 0$. Then there exists a compact subset K of $\complement A$ such that $\varepsilon_x^{\complement V}(p\, 1_{\complement A \setminus K}) < \eta$. By (2.5), there exists an open set U containing A such that $\bar{U} \cap \bar{V} = \emptyset$ and $R_p^U \leq R_p^A + \eta\, p$ on K. Since $R_p^A = p$ on A and $R_p^A = \hat{R}_p^A$ on $\complement A$ we know that $R_p^A \in \mathcal{B}(X)$. By (III.2.5), $R_p^U \in \mathcal{H}^+(\complement \bar{U})$. Therefore

$$R_p^A(x) \leq R_p^U(x) = \varepsilon_x^{\complement V}(R_p^U)$$
$$\leq \varepsilon_x^{\complement V}(R_p^A + \eta\, p + p\, 1_{\complement A \setminus K}) \leq \varepsilon_x^{\complement V}(R_p^A) + \eta\, p(x) + \eta\,.$$

Hence $R_p^A(x) \leq \varepsilon_x^{\complement V}(R_p^A)$. The converse inequality is trivial. Thus $R_p^A \in \mathcal{H}^+(\complement \bar{A})$.

The general statement now follows immediately. Indeed, there exists an increasing sequence (p_n) in P such that $\sup p_n = v$. Then $(R_{p_n}^A)$ is an increasing sequence in $\mathcal{H}^+(\complement \bar{A})$ which is majorized by w. Thus $R_v^A = \sup R_{p_n}^A \in \mathcal{H}^+(\complement \bar{A})$ by (III.2.3). ┘

2.7. REMARK. Note that in general $\hat{R}_v^A \notin \mathcal{H}^+(\complement \bar{A})$ although $\hat{R}_v^A = R_v^A$ on $\complement A$ (see example (V.9.1)).

We shall now investigate the connection between balayage on $(X, \mathcal{W})$ and balayage on open subspaces $(U, \mathcal{W}^U)$ introduced in section (V.1). For every subset A of U and function $v \in \mathcal{W}^U = {}^*\mathcal{H}^+(U)_{|U}$ let

$${}^U R_v^A = \inf \{u \in \mathcal{W}^U : u \geq v \text{ on } A\}\,.$$

2.8. LEMMA. Let A be a subset of U and $p \in P$. Then

$$R_p^{A \cup \complement U} = {}^U R_{p - H_U p}^A + H_U p\,.\ *)$$

Proof. Let $p' = p - H_U p$. Let F be a closed subset of U,

$$u = {}^U R_{p'}^{U \setminus F}\,, \quad v = R_p^{\complement F} - H_U p\,.$$

Then $u, v \in {}^*\mathcal{H}^+(U)$. Since $v = p'$ on $U \setminus F$ we have $v \geq u$. On the other hand $u + H_U p \in {}^*\mathcal{H}^+(U)$, $u + H_U p \leq p' + H_U p = p$ and $u + H_U p = p$ on $\complement F$, hence $u + H_U p \in \mathcal{W}$ by (III.4.6). Since $u' = p$ on $\complement F$ we conclude that $u' \geq R_p^{\complement F}$,

---

*) As usual we identify functions f on X satisfying $f = 0$ on $\complement U$ with $f_{|U}$. Then especially $\mathcal{W}^U \subset {}^*\mathcal{H}^+(U)$.

i.e. $u + H_U p \geq R_p^{\complement F}$, $u \geq v$. Thus

$$R_p^{\complement F} = {}^U R_{p'}^{U \setminus F} + H_U p \quad .$$

Let $x \in U$ and $\varepsilon > 0$. Using (1.2) and the fact that p' is a potential on U we may choose open subsets V,W of X such that $A \subset V \subset U$, $\complement U \subset W$,

$$R_p^{V \cup W}(x) - R_p^{A \cup \complement U}(x) < \varepsilon \ , \ {}^U R_{p'}^V(x) - {}^U R_{p'}^A(x) < \varepsilon \ , \ {}^U R_{p'}^{U \setminus W}(x) < \varepsilon \ .$$

Then $F := \complement(V \cup W)$ is a closed subset of U and

$${}^U R_{p'}^V \leq {}^U R_{p'}^{U \setminus F} = {}^U R_{p'}^{V \cup (U \setminus W)} \leq {}^U R_{p'}^V + {}^U R_{p'}^{U \setminus W} \ ,$$

hence $|{}^U R_{p'}^{U \setminus F}(x) - {}^U R_{p'}^A(x)| < 2\varepsilon$. Since $R_p^{V \cup W} = R_p^{\complement F} = {}^U R_{p'}^{U \setminus F} + H_U p$ we thus obtain that

$$|R_p^{A \cup \complement U} - {}^U R_{p'}^A|(x) < 3\varepsilon \ .$$

┘

For every subset A of U and $x \in U$ let ${}^U \varepsilon_x^A$ denote the balayage of ε_x on A with respect to $(U, \mathcal{W}^U)$.

2.9. COROLLARY. Let $A \subset U$ and $x \in U$. Then ${}^U \varepsilon_x^A = \varepsilon_x^{A \cup \complement U}|_U$.

Proof. Let $p,q \in \mathcal{P}$ such that $p = q$ on $\complement U$, and let $p' = p - H_U p$, $q' = q - H_U q$. Since $H_U p = H_U q$, we have $p' - q' = p - q$ and hence by (2.8)

$$\begin{aligned} {}^U \varepsilon_x^A(p-q) &= {}^U \varepsilon_x^A(p'-q') = {}^U \hat{R}_{p'}^A(x) - {}^U \hat{R}_{q'}^A(x) \\ &= \hat{R}_p^{A \cup \complement U}(x) - \hat{R}_q^{A \cup \complement U}(x) = \varepsilon_x^{A \cup \complement U}(p-q) \ . \end{aligned}$$

Thus ${}^U \varepsilon_x^A = \varepsilon_x^{A \cup \complement U}|_U$.

┘

2.10. PROPOSITION. Let A be a subset of X and $f : X \to \mathbb{R}$ $\mathcal{P}$-bounded and Borel measurable. Then the function $x \mapsto \varepsilon_x^A(f)$ is Borel measurable and continuous on $\complement \bar{A}$. If moreover f is continuous then $x \mapsto \varepsilon_x^A(f)$ is finely continuous and the function $x \mapsto \overset{o}{\varepsilon}{}_x^A(f)$ is harmonic on $\complement \bar{A}$ if $A \in \mathcal{B}(X)$.

Proof. By (2.2), we may assume that A is a Borel set. Then for every $p \in \mathcal{P}$, $R_p^A \in \mathcal{H}(\complement \bar{A})$ by (2.6), $\hat{R}_p^A$ is l.s.c., finely continuous and $\hat{R}_p^A = R_p^A$ on $\complement A$ by (2.3). If $f \in \mathcal{C}_{\mathcal{P}}(X)$ then there exists $p_0 \in \mathcal{P}$ such that for every $\delta > 0$ there exist $p,q \in \mathcal{P}$ satisfying

$$|f - (p - q)| \leq \delta p_0$$

and hence

$$|\varepsilon_x^A(f) - (\varepsilon_x^A(p) - \varepsilon_x^A(q))| \leq \delta\varepsilon_x^A(p_0) \leq \delta p_0(x)$$

for every $x \in X$. Using (2.6) this proves the statement if $f \in C_{\mathcal{P}}(X)$. The general case of a $\mathcal{P}$-bounded Borel measurable function now follows easily using (III.2.3) (see (III.3.2)). ┘

Let $M(\mathcal{P})$ be the set of all $\mu \in M_+(X)$ such that $\mu(p) < \infty$ for some strictly positive $p \in \mathcal{P}$. Note that every $\mu \in M_+(X)$ having compact support is contained in $M(\mathcal{P})$. Furthermore, $M_+^1(X) \subset M(\mathcal{P})$ if $1 \in \mathcal{W}$.

For every $\mu \in M(\mathcal{P})$ and every subset A of X we define the *balayage of* μ *on* A by

$$\mu^A(f) = \int \varepsilon_x^A(f)\mu(dx) \qquad (f \in \mathcal{B}^+(X)) .$$

Obviously, $\mu^A(u) \leq \mu(u)$ for every $u \in \mathcal{W}$.

<u>2.11. PROPOSITION.</u> Let $\mu \in M(\mathcal{P})$ and $A \subset X$. Then $\mu^A \in M(\mathcal{P})$, $\mu^A(\complement\bar{A}) = 0$, and

$$(*) \qquad \int u \, d\mu^A = \int \hat{R}_u^A \, d\mu$$

for every $u \in \mathcal{W}$. Moreover, μ^A is uniquely determined by (*).

Proof. (2.10), (2.1) and (I.1.5). ┘

<u>2.12. EXERCISES.</u> 1. Let $\mu \in M(\mathcal{P})$ and $A \subset X$ such that $\mu(K) = 0$ for every compact set K which is disjoint to the fine interior of A. Then $\mu^A = \mu$.

2. Suppose that $(X,\mathcal{W})$ has the local truncation property. Let $\mu \in M(\mathcal{P})$, and $A \subset X$ such that $\mu(\mathring{A}) = 0$. Then $\mu^A(\complement A^*) = 0$.

3. Suppose that $\hat{R}_p^K \in H(\complement K)$ for every compact subset K of X and every $p \in \mathcal{P}$. Then $\hat{R}_u^{A \cup B} = R_{\sup(\hat{R}_u^A, \hat{R}_u^B)}$ for all subsets A, B of X and $u \in \mathcal{W}$. (Hint: Use (III.6.6).)

3. Probabilistic Interpretation

Theorem (3.14) and theorem (3.16) will establish the fundamental probabilistic interpretation of balayage. This will be extremely useful for a deeper understanding of notions, statements, and proofs in potential theory. In particular, this connection will allow in many situations to give an analytic treatment guided by probabilistic ideas.

Let $\mathfrak{X} = (\Omega,\mathfrak{M},\mathfrak{M}_t,X_t,\Theta_t,P^x)$ be a Hunt process associated with $(X,\mathcal{W})$, i.e. let $\mathfrak{X}$ be a Hunt process on X such that $E_{\mathfrak{X}} = \mathcal{W}$. (We know by (IV.8.1) that such a process exists if $1 \in \mathcal{W}$.)

For every subset A of X_Δ we define the *first entry time* of A by

$$D_A(\omega) = \inf\{t \geq 0 : X_t(\omega) \in A\}$$

and the *first hitting time* of A by

$$T_A(\omega) = \inf\{t > 0 : X_t(\omega) \in A\}$$

where in both cases the infimum of the empty set is understood to be $+\infty$.

We intend to show that, for every $A \in \mathcal{B}(X)$, D_A and T_A are stopping times and that

$$E^x(v(X_{D_A})) = R_v^A(x) , \quad E^x(v(X_{T_A})) = \hat{R}_v^A(x)$$

for every $v \in \mathcal{W}$ and $x \in X$.

We first note some immediate consequences of the definitions.

3.1. LEMMA. For every $A \subset X_\Delta$ and $s \in \mathbb{R}_+$, $D_A \leq T_A$ and $D_A = T_A$ on $[X_0 \notin A]$,

$s + D_A \circ \Theta_s = \inf\{t \geq s : X_t \in A\}$, $s + D_A \circ \Theta_s = D_A$ on $[D_A \geq s]$,

$s + T_A \circ \Theta_s = \inf\{t > s : X_t \in A\}$, $s + T_A \circ \Theta_s = T_A$ on $[T_A > s]$.

In particular, $s + D_A \circ \Theta_s$ $(s + T_A \circ \Theta_s)$ may be called the first entry (hitting) time of A after s . It is obvious from (3.1) that these are increasing functions of s and that

$$\lim_{s \downarrow 0} (s + D_A \circ \Theta_s) = \lim_{s \downarrow 0} (s + T_A \circ \Theta_s) = T_A .$$

If D_A is a stopping time we therefore conclude by (IV.5.4) and (IV.5.10) that $s + D_A \circ \Theta_s$, T_A and $s + T_A \circ \Theta_s$ are stopping times.

3.2. LEMMA. If $A \subset B \subset X_\Delta$ then $D_A \geq D_B$ and $T_A \geq T_B$. If (A_n) is a sequence of subsets of X_Δ and $A = \bigcup_{n=1}^{\infty} A_n$ then $D_A = \inf D_{A_n}$, $T_A = \inf T_{A_n}$.

Proof. The first statement being obvious it suffices to prove the second one. Let $\omega \in \Omega$ and $D_A(\omega) < t$. Then there exists a real number $s < t$ such that $X_s(\omega) \in A$ and a natural number m such that $X_s(\omega) \in A_m$. Hence $\inf D_{A_n}(\omega) \leq D_{A_m}(\omega) < t$. Therefore $\inf D_{A_n} \leq D_A$. Similarly $\inf T_{A_n} \leq T_A$. The converse inequalities are trivial. ∎

For every stopping time T, every $\alpha \geq 0$ and $f \in \mathcal{B}^+(X)$ let

$$P_T^\alpha f(x) = E^x(e^{-\alpha T} f(X_T)) \qquad (x \in X).$$

We shall often write P_T instead of P_T^0. Note that this definition is consistent with our previous notations in the case of a constant stopping time $T = t$. Let $\mathbb{V} = (V_\alpha)_{\alpha>0}$ be the resolvent of $\mathbb{P}$.

3.3. LEMMA. For every stopping time T, every $\alpha \geq 0$, $g \in \mathcal{B}^+(X)$ and $x \in X$

$$P_T^\alpha V_\alpha g(x) = E^x\left(\int_T^\infty e^{-\alpha t} g(X_t)\,dt\right).$$

Proof. Since $\omega \longmapsto \int_0^\infty e^{-\alpha t} g(X_t(\omega))\,dt$ is $\mathfrak{M}$-measurable we obtain by the strong Markov property that

$$P_T^\alpha V_\alpha g(x) = E^x\{e^{-\alpha T} V_\alpha g(X_T)\} = E^x\{e^{-\alpha T} E^{X_T}(\int_0^\infty e^{-\alpha t} g(X_t)\,dt)\}$$

$$= E^x\{e^{-\alpha T}(\int_0^\infty e^{-\alpha t} g(X_t)\,dt) \circ \Theta_T\} = E^x\{e^{-\alpha T} \int_0^\infty e^{-\alpha t} g(X_{T+t})\,dt\}$$

$$= E^x\{\int_T^\infty e^{-\alpha t} g(X_t)\,dt\}.$$ ∎

3.4. COROLLARY. Let $v \in \mathcal{W}$ and let S,T be stopping times such that $S \leq T$. Then $P_T v \leq P_S v \leq v$.

Proof. Let $\alpha > 0$. By (II.3.11) and (II.3.12) there exists a sequence (g_n) in $\mathcal{B}^+(X)$ such that $(V_\alpha g_n)$ is increasing to v. By (3.3), we have $P_T^\alpha(V_\alpha g_n) \leq P_S^\alpha(V_\alpha g_n) \leq V_\alpha g_n$ for every $n \in \mathbb{N}$ and hence $P_T^\alpha v \leq P_S^\alpha v \leq v$. Letting α tend to zero the statement follows. ∎

3.5. LEMMA. Let $A \subset X$ and $v \in \mathcal{W}$ such that D_A is a stopping time and $P_{D_A} v \in \mathcal{B}^+(X)$. Then $P_{D_A} v \in S_{\mathfrak{X}}$ and $P_{T_A} v = \sup_{t>o} P_t P_{D_A} v \in \mathcal{W}$.

Proof. For every $t > 0$ and $x \in X$

$$P_t P_{D_A} v(x) = E^x(P_{D_A} v(X_t)) = E^x\{E^{X_t}(v \circ X_{D_A})\}$$

$$= E^x(v \circ X_{D_A} \circ \Theta_t) = E^x(v \circ X_{t+D_A \circ \Theta_t}) = P_{t+D_A \circ \Theta_t} v(x) .$$

Since $t + D_A \circ \Theta_t \geq D_A$ we conclude by (3.4) that $P_t P_{D_A} v \leq P_{D_A} v$ and hence $P_{D_A} v \in S_{\mathfrak{X}}$.

Let (q_n) be an increasing sequence in $\mathcal{P}_b$ such that $\sup q_n = v$. Since $t + D_A \circ \Theta_t \downarrow T_A$ as $t \downarrow 0$ we know by (3.4) and the right continuity of $\mathfrak{X}$ that

$$\sup_{t>o} P_t P_{D_A} v = \sup_{t>o} P_{t+D_A \circ \Theta_t} v =$$

$$= \sup_{t>o} \sup_{n} P_{t+D_A \circ \Theta_t} q_n = \sup_{n} \sup_{t>o} P_{t+D_A \circ \Theta_t} q_n = \sup_{n} P_{T_A} q_n = P_{T_A} v .$$

We conclude by (II.3.8) that $P_{T_A} v = \widetilde{P_{D_A} v} \in E_{\mathfrak{X}} = \mathcal{W}$. ┘

If $A \subset X$ such that D_A is a stopping time we shall write P_A instead of P_{D_A} and $\tilde{P}_A$ instead of P_{T_A}.

3.6. LEMMA. Let U be an open subset of X. Then $D_U = T_U$ is a stopping time and $P_U v = R_v^U$ for every $v \in \mathcal{W}$.

Proof. Since $\mathfrak{X}$ is right continuous, $D_U = T_U$ and for every $t \in \mathbb{R}_+$

$$[D_U < t] = \bigcup_{\substack{s \in \mathbb{Q}_+ \\ s<t}} [X_s \in U] \in \mathfrak{m}_t .$$

Therefore D_U is a stopping time. There exists a sequence (U_n) of relatively compact open subsets of U such that $\bar{U}_n \subset U_{n+1}$ and $\bigcup_{n=1}^{\infty} U_n = U$. Then $D_{U_n} \downarrow D_U$ by (3.2). By the right continuity of $\mathfrak{X}$, $X_{D_{U_n}} \in \bar{U}_n \subset U$ on $[D_{U_n} < \infty]$.

It clearly suffices to consider $v \in \mathcal{P}_b$. Since $R_v^U = v$ on U we conclude that for every $n \in \mathbb{N}$

$$P_{U_n} v = P_{U_n} R_v^U \leq R_v^U$$

and hence $P_U v = \lim_{n\to\infty} P_{U_n} v \leq R_v^U$. The converse inequality follows from (3.5) since

$D_U = T_U$ and $P_U v = v$ on U. ┘

3.7. LEMMA. Let K be a compact subset of X and let (U_n) be a sequence of relatively compact open subsets of X such that $\bar{U}_{n+1} \subset U_n$ and $\bigcap_{n=1}^{\infty} U_n = K$. Then $\sup D_{U_n} = D_K$ P^μ - a.s. for every $\mu \in M_+^1(X_\Delta)$.
In particular, D_A and T_A are stopping times for every closed subset A of X.

Proof. (D_{U_n}) is an increasing sequence of stopping times, hence $S = \sup D_{U_n}$ is a stopping time. Evidently, $S \leq D_K$. Let

$$\Omega_0 = [S < \infty] \cap [\lim_{n\to\infty} X_{D_{U_n}} \neq X_S] .$$

Then $\Omega_0 \in \mathfrak{M}$ by (IV.7.2.2) and $P^x(\Omega_0) = 0$ for every $x \in X_\Delta$ by the quasi-left-continuity of $\mathfrak{X}$. Hence $P^\mu(\Omega_0) = 0$ for every $\mu \in M_+^1(X_\Delta)$. For every $n \in \mathbb{N}$, $X_{D_{U_n}} \in \bar{U}_n$ on $[S < \infty]$ and hence $\lim_{n\to\infty} X_{D_{U_n}} \in K$ on $[S < \infty]$. Therefore $D_K \leq S$ on $[S < \infty] \setminus \Omega_0$, i.e. $D_K \leq S$ on $\complement \Omega_0$. Together with $S \leq D_K$ this implies that D_K is a stopping time and $D_K = S$ P^μ - a.s. for every $\mu \in M_+^1(X_\Delta)$. Using (3.2) and the preceding remark the rest of the statement follows readily. ┘

In the following let X' denote the set of all points $x \in X$ such that $P_t(x,\cdot) = \varepsilon_x$ for every $t > 0$ and $T' = D_{X'}$.

3.8. PROPOSITION. X' is closed. For every $p \in P$, $\lim_{t\to\infty} P_t p = R_p^{X'} \in P$.

Proof. Let $p \in P$. Since $t \longmapsto P_t p$ is decreasing it is clear that $q := \lim_{t\to\infty} P_t\, p \in B^+(X)$ and $P_t\, q = q$ for every $t > 0$. Hence $q, p-q \in E_{\mathfrak{X}} = W$. Thus $q \in P$. Since $q = p$ on X' it is clear that $q \geq R_p^{X'}$.

Now let $x \in X \setminus X'$ and $t > 0$ such that $P_t(x,\circ) \neq \varepsilon_x$. Since $P_t u(x) \leq u(x)$ for every $u \in W$ and $P_t q(x) = q(x)$ we conclude that $x \notin \delta(q)$. Therefore $R_p^{X'} \geq R_q^{\delta(q)} = q$ by (II.6.3) and hence $R_p^{X'} = q$. If p is a strict potential then of course $X' = \{q = p\}$ and hence X' is closed. ┘

3.9. LEMMA. T' is a stopping time such that for every $x \in X$, P^x - a.s. for every $T' \leq t < \infty$, $X_t = X_{T'} \in X'$.

Proof. By (3.7) and (3.8), T' is a stopping time. Moreover, $X_{T'} \in X'$ a.s. on $[T' < \infty]$ by the right continuity of the paths. For every $y \in X'$,

$$P^y\,[T_{X_\Delta \setminus \{y\}} < \infty] = 0$$

since $P^y\,[X_t \neq y] = 0$ for every $t \in \mathbb{R}_+$ and

$$[T_{X_\Delta \setminus \{y\}} < \infty] = \bigcup_{t \in \mathbb{Q}_+} [X_t \neq y] .$$

Now let $\{U_n : n \in \mathbb{N}\}$ be a base of X_Δ. If $\omega \in \Omega$ and $T'(\omega) < t < \infty$ such that $X_t(\omega) \neq X_{T'}(\omega)$ then there exists $n \in \mathbb{N}$ such that $X_{T'}(\omega) \in U_n$ but $X_t(\omega) \notin U_n$ and hence $(T' + T_{X_\Delta \setminus U_n} \circ \Theta_{T'})(\omega) \leq t < \infty$. By the strong Markov property we have for every $n \in \mathbb{N}$

$$P^x([X_{T'} \in U_n] \cap [T_{X_\Delta \setminus U_n} \circ \Theta_{T'} < \infty])$$

$$= E^x(1_{[X_{T'} \in U_n]}\, P^{X_{T'}}[T_{X_\Delta \setminus U_n} < \infty]) = 0$$

since for every $y \in U_n \cap X'$

$$P^y\,[T_{X_\Delta \setminus U_n} < \infty] \leq P^y\,[T_{X_\Delta \setminus \{y\}} < \infty] = 0 .$$

Thus the statement follows. ┘

<u>3.10. COROLLARY.</u> Let $q \in \mathcal{P}_b$. Then for every $x \in X$, $\lim_{t \to \infty} E^x(q(X_t)\, 1_{[T' \geq t]}) = 0$.

Proof. By (3.8),

$$\lim_{t \to \infty} E^x(q(X_t)) = R_q^{X'}(x)$$

and by (3.9)

$$\lim_{t \to \infty} E^x(q(X_t)\, 1_{[T'<t]}) = \lim_{t \to \infty} E^x(q(X_{T'})\, 1_{[T'<t]})$$

$$= E^x(q(X_{T'})\, 1_{[T'<\infty]}) = E^x(q(X_{T'})) = P_{T'}q(x) .$$

Hence it suffices to show that $P_{T'}q \geq R_q^{X'}$.

For every $n \in \mathbb{N}$, let

$$S_n = \frac{1}{n} + \inf\,\{r \in \mathbb{Q}_+ : X_r \in X'\} .$$

Then (S_n) is a decreasing sequence of stopping times with respect to $(\mathcal{F}_t^0)$ and $\lim_{n \to \infty} S_n = T' = T_{X'}$ P^x-a.s. for every $x \in X$ by (3.9). By (IV.4.1) and (IV.6.4) $P_{S_n} q \in \mathcal{B}^+(X)$ for every $n \in \mathbb{N}$, hence $P_{T'}q = \lim_{n \to \infty} P_{S_n} q \in \mathcal{B}^+(X)$ and therefore

$P_{T'}q \in \mathcal{W}$ by (3.5). Furthermore, $P_{T'}q = q$ on X' and hence $P_{T'}q \geq R_q^{X'}$. ⌟

3.11. LEMMA. Let (A_n) be a decreasing sequence of subsets of X such that every D_{A_n} is a stopping time and let $S = \sup D_{A_n}$. Then $P_S q = \lim_{n\to\infty} P_{A_n} q$ for every $q \in \mathcal{P}_b$.

Proof. Let $q \in \mathcal{P}_b$. By (3.4), $(P_{A_n} q)$ is a decreasing sequence and $P_S q \leq \lim_{n\to\infty} P_{A_n} q$.

Let $x \in X$ and $\varepsilon > 0$. Then there exists a compact subset K of X such that $R_q^{\complement K}(x) < \varepsilon$. Furthermore by (3.10) there exists $t > 0$ such that $E^x(q(X_t)\, 1_{[T' \geq t]}) < \varepsilon$. Let

$$\Omega_n = [D_{A_n} < D_{\complement K}] \quad , \quad \Omega_0 = \bigcap_{n=1}^{\infty} \Omega_n \, .$$

Obviously $X_{D_{A_n}}(\omega) \in K$ for every $\omega \in \Omega_0$ and $n \in \mathbb{N}$. Since $\mathfrak{X}$ is quasi-left-continuous we therefore obtain

$$\lim_{n\to\infty} X_{D_{A_n}} = X_S \in K \quad P^x\text{-a.s. on } \Omega_0 \cap [S < \infty]$$

and hence

$$\lim_{n\to\infty} E^x(q(X_{D_{A_n}})\, 1_{\Omega_0 \cap [S<\infty]}) = E^x(q(X_S)\, 1_{\Omega_0 \cap [S<\infty]}) \leq P_S q(x) \, .$$

Now

$$\complement(\Omega_0 \cap [S < \infty]) = (\Omega_n \setminus \Omega_0) \cup \complement\Omega_n \cup (\Omega_0 \cap [S = \infty])$$

where $\lim_{n\to\infty} P^x(\Omega_n \setminus \Omega_0) = 0$ since (Ω_n) is decreasing to Ω_0 and

$$E^x(q(X_{D_{A_n}})\, 1_{\complement\Omega_n}) \leq E^x(q(X_{D_{\complement K}})) = R_q^{\complement K}(x) < \varepsilon$$

by (3.4) and (3.6). Furthermore, if $\omega \in \Omega_0 \cap [S = \infty]$, i.e. if $D_{A_n}(\omega) < D_{\complement K}(\omega)$ for every $n \in \mathbb{N}$ but $\lim_{n\to\infty} D_{A_n}(\omega) = \infty$ then there is no $s_0 \in \mathbb{R}_+$ such that $X_s(\omega) = X_{s_0}(\omega)$ for every $s \in [s_0, \infty[$. Therefore $T' = \infty$ P^x-a.s. on $\Omega_0 \cap [S = \infty]$. Let

$$T = \begin{cases} t & \text{on } [T' \geq t] \\ \infty & \text{on } [T' < t] \, . \end{cases}$$

Then T is a stopping time and $D_{A_n} \geq T$ on $[D_{A_n} \geq t] \cap [T' = \infty]$. Therefore by (3.4)

$$E^x(q(X_{D_{A_n}})\, 1_{[D_{A_n} \geq t] \cap \Omega_0 \cap [S=\infty]}) \leq E^x(q(X_{D_{A_n}})\, 1_{[D_{A_n} \geq T]})$$

$$\leq E^x(q(X_T)) = E^x(q(X_t)\, 1_{[T' \geq t]}) < \varepsilon .$$

Finally,

$$\lim_{n\to\infty} P^x([D_{A_n} < t] \cap [S = \infty]) = 0 .$$

Combining all this we conclude that

$$\limsup_{n\to\infty} E^x(q(X_{D_{A_n}})\, 1_{[(\Omega_0 \cap [S < \infty])}) \leq 2\varepsilon .$$

and hence

$$\lim_{n\to\infty} P_{A_n} q(x) = \lim_{n\to\infty} E^x(q(X_{D_{A_n}})) \leq P_S q(x) + 2\varepsilon .$$

3.12. LEMMA. Let K be a compact subset of X and $v \in \mathcal{W}$. Then $P_K v = R_v^K$ and $\tilde{P}_K v = \hat{R}_v^K$.

Proof. It suffices to consider $v \in \mathcal{P}_b$. Let $x \in X$ and let (U_n) be a sequence of relatively compact open subsets of X such that $\bar{U}_{n+1} \subset U_n$ and $\bigcap_{n=1}^{\infty} U_n = K$. Then $\lim_{n\to\infty} D_{U_n} = D_K$ P^x-a.s. by (3.7) and hence by (3.11)

$$P_K v(x) = \lim_{n\to\infty} P_{U_n} v(x) = \lim_{n\to\infty} R_v^{U_n}(x) = R_v^K(x) .$$

Applying (3.5) we finally conclude that $\tilde{P}_K v = \widetilde{R_v^K} = \hat{R}_v^K$ since $\hat{R}_v^K$ is the greatest minorant of R_v^K in $\mathcal{W} = E_{\mathfrak{X}}$.

3.13. PROPOSITION. Let $\mu \in M_+^1(X)$. Let (K_n) be an increasing sequence of compact subsets of X and let (U_n) be a decreasing sequence of open subsets of X such that $K_n \subset U_n$ for every $n \in \mathbb{N}$, $\sup R_1^{K_n} = \inf R_1^{U_n}$ μ-a.s. and $\sup R_q^{K_n} = \inf R_q^{U_n}$ μ-a.s. for some strict potential $q \in \mathcal{P}_b$. Then

$$\inf D_{K_n} = \sup D_{U_n} \qquad P^\mu\text{-a.s.}$$

Proof. Let

$$S = \sup D_{U_n} , \quad T = \inf D_{K_n} .$$

By (3.6) and (3.7), S and T are stopping times, $S \leq T$. Applying (3.4) and (3.12) we obtain furthermore that for every $v \in \mathcal{W}$

$$\sup R_v^{K_n} = \sup P_{K_n} v \leq P_T v \leq P_S v \leq \inf P_{U_n} v = \inf R_v^{U_n} .$$

Let

$$A = \{\sup R_1^{K_n} = \inf R_1^{U_n}\} \cap \{\sup R_q^{K_n} = \inf R_q^{U_n}\}$$

and $x \in A$. Since q is a strict potential we deduce from the preceding inequalities that $P_T v(x) = P_S v(x)$ for every $v \in \mathcal{W}$. Obviously $T = S + T\circ\Theta_S$ since $S \leq T = D_{\bigcup_{n=1}^{\infty} K_n}$. Hence by the strong Markov property for all functions $f,g \in \mathcal{B}^+(X_\Delta)$

$$E^x(f(X_T)g(X_S)) = E^x(f(X_T)\circ\Theta_S\, g(X_S))$$

$$= E^x(E^{X_S}(f(X_T))g(X_S)) = E^x(P_T f(X_S)g(X_S)) \ .$$

In particular, for every $v \in \mathcal{W}$

$$E^x(v(X_S)) = P_S v(x) = P_T v(x) = E^x(v(X_T)) = E^x(P_T v(X_S)) \ .$$

Since $P_T v \leq v$ we conclude that $v(X_S) = P_T v(X_S)$ P^x-a.s. and hence for every $v \in \mathcal{W}$ and $g \in \mathcal{B}^+(X_\Delta)$

$$E^x(v(X_T)g(X_S)) = E^x(v(X_S)g(X_S)) \ .$$

This shows that $X_T = X_S$ P^x-a.s.

Furthermore, for every $n \in \mathbb{N}$, $D_{U_n} \leq S \leq T \leq D_{K_n}$ and hence

$$[D_{K_n} < \infty] \subset [T < \infty] \subset [S < \infty] \subset [D_{U_n} < \infty]$$

where

$$\lim_{n\to\infty} P^x[D_{K_n} < \infty] = \lim_{n\to\infty} R_1^{K_n}(x) = \lim_{n\to\infty} R_1^{U_n}(x) = \lim_{n\to\infty} P^x[D_{U_n} < \infty] \ .$$

Therefore $P^x[T < \infty] = P^x[S < \infty]$. Since $T\circ\Theta_T = 0$ on $[T < \infty]$ we have by the strong Markov property

$$P^x[S < T] = P^x[S < \infty\, , T\circ\Theta_S > 0] = E^x(1_{[S < \infty]}P^{X_S}[T > 0])$$

$$= E^x(1_{[T < \infty]}P^{X_T}[T > 0]) = P^x[T < \infty\, , T\circ\Theta_T > 0] = 0 \ .$$

Thus $P^\mu[S < T] = \int_A P^x[S < T]\, \mu(dx) = 0$. $\square$

<u>3.14. THEOREM.</u> Let $A \in \mathcal{B}(X)$. Then:

1. D_A and T_A are stopping times and for every $v \in \mathcal{W}$

$$P_A v = R_v^A \quad , \quad \tilde{P}_A v = \hat{R}_v^A \ .$$

2. For every $\mu \in M_+^1(X_\Delta)$ there exists an increasing sequence (K_n) of compact subsets of A and a decreasing sequence (U_n) of open subsets of X

containing A such that $\inf D_{K_n} = \sup D_{U_n} = D_A$ P^μ- a.s .

Proof. Since $D_B = \infty$ P^Δ - a.s. for any subset B of X it suffices to consider $\mu \in M_+^1(X)$. Let $q \in P_b$ be a strict potential. By (1.9) there exists an increasing sequence (K_n) of compact subsets of A and a decreasing sequence (U_n) of open subsets of X containing A such that

$$\sup R_1^{K_n} = \inf R_1^{U_n} = R_1^A \quad \mu\text{-a.s.}$$

and

$$\sup R_q^{K_n} = \inf R_q^{U_n} = R_q^A \quad \mu\text{-a.s. .}$$

Then $T := \inf D_{K_n}$ and $S := \sup D_{U_n}$ are stopping times, $S \leq D_A \leq T$ and $S = T$ P^μ-a.s. by (3.13). Thus D_A is a stopping time.

Let us now assume that $\mu = \varepsilon_x$, $x \in X$. For every $v \in W$

$$\sup R_v^{K_n} \leq R_v^A \leq \inf R_v^{U_n}$$

and by (3.4)

$$\sup P_{K_n} v \leq P_A v \leq \inf P_{U_n} v .$$

Since q is strict and $R_v^{K_n} = P_{K_n} v$, $R_v^{U_n} = P_{U_n} v$ for every $n \in \mathbb{N}$ we conclude that $R_v^A(x) = P_A v(x)$ for every $v \in W$.

Furthermore, $T_A = \lim\limits_{t \downarrow 0} (t + D_A \circ \Theta_t)$ is a stopping time. Since $R_v^A \in B^+(X)$ for every $v \in W$ we finally obtain by (3.5) that $\widetilde{P}_A v = \widetilde{R_v^A} = \hat{R}_v^A$. ┘

<u>3.15. REMARKS.</u> 1. Having established (3.14) we now may replace the compact sets K_n and open sets U_n in (3.13) by Borel sets A_n and B_n .

2. Since the life time ζ of $\mathfrak{X}$ is a stopping time and $D_{A\cup\{\Delta\}} = D_A \wedge \zeta$, $T_{A\cup\{\Delta\}} = T_A \wedge \zeta$ we obtain from (3.14) that D_A and T_A are stopping times for every $A \in B(X_\Delta)$.

<u>3.16. THEOREM.</u> For every $\mu \in M_+^1(X_\Delta)$ and all Borel subsets A and B of X ,

$$P^\mu[X_{T_A} \in B] = \mu^A(B) .$$

Proof. By (3.14) and (3.15.2) for every $v \in W$

$$E^\mu(v(X_{T_A})) = \int E^x(v(X_{T_A}))\,\mu(dx) = \int \hat{R}_v^A(x)\,\mu(dx) = \int v\,d\mu^A .$$ ┘

3.17. PROPOSITION. Let $A \in \mathcal{B}(X)$ and $x \in X$. Then $\overset{\circ}{\varepsilon}{}^A_x = \varepsilon_x$ if and only if $D_A = 0$ P^x-a.s. .

Proof. By (3.14), $\overset{\circ}{\varepsilon}{}^A_x = \varepsilon_x$ if and only if $X_{D_A} = x$ P^x-a.s. . Hence of course $\overset{\circ}{\varepsilon}{}^A_x = \varepsilon_x$ if $D_A = 0$ P^x-a.s. .

Suppose conversely that $\overset{\circ}{\varepsilon}{}^A_x = \varepsilon_x$ and let $S = D_A$. Then $S < \infty$ P^x-a.s. since $X_\infty = \Delta$, and hence $S \circ \Theta_S = 0$ P^x-a.s. by (3.1). Thus we obtain by the strong Markov property that

$$P^x[S = 0] = E^x(P^{X_S}[S = 0]) = P^x[S \circ \Theta_S = 0] = 1 .$$

It is now possible to present a probabilistic condition characterizing the balayage spaces which have the local truncation property.

3.18. THEOREM. The following conditions are equivalent:

(1) $(X,\mathcal{W})$ has the local truncation property.

(2) The paths $t \mapsto X_t$ are a.s. continuous on $[0,\zeta[$.

Proof. (1) ⇒ (2): Let $\mathcal{V}$ be a countable base of relatively compact open subsets of X closed under finite unions and let

$$\Omega_0 = \bigcup_{\substack{U,W \in \mathcal{V} \\ \bar{U} \subset W}} \; \bigcup_{\substack{s,t \in \mathbb{Q}_+ \\ s<t}} [X_s \in \bar{U}, X_t \in X \setminus W, X_r \in \bar{U} \cup (X \setminus W) \;\forall\; r \in \mathbb{Q} \cap [s,t]] .$$

Then $\Omega_0 \in \mathfrak{m}$. Let $\omega \in \Omega$ and suppose that there exists a $0 < t_0 < \zeta(\omega)$ such that $t \mapsto X_t(\omega)$ is not continuous at t_0. Then there exists $W \in \mathcal{V}$ such that

$$X_{t_0^-}(\omega) := \lim_{r \uparrow t_0, r<t_0} X_r(\omega) \in W \;, \quad X_{t_0}(\omega) \in \complement \bar{W} .$$

Furthermore, there exists $U,V \in \mathcal{V}$ such that $X_{t_0^-}(\omega) \in U$, $\bar{U} \subset V$ and $\bar{V} \subset W$. Finally, there exist $s,t \in \mathbb{Q}_+$ such that $s < t_0 < t$, $X_r(\omega) \in U$ for every $s \leq r < t_0$ and $X_r(\omega) \in \complement \bar{W}$ for every $t_0 \leq r \leq t$. Hence $\omega \in \Omega_0$. So it suffices to show that $P^x(\Omega_0) = 0$ for every $x \in X$.

So fix $x \in X$, $U, W \in \mathcal{V}$ such that $\bar{U} \subset W$ and $s,t \in \mathbb{Q}_+$ such that $s < t$. Consider

$$A = [X_s \in \bar{U}, X_t \in X \setminus W, X_r \in \bar{U} \cup (X \setminus W) \text{ for every } r \in \mathbb{Q}_+ \cap [s,t]] ,$$

choose $V \in \mathcal{V}$ such that $\bar{U} \subset V$, $\bar{V} \subset W$ and let

$$T = \inf\{r > s : X_r \in \complement V\} = s + T_{\complement V} \circ \Theta_s .$$

If $\omega \in A$ then $X_t(\omega) \in X \setminus W \subset \complement V$, hence $s \le T(\omega) \le t$ and therefore by the right continuity of the path $r \mapsto X_r(\omega)$,

$$X_T(\omega) \in (\bar{U} \cup (X \setminus W)) \cap \complement V = X \setminus W .$$

Since, for every $y \in \bar{U} \subset V$, by (3.16) and (III.8.1),

$$P^y[X_{T_{\complement V}} \in X \setminus V^*] = \varepsilon_y^{\complement V}(X \setminus V^*) = 0 ,$$

we conclude by the Markov property that

$$P^x(A) \le P^x[X_s \in \bar{U}, X_T \in X \setminus W] = P^x[X_s \in \bar{U}, X_{T_{\complement V}} \circ \Theta_s \in X \setminus W]$$

$$= \int_{[X_s \in \bar{U}]} P^{X_s}[X_{T_{\complement V}} \in X \setminus W] dP^x = 0 .$$

Thus $P^x(\Omega_0) = 0$.

(2) ⇒ (1): For every open subset U of X and every $x \in U$ obviously $X_{T_{\complement U}} \in U^* \cup \{\Delta\}$ P^x-a.s. and therefore by (3.16)

$$\varepsilon_x^{\complement U}(X \setminus U^*) = P^x[X_{T_{\complement U}} \in X \setminus U^*] = 0 .$$

So (III.8.1) yields the local truncation property. ⌟

3.19. EXAMPLES. 1. Classical potential theory. A Hunt process $\mathfrak{X}$ on $\mathbb{R}^n$, $n \ge 1$, is called *Brownian motion* if the transition function of $\mathfrak{X}$ is the Brownian semigroup $\mathbb{P}$ on $\mathbb{R}^n$, if the paths are continuous on $\mathbb{R}_+$, and if $\zeta = \infty$ on $[X_0 \in \mathbb{R}^n]$.

If $n \ge 3$ then by (III.8.4), (IV.7.5), (IV.7.6), and (3.18) there exists a Hunt process $(\tilde{W}, \tilde{M}, \tilde{M}_t, \tilde{\pi}_t, \tilde{\varphi}_t, \tilde{P}^x)$ on $\mathbb{R}^n$ with transition semigroup $\mathbb{P}$ such that the paths $t \mapsto w(t)$ are continuous a.s. on $[0, \zeta[$, and an application of (IV.7.2.1) and (IV.7.2.4) yields the existence of Brownian motion on $\mathbb{R}^n$. Considering the canonical projection of $\mathbb{R}^3$ on $\mathbb{R}^1$ ($\mathbb{R}^2$ resp.), an application of (IV.7.7) to Brownian motion on $\mathbb{R}^n$ shows the existence of Brownian motion on $\mathbb{R}^1$ ($\mathbb{R}^2$ resp.).

2. Heat equation. A Hunt process $\mathfrak{X} = (\Omega, \mathfrak{M}, \mathfrak{M}_t, X_t, \Theta_t, P^x)$ on $\mathbb{R}^{n+1}$, $n \ge 1$, is called *heat process* (or space-time Brownian motion) on $\mathbb{R}^{n+1}$ provided $\mathfrak{X}$ has

the following properties:

(1) $\mathbb{P} \otimes \mathbb{T}$ is the transition function of $\mathfrak{X}$ where $\mathbb{P}$ is the Brownian semigroup on $\mathbb{R}^n$.

(2) The paths of $\mathfrak{X}$ are continuous on $\mathbb{R}_+$.

(3) If $\omega \in \Omega$ and $X_0(\omega) = (x,s) \in \mathbb{R}^n \times \mathbb{R}$ then $X_t(\omega) \in \mathbb{R}^n \times \{s - t\}$ for every $t > 0$.

By (V.5.8), $(\mathbb{R}^{n+1}, E_{\mathbb{P}\otimes\mathbb{T}})$ is a harmonic space. Hence by (IV.7.5), (IV.7.6), (3.18) and (IV.7.2.4), there exists a Hunt process $\mathfrak{X}$ on $\mathbb{R}^{n+1}$ with transition function $\mathbb{P}\otimes\mathbb{T}$ and paths which are continuous on $[0,\zeta[$. Let π be the canonical projection of $\mathbb{R}^n \times \mathbb{R}$ on $\mathbb{R}$ and let

$$\Omega_1 = [X_0=\Delta] \cup \bigcap_{t\in\mathbb{Q}_+} [X_0 \in \mathbb{R}^{n+1}, X_t\in\mathbb{R}^{n+1}, \pi\circ X_t = \pi\circ X_0 - t] .$$

Then of course $P^\Delta(\Omega_1) = 1$ and if $(x,s) \in \mathbb{R}^n\times\mathbb{R}$ then for every $t \in \mathbb{Q}_+$

$$P^{(x,s)}[X_0 \in \mathbb{R}^{n+1}, X_t \in \mathbb{R}^{n+1}, \pi\circ X_t = \pi\circ X_0 - t]$$

$$= P^{(x,s)}[X_t \in \mathbb{R}^n \times \{s - t\}] = (P_t \otimes T_t)((x,s), \mathbb{R}^n \times \{s - t\}) = P_t(x,\mathbb{R}^n) = 1 ,$$

hence $P^{(x,s)}(\Omega_1) = 1$. If $\omega \in \Omega_1$ such that $X_0(\omega) \in \mathbb{R}^{n+1}$ then $\zeta(\omega) = \infty$ and $\pi(X_t(\omega)) = \pi(X_0(\omega)) - t$ for every $t \in \mathbb{Q}_+$, hence for every $t \in \mathbb{R}_+$ by the continuity of the paths. Another application of (IV.7.2.4) finishes the proof of the existence of the heat process.

If $\mathfrak{X}$ is a heat process on $\mathbb{R}^{n+1}$ and π denotes the canonical projection of $\mathbb{R}^n \times \mathbb{R}$ on $\mathbb{R}^n$ then $\pi(\mathfrak{X})$ is Brownian motion on $\mathbb{R}^n$.

3. Poisson semigroup. A Hunt process $\mathfrak{X}$ on $\mathbb{Z}$ is called *Poisson process* if the transition function of $\mathfrak{X}$ is the Poisson semigroup $\mathbb{P}$, if the paths are increasing by steps of size 1, and if $\zeta = \infty$ on $[X_0 \in \mathbb{Z}]$.

By (IV.7.5) and (IV.7.6) there exists a Hunt process $(\Omega,\mathfrak{M},\mathfrak{M}_t,X_t,\Theta_t,P^x)$ on $\mathbb{Z}$ with transition semigroup $\mathbb{P}$ such that $\Omega = \widetilde{W}$, $X_t(\omega) = \omega(t)$, and $\Theta_t(\omega)(s) = \omega(t+s)$. Since $E_{\mathbb{P}}$ is the set of all decreasing positive numerical functions on $\mathbb{Z}$ we obtain for every $x \in \mathbb{Z}$ that $\overset{\circ}{\varepsilon}{}_x^{\complement\{x\}} = \varepsilon_{x+1}$ and hence by (3.16), $P^x[X_{T_{\complement\{x\}}} = x+1]=1$. Let

$$\Omega_0 = \bigcup_{s\in\mathbb{Q}_+} \bigcup_{x\in\mathbb{Z}} [X_s = x , X_{T_{\complement\{x\}}}\circ\Theta_s \neq x + 1] .$$

Then $\Omega_0 \in \mathfrak{M}$ and for every $y \in \mathbb{Z}$

$$P^y[X_s = x, X_{T_{[\{x\}}} \circ \Theta_s \neq x+1] = E^y(1_{[X_s=x]} P^{X_s}[X_{T_{[\{x\}}} \neq x+1]) = 0$$

for all $s \in \mathbb{Q}_+$ and $x \in \mathbb{Z}$, hence $P^y(\Omega_0) = 0$. Moreover, obviously $P^\Delta(\Omega_0) = 0$. For every $\omega \in \Omega \setminus \Omega_0$ such that $X_0(\omega) \in \mathbb{Z}$ the path $t \mapsto X_t(\omega)$ is increasing by steps of size 1. Hence an application of (IV.7.2.1) and (IV.7.2.4) yields the existence of the Poisson process.

The following proposition shows that for every open subset U of X the canonical restriction $\mathfrak{X}^U$ of $\mathfrak{X}$ on U is associated to the open subspace $(U, \mathcal{W}^U)$.

3.20. PROPOSITION. Let U be an open subset of X and define $X_t^U : \Omega \to U_\Delta$, $t \in \overline{\mathbb{R}}_+$, by

$$X_t^U(\omega) = \begin{cases} X_t(\omega), & t < D_{\complement U}(\omega), \\ \Delta, & t \geq D_{\complement U}(\omega). \end{cases}$$

Let $\mathcal{F}_t^{oU} = \sigma(X_s^U : s \leq t)$, $\mathcal{F}^{oU} = \mathcal{F}_\infty^{oU}$. Then $\mathfrak{X}^U = (\Omega, \mathcal{F}^U, \mathcal{F}_t^U, X_t^U, \Theta_t, P^x)$ is a Hunt process on U which is associated with the restriction $(U, \mathcal{W}^U)$ of $(X, \mathcal{W})$ on U. $\mathfrak{X}^U$ is called the *restriction of* $\mathfrak{X}$ *on* U. The transition function $\mathbb{P}^U$ of $\mathfrak{X}^U$ is given by

$$P_t^U(x,B) = P^x[X_t \in B, t < D_{\complement U}] \qquad (x \in X, B \in \mathcal{B}(U)).$$

Proof. Since $D_{\complement U}$ is a stopping time of $\mathfrak{X}$ we have for every $t \in \mathbb{R}_+$ and $B \in \mathcal{B}(U)$

$$[X_t^U \in B] = [X_t \in B] \cap [t < D_{\complement U}] \in \mathfrak{M}_t.$$

Therefore $\mathcal{F}_t^{oU} \subset \mathfrak{M}_t$ for every $t \in \overline{\mathbb{R}}_+$.

Let T be a stopping time with respect to $(\mathcal{F}_{t+}^{oU})$. Then T is a stopping time of $\mathfrak{X}$. Fix $x \in U$, $s \in \mathbb{R}_+$ and $B \in \mathcal{B}(U)$. Since $D_{\complement U} = T + D_{\complement U} \circ \Theta_T$ on $[T < D_{\complement U}]$ the strong Markov property of $\mathfrak{X}$ yields that

$$\begin{aligned} P^x[X_{T+s}^U \in B] &= P^x[X_{T+s} \in B, T+s < D_{\complement U}] \\ &= P^x[X_s \circ \Theta_T \in B, T < D_{\complement U}, s < D_{\complement U} \circ \Theta_T] = E^x(1_{[T<D_{\complement U}]} P^{X_T}[X_s \in B, s < D_{\complement U}]) \\ &= E^x(P^{X_T^U}[X_s^U \in B]). \end{aligned}$$

Using (IV.6.8) and (IV.6.9) we conclude that $\mathfrak{X}^U$ is a strong Markov process and

$\mathcal{F}^U_{t+} = \mathcal{F}^U_t$ for every $t \in \mathbb{R}_+$.

Since the paths of $\mathfrak{X}^U$ are right continuous, the semigroup $(P^U_t)_{t>0}$ of $\mathfrak{X}^U$ satisfies $\lim\limits_{t\to 0} P^U_t f = f$ for every $f \in K(U)$. Let $(V^U_\lambda)_{\lambda>0}$ be the resolvent of $(P^U_t)_{t>0}$. Let S be a stopping time of $\mathfrak{X}$ such that $S \leq D_{\complement U}$, let $\alpha > 0$, $f \in \mathcal{B}^+_b(U)$, and $x \in U$. Then

$$V^U_\alpha f(x) = E^x\Big(\int_0^\infty e^{-\alpha t}\, f \circ X^U_t\, dt\Big) = E^x\Big(\int_0^{D_{\complement U}} e^{-\alpha t}\, f \circ X_t\, dt\Big)$$

and

$$P^\alpha_S\, V^U_\alpha f(x) = E^x(e^{-\alpha S}\, V^U_\alpha f(X_S)) = E^x\Big(e^{-\alpha S}\, E^{X_S}\Big(\int_0^{D_{\complement U}} e^{-\alpha t}\, f \circ X_t\, dt\Big)\Big)$$

$$= E^x\Big(e^{-\alpha S} \int_0^{D_{\complement U}\circ\Theta_S} e^{-\alpha t}\, f \circ X_t \circ \Theta_S\, dt\Big) = E^x\Big(\int_0^{D_{\complement U}\circ\Theta_S} e^{-\alpha(t+S)}\, f \circ X_{t+S}\, dt\Big)$$

$$= E^x\Big(\int_S^{D_{\complement U}} e^{-\alpha t}\, f \circ X_t\, dt\Big) = V^U_\alpha f(x) - E^x\Big(\int_0^S e^{-\alpha t}\, f \circ X_t\, dt\Big)$$

since $S + D_{\complement U}\circ\Theta_S = D_{\complement U}$.

In order to prove that $\mathcal{W}^U \subset E_{\mathfrak{X}^U}$ fix $u \in \mathcal{W}^U_b$, take $\lambda > 0$, $0 < \alpha < \varepsilon$ and define

$$g = u - \lambda V^U_{\lambda+\varepsilon} u \;, \qquad f = g - (\varepsilon-\alpha) V^U_{\lambda+\varepsilon} u \;.$$

Then by the resolvent equation

$$V^U_{\lambda+\varepsilon} u = V^U_\alpha f \;.$$

Let $x \in U$ and $A = \{f>0\}$. We choose a compact subset K of A and define $S = D_{K\cup\complement U}$. Since

$$P^\alpha_S\, V^U_\alpha f(x) = V^U_\alpha f(x) - E^x\Big(\int_0^S e^{-\alpha t}\, f \circ X_t\, dt\Big)$$

we obtain that

$$g(x) - P^\alpha_S g(x) = u(x) - P^\alpha_S u(x) - \lambda E^x\Big(\int_0^S e^{-\alpha t}\, f \circ X_t\, dt\Big) \;.$$

By (3.16) and (2.9)

$$P^\alpha_S u(x) \leq P_S u(x) = \varepsilon_x^{\circ K\cup\complement U}(u) \leq u(x) \;.$$

Moreover, $X_S \in K \cup \complement U$ on $[S < \infty]$ and $K \subset \{f > 0\} \subset \{g > 0\}$, hence $P^\alpha_S g(x) \geq 0$. Therefore

$$g(x) \geq -\lambda E^x\Big(\int_0^S e^{-\alpha t}\, f \circ X_t\, dt\Big) \;.$$

Using (3.14) we conclude that

$$g(x) \geq - \lambda E^x\left(\int_0^{D_{A\cup\complement U}} e^{-\alpha t}\, f\circ X_t\, dt\right) \geq 0$$

since $f\circ X_t \leq 0$ on $[0, D_{A\cup\complement U}[$. Letting ε tend to zero we finally obtain that $\lambda V^U_\lambda u(x) \leq u(x)$, i.e. $u \in S_{\mathfrak{X}^U}$. Since u is l.s.c., we have $u \in E_{\mathfrak{X}^U}$. Thus $\mathcal{W}^U \subset E_{\mathfrak{X}^U}$.

Suppose now conversely that $u \in E_{\mathfrak{X}^U}$ and u is bounded. If S,T are stopping times of $\mathfrak{X}$ such that $T \leq S \leq D_{\complement U}$ then $P^\alpha_S V^U_\alpha f \leq P^\alpha_T V^U_\alpha f \leq V^U_\alpha f$ for every $\alpha > 0$ and $f \in \mathcal{B}^+_b(U)$, hence $P_S u \leq P_T u \leq u$ by (II.3.12). In particular, we obtain by (3.16) that for every relatively compact open subset W of U and $x \in W$

$$H_W u(x) = P_{D_{\complement W}} u(x) \leq u(x) .$$

If $f \in \mathcal{B}^+_b(X)$ such that Vf is bounded then $V^U f = Vf - H_U Vf$ is l.s.c. . Hence u is l.s.c. by (II.3.11) if V is proper. In the general case we have to be more careful.

Fix $x \in U$ and suppose first that x is not finely isolated. Then $D_{\complement\{x\}} = 0$ P^x-a.s by (3.17). Let $\varepsilon > 0$. There exists $t > 0$ such that $P^U_t u(x) > u(x) - \varepsilon$ and a relatively compact open neighborhood W of x in U such that $\|u\|\, P^x[D_{\complement W} > t] < \varepsilon$. Let $S = \inf(t, D_{\complement U})$, $T = \inf(t, D_{\complement W})$. Then S,T are stopping times of $\mathfrak{X}$, $T \leq S \leq D_{\complement U}$ and hence $P_S u \leq P_T u$. Therefore

$$\begin{aligned} u(x) - \varepsilon < P^U_t u(x) &= P_S u(x) \leq P_T u(x) \\ &= E^x(u\circ X_t\, 1_{[t<D_{\complement W}]}) + E^x(u\circ X_{D_{\complement W}}\, 1_{[t\geq D_{\complement W}]}) \\ &\leq \varepsilon + P_{D_{\complement W}} u(x) = \varepsilon + H_W u(x) . \end{aligned}$$

Since $H_W u \leq u$ and $H_W u$ is continuous on W we obtain that

$$\hat{u}(x) \geq H_W u(x) \geq u(x) - 2\varepsilon .$$

Thus u is l.s.c. at x.

Suppose now that x is finely isolated. Let $p \in \mathcal{P}_b$ and $q = p - H_U p$ such that $q(x) = u(x)$. Since $R_p^{\{x\}\cup\complement U}$ is continuous on U we obtain by (2.8) that

$$v := {}^U R_q^{\{x\}} = R_p^{\{x\}\cup\complement U} - R_p^{\complement U} \in \mathcal{W}^U .$$

Let $S = D_{\{x\}\cup\complement U}$. Then by (2.9) and (3.16) for every $y \in U$

$$v(y) = \varepsilon_y^{\complement\{x\} \cup \complement U}(q) = P_S q(y) .$$

Therefore $v = P_S q = P_S u \leq u$, hence

$$u(x) = q(x) = v(x) = \hat{v}(x) \leq \hat{u}(x) ,$$

i.e. u is l.s.c. at x .

In total we obtain that $u \in \mathcal{W}^U$. Thus $\mathfrak{X}^U$ is associated with $(U,\mathcal{W}^U)$ and applying (IV.7.6) we finally conclude that $\mathfrak{X}^U$ is a Hunt process. (Of course the fact that $\mathfrak{X}^U$ is a standard process follows already from (IV.6.7).) ┘

3.21. EXAMPLE. Classical potential theory. Let U be a relatively compact open subset of $\mathbb{R}^2$. Then Brownian motion on U is associated with $(U,{*\mathcal{H}}^+(U))$. Indeed, let $\mathfrak{X}$ be Brownian motion on $\mathbb{R}^3$, let π denote the projection $(x_1,x_2,x_3) \mapsto (x_1,x_2)$ from $\mathbb{R}^3$ on $\mathbb{R}^2$, and take $V = U \times \mathbb{R}$. Then our definitions yield that $\tilde{\mathfrak{X}} := \pi(\mathfrak{X}^V) = (\pi(\mathfrak{X}))^U$ is Brownian motion on U and that $E_{\tilde{\mathfrak{X}}}$ is the set of all $w \in \mathcal{B}^+(U)$ such that $w \circ \pi \in E_{\mathfrak{X}^V}$. Since $E_{\mathfrak{X}^V} = {*\mathcal{H}}^+(V)$ by (3.20), we have to show that a positive function w on U is hyperharmonic on U if and only if $w \circ \pi$ is hyperharmonic on V .

So let $w : U \to \overline{\mathbb{R}}_+$. It is easily verified that w is l.s.c. on U if and only if $w \circ \pi$ is l.s.c. on V . Suppose that w is l.s.c. and consider $a = (a_1,a_2,a_3) \in \mathbb{R}^3$ and $r > 0$ such that $\bar{B}_r(a) \subset V$. For every $0 \leq \rho \leq r$ let

$$m(\rho) = \frac{1}{2\pi} \int_0^{2\pi} w(a_1 + \rho \cos \varphi , a_2 + \rho \sin \varphi) \, d\varphi .$$

Then

$$\int w \circ \pi \, d\sigma_{a,r} = \frac{1}{2} \int_{-\pi/2}^{+\pi/2} m(r \cos \vartheta) \cos \vartheta \, d\vartheta .$$

If $w \in {*\mathcal{H}}^+(U)$ then $m(\rho) \leq w(a_1,a_2)$ for every $0 \leq \rho \leq r$ and hence $\sigma_{a,r}(w\circ\pi) \leq w(a_1,a_2) = w\circ\pi(a)$. If $w\circ\pi \in {*\mathcal{H}}^+(V)$ then $\sigma_{a,r}(w\circ\pi) \leq w(a_1,a_2)$ and hence there exist $0 < \rho \leq r$ satisfying $m(\rho) \leq w\,(a_1,a_2)$. Thus the statement follows from (III.4.8).

Using the projection $(x_1,x_2,x_3) \mapsto x_1$ from $\mathbb{R}^3$ on $\mathbb{R}$ we obtain in a similar way that Brownian motion on a relatively compact open subset U of $\mathbb{R}$ is associated with $(U,{*\mathcal{H}}^+(U))$.

3.22. EXERCISES. 1. Let $\mathfrak{X}$ be a Hunt process associated with $(X,\mathcal{W})$. A subset A of X is called a *nearly Borel* set if for each $\mu \in M_+^1(X)$ there exist $B', B'' \in \mathcal{B}(X)$ such that $B' \subset A \subset B''$ and $P^\mu[D_{B''\setminus B'} < \infty] = 0$. The system $\mathcal{B}_n(X)$ of all nearly Borel subsets of X is a σ-algebra such that $\mathcal{B}(X) \subset \mathcal{B}_n(X) \subset \mathcal{B}(X)^*$. Furthermore, the assertions in (3.14) and (3.16) hold for any $A \in \mathcal{B}_n(X)$.

2. Let $A,B \in \mathcal{B}(X)$ and $\mu \in M_+(X)$ such that $B \subset A$, $\mu(A) = 0$, and $\mu^A(\complement B) = 0$. Then $\mu^A = \mu^B$. (Hint: Use 3.16.)

3. Apply (IV.7.7) to Brownian motion $\mathfrak{X}$ on $\mathbb{R}^n$ and to the mapping $\varphi : x \mapsto \|x\|$ from $\mathbb{R}^n$ onto $\mathbb{R}_+$. The resulting image of $\mathfrak{X}$ under φ is called *Bessel process.*

4. Base

This section reveals deep connections between fine topology and balayage of measures.

Let A be a subset of X. According to our convention of section (II.4) the fine interior resp. the fine closure of $A \subset X$ is denoted by $\overset{\circ}{A}{}^f$ resp. $\bar{A}^f$. We shall say that A is *thin* at a point $x \in X$ if $\varepsilon_x^A \neq \varepsilon_x$. The set $b(A)$ of all points $x \in X$ such that A is not thin at x will be called *base* of A, i.e.

$$b(A) = \{x \in X : \varepsilon_x^A = \varepsilon_x\} .$$

A subset A of X is called *basic* if $b(A) = A$.

4.1. PROPOSITION. Let $A \subset X$. Then $b(A)$ is a finely closed G_δ-set, $\overset{\circ}{A}{}^f \subset b(A) \subset \bar{A}$. For every strict $p \in \mathcal{P}$, $b(A) = \{\hat{R}_p^A = p\}$.

If $A \subset A' \subset X$ then $b(A) \subset b(A')$.

Proof. Let $p \in \mathcal{P}$ be a strict potential. We recall that for every $q \in \mathcal{P}$,

$$\varepsilon_x^A(q) = \hat{R}_q^A(x) \leq q(x) = \varepsilon_x(q)$$

and hence

$$b(A) = \{x \in A : \varepsilon_x^A(p) = p(x)\} = \{\hat{R}_p^A = p\}$$

$$= \{\hat{R}_p^A \geq p\} = \bigcap_{n=1}^{\infty} \{\hat{R}_p^A + \frac{1}{n} > p\} .$$

Furthermore, for every $x \in \complement\bar{A}$, $\varepsilon_x^A \neq \varepsilon_x$ since ε_x^A is supported by $\bar{A}$. Therefore $b(A) \subset \bar{A}$. The statement now follows immediately. ┘

4.2. COROLLARY. For all subsets A and B of X ,

$$b(A \cup B) = b(A) \cup b(B) .$$

Proof. By (4.1), $b(A) \cup b(B) \subset b(A \cup B)$. Let $p \in P$ be strict and $x \in b(A \cup B)$. Since by (1.3)

$$\hat{R}_p^{A\cup B} + \hat{R}^{A\cup B}_{\hat{R}_p^A \wedge \hat{R}_p^B} \leq \hat{R}_p^{A\cup B} + \hat{p}_{A,B} \leq \hat{R}_p^A + \hat{R}_p^B$$

we have in particular

$$p(x) + \hat{R}_p^A \wedge \hat{R}_p^B(x) \leq \hat{R}_p^A(x) + \hat{R}_p^B(x)$$

and hence $\hat{R}_p^A(x) = p(x)$ or $\hat{R}_p^B(x) = p(x)$. Thus $x \in b(A)$ or $x \in b(B)$. ┘

4.3. LEMMA. Let $A \subset X$ and $u \in W$. Then $R_u^B = R_u^A$ for every $A \subset B \subset X$ which is contained in the fine closure $\bar{A}^f$ of A .

Proof. Let $A \subset B \subset \bar{A}^f$ and $v \in W$ such that $v \geq u$ on A . Then $v \geq u$ on $\bar{A}^f$. Therefore $R_u^B \leq R_u^A$. The converse inequality is trivial. ┘

4.4. PROPOSITION. For every subset A of X , the set $A \cup b(A)$ is the fine closure $\bar{A}^f$ of A . In particular, the fine closure of every Borel measurable set and of every finely open set is Borel measurable. Moreover, every finely continuous numerical function on X is Borel measurable.

Proof. Let $p \in P$ be strict. Then by (4.3)

$$R_p^A = R_p^{\bar{A}^f} = p \quad \text{on} \quad \bar{A}^f$$

whereas by (2.3) and (4.1)

$$R_p^A = \hat{R}_p^A < p \quad \text{on} \quad \complement(A \cup b(A)) .$$

Therefore $\bar{A}^f \subset A \cup b(A)$.

Conversely, let $x \notin \bar{A}^f$. Then $x \notin A$ and there exists a compact neighborhood

K of x, a function $u \in \mathcal{W}$ and $q \in \mathcal{P}$ such that $u(x) < q(x)$ and $K \cap \{u \leq q\} \subset \complement A$, i.e.

$$A \subset \complement K \cup \{u > q\} .$$

We note first that $x \notin b(\complement K)$ since $x \in \overset{\circ}{K}$. Furthermore $x \notin b(\{u>q\})$ since $R_q^{\{u>q\}} \leq u$ and hence in particular

$$R_q^{\{u>q\}}(x) \leq u(x) < q(x) .$$

Using (4.2) we conclude that $x \notin b(A)$.

Finally, let $f : X \to \overline{\mathbb{R}}$ be finely continuous and fix $a \in \mathbb{R}$. Then for every $n \in \mathbb{N}$

$$\{f \geq a\} \subset \{f > a - \tfrac{1}{n}\} \subset b(\{f > a - \tfrac{1}{n}\}) \subset b(\{f \geq a - \tfrac{1}{n}\}) \subset \{f \geq a - \tfrac{1}{n}\} ,$$

hence $\{f \geq a\} = \bigcap_{n=1}^{\infty} b(\{f > a - \frac{1}{n}\}) \in \mathcal{B}(X)$ by (4.1). ┘

4.5. REMARK. We conclude from (4.2) and (4.4) that the property "$x \in b(A)$" is a (finely) local property: $x \in b(A)$ if and only if $x \in b(A \cap V)$ for some (every) fine neighborhood V of x. Indeed, if V is a finely open subset of X then $b(A) = b(A \cap V) \cup b(A \cap \complement V)$ where $b(A \cap \complement V) \subset \complement V$.

Moreover, if U is an open subset of X and $A \subset U$ then the base $b^U(A)$ of A with respect to the subspace $(U, \mathcal{W}^U)$ satisfies $b^U(A) = U \cap b(A)$. Indeed, let $x \in U$. Then ${}^U\varepsilon_x^A = \varepsilon_x^{A \cup \complement U}|_U$ by (2.9). If $x \in b(A)$ then $x \in b(A \cup \complement U)$, hence $\varepsilon_x^{A\cup\complement U} = \varepsilon_x$, ${}^U\varepsilon_x^A = \varepsilon_x$, i.e. $x \in b^U(A)$. Suppose now conversely that $x \in b^U(A)$, i.e. $\varepsilon_x^{A\cup\complement U}|_U = \varepsilon_x$. Then $\varepsilon_x^{A\cup\complement U} = \varepsilon_x$ since $\varepsilon_x^{A\cup\complement U}(p) \leq p(x)$ for every $p \in \mathcal{P}$. Therefore $x \in U \cap b(A \cup \complement U) = U \cap b(A)$.

4.6. PROPOSITION. For every $A \subset X$ and for every $\mu \in \mathcal{M}(\mathcal{P})$ the measure μ^A is supported by the fine closure of A, i.e. $(\mu^A)_* (\complement \bar{A}^f) = 0$.

Proof. Let K be a compact subset of $\complement \bar{A}^f$. We have to show that $\mu^A(K) = 0$. Let $p \in \mathcal{P}$ be strict such that $\mu(p) < \infty$. Since $\bar{A}^f = A \cup b(A)$ we have

$$K \subset \complement A \cap \{\hat{R}_p^A < p\} .$$

Let $n \in \mathbb{N}$. By (2.3) and (2.5), there exists an open set U_n containing A such that

$$R_p^{U_n} \leq \hat{R}_p^A + \frac{1}{n} \text{ on } K .$$

Defining $v_n = R_p^{U_n}$ we have $v_n \in \mathcal{W}$, $v_n \le p$ and $v_n = p$ on A, hence $\hat{R}_{v_n}^A = \hat{R}_p^A$ and

$$\int (p-v_n)\, d\mu^A = \int (\hat{R}_p^A - \hat{R}_{v_n}^A)\, d\mu = 0 .$$

Let

$$K_n = K \cap \{\hat{R}_p^A + \frac{2}{n} \le p\} .$$

Then $v_n + \frac{1}{n} \le p$ on K_n and hence $\mu^A(K_n) = 0$. Since $K \subset \{\hat{R}_p^A < p\}$, the compact sets K_n are increasing to K and thus $\mu^A(K) = \lim_{n\to\infty} \mu^A(K_n) = 0$. ┘

4.7. EXAMPLES. 1. Every absorbing set is basic by (4.1) since it is closed and finely open. If U is a finely open subset then $b(U)$ is basic since $b(b(U)) \subset b(U)$ by (4.4) and $U \subset b(U)$ by (4.1), hence $b(U) \subset b(b(U))$.

However the most important example is the following: For every $p \in \mathcal{P}$, the fine support $\delta(p)$ is basic. Indeed, it suffices to recall from (II.7.10) that $\delta(p)$ is a finely closed Borel set and $R_u^{\delta(p)} \in \mathcal{W}$ for every $u \in \mathcal{W}$. In section 8 we shall see that conversely every basic set is the fine support of a continuous potential.

2. Translation on $\mathbb{R}$. For every $A \subset \mathbb{R}$ the base $b(A)$ is the set A_1' of all $x \in \mathbb{R}$ such that x is the limit of a strictly increasing sequence in A. In particular, $b(\mathbb{Q}) = \mathbb{R}$.

3. Classical potential theory. Let $n \ge 3$, $a > 0$ and

$$A = \{x \in \mathbb{R}^n : x_n = 0 , \sqrt{x_2^2 + \ldots + x_{n-1}^2} \le a\, x_1\} .$$

Then A is basic. Indeed, for every $m \in \mathbb{N}$, let $a_m = 2 \circ 3^{-m}$, $b = \min(a,1)$, $r_m = b3^{-m}$ and

$$A_m = B_{r_m}(a_m) \cap (\mathbb{R}^{n-1} \times \{0\}) .$$

Then $A_m \subset \{x \in A : 3^{-m} \le \|x\| < 3^{-(m-1)}\}$. By (V.4.16.2),

$$\text{cap}\,(A_m) = c\, r_m^{n-2}$$

where $c \neq 0$. Since $3^{m(n-2)} r_m^{n-2} = b^{n-2}$ we conclude by the Wiener criterion (V.4.17) that $\complement \bigcup_{m=1}^{\infty} A_m$ is no fine neighborhood of 0. Therefore $0 \in b(\bigcup_{m=1}^{\infty} A_m) \subset b(A)$. Using translation invariance we obtain that $x \in b(x + A)$ for every $x \in \mathbb{R}^n$. If $x \in A$ then $x + A \subset A$ and therefore $x \in b(A)$. Thus $A \subset b(A)$. Moreover, $b(A) \subset A$ since A is closed. So A is basic.

Furthermore, $x \in b(x+A) \subset b(\mathbb{R}^{n-1} \times \{0\})$ for every $x \in \mathbb{R}^{n-1} \times \{0\}$, i.e. the hyperplane $\mathbb{R}^n \times \{0\}$ is basic. Using translations and rotations we conclude that every hyperplane in $\mathbb{R}^n$ is basic. More generally, we obtain the following result: Let A be a subset of $\mathbb{R}^n$ and $x \in \mathbb{R}^n$. If there exists an open cone C with vertex at x, a hyperplane H containing x, and $r > 0$ such that

$$\emptyset \neq C \cap H \cap B_r(x) \subset A$$

then $x \in b(A)$.

In particular, every closed convex subset of $\mathbb{R}^n$ which is not contained in a subspace of dimension $n-2$ is basic.

A historical example in $\mathbb{R}^3$ is *Lebesgue's spine*

$$L = \{0\} \cup \{x \in \mathbb{R}^3 : x_1 > 0 \, , \sqrt{x_2^2 + x_3^2} \leq e^{-\frac{1}{x_1}}\} \, ,$$

L is thin at 0 although L is a closed connected set such that $L = \overline{\overset{\circ}{L}}$ and $0 \in L$. Indeed, define $u \in {}^*\mathcal{H}^+(\mathbb{R}^3)$ by

$$u(x) = 4 \int_0^{\frac{1}{8}} \frac{t\,dt}{\| x - (t,0,0) \|} \, .$$

Then $u(0) = 4 \int_0^{\frac{1}{8}} dt = \frac{1}{2}$. Let $A = \{x \in L : 0 < x_1 \leq \frac{1}{8}\}$, $x \in A$ and $\alpha = x_1$, $\beta = e^{-\frac{1}{x_1}}$. Then

$$u(x) = 4 \int_0^{\frac{1}{8}} \frac{t\,dt}{\sqrt{(\alpha-t)^2 + x_2^2 + x_3^2}} \geq 4 \cdot \frac{\alpha}{2} \int_{\frac{\alpha}{2}}^{\alpha} \frac{dt}{\alpha - t + \beta} = 2\alpha \log \frac{\beta + \frac{\alpha}{2}}{\beta} \geq 2\alpha \log \frac{\alpha}{2\beta} \geq 1$$

since $\frac{1}{2\alpha} \geq 4$ implies that $e^{\frac{1}{2\alpha}} \geq 4 \cdot \frac{1}{2\alpha}$, hence $\frac{\alpha}{2\beta} = \frac{\alpha}{2} e^{\frac{1}{\alpha}} \geq e^{\frac{1}{2\alpha}}$ and $\log \frac{\alpha}{2\beta} \geq \frac{1}{2\alpha}$. Thus $R_1^A(0) \leq u(0) = \frac{1}{2} < 1$ and therefore $0 \notin b(A)$. By (4.1), $0 \notin b(\{x \in L : x_1 \geq \frac{1}{8}\})$. Moreover $b(\{0\}) = \emptyset$ since $R_1^{\{0\}} \leq \varepsilon N_0$ for every $\varepsilon > 0$ implies that $R_1^{\{0\}} = 0$ on $\complement\{0\}$, $\hat{R}_1^{\{0\}} = 0$ on $\mathbb{R}^3$. Applying (4.2) we finally conclude that L is thin at 0. In fact, the origin is the only point of the closed set L where L is thin. This follows immediately from the preceding results.

4. Riesz potentials. If $1 < \alpha < n$ then by (V.4.16.2)

$$\operatorname{cap}_\alpha(B_r(a) \cap (\mathbb{R}^{n-1} \times \{0\})) = d\, r^{n-\alpha}$$

where $d \neq 0$. Proceeding as in the classical case we obtain the same result:

All hyperplanes in $\mathbb{R}^n$ are α-basic. More generally, given $A \subset \mathbb{R}^n$ and $x \in \mathbb{R}^n$ we have $x \in b_\alpha(A)$ if there exists an open cone C with vertex at x , a hyperplane H containing x , and $r > 0$ such that

$$\emptyset \neq C \cap H \cap B_r(x) \subset A .$$

It $\alpha \leq 1$ and $\alpha < n$ then $b_\alpha(H) = \emptyset$ for every hyperplane H as we shall see in (5.4.4). However, the following holds: Given $A \subset \mathbb{R}^n$ and $x \in \mathbb{R}^n$ we have $x \in b_\alpha(A)$ if there exists $r > 0$ and an open cone C with vertex at x such that

$$\emptyset \neq C \cap B_r(x) \subset A .$$

This is easily deduced from Wiener's criterion. However, it may be interesting to see that this result is also a simple consequence of (4.2) and the symmetry properties. Indeed, there exist finitely many rotations $\rho_1, \dots, \rho_m$ of $\mathbb{R}^n$ with center x such that $\mathbb{R}^n \setminus \{x\} = \bigcup_{i=1}^{m} \rho_i(C)$. Since $x \in b_\alpha(\mathbb{R}^n \setminus \{x\})$ we obtain by (4.2) that $x \in b_\alpha(\rho_i(C))$ for some $1 \leq i \leq m$, hence $x \in b_\alpha(C)$ by the rotation invariance of $\mathbb{P}^\alpha$.

If $\beta < \alpha$ then $b_\beta(A) \subset b_\alpha(A)$ for every $A \subset \mathbb{R}^n$ since $E_{\mathbb{P}^\alpha} \subset E_{\mathbb{P}^\beta}$. If V is the set defined in (V.4.18) then $0 \in b_\alpha(\complement V) \setminus b_\beta(\complement V)$.

5. Heat equation. Let A be a subset of $\mathbb{R}$ and A_1' the base of A with respect to translation on $\mathbb{R}$ (see 4.7.2). Then by (3.19.2) and the following probabilistic characterization (4.8) of the base, we have $b(\mathbb{R}^n \times A) = \mathbb{R}^n \times A_1'$. In particular, $b(\mathbb{R}^n \times \mathbb{Q}) = \mathbb{R}^{n+1}$.

The following result states that a Borel set A is thin at x if and only if an associated process starting at x does not meet A right away.

4.8. THEOREM. If $\mathfrak{X}$ is an associated Hunt process then for every $A \in \mathcal{B}(X)$,

$$b(A) = \{x \in X : T_A = 0 \quad P^x\text{-a.s.}\} .$$

Proof. By (3.16), $x \in b(A)$ if and only if $X_{T_A} = x$ P^x-a.s. Hence of course $x \in b(A)$ if $T_A = 0$ P^x-a.s.

Suppose conversely that $x \in b(A)$. If $x \in b(A\setminus\{x\})$ then $\overset{o}{\varepsilon}{}_x^{A\setminus\{x\}} = \varepsilon_x^{A\setminus\{x\}} = \varepsilon_x$ and hence by (3.17)

$$T_A \leq T_{A\setminus\{x\}} = D_{A\setminus\{x\}} = 0 \quad P^x\text{-a.s.}$$

By (4.2) it therefore remains to consider the case $x \in A$ and $x = b(\{x\})$. Let $T = T_{\{x\}}$, $t > 0$ and $S = \inf(T,t)$. Then for every $q \in P$ by (3.16) and (3.4)

$$q(x) = P_T q(x) \leq P_S q(x) \leq q(x) ,$$

i.e. $X_S = x$ P^x-a.s. Thus $P^x[T\leq t] = P^x[T\leq S] = 1$. This shows that $T = 0$ P^x-a.s., hence $T_A = 0$ P^x-a.s. ┘

4.9. REMARKS. 1. Using (2.2) it is easily shown that for an arbitrary subset A of X

$$b(A) = \{x \in X : \forall\, B \in \mathcal{B}(X) \text{ such that } A \subset B ,\ T_B = 0 \ \ P^x\text{-a.s.}\} .$$

2. If $x \in X$ is finely isolated, i.e. if $T_{\complement\{x\}} > 0$ P^x-a.s, then $\varepsilon_x^{\complement\{x\}}(\{x\}) = 0$ by (4.6), i.e. the process jumps when leaving the point x .

4.10. COROLLARY. If $\mathfrak{X}$ is an associated Hunt process and $A \in \mathcal{B}(X)$ then $T_{\bar{A}^f} = T_A$ a.s. and $X_{T_A} \in \bar{A}^f$ a.s. on $[T_A < \infty]$.

Proof. Let $S := T_{b(A)}$. Since $\bar{A}^f = A \cup b(A)$ we have $T_{\bar{A}^f} = \inf(T_A,S)$. Let $\mu \in M_+^1(X)$. By (3.16) and (4.6) , $P^\mu[T_A < \infty , X_{T_A} \notin \bar{A}^f] = 0$ and $P^\mu[S < \infty , X_S \notin b(A)] = 0$. Since $S + T_A \circ \Theta_S = T_A$ on $[T_A > S]$ by (3.1), the strong Markov property and (4.8) imply that

$$P^\mu[T_A > S] \leq P^\mu[S < \infty , T_A \circ \Theta_S > 0] = \int_{[S<\infty]} P^{X_S}[T_A > 0]\, dP^\mu = 0 .$$ ┘

4.11. COROLLARY. If $\mathfrak{X}$ is an associated Hunt process and $V \in \mathcal{B}(X)$ then V is finely open if and only if $T_{\complement V} > 0$ P^x-a.s. for every $x \in V$.

Proof. By (4.4), V is finely open if and only if $V \cap b(\complement V) = \emptyset$. Hence the assertion follows from (4.8). ┘

4.12. PROPOSITION. Let $\mathfrak{X}$ be an associated Hunt process and let f be a numerical function on X . Then the following statements are equivalent:

(1) f is finely continuous.

(2) f is Borel measurable and $t \mapsto f \circ X_t$ is a.s. right continuous.

(3) f is Borel measurable and $\lim_{t \to 0} f \circ X_t = f(x)$ P^x-a.s. for every $x \in X$.

Proof. Using the fact that $\overline{\mathbb{R}}$ and $[-1,+1]$ are homeomorphic we may assume without loss of generality that $-1 \leq f \leq 1$.

(1) ⇒ (2): By (4.4), f is Borel measurable. Suppose that $\omega \in \Omega$ such that $t \mapsto f(X_t(\omega))$ is not right continuous. Then there exist $a,b \in \mathbb{Q}$, $t_0 \in \mathbb{R}_+$ and a sequence (t_n) in $\mathbb{R}_+$ decreasing to t_0 such that $f(X_{t_0}(\omega)) \in]a,b[$ and $f(X_{t_n}(\omega)) \in \complement[a,b]$ for every $n \in \mathbb{N}$. Let $p \in P$ be a strict potential, $V = f^{-1}(]a,b[)$, and

$$g = p - \hat{R}_p^{\complement V} .$$

Since V is finely open and the fine closure of V is contained in $f^{-1}([a,b])$, we have $g > 0$ on V and $g = 0$ on $\complement f^{-1}([a,b])$, hence

$$g(X_{t_0}(\omega)) > 0 , \quad g(X_{t_n}(\omega)) = 0$$

for every $n \in \mathbb{N}$. So $t \mapsto g(X_t(\omega))$ is not right continuous. However, $t \mapsto g \circ X_t$ is a.s. right continuous by (IV.1.4). Thus (2) follows since there are only countably many intervals with rational endpoints.

(2) ⇒ (3): Trivial.

(3) ⇒ (1): Let $a,b \in \mathbb{R}$ such that $a < b$, $V = f^{-1}(]a,b[)$ and $x \in V$. Then $T_{\complement V} > 0$ P^x-a.s. by (3). Thus V is finely open by (4.11), i.e. f is finely continuous. ┘

4.13. COROLLARY. Let $\mathfrak{X}$ be an associated Hunt process and let $u : X \to \mathbb{R}_+$ be Borel measurable. Then the following statements are equivalent:

(1) $u \in W$.

(2) For every $x \in X$, $(u \circ X_t)_{t \geq 0}$ is a supermartingale on $(\Omega, \mathfrak{M}, P^x)$ which is P^x-a.s. right continuous.

Proof. (1) ⇒ (2): (IV.7.3) and (4.12).

(2) ⇒ (1): For every $t > 0$. $P_t u(x) = E^x(u \circ X_t) \leq u(x)$, i.e. $u \in S_{\mathbb{P}}$.

Moreover, u is finely continuous by (4.12). Thus $u \in \mathcal{W}$ by (II.4.9). ┘

We conclude this section by some results concerning balayage on *subbasic sets*, i.e. subsets A of X satisfying $A \subset b(A)$ or equivalently $\hat{R}_p^A = R_p^A$ for every $p \in \mathcal{P}$. These will be used in section 6 .

4.14. LEMMA. Let A be a Borel subset of X such that $A \subset b(A)$ and let K be a compact subset of A. Then for every $p, p_o \in \mathcal{P}$ such that $\inf p_o(K) > 0$ there exists a potential $q \in \mathcal{P}$ and a compact subset L of A such that
$R_p^K \leq q \leq R_p^A$ and $q \leq R_q^L + p_o$.

Proof. Let M be a compact neighborhood of K such that $\inf p_o(M) > 0$ and define

$$\mathcal{Q} = \{q \in \mathcal{P} : R_p^K \leq q \leq R_p^A \text{ , } R_q^M = q\} .$$

$\mathcal{Q}$ is decreasingly filtered and

$$\inf \mathcal{Q} = R_p^K :$$

Indeed, let $v \in \mathcal{W}$ with $v \geq p$ on K. There exists a function $f \in C^+(X)$ such that $p\, 1_K \leq f \leq 1_M \inf (v, R_p^A)$. Then $R_f \in \mathcal{Q}$ and $R_f \leq v$. This shows that $\inf \mathcal{Q} = R_p^K$. It shows as well that $\mathcal{Q}$ is decreasingly filtered since $\inf (q_1, q_2) \in \mathcal{W}$ and $\inf (q_1, q_2) \geq p$ on K for all $q_1, q_2 \in \mathcal{Q}$.

Hence for every positive measure $\lambda \neq 0$ on M there exists a potential $q \in \mathcal{Q}$ satisfying

$$\lambda(q) < \lambda(R_p^K) + \lambda(p_o)$$

where $R_p^K = R_q^K \leq R_q^A$. Therefore if

$$F := \{(R_q^A - q + p_o)_{|M} : q \in \mathcal{Q}\}$$

then for every positive measure $\lambda \neq 0$ on M there is a function $f \in F$ with $\lambda(f) > 0$. F is a convex set of l.s.c. functions. Therefore by (I.1.9) there exists a function $f \in F$ which is strictly positive, i.e. there exists a function $q \in \mathcal{Q}$ such that

$$R_q^A + p_o > q \quad \text{on} \quad M .$$

By (1.9), there exists a compact subset L of A such that $\hat{R}_q^L + p_o > q$ on M and hence $\hat{R}_q^L + p_o \geq R_q^M = q$. ┘

4.15. LEMMA. Let (q_n) be a sequence in $\mathcal{P}$, and let (A_n) be a decreasing sequence of subsets of X, $p_0 \in \mathcal{P}$ and $B_n \subset A_n$ such that for every $n \in \mathbb{N}$

$$R_{q_n}^{B_n} \leq q_{n+1} \leq R_{q_n}^{A_n} \ , \quad q_n \leq R_{q_n}^{B_n} + 2^{-n} p_0 \ .$$

Then (q_n) converges locally uniformly to a potential $q \in \mathcal{P}$. Moreover, $R_q^{A_n} = q$ for every $n \in \mathbb{N}$.

Proof. Since $R_{q_n}^{A_n} \leq q_n$, the assumptions imply that for every $n \in \mathbb{N}$

$$0 \leq q_n - q_{n+1} \leq 2^{-n} p_0 \ .$$

Therefore the sequence (q_n) is decreasing and converges locally uniformly to a potential $q \in \mathcal{P}$. Let $n \in \mathbb{N}$. Then for every $k > n$,

$$q_k \leq q + \sum_{m=k}^{\infty} 2^{-m} p_0 = q + 2^{-k+1} p_0$$

and $B_k \subset A_k \subset A_n$, hence

$$q \leq q_k \leq R_{q_k}^{B_k} + 2^{-k} p_0 \leq R_{q+2^{-k+1} p_0}^{A_n} + 2^{-k} p_0 \leq R_q^{A_n} + 3 \circ 2^{-k} p_0 \ .$$

Thus $R_q^{A_n} = q$. ∎

4.16. PROPOSITION. Let A be a subset of X such that $A \subset b(A)$. Let K be a compact subset of A and $p \in \mathcal{P}$. Then there exists a potential $q \in \mathcal{P}$ such that $R_p^K \leq q \leq R_p^A$ and $R_q^A = q$.

Proof. Since $R_u^A = R_u^{A \cup b(A)} = R_u^{b(A)}$ for every $u \in \mathcal{W}$ we may replace A by $b(A)$. Hence we may assume that A is a Borel set. Let $p_0 \in \mathcal{P}$ be strictly positive. Starting with $K_0 = K$ and $q_0 = p$ and using (4.14) we may find potentials $q_n \in \mathcal{P}$ and compact subsets K_n of A such that for every $n \geq 0$, $K_n \subset K_{n+1}$ and

$$R_{q_n}^{K_n} \leq q_{n+1} \leq R_{q_n}^A \ , \quad q_{n+1} \leq R_{q_{n+1}}^{K_{n+1}} + \frac{p_0}{2^{n+1}} \ .$$

By the preceding lemma the decreasing sequence (q_n) converges to a potential $q \in \mathcal{P}$ satisfying $R_q^A = q$. Since $q_{n+1} = q_n$ on K_n, we have $q = p$ on K, hence $R_p^K \leq q$. The inequality $q \leq R_p^A$ is obvious from $q \leq q_1 \leq R_{q_0}^A = R_p^A$. ∎

4.17. COROLLARY. Let A be a subset of X such that $A \subset b(A)$ and let $p \in \mathcal{P}$. Then

$$R_p^A = \sup \{q \in \mathcal{P} : q \leq p \ , \ R_q^A = q\} \ .$$

Proof. (1.9) and (4.16). ┘

4.18. EXERCISES. 1. Let A be a subset of X and let $i(A)$ denote the set of all points $x \in A$ which are finely isolated in A and satisfy $x \notin b(\{x\})$. Then $\bar{A}^f = b(A) \cup i(A)$, $b(A) \cap i(A) = \emptyset$ and $i(A) = i(\bar{A}^f) \subset i(A^{*f})$.

2. Let $p \in P$ and $A \in \mathcal{B}(X)$. Then $\delta(p_A) \subset b(A)$.

3. Let A be a subbasic set. Then $(\mu^{A\cup B})^A = \mu^A$ for all $\mu \in M(P)$ and $B \subset X$.

4. If (X,W) has the local truncation property then $b(A) \cap b(\complement A) \subset b(A^*)$ for every subset A of X. (Hint: Use (3.18) and (4.8).)

5. Suppose that (X,W) has no absorbing points and that $\mathfrak{X}$ is an associated Hunt process whose transition function is a Markov semigroup. Then $H_b^+(X) = \{h \in W \cap C_b(X) : (h \circ X_t)_{t\geq 0}$ is a martingale on $(\Omega,\mathfrak{M},P^x)$ for every $x \in X\}$ and $P_b = \{p \in W \cap C_b(X) : \lim_{t\to\infty} p \circ X_t = 0 \text{ a.s.}\}$.

5. Exceptional Sets

A subset P of X is called *polar* if $\varepsilon_x^P = 0$ for every $x \in X$. The name is motivated by property (4) of the next proposition.

5.1. PROPOSITION. Let P be a subset of X. Then the following statements are equivalent:

(1) P is polar.

(2) $\hat{R}_u^P = 0$ for every $u \in W$.

(3) $\hat{R}_u^P = 0$ for some strictly positive $u \in W$.

(4) There exists a function $v \in W$ such that $v = \infty$ on P, but $\{v < \infty\}$ is dense in X.

Proof. (1) ⇒ (2) ⇒ (3) ⇒ (1): Trivial.

(4) ⇒ (3): Let $v \in W$ such that $P \subset \{v = \infty\}$ and $D = \{v < \infty\}$ is dense in X. Then $R_\infty^P \leq \frac{v}{n}$ for every $n \in \mathbb{N}$, hence $R_\infty^P = 0$ on D and $\hat{R}_\infty^P = 0$.

(2) ⇒ (4): Let $u \in \mathcal{W}$, $0 < u < \infty$. By (2.4), $\hat{R}_u^P = R_u^P$ on $\complement P$. Since $\hat{R}_u^P = 0$ we therefore know that $\complement P$ is dense in X. Let (x_n) be a sequence in $\complement P$ which is dense in X. For every $n \in \mathbb{N}$ there exists a function $v_n \in \mathcal{W}$ such that $v_n \leq u$, $v_n = u$ on P and $v_n(x_i) < 2^{-n}u(x_i)$ for every $1 \leq i \leq n$. Let $v = \sum_{n=1}^{\infty} v_n$. Then $v \in \mathcal{W}$, $v = \infty$ on P. For every $i \in \mathbb{N}$

$$v(x_i) = \sum_{n=1}^{i} v_n(x_i) + \sum_{n=i+1}^{\infty} v_n(x_i) \leq i\, u(x_i) + u(x_i) = (i+1)u(x_i) < \infty ,$$

hence $\{v < \infty\}$ is a dense subset of X. ∎

5.2. PROPOSITION. Let $\mathfrak{X}$ be an associated Hunt process and $A \in \mathcal{B}(X)$. Then A is polar if and only if $T_A = \infty$ a.s.

Proof. For every $x \in X$ by (3.16), $\varepsilon_x^A(X) = P^x[X_{T_A} \in X] = P^x[T_A < \infty]$. ∎

5.3. PROPOSITION. Every subset of a polar set is polar. Every countable union of polar sets is polar.

Proof. The first statement is trivial. So let (P_n) be a sequence of polar subsets of X and $P = \bigcup_{n=1}^{\infty} P_n$. Then for every $n \in \mathbb{N}$

$$\hat{R}_{\infty}^{\bigcup_{i=1}^{n} P_i} \leq \sum_{i=1}^{n} \hat{R}_{\infty}^{P_i} = 0$$

and hence by (1.6)

$$\hat{R}_{\infty}^{P} = \sup \hat{R}_{\infty}^{\bigcup_{i=1}^{n} P_i} = 0 .$$

So P is polar. ∎

5.4. EXAMPLES. 1. Classical potential theory. If $n \geq 3$ then the points of $\mathbb{R}^n$ are polar by (0.1.5). More generally, every affine subspace A of $\mathbb{R}^n$ having dimension $n-2$ is polar. Indeed, consider the $(n-2)$-dimensional Lebesgue measure σ on A as measure on $\mathbb{R}^n$, let $m \in \mathbb{N}$ and $\sigma_m = 1_{B_m}\sigma$. Then the potential N^{σ_m} (see(V.4.7)) is obviously finite on $\complement A$, but $N^{\sigma_m} = \infty$ on $A \cap B_m$ since for every $\varepsilon > 0$

$$\int_{\{x\in\mathbb{R}^{n-2}:\|x\|<\varepsilon\}} \frac{\lambda^{n-2}(dx)}{\|x\|^{n-2}} = \frac{2\pi^{\frac{n}{2}-1}}{\Gamma(\frac{n}{2}-1)} \int_0^{\varepsilon} \frac{r^{n-3}dr}{r^{n-2}} = \infty .$$

Hence $A = \bigcup_{m=1}^{\infty} A \cap B_m$ is polar by (5.1) and (5.3).

If U is a relatively compact subset of $\mathbb{R}^2$ then the points of U are polar (in U) since for every $x \in U$ the function $y \mapsto a - \log \|x-y\|$ is hyperharmonic on $\mathbb{R}^2$ by (0.1.5) and positive on U if $a \in \mathbb{R}_+$ is sufficiently large. If U is a relatively compact open subset of $\mathbb{R}$ then the empty set is the only polar subset.

2. Translation on $\mathbb{R}$. The empty set is the only polar subset.

3. Discrete potential theory. The empty set is the only polar subset.

4. Riesz potentials. Every polar subset with respect to classical potential theory is α-polar. If $n = 1$ and $\alpha < 1$ then the points are α-polar since $k_\alpha^x(x) = \infty$ for every $x \in \mathbb{R}$. If $n \geq 2$ and $\alpha \leq 1$ then every hyperplane in $\mathbb{R}^n$ is α-polar. Indeed, it suffices to note that for every $\varepsilon > 0$

$$\int_{\{x\in\mathbb{R}^{n-1}:\|x\|<\varepsilon\}} \frac{\lambda^{n-1}(dx)}{\|x\|^{n-\alpha}} = \frac{2\pi^{\frac{n-1}{2}}}{\Gamma(\frac{n-1}{2})} \int_0^{\varepsilon} \frac{r^{n-2}dr}{r^{n-\alpha}} = \infty$$

and to argue as in the classical case. However, if $1 < \alpha < n$ then hyperplanes in $\mathbb{R}^n$ are α-basic by (4.7.4) and hence certainly not α-polar.

5. Heat equation. For every $n \geq 1$ the points of $\mathbb{R}^n \times \mathbb{R}$ are polar. Indeed, let $z = (y,s) \in \mathbb{R}^n \times \mathbb{R}$. For every $m \in \mathbb{N}$, let $z_m = (y, s - \frac{1}{m})$ and $v_m = (\frac{m}{2\pi})^{-n/2} w^{z_m}$. Then v_m is hyperharmonic by (V.6.17), $v_m(z) = 1$ and $\lim_{m\to\infty} v_m = 0$ on $\complement\{z\}$. Therefore $\hat{R}_1^{\{z\}} = 0$, and $\{z\}$ is polar by (5.1).

If $n \geq 3$ and P is a polar subset of $\mathbb{R}^n$ with respect to the classical potential theory then $P \times \mathbb{R}$ is a polar subset of $\mathbb{R}^n \times \mathbb{R}$ with respect to the heat equation. This follows immediately from (V.6.19) and (5.1).

5.5. PROPOSITION. Let $P \in \mathcal{B}(X)$ such that every compact subset K of P is polar. Then P is polar.

Proof. Let $q \in P$, $q > 0$. Then by (1.9), $\hat{R}_q^P = \sup \{\hat{R}_q^K : K \text{ compact} \subset P\} = 0$. ┘

<u>5.6. PROPOSITION.</u> Let P be a polar subset of X and $x \in \complement P$. Then $(\varepsilon_x^A)^*(P) = 0$ for every $A \subset X$.

Proof. Let $u \in \mathcal{W}$ such that $u = \infty$ on P and $u(x) < \infty$. Then the inequality $\int u \, d\varepsilon_x^A = \hat{R}_u^A(x) \leq u(x) < \infty$ shows that $\varepsilon_x^A(\{u = \infty\}) = 0$ and hence $(\varepsilon_x^A)^*(P) = 0$. ┘

A subset T of X is called *totally thin* if it is thin at every point $x \in X$, i.e. if $b(T) = \emptyset$. A subset S of X is called *semipolar* if it is a countable union of totally thin sets. Let $\mathfrak{X}$ be an associated Hunt process and $A \in \mathcal{B}(X)$. Then by (4.8) A is totally thin if and only if $T_A > 0$ a.s. The probabilistic characterization of semipolar sets will be given in section 7.

<u>5.7. PROPOSITION.</u> 1. Every polar set is totally thin. Every totally thin set is semipolar.

2. Every subset of a totally thin (semipolar resp.) set is totally thin (semipolar resp.). Every finite union of totally thin sets is totally thin, every countable union of semipolar sets is semipolar.

3. For every totally thin set $T \subset X$ there exists a totally thin G_δ-set $T' \subset X$ such that $T \subset T'$. For every semipolar set $S \subset X$, there exists a semipolar Borel set $S' \subset X$ such that $S \subset S'$.

Proof. The first statements being obvious it suffices to prove the last two statements. So let $T \subset X$ be totally thin. By (2.2), there exists a G_δ-set $T' \supset T$ such that $\hat{R}_u^{T'} = \hat{R}_u^T$ for every $u \in \mathcal{W}$ and hence $b(T') = b(T) = \emptyset$, i.e. T' is totally thin.

If S is the countable union of a sequence of totally thin sets T_n then there exist totally thin G_δ-sets $T_n' \supset T_n$ and $S' = \bigcup_{n=1}^{\infty} T_n'$ is a semipolar Borel set containing S. ┘

<u>5.8. EXAMPLES.</u> 1. <u>Translation on $\mathbb{R}$</u>. Every singleton $\{x\}$ is totally thin but not polar since $\hat{R}_1^{\{x\}} = 1_{]x,\infty[}$. The subset $\mathbb{Q}$ is semipolar but not totally

thin since $b(\mathbb{Q}) = \mathbb{R}$ (see (4.7.2)).

2. Discrete potential theory. The empty set is the only set which is totally thin.

3. Heat equation. Every hyperplane $H_\alpha = \mathbb{R}^n \times \{\alpha\}$ is totally thin but not polar since $\hat{R}_1^{H_\alpha} = 1_{\mathbb{R}^n \times]\alpha,\infty[}$ by (V.6.14) and (V.6.19). The subset $S = \bigcup_{\alpha\in\mathbb{Q}} H_\alpha = \mathbb{R}^n \times \mathbb{Q}$ is semipolar but not totally thin since $b(S) = \mathbb{R}^{n+1}$ (see (4.7.5)).

5.9. PROPOSITION. Every totally thin set is finely closed. Every semipolar set is finely meager.

Proof. Let T be totally thin. By (4.4), the fine closure of T is the set $T \cup b(T) = T$. Therefore T is finely closed. Furthermore, the fine interior of T is empty by (4.1). ┘

5.10. COROLLARY. Let $u,v \in W$ such that the set $\{u \neq v\}$ is semipolar. Then $u = v$.

5.11. PROPOSITION. For every $V \subset W$, the set $\{\widehat{\inf V} < \inf V\}$ is semipolar.

Proof. Let $V \subset W$ and $v = \inf V$. We choose a strictly positive potential $p \in P$ and define for every $n \in \mathbb{N}$

$$T_n = \{\hat{v} + \frac{1}{n}p \leq \inf(v,np)\}.$$

Then $\bigcup_{n=1}^{\infty} T_n = \{\hat{v} < v\}$. For every $n \in \mathbb{N}$,

$$\hat{R}^{T_n}_{\hat{v}+\frac{1}{n}p} \leq \inf(v,np)$$

and hence

$$\hat{R}^{T_n}_{\hat{v}+\frac{1}{n}p} \leq \inf(\hat{v},np) < \hat{v} + \frac{1}{n}p.$$

Thus T_n is totally thin, and $\{\hat{v} < v\}$ is semipolar. ┘

5.12. COROLLARY. For every subset A of X the set $A \setminus b(A)$ is semipolar.

Proof. Let $p \in P$ be a strict potential. Then $A \setminus b(A) = \{\hat{R}^A_p < R^A_p\}$. ┘

In connection with (5.11) the following proposition gives a characterization of semipolar sets.

5.13. PROPOSITION. Let S be a semipolar subset of X. Then there exists $\mathcal{V} \subset \mathcal{W}$ such that $S = \{\widehat{\inf \mathcal{V}} < \inf \mathcal{V}\}$.

Proof. Let $p \in \mathcal{P}$ be a strict potential and let (T_n) be a sequence of totally thin sets such that $S = \bigcup_{n=1}^{\infty} T_n$. For every $n \in \mathbb{N}$, let

$$\mathcal{V}_n = \{v \in \mathcal{W} : v \leq p \text{ , } v = p \text{ on } T_n\} .$$

Then $\inf \mathcal{V}_n = R_p^{T_n}$ and hence by (2.4) and (4.1)

$$T_n = \{\widehat{\inf \mathcal{V}_n} < \inf \mathcal{V}_n\} .$$

Let $\mathcal{V} = \{\sum_{n=1}^{\infty} \frac{1}{2^n} v_n : v_n \in \mathcal{V}_n\}$. Then $\mathcal{V} \subset \mathcal{W}$, $\inf \mathcal{V} = \sum_{n=1}^{\infty} \frac{1}{2^n} \inf \mathcal{V}_n$, and $\widehat{\inf \mathcal{V}} = \sum_{n=1}^{\infty} \frac{1}{2^n} \widehat{\inf \mathcal{V}_n}$. Hence

$$S = \bigcup_{n=1}^{\infty} T_n = \bigcup_{n=1}^{\infty} \{\widehat{\inf \mathcal{V}_n} < \inf \mathcal{V}_n\} = \{\widehat{\inf \mathcal{V}} < \inf \mathcal{V}\} .$$

5.14. PROPOSITION. The fine topology is a *quasi-Lindelöf* topology, i.e. for every family $(G_i)_{i\in I}$ of finely open subsets of X there exists a countable subfamily $(G_i)_{i\in I_0}$ such that $\bigcup_{i\in I} G_i \setminus \bigcup_{i\in I_0} G_i$ is semipolar.

Proof. Let $p \in \mathcal{P}$ be a strict potential and define $F_i = \complement G_i$ $(i \in I)$. By the lemma of Choquet (I.1.8) there exists a countable subset I_0 of I such that

$$\widehat{\inf_{i\in I_0} R_p^{F_i}} = \widehat{\inf_{i\in I} R_p^{F_i}} .$$

By (5.11), we conclude that the set

$$S = \{\inf_{i\in I} R_p^{F_i} < \inf_{i\in I_0} R_p^{F_i}\}$$

is semipolar. If $x \in \bigcap_{i\in I_0} F_i$ but $x \notin F_i$ for some $i \in I$ then by (2.3)

$$R_p^{F_i}(x) = \hat{R}_p^{F_i}(x) < p(x) = \inf_{i\in I_0} R_p^{F_i}(x) ,$$

and hence $x \in S$. Thus

$$\bigcup_{i\in I} G_i \setminus \bigcup_{i\in I_0} G_i = \bigcap_{i\in I_0} F_i \setminus \bigcap_{i\in I} F_i \subset S .$$

Let us recall that the potential kernel associated with $p \in P$ (II.6.17) is the unique kernel $(x,B) \mapsto p_B(x)$ such that $p_X = p$ and $p_B \in P$, $C(p_B) \subset \bar{B}$ for every $B \in \mathcal{B}(X)$. In section 8 it will turn out that the following property is characteristic for semipolar sets.

5.15. LEMMA. Let S be a semipolar set. Then $p_S = 0$ for every $p \in P$.

Proof. Let T be a totally thin Borel set and $p \in P$. It suffices to show that $p_T = 0$. Let K be a compact subset of T and consider the convex cone

$$F = P - \mathbb{R}_+ p_K .$$

Let $x \in X$. Then $\varepsilon_x^K \neq \varepsilon_x$ since K is totally thin. Furthermore by (II.6.3)

$$\varepsilon_x^K(p_K) = \hat{R}_{p_K}^K(x) = p_K(x)$$

and hence $\varepsilon_x^K \in M_x(F)$. Therefore $Ch_F X = \emptyset$. By (I.2.2), we conclude that every function $f \in F$ is positive. In particular, $-p_K \geq 0$, i.e. $p_K = 0$. Hence $p_T = \sup \{p_K : K \text{ compact} \subset T\} = 0$. ┘

5.16. COROLLARY. Let A be a finely Borel subset of X. Then there exist Borel subsets B_1, B_2 of X such that $B_1 \subset A \subset B_2$ and $B_2 \setminus B_1$ is semipolar. In particular, A is measurable with respect to every measure $B \to p_B(x)$, $x \in X$.

Proof. Since countable unions of semipolar sets are semipolar the set of all $A \subset X$ having the stated property is a σ-algebra. Hence it suffices to consider a finely open subset A of X. For every $x \in A$ there exists a compact fine neighborhood of x which is contained in A. Using (5.14) we therefore obtain a sequence (K_n) of compact subsets of A such that $S = A \setminus \bigcup_{n=1}^{\infty} K_n$ is semipolar. By (5.7), there exists a semipolar set $S' \in \mathcal{B}(X)$ containing S. Obviously, $\bigcup_{n=1}^{\infty} K_n \subset A \subset S' \cup \bigcup_{n=1}^{\infty} K_n$. Thus the first part of the proposition is proved. The second part follows immediately by (5.15). ┘

Using (5.15) we now may determine the harmonic kernels associated with the balayage space $(X, \mathcal{W}^p)$ introduced in (V.2.7). Let us recall that given $p \in P_b$, we have $\mathcal{W}_b^p = (I + K_p)^{-1} \mathcal{W}_b$ where K_p denotes the potential kernel associated with p . If U is an open subset of X then by (V.1.1), the function $p^U = p - H_U p$ is a potential on U and it is easily verified that $K_p^U := K_p - H_U K_p$ is the potential kernel associated with p^U .

5.17. PROPOSITION. Suppose that $1 \in \mathcal{W}$, let $p \in \mathcal{P}_b$, and let $(H_U)_{U \in \mathcal{U}}$ be a family of harmonic kernels associated with $(X,\mathcal{W})$. Then $((I + K_p^U)^{-1} H_U)_{U \in \mathcal{U}}$ is a family of harmonic kernels associated with $(X,\mathcal{W}^p)$.

Proof. Let U be a relatively compact open subset of X, $q \in \mathcal{P}_b$, and

$$g = \inf \{u \in \mathcal{W}^p : u \geq q \text{ on } \complement U\} .$$

We have to show that $g = (I + K_p^U)^{-1} R_q^{\complement U}$.

There exists a decreasing sequence (u_n) in $\mathcal{W}_b^p$ such that $\inf u_n = g$. Fix $n \in \mathbb{N}$ and let $v_n = u_n + K_p^U u_n$. Then $v_n = u_n + K_p u_n - H_U K_p u_n \in {}^*\mathcal{H}^+(U)$. Moreover, v_n is l.s.c. on X and $v_n \geq q$ on $\complement U$, hence $w_n := \inf (v_n, q) \in \mathcal{W}$ by (III.4.6). Therefore

$$(I + K_p^U)g = \inf v_n \geq \inf w_n \geq R_q^{\complement U} .$$

Fix $x \in X$ and $\varepsilon > 0$. Then there exists an open set V such that $\bar{V} \subset U$ and $R_{K_p g}^{\complement V}(x) < R_{K_p g}^{\complement U}(x) + \varepsilon$. Let $w \in \mathcal{W}_b$ such that $w \leq q$ and $w = q$ on $\complement U$. Define

$$u = (I + K_p^V)^{-1} w , \qquad s = K_p u .$$

Then $u \in \mathcal{B}_b(X)$ and $u + K_p u = w + R_s^{\complement V}$, hence $\hat{u} + K_p u = w + \hat{R}_s^{\complement V} \in \mathcal{W}$. Moreover, the set $\{\hat{u} < u\} = \{\hat{R}_s^{\complement V} < R_s^{\complement V}\}$ is a semipolar subset of V^* and hence $K_p u = K_p \hat{u}$ by (5.15). Therefore $\hat{u} + K_p \hat{u} \in \mathcal{W}_b$, i.e. $\hat{u} \in \mathcal{W}_b^p$. Since $\hat{u} = u = w = q$ on $\complement U$ we obtain that $\hat{u} \geq g$ and hence $u \geq g$. So

$$w = (I + K_p^V)u \geq (I + K_p^V)g ,$$

and we conclude that $R_q^{\complement U} \geq (I + K_p^V)g$. In particular,

$$\begin{aligned} R_q^{\complement U}(x) &\geq g(x) + K_p g(x) - H_V K_p g(x) \\ &> g(x) + K_p g(x) - H_U K_p g(x) - \varepsilon = (I + K_p^U)g(x) - \varepsilon . \end{aligned}$$

⌟

5.18. REMARK. Every subset of $\mathbb{R}^n$ which is semipolar with respect to the classical theory or the Riesz potentials is a null set with respect to Lebesgue measure. Indeed, if K is a compact subset of $\mathbb{R}^n$ such that $\lambda^n(K) > 0$ then

$p := k_\alpha * 1_K \neq 0$ and by (V.4.7) p is a continuous real α-potential which is α-harmonic on $\complement K$, hence K is not semipolar by (5.15) and (II.6.18).

Let us note that this fact implies that for every $A \in \mathcal{B}(X)$ and every bounded α-hyperharmonic function $v \geq 0$ the function ${}^\alpha\hat{R}_v^A$ is α-harmonic on $\complement \bar{A}$: If $\alpha = 2$ (classical theory) we have a harmonic space and the statement is immediate from (2.6) and (III.8.1). So let $\alpha < 2$ (Riesz potentials). By (2.6), ${}^\alpha R_v^A \in {}^\alpha\mathcal{H}(\complement \bar{A})$. If U is an open ball in $\mathbb{R}^n$ such that $\bar{U} \cap \bar{A} = \emptyset$ and $x \in U$ then ${}^\alpha H_U\, {}^\alpha\hat{R}_v^A(x) = {}^\alpha H_U\, {}^\alpha R_v^A(x) = {}^\alpha R_v^A(x) = {}^\alpha\hat{R}_v^A(x)$ since ${}^\alpha H_U(x,\cdot)$ has a density with respect to Lebesgue measure and the set $\{{}^\alpha\hat{R}_v^A < {}^\alpha R_v^A\}$ is α-semipolar.

Thus in connection with (V.4.12) the following proposition shows that in the case of classical potential theory or Riesz potentials the notions polar and semipolar coincide.

5.19. PROPOSITION. Suppose that for every compact subset K of X and every $p \in \mathcal{P}$ the potential $\hat{R}_p^K$ is a countable sum of continuous real potentials and harmonic on $\complement K$. Then every semipolar subset S of X is polar.

Proof. By (5.7) there is no loss of generality in assuming that S is a semipolar Borel set. Let K be a compact subset of S and $p \in \mathcal{P}$, $p > 0$. By assumption $\hat{R}_p^K \in \mathcal{H}(\complement K)$ and there exists a sequence (p_n) in $\mathcal{P}$ such that $\sum_{n=1}^{\infty} p_n = \hat{R}_p^K$. Then $C(p_n) \subset K$ and hence $p_n = 0$ for every $n \in \mathbb{N}$ by (II.6.20) and (5.15). Therefore $\hat{R}_p^K = 0$, i.e. K is polar. Thus S is polar by (5.5). ⌟

5.20. REMARK. Example (V.9.1) shows that in general the assumption that $\hat{R}_p^K \in \mathcal{H}(\complement K)$ cannot be omitted since $\{z\}$ is totally thin but not polar.

In general, we shall say that the balayage space $(X,\mathcal{W})$ satisfies the *axiom of polarity* if every semipolar subset of X is polar.

5.21. PROPOSITION. The following properties are equivalent:

(1) $(X,\mathcal{W})$ satisfies the axiom of polarity.

(2) For every subset A of X, the set $A \setminus b(A)$ is polar.

(3) For every subset A of X and every $\mu \in \mathcal{M}(\mathcal{P})$, the measure μ^A is

supported by $b(A)$.

(4) For every subset A of X and every $\mu \in M(P)$, $(\mu^A)^A = \mu^A$.

Proof. (1) ⇒ (2): (5.12).

(2) ⇒ (3): Let $A \subset X$ and $\mu \in M(P)$. Since $\bar{A}^f = A \cup b(A)$ and $A \setminus b(A)$ is polar, we have for every $p \in P$

$$\hat{R}_p^{b(A)} \leq \hat{R}_p^{\bar{A}^f} = \hat{R}_p^A \leq \hat{R}_p^{b(A)} + \hat{R}_p^{A \setminus b(A)} = \hat{R}_p^{b(A)} .$$

Therefore $\mu^A = \mu^{b(A)}$ and using (4.6) we obtain that μ^A is supported by $\overline{b(A)}^f = b(A)$.

(3) ⇒ (4): Let $A \subset X$, $\mu \in M(P)$, and $p \in P$. Then $\hat{R}_p^A = p$ on $b(A)$ and hence

$$(\mu^A)^A(p) = \int \hat{R}_p^A \, d\mu^A = \int p \, d\mu^A .$$

This shows that $(\mu^A)^A = \mu^A$.

(4) ⇒ (1): Suppose that (1) does not hold. Then there exists a subset T of X which is totally thin but not polar. There exists a point $x \in X$ such that $\varepsilon_x^T \neq 0$. Let $p \in P$ be a strict potential. Since $\hat{R}_p^T < p$ we obtain that

$$(\varepsilon_x^T)^T(p) = \int \hat{R}_p^T \, d\varepsilon_x^T < \int p \, d\varepsilon_x^T ,$$

i.e. $(\varepsilon_x^T)^T \neq \varepsilon_x^T$. Thus (4) does not hold. ⌟

Next we show that the classical capacities cap_α are actually capacities in the sense of section (I.3) and prove formulas which are useful for estimates of capacities.

5.22. EXAMPLE. Riesz potentials ($n \geq 1$, $0 < \alpha \leq 2$, $\alpha < n$). The α-capacity defined in section (V.4) can be (uniquely) extended to a capacity $\mathrm{cap}_\alpha : \mathfrak{P}(\mathbb{R}^n) \to \overline{\mathbb{R}}_+$. This capacity is *strongly subadditive*, i.e.

$$\mathrm{cap}_\alpha(A \cup B) + \mathrm{cap}_\alpha(A \cap B) \leq \mathrm{cap}_\alpha(A) + \mathrm{cap}_\alpha(B)$$

for all $A,B \in \mathfrak{P}(\mathbb{R}^n)$, and $\mathrm{cap}_\alpha(A) = 0$ if and only if A is α-polar. Moreover, $\mathrm{cap}_\alpha(A) < \infty$ if and only if there exists a (unique) finite measure μ_A on $\mathbb{R}^n$ such that ${}^\alpha\hat{R}_1^A = N_\alpha^{\mu_A}$, and then $\mathrm{cap}_\alpha(A) = \mu_A(\mathbb{R}^n)$. For every $A \in \mathcal{B}(\mathbb{R}^n)$,

$$\operatorname{cap}_\alpha(A) = \sup \{\mu(\mathbb{R}^n) : \mu \in M_+(\mathbb{R}^n),\ N_\alpha^\mu \leq 1,\ \mu(\complement A) = 0\}$$
$$= \inf \{\mu(\mathbb{R}^n) : \mu \in M_+(\mathbb{R}^n),\ N_\alpha^\mu \geq 1 \text{ on } A\}.$$

Proof. Fix a relatively compact subset A_0 of $\mathbb{R}^n$ and choose $\nu \in M_+(\mathbb{R}^n)$ such that N_α^ν is a continuous real α-potential and $N_\alpha^\nu = 1$ on $\bar{A}_0$. Then for every $A \subset A_0$

$$\operatorname{cap}_\alpha(A) = \mu_A(\mathbb{R}^n) = \int N_\alpha^\nu \, d\mu_A = \int N_\alpha^{\mu_A} d\nu = \int \hat{R}_1^A \, d\nu .$$

If (A_m) is an increasing sequence of subsets of A_0, then we conclude from (1.7) that the sequence $(\operatorname{cap}_\alpha(A_m))$ is increasing to $\operatorname{cap}_\alpha(\bigcup_{m=1}^\infty A_m)$. Consider now a decreasing sequence (K_m) of compact subsets of A_0 and let $K = \bigcap_{m=1}^\infty K_m$. Then $R_1^K = \inf R_1^{K_m}$ by (1.2), hence the set $E = \{\hat{R}_1^K < \inf \hat{R}_1^{K_m}\} \subset \{\hat{R}_1^K < R_1^K\}$ is α-semi-polar by (5.11) (and therefore α-polar). By (5.15) and (V.4.10), $\nu(E) = 0$. Thus

$$\operatorname{cap}_\alpha(K) = \int \hat{R}_1^K \, d\nu = \inf \int \hat{R}_1^{K_m} \, d\nu = \inf \operatorname{cap}_\alpha(K_m) .$$

In particular, $\operatorname{cap}_\alpha(A) = \sup_{r>0} \operatorname{cap}_\alpha(A \cap B_r(0))$ for every relatively compact subset A of $\mathbb{R}^n$. Defining the α-capacity of an arbitrary subset A of $\mathbb{R}^n$ by

$$\operatorname{cap}_\alpha(A) = \sup_{r>0} \operatorname{cap}_\alpha(A \cap B_r(0))$$

we obtain that $\operatorname{cap}_\alpha$ is a capacity on $\mathbb{R}^n$. The strong subadditivity of $\operatorname{cap}_\alpha$ follows from (1.7) and the inequality $\hat{R}_1^{A\cup B} + \hat{R}_1^{A\cap B} \leq \hat{R}_1^A + \hat{R}_1^B$ in (1.3). Moreover, clearly $\operatorname{cap}_\alpha(A) = 0$ if and only if A is α-polar.

Given an α-polar set P in $\mathbb{R}^n$ there exists a finite measure σ on $\mathbb{R}^n$ such that $N_\alpha^\sigma = \infty$ on P. Indeed, fix $m \in \mathbb{N}$, let $P_m = P \cap B_m$, and $x_m \in \mathbb{R}^n$ such that $\|x_m\| = 2m$. Since $R_1^{P_m} = 0$ on $\complement P_m$, there exists an α-potential q_m on $\mathbb{R}^n$ such that $q_m \geq 1$ on P_m and $q_m(x_m) < (3m)^{\alpha-n} 2^{-m}$. Let $\sigma_m \in M_+(\mathbb{R}^n)$ such that $N_\alpha^{\sigma_m} = q_m$. Then $q_m(x_m) \geq (3m)^{\alpha-n} \sigma_m(\mathbb{R}^n)$, hence $\sigma_m(\mathbb{R}^n) < 2^{-m}$. It now suffices to take $\sigma = \sum_{m=1}^\infty \sigma_m$.

Now fix $A \in B(X)$ and let

$$a = \sup \{\mu(\mathbb{R}^n) : \mu \in M_+(\mathbb{R}^n),\ N_\alpha^\mu \leq 1,\ \mu(\complement A) = 0\},$$
$$b = \inf \{\mu(\mathbb{R}^n) : \mu \in M_+(\mathbb{R}^n),\ N_\alpha^\mu \geq 1 \text{ on } A\}.$$

If K is a compact subset of A then $N_\alpha^{\mu_K} = \hat{R}_1^K \leq 1$ and $\mu_K(\complement A) = 0$, hence

$\mathrm{cap}_\alpha(K) = \mu_K(\mathbb{R}^n) \leq a$. Therefore $\mathrm{cap}_\alpha(A) \leq a$. If $\mu \in M_+(\mathbb{R}^n)$ such that $N_\alpha^\mu \geq 1$ on A and if $\nu \in M_+(\mathbb{R}^n)$ such that $N_\alpha^\nu \leq 1$ and $\nu(\complement A) = 0$, then

$$\nu(\mathbb{R}^n) \leq \int N_\alpha^\mu \, d\nu = \int N_\alpha^\nu \, d\mu \leq \mu(\mathbb{R}^n) ,$$

hence $a \leq b$. Thus

$$\mathrm{cap}_\alpha(A) \leq a \leq b .$$

Suppose next that ${}^\alpha\hat{R}_1^A$ is an α-potential, i.e. ${}^\alpha\hat{R}_1^A = N_\alpha^{\mu_A}$, $\mu_A \in M_+(\mathbb{R}^n)$. Let σ be a finite measure on $\mathbb{R}^n$ such that $N_\alpha^\sigma = \infty$ on the α-polar set $\{{}^\alpha\hat{R}_1^A < {}^\alpha R_1^A\}$. Then $N_\alpha^{\mu_A+\varepsilon\sigma} \geq 1$ on A for every $\varepsilon > 0$ and therefore

$$b \leq \mu_A(\mathbb{R}^n) .$$

So $\mathrm{cap}_\alpha(A) = a = b$ if A is relatively compact.

In order to show that even under our general assumption $\mu_A(\mathbb{R}^n) \leq \mathrm{cap}_\alpha(A)$ we may suppose without loss of generality that A is finely closed since $\hat{R}_1^B = \hat{R}_1^{\bar{B}^f}$ for every subset B of $\mathbb{R}^n$. Then $\mu_A(\complement A) = 0$. Indeed, by (V.4.12) and (V.4.9) there exist $\mu_m \in M_+(\mathbb{R}^n)$ such that $q_m := N_\alpha^{\mu_m} \in C(\mathbb{R}^n)$ for every $m \in \mathbb{N}$ and $\sum_{m=1}^\infty \mu_m = \mu_A$. Since ${}^\alpha\hat{R}^A_{{}^\alpha\hat{R}_1^A} = {}^\alpha\hat{R}_1^A$, we obtain that ${}^\alpha R^A_{q_m} = q_m$ for every $m \in \mathbb{N}$, hence $\delta(q_m) \subset A$ and therefore $\mu_m(\complement A) = 0$ by (II.6.20). Now let K be a compact subset of A and $\mu = 1_K\mu_A$. Then $N_\alpha^\mu \leq N_\alpha^{\mu_A} \leq 1$ and $\mu(\complement A) = 0$, hence $\mu(K) \leq \mathrm{cap}_\alpha(K) \leq \mathrm{cap}_\alpha(A)$. Thus $\mu_A(\mathbb{R}^n) \leq \mathrm{cap}_\alpha(A)$.

For every subset A of $\mathbb{R}^n$ there exists $A' \in \mathcal{B}(\mathbb{R}^n)$ such that $A \subset A'$ and $A' \setminus A$ is α-polar. Then $\hat{R}_1^A = \hat{R}_1^{A'}$ and $\mathrm{cap}_\alpha(A) \leq \mathrm{cap}_\alpha(A') \leq \mathrm{cap}_\alpha(A) + \mathrm{cap}_\alpha(A' \setminus A) = \mathrm{cap}_\alpha(A)$. Hence it remains to show that $\hat{R}_1^A$ is an α-potential if $A \in \mathcal{B}(X)$ such that $\mathrm{cap}_\alpha(A) < \infty$. So let $A \in \mathcal{B}(\mathbb{R}^n)$ and $\mathrm{cap}_\alpha(A) < \infty$. Let $(r_m) \subset [1,\infty[$ be an increasing sequence such that $r_1 = 1$, $\lim_{m\to\infty} r_m = \infty$, and $\sum_{m=1}^\infty r_m^{\alpha-n} < \infty$. For every $m \in \mathbb{N}$ let $A_m = A \cap (B_{r_{m+1}} \setminus B_{r_m})$. Then ${}^\alpha\hat{R}_1^{A_m} = N_\alpha^{\mu_m}$ where

$$N_\alpha^{\mu_m}(0) \leq r_m^{\alpha-n} \cdot \mathrm{cap}_\alpha(A_m) \leq r_m^{\alpha-n} \cdot \mathrm{cap}_\alpha(A) .$$

Defining $\mu = \sum_{m=1}^\infty \mu_m$ we hence have $\mu \in M_+(\mathbb{R}^n)$ and

$$N_\alpha^\mu(0) \leq \left(\sum_{m=1}^\infty r_m^{\alpha-n}\right) \mathrm{cap}_\alpha(A) < \infty .$$

So N_α^μ is an α-potential on $\mathbb{R}^n$. Let $A_0 = A \cap B_1$. Then

$$ {}^{\alpha}R_1^A \leq \sum_{m=0}^{\infty} {}^{\alpha}\hat{R}_1^{A_m} = {}^{\alpha}\hat{R}_1^{A_0} + N_\alpha^\mu $$

where ${}^{\alpha}\hat{R}_1^{A_0} + N_\alpha^\mu$ is an α-potential on $\mathbb{R}^n$. Thus ${}^{\alpha}\hat{R}_1^A$ is an α-potential. ┘

An application of the next proposition will give a nice criterion of thinness in classical potential theory (which of course follows as well from Wiener's criterion).

5.23. PROPOSITION. Let A be a subset of X, x a polar point of X and $s \in \mathcal{W} \cap C(X)$, $s > 0$. Then the following statements are equivalent:

(1) A is not thin at x.

(2) For every open neighborhood V of x, $\hat{R}_s^{A\cap V}(x) = \hat{R}_s^A(x) > 0$.

(3) For every open neighborhood V of x, $\hat{R}^A_{R_s^V}(x) = \hat{R}_s^A(x) > 0$.

Proof. (1) ⇒ (2): For every open neighborhood V of x, the set $A \cap V$ is not thin at x, hence $\hat{R}_s^{A\cap V}(x) = s(x) = \hat{R}_s^A(x)$.

(2) ⇒ (3): It suffices to note that for every open set V obviously

$$ R_s^{A\cap V} = R^{A\cap V}_{R_s^V} \leq R^A_{R_s^V} \leq R_s^A . $$

(3) ⇒ (1): Let (V_n) be a decreasing sequence of open sets which is a fundamental system of neighborhoods of x. Then for every $n \in \mathbb{N}$

$$ \int R_s^{V_n} \, d\varepsilon_x^A = \hat{R}^A_{R_s^{V_n}}(x) = \hat{R}_s^A(x) , $$

hence by (1.2)

$$ \int R_s^{\{x\}} \, d\varepsilon_x^A = \lim_{n\to\infty} \int R_s^{V_n} \, d\varepsilon_x^A = \hat{R}_s^A(x) > 0 . $$

Since x is polar, $R_s^{\{x\}} = 0$ on $\complement\{x\}$. Hence the preceding equality yields that $\varepsilon_x^A(\{x\}) > 0$. On the other hand $\varepsilon_x^A = \varepsilon_x^{A\setminus\{x\}}$ and hence $\varepsilon_x^A(\{x\}) = 0$ by (4.6) if $x \notin b(A)$. Thus $x \in b(A)$, i.e. A is not thin at x. ┘

5.24. EXAMPLE. Classical potential theory. Let U be a convex open subset of $\mathbb{R}^n$, relatively compact if $n \leq 2$, such that $0 \in U$. Let A be a subset of U such that $\gamma A \subset A$ for some $0 < \gamma < 1$. Then A is thin at 0 if and only if A is polar. In particular:

1. If $B \subset U$ is not polar and $0 < \gamma < 1$ then $\bigcup_{m=1}^{\infty} \gamma^m B$ is not thin at 0.

2. If $C \subset \mathbb{R}^n$ is a cone with vertex at 0 then $C \cap U$ is thin at 0 if and only if $C \cap U$ is polar.

Proof. Suppose that A is not polar. Then there exists $R > 0$ such that $A' = A \cap B_R(0)$ is not polar and hence $\hat{R}_1^{A'}(0) > 0$ by (V.1.4.1). Fix $0 < r < R$ such that $V = B_r(0) \subset U$ and choose $m \in \mathbb{N}$ such that $\gamma^m < \frac{r}{R}$. Then $\gamma^m A' \subset A' \cap V$. Consider now $u \in {}^*\mathcal{H}^+(U)$ such that $u \geq 1$ on $A' \cap V$ and define $v : U \to \overline{\mathbb{R}}_+$ by $v(x) = u(\gamma^m x)$. Then $v \in {}^*\mathcal{H}^+(U)$ and $v \geq 1$ on A'. Hence for every $x \in U$

$$R_1^{A'}(x) \leq R_1^{A' \cap V}(\gamma^m x) .$$

In particular, $\hat{R}_1^{A'}(0) \leq \hat{R}_1^{A' \cap V}(0)$, i.e. $\hat{R}_1^{A' \cap V}(0) = \hat{R}_1^{A'}(0) > 0$. Applying (5.23) we conclude that A' and thus A is not thin at 0. The converse is trivial. $\lrcorner$

5.25. EXERCISES. 1. Let U be an open subset of X and $A \subset U$. Then A is semipolar if and only if A is a semipolar subset in the balayage space $(U, \mathcal{W}^U)$. The corresponding statement for polar sets holds for harmonic spaces, but not for balayage spaces in general. (Hint: See example (V.9.1).)

2. For every balayage space $(X, \mathcal{W})$ the following properties are equivalent:

(1) Every semipolar subset of X is countable.

(2) The fine topology is a Lindelöf topology.

(3) The fine topology is normal.

(Hint: If the fine topology is normal then every subset of a compact totally thin set is a G_δ-set. Note however that no countable dense subset of an uncountable compact subset K of X is a G_δ-set.)

3. Let $\mathfrak{X}$ be an associated Hunt process. Then the system of all $A \in \mathcal{B}(X)^*$ such that $A \Delta B$ is polar for some Borel subset B of X is a σ-algebra contained in $\mathcal{B}_n(X)$.

4. Suppose that X is connected and that $(X, \mathcal{W})$ has the local truncation property. Then the complement of every polar subset is connected.

5. For every subset A of $\mathbb{R}^n$ and every $\gamma > 0$, $\text{cap}_\alpha(\gamma A) = \gamma^{n-\alpha}\, \text{cap}_\alpha(A)$

6. Let $x \in \mathbb{R}^n$, $A \subset \mathbb{R}^n$, and $\lambda > 1$. Then the following statements are

equivalent:

(1) A is α-thin at x .

(2) $\sum_{m=1}^{\infty} \lambda^{m(n-\alpha)} \mathrm{cap}_\alpha(\{y \in A : \lambda^{-(m+1)} \leq \|y-x\| < \lambda^{-m}\}) < \infty$.

(3) $\sum_{m=1}^{\infty} \lambda^{m(n-\alpha)} \mathrm{cap}_\alpha(A \cap B_{\lambda^{-m}}(x)) < \infty$.

(Hint: Use (V.4.17) and exercise (I.3.12.3).)

7. Let F_a denote the rectangle $[-a,+a] \times [-\frac{1}{2},+\frac{1}{2}] \times \{0\}$ in $\mathbb{R}^3$, $a \geq 1$. Then $\frac{a}{4+\log a} \leq \mathrm{cap}(F_a) \leq \frac{2a}{\log(2a+1)}$. (Hint: Show that the surface measure μ on F_a satisfies $\log(2a+1) \leq N^\mu((-a,\frac{1}{2})) \leq N^\mu \leq N^\mu(0) \leq 8 + 2\log a$ on F_a.)

8. Let $k \geq 1$ and $A = \{x \in \mathbb{R}^3 : x_1 > 0 , |x_2| < x_1^k , x_3 = 0\}$. Then A is not thin at 0 (with respect to the classical situation on $\mathbb{R}^n$). (Hint: Use Wiener's criterion and the preceding exercise.)

9. Let $0 < \alpha \leq 2$, $n > \alpha + 1$, $k > 1$, and $A = \{x \in \mathbb{R}^n : x_1 \geq 0 , \sqrt{x_2^2 + \ldots + x_n^2} \leq x_1^k\}$. Then A is α-thin at 0 .

6. Essential Base

For every subset A of X let $\beta(A)$ be the set of all points $x \in X$ such that A is not "semipolar at x", i.e. such that for every fine neighborhood V of x the set $A \cap V$ is not semipolar. The set $\beta(A)$ will be called *essential base* of A. It is obvious from the definition that " $x \in \beta(A)$ " is a (finely) local property. Furthermore, it is easily seen that $\beta(A)$ is a finely closed set, $\mathring{A}^f \subset \beta(A) \subset \bar{A}^f$, and

$$\beta(A \cup B) = \beta(A) \cup \beta(B)$$

for all $A,B \subset X$.

6.1. PROPOSITION. For every subset A of X , the essential base $\beta(A)$ is the smallest finely closed subset F of X such that $A \setminus F$ is semipolar. In particular, $\beta(A) \subset b(A)$ and $\beta(A) = \emptyset$ if and only if A is semipolar.

Proof. By definition, $\complement\beta(A)$ is the union of all finely open subsets V of X such that $A \cap V$ is semipolar. Hence by (5.14) there exists a sequence (V_n) of finely open subsets of X such that $A \cap V_n$ is semipolar for every $n \in \mathbb{N}$ and $S := (\complement\beta(A)) \setminus \bigcup_{n=1}^{\infty} V_n$ is semipolar. Since

$$A\setminus\beta(A) \subset S \cup \bigcup_{n=1}^{\infty} (A \cap V_n)$$

we conclude that $A \setminus \beta(A)$ is semipolar. If conversely F is a finely closed set such that $A \setminus F = A \cap \complement F$ is semipolar then of course the finely open set $\complement F$ is contained in $\complement\beta(A)$, i.e. $\beta(A) \subset F$. Since $A \setminus b(A)$ is semipolar and $b(A)$ is finely closed we have in particular $\beta(A) \subset b(A)$. Of course the inclusion $\beta(A) \subset b(A)$ can be seen in a more direct way: If $x \in A$ and x is not totally thin then $x \in b(A)$. So let $x \in \beta(A)$ and suppose that $A \cap \{x\}$ is semipolar. Then for every fine neighborhood V of x the set $(A\setminus\{x\}) \cap V$ is not semipolar and hence $(A\setminus\{x\}) \cap V \neq \emptyset$. Therefore $x \in \overline{A\setminus\{x\}}^f$ and hence $x \in b(A)$ by (4.4). ┘

6.2. COROLLARY. $\beta^2 = b\beta = \beta$, i.e. $\beta(\beta(A)) = b(\beta(A)) = \beta(A)$ for every subset A of X. In particular, $\beta(A)$ is a basic set.

If (A_n) is a sequence of subsets of X, $A = \bigcup_{n=1}^{\infty} A_n$ and $A' = \bigcup_{n=1}^{\infty} \beta(A_n)$ then $\beta(A) = \bar{A}'^f = b(A') = \beta(A')$.

Proof. Let A be a subset of X. Then on one hand $\beta(\beta(A)) \subset b(\beta(A)) \subset \beta(A)$. On the other hand the sets $A\setminus\beta(A)$ and $\beta(A) \setminus \beta(\beta(A))$ are semipolar, hence $A\setminus\beta(\beta(A))$ is semipolar. Furthermore $\beta(\beta(A))$ is finely closed. Hence $\beta(A) \subset \beta(\beta(A))$ by (6.1).

Suppose now that $A = \bigcup_{n=1}^{\infty} A_n$, $A' = \bigcup_{n=1}^{\infty} \beta(A_n)$. Then evidently $A' \subset \beta(A)$ and hence

$$\beta(A') \subset b(A') \subset \bar{A}'^f \subset \beta(A) .$$

Moreover, $A \setminus A' \subset \bigcup_{n=1}^{\infty} (A_n\setminus\beta(A_n))$, hence $A\setminus A'$ is semipolar and therefore $\beta(A) \subset \beta(A')$. ┘

6.3. REMARKS. 1. The map $A \mapsto A \cup \beta(A)$ defines the closure operation of a topology on X which clearly is finer that the fine topology. However every numerical function f on X which is continuous with respect to this topology is already finely continuous. Indeed let $a \in \mathbb{R}$. Then for every $n \in \mathbb{N}$

$$\{f \geq a\} \subset \{f > a - \tfrac{1}{n}\} \subset \beta(\{f > a - \tfrac{1}{n}\}) \subset \beta(\{f \geq a - \tfrac{1}{n}\}) \subset \{f \geq a - \tfrac{1}{n}\},$$

hence $\{f \geq a\} = \bigcap_{n=1}^{\infty} \beta(\{f > a - \frac{1}{n}\})$ is finely closed.

2. We note that in general $\beta \neq b$. Indeed, consider translation on $\mathbb{R}$. Then $b(\mathbb{Q}) = \mathbb{R}$ whereas $\beta(\mathbb{Q}) = \emptyset$ (see (4.7.2)). However, $\beta = b$ in our examples classical potential theory and Riesz potentials since the following conditions are equivalent:

(a) $\beta = b$.

(b) $(X,\mathcal{W})$ satisfies the *axiom of thinness*, i.e. every semipolar set is totally thin.

Proof. (a) ⇒ (b): If $S \subset X$ is semipolar then $b(S) = \beta(S) = \emptyset$.

(b) ⇒ (a): Let $A \subset X$. Then the semipolar set $A \setminus \beta(A)$ is totally thin, hence $b(A) \subset b(A \setminus \beta(A)) \cup b(\beta(A)) = b(\beta(A)) = \beta(A) \subset b(A)$. ┘

<u>6.4. LEMMA.</u> Let $q \in P$ and let F be a finely closed subset of X. Then:

$$q_F = q \iff R_q^{\beta(F)} = q \iff R_q^F = q \iff \delta(q) \subset F.$$

Proof. For every compact subset K of $\beta(F)$, $q_K = R_{q_K}^K \leq R_q^{\beta(F)}$. Moreover, $q_{F \setminus \beta(F)} = 0$ by (5.15) since $F \setminus \beta(F)$ is semipolar. If $q_F = q$ we therefore obtain that

$$q = q_F = q_{\beta(F)} = \sup\{q_K : K \text{ compact} \subset \beta(F)\} \leq R_q^{\beta(F)},$$

i.e. $R_q^{\beta(F)} = q$. If $R_q^{\beta(F)} = q$ then of course $R_q^F = q$ since $\beta(F) \subset F$. If $R_q^F = q$ then $\varepsilon_x^F(q) = q(x)$ and $\varepsilon_x^F \neq \varepsilon_x$ for every $x \in \complement F$, hence $\complement F \cap \delta(q) = \emptyset$, $\delta(q) \subset F$. Finally, if $\delta(q) \subset F$ then $q_F = q$ by (II.6.20). ┘

Using the deep result (4.17) we obtain that basic sets and essentially basic sets coincide and that the essential base of a finely closed set A is the greatest basic subset of A.

<u>6.5. PROPOSITION.</u> Let A be a subset of X. Then $b(A) = A$ if and only if $\beta(A) = A$.

Proof. Assume that $b(A) = A$ and let $p \in P$ be a strict potential. Then by (4.17) and (6.4),

$$R_p^A = \sup \{q \in P : q \leq p \ , \ R_q^A = q\}$$
$$= \sup \{q \in P : q \leq p \ , \ R_q^{\beta(A)} = q\} = R_p^{\beta(A)} \ .$$

Therefore $b(A) = b(\beta(A))$, i.e. $A = \beta(A)$. If conversely $\beta(A) = A$ then of course $b(A) = b(\beta(A)) = \beta(A) = A$. ┘

6.6. COROLLARY. Let A be a finely closed subset of X . Then $\beta(A)$ is the greatest subset B of A such that $B \subset b(B)$.

Proof. By (6.2) and (6.1), $\beta(A)$ is a basic subset of A . Let B be an arbitrary subset of A such that $B \subset b(B)$. Then $b(B) \subset b(b(B)) \subset b(B)$, hence $\beta(b(B)) = b(B)$ by (6.5) and therefore $B \subset b(B) = \beta(b(B)) \subset \beta(b(A)) \subset \beta(A)$. ┘

Let $A \subset X$ and $u,v \in W$. We shall say that v is *essentially equal* to u on A ($v = u$ ess. on A) if the set of all $x \in A$ such that $v(x) \neq u(x)$ is semipolar. The *essential balayage* of u on A is defined by

$$B_u^A = \inf \{v \in W : v = u \ \text{ess. on } A\} \ .$$

The following result is almost trivial.

6.7. PROPOSITION. $B_u^A = R_u^{\beta(A)}$.

Proof. $R_u^{\beta(A)} \in W$ and $R_u^{\beta(A)} = u$ ess. on A since $A \setminus \beta(A)$ is semipolar. Therefore $B_u^A \leq R_u^{\beta(A)}$.

Conversely, let $v \in W$ such that $A \cap \{v \neq u\}$ is semipolar. Then $F = \{v=u\}$ is finely closed and $A \setminus F$ is semipolar. Hence $\beta(A) \subset F$ and $v \geq R_u^{\beta(A)}$. Thus conversely $B_u^A \geq R_u^{\beta(A)}$. ┘

So essential balayage on A is simply balayage on the essential base of A . Unless (X,W) satisfies the axiom of polarity the mappings $A \mapsto R_u^{\beta(A)}$ are far from furnishing capacities (see (6.18.2)). Nevertheless, essential balayage on Borel sets can be approximated by essential balayage on compact subsets. Let us note that in proving this important property we shall use the fact that Borel sets are hypercapacitable for the first time in full generality.

6.8. THEOREM. For every $A \in \mathcal{B}(X)$ and every $u \in \mathcal{W}$,

$$B_u^A = \sup \{B_u^K : K \text{ compact}, K \subset A\} .$$

Proof. Choosing an exhaustion (L_n) of X and an increasing sequence (p_n) in P such that $u = \sup p_n$ we obtain by (6.7) and (1.7) that

$$B_u^A = R_u^{\beta(A)} = \sup R_{p_n}^{\beta(A\cap L_n)} = \sup B_{p_n}^{A\cap L_n} .$$

Hence we may suppose that A is relatively compact and consider $p \in P$ instead of $u \in \mathcal{W}$. By (4.17) and (6.4)

$$B_p^A = \sup \{q \in P : q \leq p , B_q^A = q\} .$$

Let $A_1 = A$ and $q_1 \in P$ such that $q_1 \leq p$ and $B_{q_1}^A = q_1$. Let $p_0 \in P, p_0 > 0$. In order to prove the theorem it clearly suffices to find a compact subset K of A such that $B_p^K \geq q_1 - p_0$.

Let $n \in \mathbb{N}$ and assume that a subset A_n of A and a potential $q_n \in P$ such that $C(q_n) \subset \bar{A}$ and

$$B_{q_n}^{A_n} > q_n - \frac{p_0}{2^n}$$

are already chosen. Then there exists a compact subset K_n of $\beta(A_n)$ such that

$$\hat{R}_{q_n}^{K_n} > q_n - \frac{p_0}{2^n}$$

on $\bar{A}$ and hence on X since $C(q_n) \subset \bar{A}$. By (4.16), there exists a potential $q_{n+1} \in P$ such that

$$R_{q_n}^{K_n} \leq q_{n+1} \leq B_{q_n}^{A_n} \quad \text{and} \quad B_{q_{n+1}}^{A_n} = q_{n+1} .$$

Then $C(q_{n+1}) \subset \bar{A}_n \subset \bar{A}$. We take

$$C_n = \{A' \subset X : B_{q_{n+1}}^{A'} > q_{n+1} - \frac{p_0}{2^{n+1}}\}$$

and choose $A_{n+1} \in C_n$, $A_{n+1} \subset A_n$.

We note that C_n is a capacitance, i.e. if $A' \subset A'' \subset X$ and $A' \in C_n$ then $A'' \in C_n$ and if $A'_m \uparrow A'$, $A' \in C_n$ then $A'_m \in C_n$ for some m. Hence by (I.3.8) and (I.3.4) the sequence (A_n) can be chosen in such a way that the compact set $K = \bigcap_{n=1}^{\infty} \bar{A}_n$ is contained in A.

By (4.15) the corresponding decreasing sequence (q_n) converges to a potential $q \in P$ such that $R_q^{A_n} = q$ for every $n \in \mathbb{N}$ and hence

$$R_q^K = \lim_{n\to\infty} R_q^{\bar{A}_n} = q \ .$$

Therefore $B_q^K = q$ by (6.4). Since $q_1 \leq p$ and $q \leq q_1 < q+p_o$ we finally conclude that $B_p^K \geq B_q^K = q \geq q_1 - p_o$. ┘

6.9. PROPOSITION. Let A be a Borel subset of X . Then there exists a sequence (K_n) of compact subsets of A such that $\beta(\bigcup_{n=1}^{\infty} \beta(K_n)) = \beta(\bigcup_{n=1}^{\infty} K_n) = \beta(A)$.

Proof. Let $p \in P$ be a strict potential. By (6.7) and (6.8) there exists a sequence (K_n) of compact subsets of A such that

$$\sup R_p^{\beta(K_n)} = R_p^{\beta(A)} \ .$$

Let $A' = \bigcup_{n=1}^{\infty} \beta(K_n)$, $A'' = \bigcup_{n=1}^{\infty} K_n$. Then $A' \subset A'' \subset A$, hence $\beta(A') \subset \beta(A'') \subset \beta(A)$. Using (6.2) we have on the other hand

$$R_p^{\beta(A)} = \sup R_p^{\beta(K_n)} \leq R_p^{A'} = R_p^{\bar{A}'^f} = R_p^{\beta(A')}$$

and hence $\beta(A) \subset \beta(A')$. ┘

6.10. COROLLARY. Let A be a Borel subset of X . Then $\beta(A)$ is the smallest finely closed set containing $\beta(K)$ for every compact subset K of A . In particular, if every compact subset K of A is semipolar then A is semipolar.

6.11. PROPOSITION. Let A be a Borel subset of X and let $A_o \subset \beta(A)$ be a closed (compact resp.) set. Then there exists a closed (compact resp.) set $A_1 \subset (A \cap \beta(A)) \cup A_o$ such that $A_o \subset \beta(A \cap A_1)$.

Proof. Let $p \in P$ be a strict potential and let (L_n) be an exhaustion of X . Let (U_n) be a sequence of open neighborhoods of A_o such that $\bar{U}_{n+1} \subset U_n$ and $\bigcap_{n=1}^{\infty} U_n = A_o$. Fix $n \in \mathbb{N}$. Then $A_o \subset \beta(A) \cap U_n \subset \beta(A \cap U_n) = \beta(A \cap \beta(A) \cap U_n)$. Therefore by (6.8) there exists a compact subset K_n of $A \cap \beta(A) \cap U_n$ such that

$$B_p^{K_n}(x) > p(x) - \frac{1}{n}$$

for every $x \in A_o \cap L_n$. We take

$$A_1 = A_o \cup \bigcup_{n=1}^{\infty} K_n \ .$$

Then A_1 is a closed subset of $(A \cap \beta(A)) \cup A_0$, and for every $x \in A_0$

$$B_p^{A \cap A_1}(x) \geq \sup B_p^{K_n}(x) \geq p(x) .$$

Hence $A_0 \subset \beta(A \cap A_1)$. If A_0 is compact, we choose a compact neighborhood L of A_0 and replace A_1 by $A_1 \cap L$. ┘

6.12. COROLLARY. Let A be a Borel subset of X and let $A_0 \subset A \cap \beta(A)$ be a closed (compact resp.) set. Then there exists a closed (compact resp.) set $A_1 \subset A \cap \beta(A)$ such that A_0 is contained in the basic subset $\beta(A_1)$ of A_1.

6.13. THEOREM. Let A be a Borel subset of X and let $A_0 \subset A \cap \beta(A)$ be a closed set. Then there exists an increasing family $(A_t)_{0<t<1}$ of closed sets contained in $A \cap \beta(A)$ such that $A_0 = \bigcap_{0<t<1} A_t$, $\beta(A) = \beta(\bigcup_{0<t<1} A_t)$, and $A_s \subset \beta(A_t)$ for all $0 < s < t < 1$.

Proof. Let (U_n) be a sequence of open subsets of X such that $A = \bigcap_{n=1}^{\infty} U_n$. By (6.9) there exists an increasing sequence (K_n) of compact subsets of $A \cap \beta(A)$ such that $\beta(A) = \beta(\bigcup_{n=1}^{\infty} K_n)$.

By (6.12) there exists a closed subset $A_{\frac{1}{2}}$ of $A \cap \beta(A)$ such that $A_0 \subset \beta(A_{\frac{1}{2}})$. Again by (6.12) there exists a closed subset $A_{\frac{1}{4}}$ of $\beta(A_{\frac{1}{2}}) \cap U_1$ such that $A_0 \subset \beta(A_{\frac{1}{4}})$ and a closed subset $A_{\frac{3}{4}}$ of $A \cap \beta(A)$ such that $A_{\frac{1}{2}} \cup K_1 \subset \beta(A_{\frac{3}{4}})$. Continuing in this way and using $D = \{i\, 2^{-n} : i,n \in \mathbb{N}, i < 2^n\}$ we obtain a family $(A_d)_{d \in D}$ of closed sets contained in A such that $A_d \subset \beta(A_{d'}) \subset A_{d'}$ for all $d,d' \in D$ satisfying $d < d'$ and $A_{2^{-(n+1)}} \subset U_n$, $K_n \subset \beta(A_{1-2^{-(n+1)}})$ for every $n \in \mathbb{N}$. Defining

$$A_t = \bigcap_{d \in D, d \geq t} A_d \qquad (0 < t < 1)$$

the family $(A_d)_{d \in D}$ is extended to an increasing family $(A_t)_{0<t<1}$ of closed subsets having the desired properties. ┘

Changing the proof of (6.13) in an obvious way or using (6.13) and replacing A_t by $A_t' = A_{t/2}$ we obtain the following result.

6.14. COROLLARY. Let A be a Borel subset of X and let $A_0 \subset A \cap \beta(A)$ be a closed set. Then there exists an increasing family $(A_t)_{o<t\leq 1}$ of closed sets contained in $A \cap \beta(A)$ such that $A_o = \bigcap_{o<t<1} A_t$ and $A_s \subset \beta(A_t)$ for all $o < s < t \leq 1$.

<u>6.15. REMARK.</u> Let us finally note that balayage and essential balayage coincide if and only if $(X,\mathcal{W})$ satisfies the axiom of polarity. Indeed, fix $u \in \mathcal{W}$, $u > 0$. If S is a semipolar subset of X such that $\hat{R}_u^S = B_u^S$ then S is polar since $B_u^S = 0$ by (6.7). Suppose now conversely that $(X,\mathcal{W})$ satisfies the axiom of polarity and let $A \subset X$. Then $A \setminus \beta(A)$ is polar and hence $\hat{R}_u^A = \hat{R}_u^{A\cap\beta(A)} \leq R_u^{\beta(A)} = B_u^A$, i.e. $\hat{R}_u^A = B_u^A$.

Finally, we shall consider restrictions of $(X,\mathcal{W})$ on closed basic subsets. Since every absorbing set is closed and basic, the following result generalizes (V.1.5).

<u>6.16. PROPOSITION.</u> Let A be a closed basic subset of X . Then $(A,\mathcal{W}_{|A})$ is a balayage space such that $\mathcal{W}_{|A} = S(\{p_{|A} : p \in \mathcal{P}, C(p) \subset A\})$. For every $x \in A$ and $B \subset A$ the measure ε_x^B is the balayage of ε_x on B with respect to $(A,\mathcal{W}_{|A})$.

Proof. Clearly, $\mathcal{W}_{|A}$ is a convex cone of l.s.c. positive numerical functions on the locally compact space A . Let us note that the mapping

$$v \longmapsto R_v^A = \inf\{u \in \mathcal{W} : u \geq v \text{ on } A\}$$

defines a bijection from $\mathcal{W}_{|A}$ on the set of all functions $u \in \mathcal{W}$ such that $R_u^A = u$. This bijection is additive, positively homogeneous and increasing. In particular, if (v_n) is an increasing sequence in $\mathcal{W}_{|A}$ then $\sup v_n = \sup (R_{v_n}^A{}_{|A}) = (\sup R_{v_n}^A)_{|A} \in \mathcal{W}_{|A}$, i.e. (B_1) holds.

Next let $\mathcal{V} \subset \mathcal{W}_{|A}$ and $v_o = \inf\{R_v^A : v \in \mathcal{V}\}$. Then $v_{o|A} = \inf \mathcal{V}$ and $\hat{v}_o = \hat{v}_o^f \in \mathcal{W}$. Fix $x \in A$ and $\varepsilon > 0$. Then there exists a fine neighborhood V of x such that $\hat{v}_o \leq \hat{v}_o(x) + \varepsilon$ on V . Since $x \in A = \beta(A)$ by (6.5), the set $A \cap V$ is not semipolar, whereas the set $\{\hat{v}_o < v_o\}$ is semipolar by (5.11). Hence there exists $y \in A \cap V$ such that $\hat{v}_o(y) = v_o(y)$ and therefore $v_o(y) \leq \hat{v}_o(x) + \varepsilon$.

This shows that $f - \liminf_{y \to x, y \in A} v_0(y) \leq \hat{v}_0(x)$ and hence

$$f - \liminf_{y \to x, y \in A} v_0(y) = \hat{v}_0(x)$$

since $\hat{v}_0(x) = f - \liminf_{y \to x} v_0(y)$. Thus (B_2) holds.

Now let $u, v', v'' \in \mathcal{W}_{|A}$ such that $u \leq v' + v''$. Then $R_u^A \leq R_{v'}^A + R_{v''}^A$, hence there exist $w', w'' \in \mathcal{W}$ such that $R_u^A = w' + w''$, $w' \leq R_{v'}^A$, $w'' \leq R_{v''}^A$. Defining $u' = w'_{|A}$, $u'' = w''_{|A}$ we have $u = u' + u''$, $u' \leq v'$, $u'' \leq v''$. Hence (B_3) holds. Moreover, $\mathcal{P}_{|A}$ is a function cone on A by (I.1.1.1) and obviously $S(\mathcal{P}_{|A}) = \mathcal{W}_{|A}$, i.e. (B_4) holds.

So $(A, \mathcal{W}_{|A})$ is a balayage space. Given $p \in \mathcal{P}$, we have $R_p^A \in \mathcal{P}$ and $C(R_p^A) \subset A$ by (III.2.6). Hence $\mathcal{P}_{|A} = \{q_{|A} : q \in \mathcal{P}, C(q) \subset A\}$.

Finally, let $x \in A$, $B \subset A$ and $p \in \mathcal{P}$. It is immediately verified that ${}^A R^B_{p_{|A}} = (R_p^B)_{|A}$ and hence

$${}^A\varepsilon_x^B(p_{|A}) = f - \liminf_{y \to x, y \in A} {}^A R^B_{p_{|A}}(y) = \hat{R}_p^B(x) = \varepsilon_x^B(p) = \varepsilon_x^B(p_{|A}) .$$

Therefore ${}^A\varepsilon_x^B = \varepsilon_x^B$. ┘

<u>6.17. REMARK.</u> If $(X, \mathcal{W})$ is a harmonic space and A is absorbing then $(A, \mathcal{W}_{|A})$ is a harmonic space. Note, however, that in general this is no longer true if A is not absorbing. Indeed, for every relatively compact open subset U of X such that $A \cap U \neq 0$ the harmonic kernel of $A \cap U$ with respect to A is given by

$${}^A H_U(x, \cdot) = \begin{cases} \varepsilon_x^{A \setminus U} , & x \in A \cap U , \\ \varepsilon_x , & x \in A \setminus U . \end{cases}$$

Hence all we can deduce in the situation of a harmonic space $(X, \mathcal{W})$ is that the support of ${}^A H_U(x, \cdot)$, $x \in A \cap U$, is contained in $(U^* \cap A) \cup (A^* \setminus U)$.

For example, if we restrict the harmonic space $(\mathbb{R}^n, {}^*\mathcal{H}^+(\mathbb{R}^n))$ of classical potential theory, $n \geq 3$, to the hyperplane $\mathbb{R}^{n-1} \times \{0\}$ then we obtain the balayage space of Riesz potentials of index 1 on $\mathbb{R}^{n-1}$! This follows immediately from (6.16) and (V.4.9). (Restricting Riesz potentials of index α, $1 < \alpha < 2$, from $\mathbb{R}^n$ on $\mathbb{R}^{n-1} \times \{0\}$ we obtain Riesz potentials of order $\alpha-1$.)

<u>6.18. EXERCISES.</u> 1. Let $p \in \mathcal{P}$ and $A \subset X$. Then $R_p^A = p$ if and only if $\delta(p) \subset b(A)$.

2. Let (A_n) be an increasing sequence of subsets of X, $A = \bigcup_{n=1}^{\infty} A_n$ and $u \in \mathcal{W}$. Then $\sup B_u^{A_n} = B_u^A$.

3. Let u be a strictly positive function in $\mathcal{W} \cap C(X)$. Then $\widehat{\inf B_u^{K_n}} = B_u^{\cap K_n}$ for every decreasing sequence (K_n) of compact subsets of X if and only if $(X,\mathcal{W})$ satisfies the axiom of polarity.

4. The following statements are equivalent:

(1) The fine closure of every semipolar set is semipolar.

(2) $\beta = \beta b$.

5. The following statements are equivalent:

(1) $b(S)$ is polar for every semipolar set S.

(2) $B_p^A = \hat{R}_p^{b(A)}$ for every $A \subset X$ and $p \in \mathcal{P}$.

6. Let $X =]-1,3[$ and $F = \{0\} \cup \{\frac{1}{n} + \frac{1}{m} : n,m \in \mathbb{N}\}$. Let $\mathcal{W}$ be the set of all l.s.c. positive numerical functions on X which are locally concave on $\complement F$ and decreasing on $A = [0,3[$. Then $(X,\mathcal{W})$ is a harmonic space having the properties of exercise 4 but not of exercise 5. The harmonic space $(A,\mathcal{W}_{|A})$ (A is an absorbing subset of X !) has the properties of exercise 5 but does not satisfy the axiom of thinness. (Hint: $S = \{0\} \cup \{\frac{1}{n} : n \in \mathbb{N}\}$ is the greatest semipolar subset of X.)

7. Penetration Time

Let $\mathfrak{X}$ be a Hunt process associated with $(X,\mathcal{W})$. For every subset A of X we define the *penetration time* τ_A of A by

$$\tau_A = \inf \{t > 0 : \{s \in [0,t] : X_s \in A\} \text{ is uncountable}\} .$$

We intend to show that $\tau_A = \infty$ a.s. if and only if A is semipolar. More generally, we shall see that, for every $A \in \mathcal{B}(X)$,

$$\tau_A = T_{\beta(A)} \quad \text{a.s.}$$

<u>7.1. PROPOSITION.</u> Let A be a semipolar subset of X. Then $\tau_A = \infty$ a.s., i.e. a.s. the set $\{t \geq 0 : X_t \in A\}$ is at most countable.

Proof. By (5.7) it suffices to consider a set $A \in \mathcal{B}(X)$ which is totally thin. Let $q \in \mathcal{P}_b$ be a strict potential. Then $\hat{R}_q^A < q$ on X. Let K be a compact subset of X, $0 < \alpha < 1$ and

$$B = A \cap K \cap \{\hat{R}_q^A \leq \alpha q\} .$$

Then B is a finely closed Borel subset of A and $\hat{R}_q^B \leq \alpha q$ on B. Defining recursively

$$T_1 = T_B , \; T_{n+1} = T_n + T_B \circ \Theta_{T_n}$$

we obtain an increasing sequence (T_n) of stopping times. We note that for every $\omega \in \Omega$, every $n \in \mathbb{N}$ and every $t \in]T_{n-1}(\omega) , T_n(\omega)[$, $X_t(\omega) \notin B$ (where $T_0 := 0$). Hence it suffices to show that $\lim_{n\to\infty} T_n = \infty$ P^x-a.s. for every $x \in X$.

So let $x \in X$. For every $n \in \mathbb{N}$ and every $f \in \mathcal{B}_b^+(X)$

$$E^x(f(X_{T_{n+1}})) = E^x(f \circ X_{T_B} \circ \Theta_{T_n}) = E^x(E^{X(T_n)}(f \circ X_{T_B})) .$$

For every $y \in X$, $P^y[X_{T_B} \in X \setminus B] = \varepsilon_y^B(X \setminus B) = 0$ by (3.16) and (4.6). Choosing $f = 1_{X \setminus B}$ we hence obtain that $P^x[X_{T_n} \in X \setminus B] = 0$ for every $n \in \mathbb{N}$. Therefore in particular,

$$E^x(q(X_{T_{n+1}})) = E^x(E^{X(T_n)}(q \circ X_{T_B})) = E^x(\hat{R}_q^B(X_{T_n})) \leq \alpha \, E^x(q(X_{T_n}))$$

and hence

$$\lim_{n\to\infty} E^x(q(X_{T_n})) = 0 .$$

Let $T = \lim_{n\to\infty} T_n$. For every $n \in \mathbb{N}$, $X_{T_n} \in \bar{B} \subset K$ on $[T < \infty]$ and therefore

$$\inf q(K) \cdot P^x[T < \infty] \leq E^x(q(X_{T_n})) .$$

Thus $P^x[T < \infty] = 0$. ┘

<u>7.2. LEMMA.</u> Let M be a non-empty set of positive real numbers such that $\{\inf t_n : (t_n) \subset M , t_{n+1} < t_n$ for every $n\} = M$. Then M is uncountable.

Proof. Let (a_n) be a sequence in M. Then there exist sequences (s_n), (t_n) in M such that $s_n < s_{n+1} < t_{n+1} < t_n$ and $a_n \notin [s_{n+1} , t_{n+1}]$. Indeed, there exist $s_1 , t_1 \in M$ such that $s_1 < t_1$. Let $n \in \mathbb{N}$ and suppose that $s_n , t_n \in M$ satisfying $s_n < t_n$ are already chosen. By assumption, s_n is the infimum of a strictly decreasing sequence in M. Hence there exist $s_{n+1} , t_{n+1} \in M$

such that $s_n < s_{n+1} < t_{n+1} < t_n$ and $a_n \notin [s_{n+1}, t_{n+1}]$.

Let $t = \inf t_n$. Then $t \in M$ and $t \neq a_n$ for every $n \in \mathbb{N}$ since $t \in [s_{n+1}, t_{n+1}]$. This shows that M is uncountable. ┘

7.3. PROPOSITION. Let A be a basic subset of X . Then $\tau_A = T_A$ a.s.

Proof. We define a decreasing sequence (S_n) of stopping times by

$$S_n = \frac{1}{n} + T_A \circ \Theta_{\frac{1}{n}} .$$

Since $b(A) = A$ we obtain by (4.8) that $\inf S_n = T_A = 0$ P^y-a.s. for every $y \in A$.

Let $\mathcal{T}$ be the set of all stopping times T such that $X_T \in A$ a.s. on $[T < \infty]$. By (4.10), $T_A \in \mathcal{T}$. More generally, if S is an arbitrary stopping time then for every $x \in X_\Delta$

$$P^x[X_{S+T_A\circ\Theta_S} \notin A, S + T_A\circ\Theta_S < \infty] = P^x[X_{T_A}\circ\Theta_S \notin A, T_A\circ\Theta_S < \infty]$$
$$= E^x(P^{X_S}[X_{T_A} \notin A, T_A < \infty]) = 0$$

and hence $S + T_A\circ\Theta_S \in \mathcal{T}$. For every $T \in \mathcal{T}$ and $n \in \mathbb{N}$ let

$$T_n = T + S_n\circ\Theta_T = (T + \tfrac{1}{n}) + T_A\circ\Theta_{T+\frac{1}{n}} .$$

Then (T_n) is a decreasing sequence in $\mathcal{T}$ such that $T + \frac{1}{n} \leq T_n$ for every $n \in \mathbb{N}$. Furthermore for every $x \in X_\Delta$

$$P^x[\inf T_n > T] = P^x[\inf S_n\circ\Theta_T > 0, T < \infty] = E^x\{1_{[T<\infty]} P^{X_T}[\inf S_n > 0]\} = 0 ,$$

i.e. $\inf T_n = T$ a.s. . Let

$$\mathcal{T}_0 = \{(\dots ((T_A)_{n_1})_{n_2} \dots)_{n_k} : k \in \mathbb{N}, n_i \in \mathbb{N}\} ,$$

i.e. $\mathcal{T}_0$ is the smallest subset of $\mathcal{T}$ such that $T_A \in \mathcal{T}_0$ and $(T_n) \subset \mathcal{T}_0$ for every $T \in \mathcal{T}_0$. $\mathcal{T}_0$ is countable and hence

$$\Omega_0 := \bigcup_{T\in\mathcal{T}_0} [\inf T_n > T] \cup [X_T \notin A, T < \infty] \in \mathfrak{M} ,$$

$P^x(\Omega_0) = 0$ for every $x \in X_\Delta$.

Let $p \in \mathcal{P}$ be a strict potential and $f := p - R_p^A$. Since $A = b(A)$ we have $R_p^A \in \mathcal{W}$ and $A = \{f = 0\}$. Fix $x \in X_\Delta$. Then by (4.12) there exists a set $\Omega_1 \in \mathfrak{M}$ such that $P^x(\Omega_1) = 0$ and $t \mapsto f(X_t(\omega))$ is right continuous for every $\omega \in \complement\Omega_1$.

Consider now $\omega \in \complement(\Omega_0 \cup \Omega_1)$. We claim that $T_A(\omega) = \tau_A(\omega)$. Obviously $T_A(\omega) \leq \tau_A(\omega)$. So assume that $T_A(\omega) < \infty$, take $\varepsilon > 0$, $\mathcal{T}_0(\omega) = \{T(\omega) : T \in \mathcal{T}_0\}$, and let

$$M = \{\inf t_n : (t_n) \subset \mathcal{T}_0(\omega) \cap [T_A(\omega), T_A(\omega) + \varepsilon[\} .$$

Then M is uncountable by (7.2). Furthermore, $X_t(\omega) \in A$ for every $t \in M$. Indeed, let (t_n) be a decreasing sequence in $\mathcal{T}_0(\omega)$ and $t = \inf t_n$. Then $X_{t_n}(\omega) \in A$ for every $n \in \mathbb{N}$, hence $f(X_t(\omega)) = \lim_{n\to\infty} f(X_{t_n}(\omega)) = 0$, i.e. $X_t(\omega) \in A$. Therefore $\tau_A(\omega) \leq T_A(\omega) + \varepsilon$ and hence $\tau_A(\omega) = T_A(\omega)$. ┘

7.4. REMARK. (7.1) and (7.3) yield a probabilistic proof of (6.5). Indeed, let A be a basic subset of X and $x \in A$. By (7.1), $\tau_{A\setminus\beta(A)} = \infty$ P^x-a.s. and hence $T_{\beta(A)} = \tau_{\beta(A)} \leq \tau_A = T_A = 0$ P^x-a.s. by (7.3). Therefore $x \in \beta(A)$.

7.5. THEOREM. Let A be a Borel subset of X . Then $\tau_A = T_{\beta(A)}$ a.s. In particular, $\beta(A) = \{x \in X : \tau_A = 0 \;\; P^x\text{-a.s.}\}$.

Proof. Since $A\setminus\beta(A)$ is semi-polar we know by (7.1) that $\tau_{A\setminus\beta(A)} = \infty$ a.s. Therefore by (7.3)

$$\tau_A \geq \tau_{\beta(A) \cup (A\setminus\beta(A))} = \inf(\tau_{\beta(A)}, \tau_{A\setminus\beta(A)}) = \tau_{\beta(A)} = T_{\beta(A)} \quad \text{a.s.}$$

On the other hand by (6.9) and (6.2) there exists a sequence (A_n) of basic subsets of A such that $\beta(A)$ is the fine closure of $A' = \bigcup_{n=1}^{\infty} A_n$. Hence by (4.10) and (7.3)

$$T_{\beta(A)} = T_{A'} = \inf T_{A_n} = \inf \tau_{A_n} \geq \tau_A \quad \text{a.s.}$$ ┘

7.6. COROLLARY. For every $A \in \mathcal{B}(X)$ the following statements are equivalent:

(1) A is semipolar.

(2) $\tau_A = \infty$ a.s.

(3) $\tau_A > 0$ a.s.

We shall close this section by giving a probabilistic interpretation for $\delta(p_A)$ if p is a strict potential. To that end let $A \in \mathcal{B}(X)$ define

$$\tilde{\tau}_A = \inf\{t > 0 : \lambda(\{s \in [0,t] : X_s \in A\}) > 0\}$$

where λ denotes Lebesgue measure on $\mathbb{R}$. Then for every $t > 0$

$$[\tilde{\tau}_A < t] = \{\omega \in \Omega : \int_0^t 1_A(X_s(\omega))ds > 0\}$$

hence $\tilde{\tau}_A$ is a stopping time. For every $x \in X$,

$$V1_A(x) = E^x(\int_0^\infty 1_A \circ X_s \, ds)$$

and hence $\tilde{\tau}_A = \infty$ P^x-a.s. if and only if $V1_A(x) = 0$. Evidently $\tau_A \leq \tilde{\tau}_A$, and in many cases τ_A is strictly less then $\tilde{\tau}_A$ (see (7.8)).

7.7. PROPOSITION. Let A be a Borel subset of X and suppose that $p := V1 \in P_b$. Then $\tilde{\tau}_A = T_{\delta(p_A)}$ a.s. In particular, $\delta(p_A) = \{x \in X : \tilde{\tau}_A = 0 \; P^x\text{-a.s.}\}$.

Proof. By (II.7.8) and (6.4),

$$V1_{A \setminus \delta(p_A)} = p_{A \setminus \delta(p_A)} = (p_A)_{\complement \delta(p_A)} = 0 .$$

Hence, for every $x \in X$,

$$\tilde{\tau}_A \geq \tilde{\tau}_{\delta(p_A)} \geq T_{\delta(p_A)} \qquad P^x\text{-a.s.}$$

In the following let $T = \tilde{\tau}_A$. Since $t + T \circ \Theta_t$ decreases to T as t decreases to zero, we obtain that

$$\sup_{t>0} E^x(e^{-t} E^{X_t}(e^{-T})) = \sup_{t>0} E^x(e^{-(t+T\circ\Theta_t)}) = E^x(e^{-T})$$

for every $x \in X$. So the function $x \mapsto E^x(e^{-T})$ is 1-excessive and hence it is finely continuous by (V.2.3). In particular, the set

$$F := \{x \in X : T = 0 \; P^x\text{-a.s.}\} = \{x \in X : E^x(e^{-T}) = 1\}$$

is finely closed.

We claim that $\delta(p_A) \subset F$. Indeed, let $x \in \delta(p_A)$ and let K be a compact fine neighborhood of x. Then $x \in \delta(p_{A \cap K})$. Therefore $0 < p_{A \cap K}(x) = V1_{A \cap K}(x)$ and hence $P^x[\tilde{\tau}_{A \cap K} < \infty] > 0$. This is sufficient to show the existence of a point $y \in F \cap K$: Let $S = \tilde{\tau}_{A \cap K}$. It is easily seen that $S = S + S \circ \Theta_S$ and hence $S \circ \Theta_S = 0$ on $[S < \infty]$. Therefore by the strong Markov property

$$P^x[S < \infty] = P^x[S < \infty, S \circ \Theta_S = 0] = \int_{[S < \infty]} P^{X_S}[S=0] \, dP^x ,$$

i.e. $P^{X_S}[S=0] = 1$ P^x-a.s. on $[S < \infty]$. Since obviously $X_S \in K$ on $[S < \infty]$ and

$P^x[S<\infty] > 0$ we conclude that there exists a point $y \in K$ such that $P^y[S=0] = 1$. Since of course $T = \tilde{\tau}_A \leq \tilde{\tau}_{A\cap K} = S$ we then know that $T = 0$ P^y-a.s., i.e. $y \in F$. As noted above F is finely closed. K being an arbitrary compact fine neighborhood of x we thus have proved that $x \in F$.

Having $T = 0$ P^x-a.s. for every $x \in \delta(p_A)$ it is now easy to complete the proof. Let $x \in X$ and $R = T_{\delta(p_A)}$. Since $T \geq R$ P^x-a.s. we obviously have $T = R + T\circ\Theta_R$ P^x-a.s. and hence again by the strong Markov property

$$P^x[R<T] = P^x[R<T\,,\,T\circ\Theta_R > 0] = \int_{[R<T]} P^{X_R}[T>0]\,dP^x = 0$$

since $X_R \in \delta(p_A)$ P^x-a.s. on $[R<\infty]$. Thus $R = T$ P^x-a.s. ┘

7.8. EXERCISE. Let $\mathfrak{X}$ be Brownian motion on $\mathbb{R}^3$ and let $A = \{x \in \mathbb{R}^3 : x_1^2 + x_2^2 \leq 1\,,\, x_3 = 0\}$. Then $\tilde{\tau}_A = \infty$ P^x-a.s. for every $x \in \mathbb{R}^3$, $\tau_A = 0$ P^x-a.s. for every $x \in A$.

8. Fine Support of Potentials

Let us recall that by definition the *fine support* $\delta(q)$ of a potential $q \in P$ is the Choquet boundary of X with respect to the convex cone $P + \mathbb{R}q$, i.e. $\delta(q)$ is the set of all points $x \in X$ such that ε_x is the only measure μ on X satisfying $\mu(p) \leq p(x)$ for every $p \in P$ and $\mu(q) = q(x)$. By (4.7.1) the fine support $\delta(q)$ is a basic set and by (6.4) it is the smallest finely closed subset F of X such that $q_F = q$ ($R_q^F = q$ resp.).

8.1. LEMMA. Let (q_n) be a sequence in P such that $q = \sum_{n=1}^{\infty} q_n \in P$. Then $\delta(q) = b\,(\bigcup_{n=1}^{\infty} \delta(q_n))$. In particular, $\delta(q_1+q_2) = \delta(q_1) \cup \delta(q_2)$ for all $q_1, q_2 \in P$.

Proof. By (II.6.1), $\delta(q_n) \subset \delta(q)$ for every $n \in \mathbb{N}$ and hence

$$F := b(\bigcup_{n=1}^{\infty} \delta(q_n)) \subset b(\delta(q)) = \delta(q)\,.$$

On the other hand, $R_{q_n}^F = q_n$ for every $n \in \mathbb{N}$, hence $R_q^F = q$ and $\delta(q) \subset F$. ┘

The following result shows that $\delta(p)$ is really the fine support of p in any possible sense of the word.

8.2. PROPOSITION. Let $p \in P$ and $x \in X$. Then the following statements are equivalent:

(1) $x \in \delta(p)$.

(2) $R_p^{\complement V}(x) < p(x)$ for every fine neighborhood V of x.

(3) $p_V \neq 0$ for every Borel fine neighborhood V of x.

(4) $p_V(x) \neq 0$ for every Borel fine neighborhood V of x.

Proof. (1) ⇒ (2): If V is a fine neighborhood of a point $x \in \delta(p)$ then $R_p^{\complement V}(x) < p(x)$ since $\varepsilon_x^{\complement V} \neq \varepsilon_x$.

(2) ⇒ (1): $V = \complement\delta(p)$ is finely open and $R_p^{\complement V} = R_p^{\delta(p)} = p$.

(1) ⇒ (3): Let $V \in B(X)$ be a fine neighborhood of x such that $p_V = 0$. Then $\delta(p) = \delta(p_{\complement V})$ is contained in the fine closure of $\complement V$ and hence $x \notin \delta(p)$.

(3) ⇒ (1): $V = \complement\delta(p)$ is finely open and $p_V = p - p_{\delta(p)} = 0$.

(3) ⇒ (4): Let $V \in B(X)$ be a fine neighborhood of x such that $p_V(x) = 0$. Since $\{p_V = 0\}$ is finely open by (V.1.3.3) there exists a compact fine neighborhood K of x such that $K \subset \{p_V = 0\} \cap V$. Then $p_K \leq p_V = 0$ on K and therefore $p_K = 0$.

(4) ⇒ (3): Trivial. ┘

8.3. REMARK. Let U be an open subset of X and $p \in P$. Then $p^U = p - R_p^{\complement U}$ is a potential on the subspace U (see (V.1.1)) such that $\delta^U(p^U) = \delta(p) \cap U$ where δ^U denotes the fine support with respect to (U, W^U). This follows immediately from (8.2) and (2.8). In particular, p^U is strict if p is strict.

A kernel D on X is called a *dilation* if $Dp \leq p$ for every $p \in P$. A dilation D is called a W-*dilation* (P-*dilation* resp.) provided $D(W) \subset W$ ($D(P) \subset P$ resp.). If D is a P-dilation then D is a W-dilation and $D(C_P(X)) \subset C_P(X)$.

For any W-dilation let

$$b(D) = \{x \in X : D(x,\cdot) = \varepsilon_x\} .$$

If $p \in P$ is strict then $b(D) = \{Dp = p\}$, in particular $b(D)$ is a finely closed G_δ-set. For example, for every subset A of X the kernel $D_A : (x,B) \mapsto \varepsilon_x^A(B)$ is a W-dilation, $b(D_A) = b(A)$.

We shall say that a P-dilation D is *supported* by a finely closed Borel set F if all measures $D(x,\cdot)$, $x \in X$, are supported by F and if $\delta(Dp) \subset F$ for every $p \in P$.

In chapter VII P-dilations will be used to solve the weak Dirichlet problem. They can be obtained by a "smearing" of balayage on suitable families of closed sets.

8.4. PROPOSITION. Let $(A_t)_{0<t<1}$ be an increasing family of closed subsets of X such that $A_s \subset \beta(A_t)$ for all $0<s<t<1$, and let $A_0 = \bigcap_{0<t<1} A_t$, $A_1 = \bigcup_{0<t<1} A_t$. Then

$$D : (x,B) \longmapsto \int_0^1 \varepsilon_x^{A_t}(B)\, dt$$

is a P-dilation such that $b(D) = A_0$, $D1_{\complement A_1} = 0$. Moreover $(Dp)_{\complement A_1} = 0$ and $\delta(p) \cap \beta(A_1) \subset \delta(Dp) \subset \beta(A_1)$ for every $p \in P$. In particular, D is supported by $\beta(A_1)$ and $\delta(Dp) = \beta(A_1)$ for every strict potential $p \in P$.

Proof. Let us note first that $A_1 \subset \beta(A_1)$ and hence $\beta(A_1)$ is the fine closure of A_1 . Let $p \in P$. Obviously the mapping $t \mapsto \hat{R}_p^{A_t}$ is increasing and $R_p^{A_0} \leq \hat{R}_p^{A_t} \leq R_p^{\beta(A_1)}$ for every $0<t<1$. We define

$$Dp = \int_0^1 \hat{R}_p^{A_t}\, dt .$$

Then $Dp \in W$ and

$$R_p^{A_0} \leq Dp \leq R_p^{\beta(A_1)} .$$

The mapping $p \mapsto Dp$ is additive, positively homogeneous, and increasing. Hence by (II.1.2) D can be uniquely extended to a kernel on X which will again be denoted by D . Obviously, D is a W-dilation. Moreover, by the monotone class theorem, for every $f \in B^+(X)$ and $x \in X$,

$$Df(x) = \int_0^1 \varepsilon_x^{A_t}(f)\, dt .$$

In particular,

$$D1_{\complement A_1}(x) = \int_0^1 \varepsilon_x^{A_t}(\complement A_1)\, dt = 0 .$$

Again let $p \in P$. If $0 < s < t < 1$ then

$$\hat{R}_p^{A_s} \leq R_p^{A_s} \leq R_p^{\beta(A_t)} \leq \hat{R}_p^{A_t}$$

and hence

$$Dp = \int_0^1 R_p^{A_t}\, dt .$$

(In fact, for every $x \in X$ there is at most one $t \in]0,1[$ such that $\hat{R}_p^{A_t}(x) \neq R_p^{A_t}(x)$.) By (2.6), $R_p^{A_t}$ is u.s.c. for every $0 < t < 1$. Hence Dp is u.s.c. as well. This shows that D is a P-dilation.

If $x \in A_0$ then $Dp(x) \geq R_p^{A_0}(x) = p(x)$. If $x \in \complement A_0$ then there exists $s \in]0,1[$ such that $x \in \complement A_t$ for all $s < t < 1$ and therefore $Dp(x) < p(x)$ whenever p is strict. Thus $b(D) = A_0$.

Fix $n \in \mathbb{N}$ and let $s_n = 1 - \frac{1}{n}$,

$$p_n = \int_0^{s_n} \hat{R}_p^{A_t}\, dt = \int_0^{s_n} R_p^{A_t}\, dt , \quad q_n = \int_{s_n}^1 \hat{R}_p^{A_t}\, dt = \int_{s_n}^1 R_p^{A_t}\, dt .$$

Then $p_n , q_n \in P , p_n + q_n = Dp$. For every relatively compact open subset U of $\complement A_{s_n}$ we obtain by Fubini's theorem and (2.6) that

$$H_U p_n = \int_0^{s_n} H_U R_p^{A_t}\, dt = \int_0^{s_n} R_p^{A_t}\, dt = p_n ,$$

i.e. $C(p_n) \subset A_{s_n}$, $(p_n)_{\complement A_{s_n}} = 0$. Using $A_{s_n} \subset A_1$ we obtain that

$$Dp \geq (Dp)_{A_1} = (p_n)_{A_1} + (q_n)_{A_1} \geq p_n .$$

Letting n tend to infinity we conclude that $Dp = (Dp)_{A_1}$. Since $A_1 \subset \beta(A_1)$ this implies by (6.4) that $\delta(Dp) \subset \beta(A_1)$.

It remains to show that $\delta(p) \cap \beta(A_1) \subset \delta(Dp)$. So fix $x \in \delta(p) \cap \beta(A_1)$ and consider $\mu \in M_+(X)$ such that $\mu(q) \leq q(x)$ for every $q \in P$ and $\mu(Dp) = Dp(x)$. Then obviously $\mu(q_n) = q_n(x)$ for every $n \in \mathbb{N}$. Since

$$nq_n = \int_0^1 \hat{R}_p^{A_{1-\frac{t}{n}}}\, dt ,$$

the sequence (nq_n) is increasing. Furthermore, for every $n \in \mathbb{N}$,

$$R_p^{A_{s_n}} \leq nq_n \leq R_p^{\beta(A_1)}$$

and hence
$$R_p^{\beta(A_1)} = R_p^{A_1} = \sup R_p^{A_{s_n}} \leq \sup nq_n \leq R_p^{\beta(A_1)} .$$
Therefore
$$p(x) = R_p^{\beta(A_1)}(x) = \sup nq_n(x) = \sup \mu(nq_n) = \mu(R_p^{\beta(A_1)}) \leq \mu(p) \leq p(x) .$$
Thus $\mu = \varepsilon_x$, i.e. $x \in \delta(Dp)$. ┘

8.5. COROLLARY. Let A be a Borel subset of X and let $A_o \subset A \cap \beta(A)$ be a closed set. Then there exists a P-dilation D such that $b(D) = A_o$, $D1_{\complement A} = 0$, and $(Dp)_{\complement A} = 0$, $\delta(p) \cap \beta(A) \subset \delta(Dp) \subset \beta(A)$ for every $p \in P$.

Moreover there exists a sequence (D_n) of P-dilations such that $A_o = b(D_1)$, every D_n is supported by a closed subset of $A \cap \beta(A)$, and such that, for every $p \in P$, the sequence $(D_n p)$ is increasing to $R_p^{\beta(A)}$.

Proof. By (6.13), there exists an increasing family $(A_t)_{0<t<1}$ of closed sets contained in $A \cap \beta(A)$ such that $A_o = \bigcap_{0<t<1} A_t$, $\beta(A) = \beta(\bigcup_{0<t<1} A_t)$, and $A_s \subset \beta(A_t)$ for all $0<s<t<1$. Then $\lim_{t\uparrow 1} \hat{R}_p^{A_t} = R_p^{\beta(A)}$ for every $p \in P$. Using (8.4) and defining
$$D(x,B) = \int_0^1 \varepsilon_x^{A_t}(B)\, dt \quad , \quad D_n(x,B) = 2^n \int_{1-2^{-n+1}}^{1-2^{-n}} \varepsilon_x^{A_t}(B)\, dt$$
we obtain P-dilations having the desired properties. ┘

8.6. COROLLARY. For every subset A of X the following statements are equivalent:

(1) A is basic.

(2) There exists a potential $q \in P$ such that $\delta(q) = A$.

Proof. (1) ⇒ (2): By (8.5), there exists a P-dilation D on X such that $q = Dp$ satisfies $\delta(q) = \beta(A)$ if $p \in P$ is strict. By (6.5), $\beta(A) = A$.

(2) ⇒ (1): (4.7.1). ┘

8.7. REMARK. Instead of applying the strong result (8.5) we could go back directly to (4.17). Indeed, let A be a basic set and let $p \in P$ be a strict potential. Since $R_p^A = \sup \{q \in P : q \leq p , R_q^A = q\}$ there exists an increasing sequence (q_n) in P such that $\sup q_n = R_p^A$ and $q_n \leq p$, $R_{q_n}^A = q_n$. Choosing $q = \sum_{n=1}^{\infty} 2^{-n} q_n$ it is easily shown that $q \in P$ and $\delta(q) = A$.

8.8. COROLLARY. For every Borel subset A of X the following statements are equivalent:

(1) A is semipolar.

(2) $q_A = 0$ for every $q \in P$.

(3) The constant zero is the only potential $q \in P$ such that $C(q) \subset A$ ($\delta(q) \subset A$ resp.).

Proof. (1) ⇒ (2): (5.15).

(2) ⇒ (1): Suppose that A is not semipolar. Then there exists a compact subset K of A which is not semipolar and a potential $q \in P$ such that $\delta(q) = \beta(K)$. Then $q_A = q \neq 0$.

(2) ⇒ (3): If $q \in P$, $q \neq 0$, such that $\delta(q) \subset A$ then $q_A = q \neq 0$.

(3) ⇒ (2): If $q \in P$ such that $q_A \neq 0$ then there exists a compact subset K of A such that $q_K \neq 0$ and it suffices to note that $q_K \in P$ and $C(q_K) \subset K \subset A$. ⌟

8.9. PROPOSITION. Let A be a Borel subset of X. Then there exists a measure μ on X such that $\mu(\complement A) = 0$, $\mu(S) = 0$ for every semipolar subset S of X and $\beta(A)$ is the fine support of μ (i.e. $\beta(A)$ is the set of all points $x \in X$ such that $\mu(V) > 0$ for every fine neighborhood $V \in \mathcal{B}(X)$ of x).

Proof. Choosing a strict potential $p \in P$ and applying (8.5) we obtain a potential $q = Dp \in P$ such that $q_{\complement A} = 0$ and $\delta(q) = \beta(A)$. Let (x_n) be a dense sequence in X and (γ_n) a sequence of strictly positive real numbers such that $\sum_{n=1}^{\infty} \gamma_n q(x_n) < \infty$. Define μ by

$$\mu(B) = \sum_{n=1}^{\infty} \gamma_n q_B(x_n) \qquad (B \in \mathcal{B}(X)).$$

Of course, μ is a finite measure on X such that $\mu(\complement A) = 0$ and $\mu(S) = 0$ for every semipolar subset S of X. Hence $\mu(\complement\beta(A)) = 0$. Finally, for every $x \in \beta(A)$ and every compact fine neighborhood K of x, $q_K \neq 0$ since $x \in \delta(q)$, and hence

$\mu(K) > 0$. ┘

8.10. EXERCISES. 1. Let V be the potential kernel associated with $p \in P$. Then $\delta(p) = X$ if and only if for every finely continuous bounded function f the equality $Vf = 0$ implies that $f = 0$.

2. For every $A \in \mathcal{B}(X)$ there exists a strict $p \in P$ such that $\delta(p_A) = \beta(A)$.

3. If $1 \in W$ then for every $A \in \mathcal{B}(X)$ there exists an associated Hunt process such that $\tilde{\tau}_A = \tau_A$ a.s.

4. Let A be a Borel subset of $\mathbb{R}$ and let B denote the set of all condensation points of A , i.e. the set of all $x \in \mathbb{R}$ such that $A \cap U$ is uncountable for every neighborhood of x . Then there exists a measure μ on $\mathbb{R}$ such that $\mu(\complement A) = 0$, $\mu(\{x\}) = 0$ for every $x \in \mathbb{R}$ and B is the support of μ . (Hint: Consider translation on $\mathbb{R}$ and apply (8.9) to A and $-A$.)

9. Fine Properties of Balayage

9.1. LEMMA. Let B be a Borel subset of X and $B \subset A \subset X$. Then for every $x \in X$ and every $p \in P$, $\hat{R}_p^B(x) \leq \int R_p^B \, d\varepsilon_x^A \leq R_p^B(x)$.

Proof. Let $p \in P$. Then by (1.2)

$$\hat{R}_p^B = \hat{R}^A_{R_p^B} \leq \inf\{\hat{R}^A_{R_p^U} : U \text{ open } \supset B\} \leq \inf\{R_p^U : U \text{ open } \supset B\} = R_p^B$$

where by (1.9) for every $x \in X$

$$\inf\{\hat{R}^A_{R_p^U}(x) : U \text{ open } \supset B\} = \inf\{\int R_p^U \, d\varepsilon_x^A : U \text{ open } \supset B\} = \int R_p^B \, d\varepsilon_x^A .$$ ┘

9.2. PROPOSITION. Let A be a subset of X , $x \in X$ and $\mu \in M(P)$ such that $\mu^*(A \setminus \{x\}) = 0$. Then

$$\mu^A = \mu^{A \setminus \{x\}} + \mu^A(\{x\})(\varepsilon_x - \varepsilon_x^{A \setminus \{x\}}) .$$

In particular,

$$\varepsilon_x^A = \varepsilon_x^A(\{x\}) \, \varepsilon_x + (1 - \varepsilon_x^A(\{x\})) \, \varepsilon_x^{A \setminus \{x\}} .$$

Proof. By (2.2), we may assume that A is a Borel set. Let $p \in P$. By (2.3),

$\hat{R}_p^{A\setminus\{x\}} = R_p^{A\setminus\{x\}}$ on $\complement(A\setminus\{x\})$. Hence by (9.1)

$$\int p \, d\mu^{A\setminus\{x\}} = \int \hat{R}_p^{A\setminus\{x\}}(y) \, \mu(dy) = \int \left(\int R_p^{A\setminus\{x\}} \, d\varepsilon_y^A\right) \mu(dy) = \int R_p^{A\setminus\{x\}} \, d\mu^A .$$

By (4.6), $\mu^A(\complement\bar{A}^f) = 0$. Since $\bar{A}^f \subset \{x\} \cup \overline{A\setminus\{x\}}^f$, $R_p^{A\setminus\{x\}}(x) = \hat{R}_p^{A\setminus\{x\}}(x)$ and $R_p^{A\setminus\{x\}} = p$ on $\overline{A\setminus\{x\}}^f$ we therefore have

$$\int R_p^{A\setminus\{x\}} \, d\mu^A = \mu^A(\{x\}) \, \hat{R}_p^{A\setminus\{x\}}(x) + \int_{\overline{A\setminus\{x\}}^f\setminus\{x\}} p \, d\mu^A$$

$$= \mu^A(\{x\}) \, \hat{R}_p^{A\setminus\{x\}}(x) + \int p \, d\mu^A - \mu^A(\{x\}) \, p(x) .$$

This proves the statement. ┘

9.3. PROPOSITION. Let A and B be subsets of X and $\mu \in M(P)$. Then $\mu^{A\cup B} \leq \mu^A + \mu^B$.

Proof. By (1.3), for every $p \in P$, $\hat{R}_p^{A\cup B} + \hat{p}_{A,B} = \hat{R}_p^A + \hat{R}_p^B$ where the mapping $p \mapsto \hat{p}_{A,B}$ is additive, positively homogeneous and increasing. Hence by (I.1.4), $\varepsilon_x^{A\cup B} \leq \varepsilon_x^A + \varepsilon_x^B$ for every $x \in X$ and therefore $\mu^{A\cup B} \leq \mu^A + \mu^B$. ┘

9.4. PROPOSITION. Let B be a Borel subset of X , $B \subset A \subset X$ and $x \in X$. Then the following statements are equivalent:

(1) $x \in \complement(B \cap (b(A) \setminus b(B)))$.

(2) For every $p \in P$, $\int R_p^B \, d\varepsilon_x^A = \hat{R}_p^B(x)$.

(3) $\varepsilon_x^B = \varepsilon_x^A|_B + (\varepsilon_x^A|_{\complement B})^B$.

(4) $\varepsilon_x^A|_B \leq \varepsilon_x^B$.

Proof. By (2.3), for every $p \in P$

$$\int R_p^B \, d\varepsilon_x^A = \int_B p \, d\varepsilon_x^A + \int_{\complement B} \hat{R}_p^B \, d\varepsilon_x^A = (\varepsilon_x^A|_B + (\varepsilon_x^A|_{\complement B})^B)(p) .$$

Therefore (2) and (3) are equivalent. Obviously (3) implies (4). Moreover, if $x \in B \cap (b(A) \setminus b(B))$ then $\varepsilon_x^A|_B = \varepsilon_x$ whereas $\int p \, d\varepsilon_x^B < p(x)$ for some $p \in P$. Therefore (4) implies (1). Thus it remains to show that (1) implies (2).

By (9.1), it suffices to consider $x \in B \setminus b(A)$. By (2.2) and (4.3) we may assume that A and B are finely closed Borel sets. Let $p \in P$ and $\varepsilon > 0$. By (4.6) there exists a compact subset K of $A\setminus B$ such that defining $C = B \cup K$

we have

$$\int_{A\setminus C} p \, d\varepsilon_x^A < \varepsilon .$$

We choose a potential $q \in P$ such that $q > 0$ and $q(x) < \varepsilon$. By (2.5) there exists an open set U containing B such that

$$u := R_p^U \leq R_p^B + q \quad \text{on} \quad K ,$$

and hence $u \leq R_p^B + q$ on C. This implies that $\hat{R}_u^C \leq \hat{R}_p^B + q$, in particular

$$\hat{R}_u^C(x) \leq \hat{R}_p^B(x) + \varepsilon .$$

Since $x \notin \overline{A\setminus C}^f$ we know by (9.3), and (4.6) that

$$\alpha := \varepsilon_x^A(\{x\}) \leq \varepsilon_x^C(\{x\}) + \varepsilon_x^{A\setminus C}(\{x\}) = \varepsilon_x^C(\{x\}) =: \gamma .$$

Moreover by (9.1) and the first part of our proof

$$\varepsilon_x^{A\setminus\{x\}}\big|_{C-\{x\}} \leq \varepsilon_x^{C\setminus\{x\}} .$$

Using (9.2) we therefore obtain that

$$\int_C u \, d\varepsilon_x^A = \alpha u(x) + (1-\alpha) \int_{C\setminus\{x\}} u \, d\varepsilon_x^{A\setminus\{x\}} \leq \alpha u(x) + (1-\alpha) \hat{R}_u^{C\setminus\{x\}}(x)$$

$$= \hat{R}_u^{C\setminus\{x\}}(x) + \alpha(u(x) - \hat{R}_u^{C\setminus\{x\}}(x)) \leq \hat{R}_u^{C\setminus\{x\}}(x) + \gamma(u(x) - \hat{R}_u^{C\setminus\{x\}}(x)) = \hat{R}_u^C(x) .$$

Hence

$$\int R_p^B \, d\varepsilon_x^A \leq \int u \, d\varepsilon_x^A = \int_C u \, d\varepsilon_x^A + \int_{A\setminus C} u \, d\varepsilon_x^A \leq \hat{R}_u^C(x) + \int_{A\setminus C} p \, d\varepsilon_x^A < \hat{R}_p^B(x) + 2\varepsilon .$$

Thus $\int R_p^B \, d\varepsilon_x^A = \hat{R}_p^B(x)$. ┘

<u>9.5. COROLLARY.</u> Let B be a Borel subset of X, $B \subset A \subset X$ and $\mu \in M(P)$. Then the following statements are equivalent:

(1) $\mu^B = \mu^A|_B + (\mu^A|_{\complement B})^B$.

(2) $\mu(b(A) \cap (B\setminus b(B))) = 0$.

<u>9.6. COROLLARY.</u> Let $B \subset A \subset X$ and $\mu \in M(P)$. Then the following assertions are equivalent:

(1) $\mu^A = \mu^B$.

(2) $(\mu^A)_* (\complement \bar{B}^f) = 0$ and $\mu(b(A)\setminus b(B)) = 0$.

(3) $(\mu^A)_* (\complement \bar{B}^f) = 0$ and $\mu^*(b(A) \cap (B\setminus b(B))) = 0$.

Proof. (1) ⇒ (2): If $(\mu^A)_*(\complement\bar{B}^f) \neq 0$ then $\mu^A \neq \mu^B$ since $(\mu^B)_*(\complement\bar{B}^f) = 0$ by (4.6). If $p \in P$ is strict then $\hat{R}_p^A = p > \hat{R}_p^B$ on $b(A) \setminus b(B)$ and hence

$$\mu^A(p) = \int \hat{R}_p^A \, d\mu > \int \hat{R}_p^B \, d\mu = \mu^B(p)$$

if $\mu(b(A) \setminus b(B)) > 0$.

(2) ⇒ (3): Obvious.

(3) ⇒ (1): By (2.2) we may assume that B is a Borel set. Since $\bar{B}^f = B \cup b(B)$ and $b(\bar{B}^f) = b(B)$ we have $\bar{B}^f \setminus b(\bar{B}^f) = B \setminus b(B)$. Hence we may even assume that B is finely closed. Then $\mu^A|_B = \mu^A$, $\mu^A|_{\complement B} = 0$. Thus $\mu^B = \mu^A$ by (9.5). ┘

9.7. COROLLARY. Let A and B be finely closed Borel subsets of X and $\mu \in M(P)$. Then the following statements are equivalent:

(1) $\mu^A + \mu^B = \mu^{A\cup B} + \mu^{A\cup B}|_{A\cap B} + (\mu^{A\cup B}|_{\complement A})^A + (\mu^{A\cup B}|_{\complement B})^B$.

(2) $\mu[(A \setminus b(A)) \cap b(B)] = 0$ and $\mu[(B \setminus b(B)) \cap b(A)] = 0$.

Proof. Let $\nu_A = \mu^{A\cup B}|_A + (\mu^{A\cup B}|_{\complement A})^A$ and $\nu_B = \mu^{A\cup B}|_B + (\mu^{A\cup B}|_{\complement B})^B$. Then the right hand side of (1) is equal to $\nu_A + \nu_B$. By (9.1) we have for every $p \in P$,

$$\int p \, d\mu^A = \int \hat{R}_p^A \, d\mu \leq \int R_p^A \, d\mu^{A\cup B} = \int p \, d\nu_A$$

and analogously $\int p \, d\mu^B \leq \int p \, d\nu_B$. Therefore (1) holds if and only if $\mu^A = \nu_A$ and $\mu^B = \nu_B$. Thus the equivalence of (1) and (2) follows immediately from (9.5). ┘

Given $\mu, \nu \in M(P)$ we shall write $\mu \prec \nu$ if $\mu(p) \leq \nu(p)$ for every $p \in P$. For example, if $B \subset A \subset X$ and $\mu \in M(P)$ then obviously $\mu^B \prec \mu^A$. By (I.1.5), "$\prec$" is an order relation on $M(P)$. It is called *specific order*.

9.8. PROPOSITION. Let $A \subset X$, $B \in \mathcal{B}(X)$ and $\mu \in M(P)$ such that $\mu(b(A) \cap (B \setminus b(B))) = 0$. Then

$$\mu^{A\cup B} + \mu^A|_B + (\mu^A|_{\complement B})^B \prec \mu^A + \mu^B .$$

Proof. We may assume that B is finely closed. By (9.3), $\mu^{A\cup B} \leq \mu^A + \mu^B$, hence

$$\nu := (\mu^A + \mu^B - \mu^{A\cup B})|_{\complement B} = \mu^A|_{\complement B} - \mu^{A\cup B}|_{\complement B}$$

is a positive measure on X. Therefore $\nu^B \prec \nu$, i.e.

$$(\mu^A|_{\complement B})^B + \mu^{A \cup B}|_{\complement B} \prec \mu^A|_{\complement B} + (\mu^{A \cup B}|_{\complement B})^B .$$

Adding $\mu^A|_B + \mu^{A \cup B}|_B$ and applying (9.5) which is possible since

$b(A \cup B) \diagdown b(B) = b(A) \diagdown b(B)$ we thus obtain that

$$\mu^{A \cup B} + \mu^A|_B + (\mu^A|_{\complement B})^B \prec \mu^A + \mu^{A \cup B}|_B + (\mu^{A \cup B}|_{\complement B})^B = \mu^A + \mu^B .$$ ⌐

9.9. PROPOSITION. Let A and B be subsets of X and $x \in X$. Then $\varepsilon_x^B \prec \varepsilon_x^A$

if and only if $\varepsilon_x^{A \cup B} = \varepsilon_x^A$.

Proof. If $\varepsilon_x^{A \cup B} = \varepsilon_x^A$ then $\varepsilon_x^B \prec \varepsilon_x^A$ since trivially $\varepsilon_x^B \prec \varepsilon_x^{A \cup B}$. If $\varepsilon_x^A = \varepsilon_x$

then of course $\varepsilon_x^{A \cup B} = \varepsilon_x$. So let $x \in \complement b(A)$ and $\varepsilon_x^B \prec \varepsilon_x^A$. We may suppose that

B is a finely closed Borel set. Then by (1.9) for every $p \in P$,

$$\int p \, d\varepsilon_x^B = \int R_p^B \, d\varepsilon_x^B = \inf_{U \text{ open } \supset B} \int R_p^U \, d\varepsilon_x^B$$

$$\leq \inf_{U \text{ open } \supset B} \int R_p^U \, d\varepsilon_x^A = \int R_p^B \, d\varepsilon_x^A ,$$

i.e. $\varepsilon_x^B \prec \varepsilon_x^A|_B + (\varepsilon_x^A|_{\complement B})^B$. Thus $\varepsilon_x^{A \cup B} \prec \varepsilon_x^A$ by (9.8) and hence $\varepsilon_x^{A \cup B} = \varepsilon_x^A$. ⌐

9.10. PROPOSITION. Let $\mathbf{X}$ be an associated Hunt process and let $A, B \in B(X)$,

$x \in X$. Then $\varepsilon_x^B \prec \varepsilon_x^A$ if and only if $T_A \leq T_B$ P^x-a.s. In particular, $\varepsilon_x^A = \varepsilon_x^B$

if and only if $T_A = T_B$ P^x-a.s.

Proof. If $T_A \leq T_B$ P^x-a.s. then for every $q \in P$ by (3.16) and (3.4)

$$\varepsilon_x^B(q) = P_{T_B} q(x) \leq P_{T_A} q(x) = \varepsilon_x^A(q) ,$$

i.e. $\varepsilon_x^B \prec \varepsilon_x^A$.

Suppose now conversely that $\varepsilon_x^B \prec \varepsilon_x^A$ and let $C = A \cup B$. Then $\varepsilon_x^C = \varepsilon_x^A$ by

(9.9). Since $T_C = \inf(T_A , T_B)$ we have to show that $T_A = T_C$ P^x-a.s. If $x \in b(A)$

then $T_C \leq T_A = 0$ P^x-a.s. So let us assume that $x \notin b(A)$. Then $\varepsilon_x^C(\{x\}) = \varepsilon_x^A(\{x\})$

< 1 and hence $\varepsilon_x^{C \diagdown \{x\}} = \varepsilon_x^{A \diagdown \{x\}}$ by (9.2). Using (3.13) and (3.15.1) we obtain

that

$$D_{C \diagdown \{x\}} = D_{A \diagdown \{x\}} \quad P^x\text{-a.s.}$$

Hence $T_C = D_{C \diagdown \{x\}} = D_{A \diagdown \{x\}} = T_A$ P^x-a.s. if $x \notin C$. If $x \in A$ then

$$T_C = \inf(D_{C \diagdown \{x\}}, T_{\{x\}}) = \inf(D_{A \diagdown \{x\}}, T_{\{x\}}) = T_A \quad P^x\text{-a.s.}$$

Therefore it remains to consider the case $x \in C \setminus (A \cup b(A))$. Then $\varepsilon_x^C(\{x\}) = \varepsilon_x^A(\{x\}) = 0$ by (4.6), hence $P^x[X_{T_C} = x] = 0$ by (3.16) and thus $P^x[T_C = T_{\{x\}}] = 0$,

$$T_C = \inf(D_{C \setminus \{x\}}, T_{\{x\}}) = D_{C \setminus \{x\}} = D_{A \setminus \{x\}} = T_A \quad P^x\text{-a.s.}$$

9.11. EXERCISES. 1. For all subsets A,B of X and every $\mu \in M(P)$, $\mu^{A \cup B} \leq \sup(\mu^A, \mu^B)$. (Hint: Use (9.4) or (3.16).)

2. Let A be an absorbing subset of X , $E \subset X$, and $\mu \in M(P)$. Then $\mu^{E \setminus A} = \mu^E|\complement A$ and $\mu^{E \setminus A} + \mu^{E \cap A} = \mu^E + (\mu^{E \setminus A})^{E \cap A}$. (Hint: Use (9.5), (9.7).)

3. Suppose that $(X,\mathcal{W})$ has the local truncation property. Let $A \subset X$ and $x \in \complement \bar{A}^f$. Then $\varepsilon_x^A = \varepsilon_x^{\bar{A}^f \cap A^*}$.

4. Give an analytic proof for the statement of (4.18.4). (Hint: Given $x \in b(\mathring{A}) \cap b(\complement \bar{A})$, choose an exhaustion (K_n) of $\complement \bar{A}$ and show that $\lim_{n\to\infty} \varepsilon_x^{K_n}(R_p^{\mathring{A}}) \leq \hat{R}_p^{A^*}(x)$ for every $p \in P$.)

5. Let $(\mathbb{R}^{n+1},\mathcal{W})$ be the harmonic space associated with the heat equation, let $B \in \mathcal{B}(X)$, and $\tau \in \mathbb{R}$. Then $B \times \{\tau\}$ is polar if and only if $\lambda^n(B) = 0$. (Hint: Use (9.4).)

10. Convergence of Balayage Measures

For every $\mu \in M(P)$ let P_μ denote the convex cone of all $p \in P$ such that $\mu(p) < \infty$. Then P_μ is a function cone.

Indeed, by definition of $M(P)$ there exists a strictly positive $p_0 \in P$ such that $\mu(p_0) < \infty$. Moreover, P_μ is linearly separating since P is linearly separating and for every $q \in P$, the sequence $(\inf(q,np_0))$ in P_μ is increasing to q . Finally, given $p \in P_\mu$, we may choose a decreasing sequence (p_n) in P such that $p_n \leq p$ and $\{p_n < p\}$ is relatively compact for every $n \in \mathbb{N}$, and $\inf p_n = 0$. If (K_n) is an exhaustion of X we may assume without loss of generality that $p_n < 2^{-n}$ on K_n and $\mu(p_n) < 2^{-n}$ for every $n \in \mathbb{N}$. Then $q = \sum_{n=1}^{\infty} p_n \in P_\mu$ and $p \in o(q)$.

Obviously, P_μ contains all $p \in P$ with compact support $C(p)$ and by (I.1.5) there are strict potentials $p \in P_\mu$.

10.1. LEMMA. Let P' be a function cone on X and let F be a filter on $M_+(X)$ such that $\lim_{\mu,F} \mu(p)$ exists for every $p \in P'$. Then there exists a unique measure $\nu \in M_+(X)$ such that $\lim_{\mu,F} \mu(p) = \nu(p)$ for every $p \in P'$. Moreover, $\lim_{\mu,F} \mu(f) = \nu(f)$ for every $f \in C_{P'}(X)$.

Proof. The first part of the statement follows immediately from (I.1.4) and the second part by (I.1.3). ⌟

10.2. PROPOSITION. Let A be a subset of X and $\mu \in M(P)$. Let (A_n) be an increasing sequence of subsets of A or a decreasing sequence of subsets of X containing A such that $\lim_{n\to\infty} \mu^{A_n}(p) = \mu^A(p)$ for some strict potential $p \in P_\mu$. Then $\lim_{n\to\infty} \mu^{A_n}(f) = \mu^A(f)$ for every $f \in C_{P_\mu}(X)$.

Proof. By (10.1) there exists a measure $\nu \in M(P)$ such that $\lim_{n\to\infty} \mu^{A_n}(f) = \nu(f)$ for every $f \in C_{P_\mu}(X)$. Obviously the assumption implies that $\nu(q) \leq \mu^A(q)$ for every $q \in P$ or $\nu(q) \geq \mu^A(q)$ for every $q \in P$. Moreover $\nu(p) = \mu(p)$. Thus $\nu = \mu$ by definition of a strict potential. ⌟

10.3. COROLLARY. Let (A_n) be an increasing sequence of subsets of X, $A = \bigcup_{n=1}^{\infty} A_n$ and $\mu \in M(P)$. Then $\lim_{n\to\infty} \mu^{A_n}(f) = \mu^A(f)$ for every $f \in C_{P_\mu}(X)$.

Proof. For every $p \in P_\mu$, $\lim_{n\to\infty} \mu^{A_n}(p) = \lim_{n\to\infty} \int \hat{R}_p^{A_n}\, d\mu = \int \hat{R}_p^A\, d\mu = \mu^A(p)$. ⌟

10.4. COROLLARY. Let A be a Borel subset of X. Then there exists an increasing sequence (K_n) of compact subsets of A such that, for every $\mu \in M(P)$ and every $f \in C_{P_\mu}(X)$, $\lim_{n\to\infty} \mu^{K_n}(f) = \mu^{\bigcup_{n=1}^{\infty} K_n}(f) = \mu^A(f)$.

Proof. Let $p \in P$ be a strict potential. By (1.9) there exists an increasing sequence (K_n) of compact subsets of A such that $\lim_{n\to\infty} \hat{R}_p^{K_n} = \hat{R}_p^A$. Let $A' = \bigcup_{n=1}^{\infty} K_n$. Using (10.2) we conclude that $\lim_{n\to\infty} \hat{R}_q^{K_n} = \hat{R}_q^A$, hence $\lim_{n\to\infty} \mu^{K_n}(q) = \mu^{A'}(q) = \mu^A(q)$ for every $q \in P$, and finally $\lim_{n\to\infty} \mu^{K_n}(f) = \mu^{A'}(f) = \mu^A(f)$ for every $f \in C_{P_\mu}(X)$. ⌟

10.5. COROLLARY. Let A be a subset of X and $\mu \in M(P)$ such that $\mu^*(A \setminus b(A)) = 0$. Then there exists a decreasing sequence (U_n) of open neighborhoods of A such that $\lim_{n\to\infty} \mu^{U_n}(f) = \mu^A(f)$ for every $f \in C_{P_\mu}(X)$.

Proof. By (2.2) and (4.1) we may assume that A is a Borel set. Let $p \in P_\mu$ be a strict potential. Then by (1.9) there exists a decreasing sequence (U_n) of open neighborhoods of A such that

$$\lim_{n\to\infty} \int R_p^{U_n}\, d\mu = \int R_p^A\, d\mu .$$

Since $R_p^A = \hat{R}_p^A$ on $b(A) \cup \complement A$ and $\mu(A\setminus b(A)) = 0$ we obtain $\lim_{n\to\infty} \mu^{U_n}(p) = \mu^A(p)$. It now suffices to apply (10.2). ∎

10.6. PROPOSITION. Let A be a subset of X, (A_n) a decreasing sequence of subsets of X and $\mu \in M(P)$ such that $\inf \mu^{A\cup A_n}(p) = \mu^A(p)$ for some strict potential $p \in P_\mu$. Then $\lim_{n\to\infty} \mu^{A\cup A_n}(f) = \mu^A(f)$ for every $f \in C_{P_\mu}(X)$ and the sequence $(\mu^{A\cup A_n}|\complement\bar{A}_1)$ increases to $\mu^A|\complement\bar{A}_1$.

Proof. For every $q \in P_\mu$ the sequence $(\mu^{A\cup A_n}(q))$ is decreasing and $\lim_{n\to\infty} \mu^{A\cup A_n}(q) \geq \mu^A(q)$. Hence by (10.1) there exists a measure $\nu \in M(P)$ such that $\lim_{n\to\infty} \mu^{A\cup A_n}(f) = \nu(f)$ for every $f \in C_{P_\mu}(X)$. In particular, $\mu^A \prec \nu$ and $\mu^A(p) = \nu(p)$. Hence $\mu^A = \nu$.

Let B be a Borel subset of $\complement\bar{A}_1$ and let $C \subset D \subset A_1$. Then by (9.3)

$$\mu^{A\cup D}(B) \leq \mu^{A\cup C}(B) + \mu^{D\setminus(A\cup C)}(B) = \mu^{A\cup C}(B)$$

since $\mu^{D\setminus(A\cup C)}$ is supported by $\bar{A}_1$. Hence the sequence $(\mu^{A\cup A_n}(B))$ is increasing and majorized by $\mu^A(B)$. Since $\lim_{n\to\infty} \mu^{A\cup A_n}(f) = \mu^A(f)$ for every $f \in C_{P_\mu}(X)$ we conclude easily that $\lim_{n\to\infty} \mu^{A\cup A_n}(B) = \mu^A(B)$. ∎

10.7. COROLLARY. Let A be a Borel subset of X and $\mu \in M(P)$ such that $\mu(A\setminus b(A)) = 0$. Let U be an open subset of X. Then there exists a decreasing sequence (U_n) of open subsets of X having the following properties:

(1) $A \cap U \subset U_n \subset U$ for every $n \in \mathbb{N}$.

(2) $\lim_{n\to\infty} \mu^{A\cup U_n}(f) = \mu^A(f)$ for every $f \in C_{P_\mu}(X)$.

(3) The sequence $(\mu^{A\cup U_n}|\complement\bar{U})$ increases to $\mu^A|\complement\bar{U}$.

Proof. Let $p \in P_\mu$ be strict. By (10.5) there exists a decreasing sequence (V_n) of open neighborhoods of A such that $\lim_{n\to\infty} \mu^{V_n}(p) = \mu^A(p)$. For every $n \in \mathbb{N}$ let $U_n = U \cap V_n$. Then $A \cap U \subset U_n \subset U$ and $A \subset A \cup U_n \subset V_n$. Hence

$\inf \mu^{A \cup U_n}(p) = \mu^A(p)$, and (2) and (3) follow from (10.2) and (10.6). ∎

10.8. PROPOSITION. Let (A_n) be a decreasing sequence of Borel subsets of X and $A \subset \bigcap_{n=1}^{\infty} A_n$ such that $\lim_{n\to\infty} R_p^{A_n} = R_p^A$ for some strict $p \in P$. Let $\mu \in M(P)$ such that $\mu(\bigcap_{n=1}^{\infty} A_n \cap (b(A_n) \setminus b(A))) = 0$. Then $\lim_{n\to\infty} \mu^{A_n}(f) = \mu^A(f)$ for every $f \in C_{P_\mu}(X)$.

Proof. Let $p \in P$ be a strict potential such that $\lim_{n\to\infty} R_p^{A_n} = R_p^A$ and let $A' = \bigcap_{n=1}^{\infty} A_n$. Then $R_p^A = R_p^{A'}$. In particular, $b(A) = b(A')$. If $m \in \mathbb{N}$ and $x \in \complement(A_m \cap (b(A_m) \setminus b(A)))$ then by (9.4)

$$\lim_{n\to\infty} \hat{R}_p^{A_n}(x) = \lim_{n\to\infty} \int R_p^{A_n} \, d\varepsilon_x^{A_m} = \int R_p^{A'} \, d\varepsilon_x^{A_m} = \hat{R}_p^{A'}(x) = \hat{R}_p^A(x) .$$

hence $\lim_{n\to\infty} \hat{R}_q^{A_n}(x) = \hat{R}_q^A(x)$ for every $q \in P$ by (10.2). Therefore

$$\lim_{n\to\infty} \int q \, d\mu^{A_n} = \lim_{n\to\infty} \int \hat{R}_q^{A_n} \, d\mu = \int \hat{R}_q^A \, d\mu = \int q \, d\mu^A$$

for every $q \in P_\mu$. Another application of (10.2) finishes the proof. ∎

10.9. REMARK. By (1.2) the first condition is satisfied if for every neighborhood U of A there exists $n \in \mathbb{N}$ such that $A_n \subset U$.

10.10. EXERCISE. Let (A_n) be an increasing sequence of subsets of X and $x \in b(\bigcup_{n=1}^{\infty} A_n)$. Then $\lim_{n\to\infty} (\varepsilon_x^{A_n})^B = \varepsilon_x^B$ for every $B \subset X$.

11. Accumulation Points of Balayage Measures

Throughout this section let A be a subset of X and $z \in X$. We intend to study the behavior of ε_x^A as x tends to z . Let us first consider the simple case $z \in \complement\bar{A}$ or $z \in b(A)$.

11.1. PROPOSITION. If $z \in \complement\bar{A}$ then $\lim_{x\to z} \varepsilon_x^A(f) = \varepsilon_z^A(f)$ for every $f \in C_P(X)$.

Proof. By (2.3), $\lim_{x\to z} \varepsilon_x^A(p) = \varepsilon_z^A(p)$ for every $p \in P$. It now suffices to apply (10.1). ∎

11.2. PROPOSITION. The following statements are equivalent:

(1) A is not thin at z .

(2) $\lim_{x \to z} \varepsilon_x^A(f) = f(z)$ for every $f \in C_p(X)$.

(3) $z \in \overset{\circ}{A}$ or $\lim_{x \to z, x \in \complement A} \varepsilon_x^A(f) = f(z)$ for every $f \in C_p(X)$.

Proof. (1) ⇒ (2): If $\varepsilon_z^A = \varepsilon_z$ then for every $p \in P$,

$$p(z) = \hat{R}_p^A(z) = \liminf_{x \to z} \hat{R}_p^A(x) \leq \limsup_{x \to z} \hat{R}_p^A(x) \leq \limsup_{x \to z} p(x) = p(z)$$

and hence $\lim_{x \to z} \varepsilon_x^A(p) = p(z)$. Using (10.1) we obtain (2).

(2) ⇒ (3): Obvious.

(3) ⇒ (1): If $z \in \overset{\circ}{A}$ then A is not thin at z . So let $z \in \complement\overset{\circ}{A}$ and $\lim_{x \to z, x \in \complement A} \varepsilon_x^A(f) = f(z)$ for every $f \in C_p(X)$. Let $p \in P$. Then by (2.3)

$$\lim_{x \to z, x \in \complement A} R_p^A(x) = \lim_{x \to z, x \in \complement A} \hat{R}_p^A(x) = p(z) .$$

Since $R_p^A = p$ on A we therefore have

$$\lim_{x \to z} R_p^A(x) = p(z) ,$$

i.e. $\hat{R}_p^A(z) = p(z)$. Thus A is not thin at z . ┘

If F is a filter on X converging to z and $p \in P$ such that $\lim_F \hat{R}_p^A$ exists then of course

$$\hat{R}_p^A(z) = \liminf_{x \to z} \hat{R}_p^A(x) \leq \lim_F \hat{R}_p^A \leq \lim_F p = p(z) ,$$

and hence obviously

$$\lim_F \hat{R}_p^A = \alpha\, p(z) + (1 - \alpha)\, \hat{R}_p^A(z)$$

for some $\alpha \in [0,1]$. If $z \notin b(A)$ and p is strict then α is uniquely determined by the preceding equality. The following two propositions will establish the surprising fact that α does not depend on the choice of p .

11.3. THEOREM. Let F be an ultrafilter on X converging to z . Then there exists $\alpha \in [0,1]$ such that for every $f \in C_p(X)$,

$$\lim_{x,F} \varepsilon_x^A(f) = (\alpha\varepsilon_z + (1 - \alpha)\varepsilon_z^A)(f) .$$

Proof. For every $q \in P$

$$0 \leq \lim_{x,F} \varepsilon_x^A(q) \leq \lim_{x,F} q(x) = q(z) < \infty .$$

Hence by (10.1) there exists a measure $\lambda \in M(P)$ such that

$$\lim_{X,F} \varepsilon_x^A(f) = \lambda(f)$$

for every $f \in C_p(X)$. Let $\beta = \lambda(\{z\})$ and

$$\mu = \lambda - \beta\varepsilon_z .$$

We intend to show that $\mu = (1 - \beta)\, \varepsilon_z^{A\setminus\{z\}}$. Let us note first that obviously

$$\int u \, d\mu = \int u \, d\lambda - \beta u(z) \leq (1 - \beta)\, u(z)$$

for every $u \in P$ and hence for every $u \in W$. In particular $\beta \leq 1$.

Let V be a closed neighborhood of z . Then for every $f \in K^+(X)$ such that $f = 0$ on V and every $x \in X$ by (9.3)

$$\varepsilon_x^A(f) \leq \varepsilon_x^{A\setminus V}(f) + \varepsilon_x^{A\cap V}(f) = \varepsilon_x^{A\setminus V}(f)$$

and hence

$$\mu(f) = \lambda(f) = \lim_{X,F} \varepsilon_x^A(f) \leq \lim_{X,F} \varepsilon_x^{A\setminus V}(f) = \varepsilon_z^{A\setminus V}(f) .$$

Therefore

$$\mu|\complement V \leq \varepsilon_z^{A\setminus V} .$$

Consider now $p \in P$, $\varepsilon > 0$ and $u \in W$ such that $u = p$ on $A\setminus\{z\}$, $u \leq p$ and $u(z) < R_p^{A\setminus\{z\}}(z) + \varepsilon$. We note that $R_p^{A\setminus\{z\}}(z) = \hat{R}_p^{A\setminus\{z\}}(z)$ by (2.3). Let (V_n) be a decreasing sequence of closed neighborhoods of z such that $\bigcap_{n=1}^{\infty} V_n = \{z\}$. Then for every $n \in \mathbb{N}$

$$\int_{\complement V_n} (p - u)\, d\mu \leq \int_{\complement V_n} (p - u)\, d\varepsilon_z^{A\setminus V_n}$$

$$\leq \int (p - u)\, d\varepsilon_z^{A\setminus V_n} = \hat{R}_p^{A\setminus V_n}(z) - \hat{R}_u^{A\setminus V_n}(z) = 0 .$$

Therefore

$$\int p \, d\mu \leq \int u \, d\mu \leq (1 - \beta)\, u(z) \leq (1 - \beta)\, \hat{R}_p^{A\setminus\{z\}}(z) + \varepsilon .$$

On the other hand let (p_n) be an increasing sequence in P such that $u = \sup p_n$. Then by (2.3) for every $n \in \mathbb{N}$

$$\hat{R}_{p_n}^{A\setminus V_n}(z) = \lim_F \hat{R}_{p_n}^{A\setminus V_n} \leq \lim_F \hat{R}_{p_n}^A = \int p_n \, d\lambda$$

$$= \beta\, p_n(z) + \int p_n \, d\mu \leq \beta\, u(z) + \int p \, d\mu$$

and hence

$$\hat{R}_p^{A\setminus\{z\}}(z) = \hat{R}_u^{A\setminus\{z\}}(z) = \sup \hat{R}_{p_n}^{A\setminus V_n}(z)$$

$$\leq \beta\, u(z) + \int p\, d\mu \leq \beta\, \hat{R}_p^{A\setminus\{z\}}(z) + \int p\, d\mu + \varepsilon ,$$

i.e. $(1-\beta)\, \hat{R}_p^{A\setminus\{z\}}(z) \leq \int p\, d\mu + \varepsilon$. Thus

$$\int p\, d\mu = (1-\beta)\, \hat{R}_p^{A\setminus\{z\}}(z) ,$$

$$\mu = (1-\beta)\, \varepsilon_z^{A\setminus\{z\}} , \quad \lambda = \beta\varepsilon_z + (1-\beta)\, \varepsilon_z^{A\setminus\{z\}} .$$

If $\beta = 1$ then $\lambda = \varepsilon_z = 1\cdot\varepsilon_z + 0\cdot\varepsilon_z^A$. So let $\beta < 1$. Then $\lambda \neq \varepsilon_z$ and hence $\varepsilon_z^{A\setminus\{z\}} \neq \varepsilon_z$. We choose a potential $q \in P$ such that $\hat{R}_q^{A\setminus\{z\}}(z) < q(z)$. Defining $\beta_o = \varepsilon_z^A(\{z\})$ we have

$$\varepsilon_z^A = \beta_o\, \varepsilon_z + (1-\beta_o)\, \varepsilon_z^{A\setminus\{z\}}$$

by (9.2) (or by the first part of the proof using the ultrafilter containing $\{z\}$).

Since

$$\beta_o(q(z) - \hat{R}_q^{A\setminus\{z\}}(z)) + \hat{R}_q^{A\setminus\{z\}}(z) = \hat{R}_q^A(z) \leq \lim_F \hat{R}_q^A = \int q\, d\lambda$$

$$= \beta(q(z) - \hat{R}_q^{A\setminus\{z\}}(z)) + \hat{R}_q^{A\setminus\{z\}}(z)$$

we conclude that $\beta_o \leq \beta < 1$. Thus $\alpha := \dfrac{\beta - \beta_o}{1 - \beta_o} \in [0,1[$ and

$$\alpha\, \varepsilon_z + (1-\alpha)\, \varepsilon_z^A = \frac{1}{1-\beta_o}\, ((\beta - \beta_o)\varepsilon_z + (1-\beta)[\beta_o\varepsilon_z + (1-\beta_o)\varepsilon_z^{A\setminus\{z\}}])$$

$$= \beta\varepsilon_z + (1-\beta)\, \varepsilon_z^{A\setminus\{z\}} = \lambda . \qquad \lrcorner$$

<u>11.4. PROPOSITION.</u> Let F be a filter on X converging to z and $\alpha \in [0,1]$. Then the following statements are equivalent:

(1) $\lim_F \hat{R}_p^A = \alpha p(z) + (1-\alpha)\, \hat{R}_q^A(z)$ for some strict $p \in P$.

(2) $\lim_{x,F} \varepsilon_x^A(f) = (\alpha\varepsilon_z + (1-\alpha)\varepsilon_z^A)(f)$ for every $f \in C_P(X)$.

(3) $\lim_{x,F} \varepsilon_x^A = \alpha\varepsilon_z + (1-\alpha)\varepsilon_z^A$.

Proof. Let $\lambda = \alpha\varepsilon_z + (1-\alpha)\varepsilon_z^A$ and let F' be an ultrafilter on X which is finer than F . By (11.3) there exists $\alpha' \in [0,1]$ such that $\lambda' = \alpha'\varepsilon_z + (1-\alpha')\varepsilon_z^A$ satisfies $\lim_{x,F'} \varepsilon_x^A(f) = (\alpha'\varepsilon_z + (1-\alpha')\varepsilon_z^A)(f)$ for every $f \in C_P(X)$. In particular, $\lim_{x,F'} \varepsilon_x^A = \lambda'$ and $\lim_{F'} \hat{R}_p^A = \lambda'(p)$ for every $p \in P$. Moreover, $\lambda \prec \lambda'$ if $\alpha \leq \alpha'$ and $\lambda' \prec \lambda$ if $\alpha' \leq \alpha$. Thus $\lambda = \lambda'$ if (1) or (3) holds, i.e. (1) implies (2) and (3) implies (2). Conversely, (1) and (3) are trivial consequences of (2). ┘

11.5. REMARK. If $\alpha = 0$ or $\alpha = 1$ the proof of the equivalences in (11.4) does not require (11.3). Indeed, if F' is an ultrafilter which is finer than F then by (10.1) there exists $\lambda' \in M(P)$ such that $\lim_{x,F} \varepsilon_x^A(f) = \lambda'(f)$ for every $f \in C_P(X)$. Since obviously $\varepsilon_z^A \prec \lambda' \prec \varepsilon_z$ we may conclude as before. Thus the following two corollaries can easily be obtained without using the strong result (11.3).

11.6. COROLLARY. If $z \in \overset{\circ}{\complement A}$ then there exists a sequence (x_n) in $\complement A$ such that $\lim\limits_{n\to\infty} x_n = z$ and $\lim\limits_{n\to\infty} \varepsilon_{x_n}^A = \varepsilon_z^A$.

Proof. Let $p \in P$ be a strict potential. Since $R_p^A = p$ on A and $\lim\limits_{x\to z} p(x) = p(z) \geq \hat{R}_p^A(z)$ we have

$$\hat{R}_p^A(z) = \liminf_{x\to z} R_p^A(x) = \liminf_{x\to z, x\in\complement A} R_p^A(x) = \liminf_{x\to z, x\in\complement A} \hat{R}_p^A(x) .$$

Hence there exists a sequence (x_n) in $\complement A$ such that $\lim\limits_{n\to\infty} x_n = z$ and $\hat{R}_p^A(z) = \lim\limits_{n\to\infty} \hat{R}_p^A(x_n)$. Thus the statement follows from (11.4). ┘

11.7. COROLLARY. Let $B \subset X$ such that $z \in \bar{B}$ and $\limsup\limits_{x\to z, x\in B} \hat{R}_p^A(x) = p(z)$ for some strict $p \in P$. Then there exists a sequence (x_n) in B such that $\lim\limits_{n\to\infty} \varepsilon_{x_n}^A = \varepsilon_z$.

Proof. It suffices to choose a sequence (x_n) in B such that $\lim\limits_{n\to\infty} x_n = z$ and $\lim\limits_{n\to\infty} \hat{R}_p^A(x_n) = p(z)$ and to apply (11.4). ┘

11.8. PROPOSITION. Let B be a subset of X such that $A \cap U = B \cap U$ for some neighborhood U of z. Let (x_n) be a sequence in X converging to z and $\alpha \in [0,1]$. Then $(\varepsilon_{x_n}^A)$ converges to $\alpha\varepsilon_z + (1-\alpha)\varepsilon_z^A$ if and only if $(\varepsilon_{x_n}^B)$ converges to $\alpha\varepsilon_z + (1-\alpha)\varepsilon_z^B$.

Proof. We may suppose that A and B are finely closed Borel sets. Let U be an open neighborhood of z such that $A \cap U = B \cap U$ and let $C = A \cup \complement U = B \cup \complement U$.

Let $p \in P$ be a strict potential and let (y_n) be a subsequence of (x_n) such that the sequence $(\hat{R}_p^C(y_n))$ is convergent. Then by (11.4) there exists

$\gamma \in [0,1]$ such that defining

$$\lambda = \gamma\, \varepsilon_z + (1-\gamma)\varepsilon_z^C$$

we have $\lim_{n\to\infty} \varepsilon_{y_n}^C(f) = \lambda(f)$ for every $f \in C_p(X)$. In order to prove the proposition it suffices to show that

$$\lim_{n\to\infty} \hat{R}_p^A(y_n) = \gamma p(z) + (1-\gamma)\hat{R}_p^A(z)$$

and

$$\lim_{n\to\infty} \hat{R}_p^B(y_n) = \gamma p(z) + (1-\gamma)\hat{R}_p^B(z) .$$

Let $\delta > 0$ and let K be a compact neighborhood of z in U . There exist a compact subset L of A and a compact subset M of $\complement(A \cup U)$ such that

$$\int (\hat{R}_p^A - \hat{R}_p^L)\, d\varepsilon_z^{C\setminus K} + \int_{\complement(L\cup M)} p\, d\varepsilon_z^{C\setminus K} < \delta .$$

Since by (2.10) the function

$$h : x \longmapsto \int (\hat{R}_p^A - \hat{R}_p^L)\, d\varepsilon_x^{C\setminus K} + \int_{\complement(L\cup M)} p\, d\varepsilon_x^{C\setminus K}$$

is continuous on $\overset{\circ}{K}$ there exists a neighborhood V of z such that $V \subset \overset{\circ}{K}$ and $h(x) < \delta$ for every $x \in V$ and hence by (9.4)

$$\int_{\complement K} (\hat{R}_p^A - \hat{R}_p^L)\, d\varepsilon_x^C + \int_{\complement(K\cup L\cup M)} p\, d\varepsilon_x^C \leq h(x) < \delta .$$

Since $\hat{R}_p^L$ is continuous on M there exists a function $f \in C_p(X)$ such that $0 \leq f \leq p$, $f = \hat{R}_p^L$ on M and $f = p$ on $K \cup L$. Let $x \in V$. Then by (9.4) $\hat{R}_p^A(x) = \int R_p^A\, d\varepsilon_x^C$. Since

$$|(f - R_p^A)1_C| \leq p\, 1_{\complement(K\cup L\cup M)} + (\hat{R}_p^A - \hat{R}_p^L)1_M$$

we obtain that

$$|\hat{R}_p^A(x) - \varepsilon_x^C(f)| < \delta .$$

Now there exists $n_0 \in \mathbb{N}$ such that for every $n \geq n_0$, $y_n \in V$ and

$$|\varepsilon_{y_n}^C(f) - \lambda(f)| < \delta ,$$

hence

$$|\hat{R}_p^A(y_n) - (\gamma p(z) + (1-\gamma)\hat{R}_p^A(z))|$$

$$\leq |\varepsilon_{y_n}^C(f) - (\gamma p(z) + (1-\gamma)\varepsilon_z^C(f))| + 2\delta = |\varepsilon_{y_n}^C(f) - \lambda(f)| + 2\delta < 3\delta .$$

Thus

$$\lim_{n\to\infty} \hat{R}_p^A(y_n) = \gamma p(z) + (1-\gamma)\hat{R}_p^A(z) .$$

Analogously

$$\lim_{n\to\infty} \hat{R}_p^B(y_n) = \gamma p(z) + (1-\gamma)\hat{R}_p^B(z) .$$

┘

<u>11.9. COROLLARY.</u> Let U be an open neighborhood of z, let (x_n) be a sequence in U converging to z and $\alpha \in [0,1]$. Then $\lim_{n\to\infty} \varepsilon_{x_n}^A = \alpha\,\varepsilon_z + (1-\alpha)\varepsilon_z^A$ if and only if $\lim_{n\to\infty} {}^U\varepsilon_{x_n}^{A\cap U} = \alpha\,\varepsilon_z + (1-\alpha)\,{}^U\varepsilon_z^{A\cap U}$.

Proof. Let $p \in P$ be a strict potential. Then by (8.3) $p' = p - R_p^{\complement U}$ is a strict potential on U and by (2.8)

$$\hat{R}_p^{A\cup\complement U} = {}^U\hat{R}_{p'}^{A\cap U} + R_p^{\complement U} .$$

Since $R_p^{\complement U}$ is continuous on U the statement follows easily from (11.8) and (11.4).

┘

<u>11.10. REMARK.</u> 1. Let (x_n) be a sequence in X converging to z. We might say that (x_n) is 1-*regular* (or simply *regular*) with respect to A if $\lim_{n\to\infty} \varepsilon_{x_n}^A = \varepsilon_z$ and α-*regular* with respect to A, $0 \le \alpha < 1$, if $\varepsilon_z \neq \varepsilon_z^A$ and $\lim_{n\to\infty} \varepsilon_{x_n}^A = \alpha\,\varepsilon_z + (1-\alpha)\varepsilon_z^A$. Then the preceding two results state that the α-regularity of (x_n) is a local property of A at z.

For every subset B of X we denote by $\Lambda_B^A(z)$ the set of all measures λ on X such that $(\varepsilon_{x_n}^A)$ converges (vaguely) to λ for some sequence (x_n) in B converging to z. We shall write $\Lambda^A(z)$ instead of $\Lambda_X^A(z)$. If $z \in \complement\bar{A}$ then $\Lambda^A(z) = \{\varepsilon_z^A\}$ by (11.1) and if $z \in b(A)$ then $\Lambda^A(z) = \{\varepsilon_z\}$ by (11.2). Moreover $\Lambda^A(z) \subset \{\alpha\varepsilon_z + (1-\alpha)\varepsilon_z^A : 0 \le \alpha \le 1\}$ by (11.3).

<u>11.11. COROLLARY.</u> If $z \in \complement\mathring{A}$ then $\Lambda_{\complement A}^A(z) = \Lambda_{\complement\mathring{A}}^A(z)$. In particular, $\Lambda^A(z) = \Lambda_{\complement A}^A(z)$ if $z \in \complement\bar{\mathring{A}}$, and $\Lambda^A(z) = \{\varepsilon_z\} \cup \Lambda_{\complement A}^A(z)$ if $z \in \bar{\mathring{A}}$.

Proof. Suppose first that $z \in \complement\mathring{A}$ and let $p \in P$ be a strict potential. Let $\lambda \in \Lambda_{\complement\mathring{A}}^A(z)$ and let (z_n) be a sequence in $\complement\mathring{A}$ such that $\lim_{n\to\infty} z_n = x$ and $\lim_{n\to\infty} \hat{R}_p^A(z_n) = \lambda(p)$. Choosing a decreasing sequence (U_n) of relatively compact open neighborhoods U_n of z such that $\bar{U}_{n+1} \subset U_n$ and $\bigcap_{n=1}^{\infty} U_n = \{z\}$ we may assume without loss of generality that $z_n \in U_n$ for every $n \in \mathbb{N}$. By (11.6), there exist $x_n \in U_n \cap \complement A$ such that $|\hat{R}_p^A(z) - \hat{R}_p^A(x_n)| < \frac{1}{n}$ for every $n \in \mathbb{N}$. Then obviously $\lim_{n\to\infty} x_n = z$ and $\lim_{n\to\infty} \hat{R}_p^A(x_n) = \lim_{n\to\infty} \hat{R}_p^A(z_n) = \lambda(p)$. Hence $\lim_{n\to\infty} \varepsilon_{x_n}^A = \lambda$ by (11.4), i.e. $\lambda \in \Lambda_{\complement A}^A(z)$. Thus $\Lambda_{\complement \mathring{A}}^A = \Lambda_{\complement A}^A(z)$.

In particular, if $z \in \complement \bar{\mathring{A}}$ then $\Lambda^A(z) = \Lambda_{\complement \mathring{A}}^A(z) = \Lambda_{\complement A}^A(z)$. Suppose finally that $z \in \bar{\mathring{A}}$. Then obviously $\Lambda_{\complement \mathring{A}}^A(z) = \{\varepsilon_z\}$ and hence $\Lambda^A(z) = \{\varepsilon_z\} \cup \Lambda_{\complement \mathring{A}}^A(z)$. Therefore $\Lambda^A(z) = \{\varepsilon_z\} \cup \Lambda_{\complement A}^A(z)$ by the first part of the proof since $\Lambda_{\complement \mathring{A}}^A(z) = \Lambda_{\complement A}^A(z) = \emptyset$ if $z \in \mathring{A}$. ┘

Let

$$T(A) := \overline{\bigcup_{x \in \complement A} \operatorname{supp}(\varepsilon_x^A)} .$$

For the remainder of this section let us assume that $(X,\mathcal{W})$ *has the local truncation property.*

Then $T(A)$ is a closed subset of the boundary A^* of A by the following proposition.

11.12. PROPOSITION. Let A be a subset of X and $\mu \in M(P)$ such that $\mu(\mathring{A}) = 0$. Then $\mu^A(\complement A^*) = 0$. Moreover, if even $\mu(b(A)) = 0$ then $\mu^A = \mu^{A^* \cap (A \cup b(A))}$.

Proof. Let K be a compact subset of $\mathring{A}$ and $p,q \in P$ such that $\mu(p) < \infty$, $q \leq p$, and $\{q < p\} \subset K$. Let $u \in \mathcal{W}$ such that $u \leq q$ on X, $u = q$ on A and define v by

$$v = \begin{cases} p & \text{on } K \\ u & \text{on } \complement K . \end{cases}$$

Then $v = p$ on A, hence $v \in {}^*\mathcal{H}^+(\mathring{A})$ and $v \in {}^*\mathcal{H}^+(\complement K)$ by (III.8.1). Therefore $v \in \mathcal{W}$. This shows that $R_p^A = R_q^A$ on $\complement K$ and hence

$$\mu^A(p-q) = \int (\hat{R}_p^A - \hat{R}_q^A)\, d\mu = 0 .$$

Therefore $\mu^A(\mathring{A}) = 0$. By (2.11), $\mu^A(\complement \bar{A}) = 0$. Hence $\mu^A(\complement A^*) = 0$. Since $\mu^A = \mu^{A \cup b(A)}$ and $(\mu^{A \cup b(A)})_* (\complement(A \cup b(A))) = 0$ by (4.6) we now conclude by (9.6) that $\mu^A = \mu^{A^* \cap (A \cup b(A))}$ if $\mu(b(A)) = 0$. ┘

11.13. COROLLARY. Let A be a subset of X. Then A is an absorbing set if and only if A is closed and finely open.

Proof. Suppose that A is closed and finely open. Fix $x \in A$. By (11.12), $\varepsilon_x^{\complement A}(\complement(\complement A)^*) = 0$ and hence $\varepsilon_x^{\complement A}(\complement A) = 0$ since $(\complement A)^* = A^* \subset A$. By (4.6), $\varepsilon_x^{\complement A}(A) = 0$ since $\complement A$ is a finely closed Borel set. Therefore $\overset{o}{\varepsilon}{}_x^{\complement A} = \varepsilon_x^{\complement A} = 0$. Thus A is an absorbing set. ∎

The following characterisation of $T(A)$ will prepare our main theorem.

<u>11.14. PROPOSITION.</u> If $z \in \overset{o}{\complement A}$ the following statements are equivalent:

(1) $\Lambda_{\complement A}^A(z) = \{\varepsilon_z^A\}$, $\varepsilon_z^A \neq \varepsilon_z$.

(2) $\varepsilon_z \notin \Lambda_{\complement A}^A(z)$.

(3) $z \notin T(A)$.

Proof. (1) ⇒ (2): Trivial.

(2) ⇒ (3): Let $p \in P$ be a strict potential. Then $\limsup_{x \to z, x \in \complement A} \hat{R}_p^A(x) < p(z)$ by (11.7). Hence there exists $\delta > 0$ and an open neighborhood V of z such that

$$\hat{R}_p^A \leq p - \delta \quad \text{on} \quad V \setminus A .$$

Let $x \in \complement A$. We intend to show that $\varepsilon_x^A(V) = 0$. To that end we may assume that A is a Borel set. By (10.7), there exists a decreasing sequence (U_n) of open neighborhoods of A such that $A \subset U_n \subset X \setminus \{x\}$ and $\lim_{n\to\infty} \varepsilon_x^{U_n}(f) = \varepsilon_x^A(f)$ for every $f \in C_P(X)$. In particular,

$$\liminf_{n \to \infty} \varepsilon_x^{U_n}(V) \geq \varepsilon_x^A(V) .$$

By (9.4), we have for every $n \in \mathbb{N}$

$$\varepsilon_x^A(p) = \int R_p^A \, d\varepsilon_x^{U_n} \leq \int p \, d\varepsilon_x^{U_n} - \delta\varepsilon_x^{U_n}(V \setminus A) = \varepsilon_x^{U_n}(p) - \delta\varepsilon_x^{U_n}(V)$$

since $\varepsilon_x^{U_n}(U_n) = 0$. Letting n tend to infinity we conclude that $\lim_{n\to\infty} \varepsilon_x^{U_n}(V) = 0$ and hence $\varepsilon_x^A(V) = 0$. Thus $z \notin T(A)$.

(3) ⇒ (1): Since obviously $\overline{b(A) \setminus A} \subset T(A)$ there exists a neighborhood U of z such that $U \cap (b(A) \setminus A) = \emptyset$. Let $x \in U \setminus A$ and $B = \bar{A}^f$. Then $x \notin b(A) = b(B)$, hence $\varepsilon_x^A = \varepsilon_x^B = \varepsilon_x^{B \cap T(A)}$ by (9.6). Therefore

$$\Lambda_{\complement A}^A(z) \subset \Lambda^{B \cap T(A)}(z) = \{\varepsilon_z^{B \cap T(A)}\} .$$

Since $\varepsilon_z^A \in \Lambda_{\complement A}^A(z)$ by (11.6) we conclude that $\Lambda_{\complement A}^A(z) = \{\varepsilon_z^A\}$ and

$\varepsilon_z^A = \varepsilon_z^{B \cap T(A)} \neq \varepsilon_z$. ∎

<u>11.15. COROLLARY.</u> Let B be a subset of X such that $A \cap U = B \cap U$ for some neighborhood U of z . Then $z \in T(A)$ if and only if $z \in T(B)$.

<u>11.16. THEOREM.</u> $\Lambda^A(z) \subset \{\varepsilon_z, \varepsilon_z^A\}$ or $\Lambda^A(z) = \{\alpha\varepsilon_z + (1-\alpha)\varepsilon_z^A : 0 \leq \alpha \leq 1\}$. More precisely, if $\Lambda^A(z) \neq \{\alpha\varepsilon_z + (1-\alpha)\varepsilon_z : 0 \leq \alpha \leq 1\}$ then there exist an open neighborhood U of z and an absorbing set W in the subspace U such that $\Lambda_W^A(z) = \{\varepsilon_z^A\}$ and $\Lambda_{\complement W}^A(z) \subset \{\varepsilon_z\}$.

Proof. We may suppose that A is a finely closed Borel set. If $z \in b(A)$ then $\Lambda^A(z) = \{\varepsilon_z\}$. So let $z \in \complement b(A)$. Then $\varepsilon_z^A \neq \varepsilon_z$ and $\varepsilon_z^A \in \Lambda^A(z)$ by (11.6). Let us assume that there exists an $\alpha \in]0,1[$ such that

$$\lambda = \alpha\,\varepsilon_z + (1-\alpha)\varepsilon_z^A \notin \Lambda^A(z) .$$

We have to show that $\Lambda^A(z) \subset \{\varepsilon_z, \varepsilon_z^A\}$.

Let $p \in P$ be a strict potential such that $p(z) = 1$. Then $\hat{R}_p^A(z) < \lambda(p) < 1$. By (11.4) there exists a relatively compact open neighborhood U of z such that $\hat{R}_p^A \neq \lambda(p)$ on U . Let

$$V = U \cap \{\hat{R}_p^A > \lambda(p)\} , \quad W = U \cap \{\hat{R}_p^A < \lambda(p)\} .$$

Then V is open, W is finely open and $W = U \setminus V$. Hence W is an absorbing set in the subspace U by (11.13). Moreover, $z \in \overline{W \setminus A}$ since $\liminf_{x \to z} R_p^A(x) = \hat{R}_p^A(z) < \lambda(p)$ whereas $R_p^A > \lambda(p)$ on V .

Let $B = A \cup \complement U$. By (11.8) and (11.4)

$$\limsup_{x \to z, x \in W} \hat{R}_p^B(x) \leq \alpha p(z) + (1-\alpha)\hat{R}_p^B(z) < p(z) .$$

Let $x \in W$. Then $\varepsilon_x^{B \cup V}(V) = 0$ and $\varepsilon_x^{B \cup V}$ is supported by $\overline{B \cup V}^f = B \cup b(V)$. Since V is finely closed in U we have $b(V) \setminus V \subset U^* \subset B$. Hence $\varepsilon_x^{B \cup V} = \varepsilon_x^B$ by (9.6). In particular,

$$\limsup_{x \to z, x \in W} \hat{R}_p^{B \cup V}(x) = \limsup_{x \to z, x \in W} \hat{R}_p^B(x) < p(z) .$$

Since $W = \complement(B \cup V)$ we conclude by (11.7) and (11.14) $\Lambda_W^{B \cup V}(z) = \{\varepsilon_z^{B \cup V}\}$. Therefore

$$\lim_{x\to z, x\in W} \varepsilon_x^B = \lim_{x\to z, x\in W} \varepsilon_x^{B\cup V} = \varepsilon_z^{B\cup V} = \varepsilon_z^B .$$

Using (11.8) again we obtain that

$$\lim_{x\to z, x\in W} \varepsilon_x^A = \varepsilon_z^A .$$

If $z \notin \bar{V}$ then $\Lambda^A(z) = \Lambda_W^A(z) = \{\varepsilon_z^A\}$. So let us assume that $z \in \bar{V}$ and let

$$a = \liminf_{x\to z, x\in V} \hat{R}_p^A(x) \quad , \quad b = \hat{R}_p^A(z) .$$

Then

$$1 \geq a > \lambda(p) > b \geq 0 .$$

We intend to show that $a = 1$, i.e. $\lim_{x\to z, x\in V} \hat{R}_p^A(x) = p(z)$ and hence $\Lambda_V^A(z) = \{\varepsilon_z\}$.

Let $C = A \cup W$ and suppose for a moment that A is closed. Then $\hat{R}_p^A$ is continuous on $\complement A$, hence $W\setminus A$ is open and $W^* \subset A \cup U^*$, $b(W)\setminus C \subset U^*$. Choosing a function $\varphi \in C_p^+(X)$ such that $\varphi \leq p$, $\varphi(z) = 1$ and $S(\varphi) \subset U$ we would obtain that for every $x \in V$

$$\varepsilon_x^C(\varphi) = \varepsilon_x^{A\cup b(W)}(\varphi) = \varepsilon_x^{A\cup(b(W)\setminus W)}(\varphi)$$

$$\leq \varepsilon_x^A(\varphi) + \varepsilon_x^{b(W)\setminus C}(\varphi) = \varepsilon_x^A(\varphi) \leq \varepsilon_x^A(p) = \hat{R}_p^A(x) \leq 1 .$$

Moreover, $\lim_{x\to z} \varepsilon_x^C(\varphi) = \varphi(z) = 1$ since $z \in W \subset b(C)$. Therefore

$$a = \liminf_{x\to z, x\in V} \hat{R}_p^A(x) = 1$$

and the proof would be finished.

However, in our general situation we have to be more careful. Let $\delta > 0$. Then there exists an open neighborhood U' of z in U such that

$$|p - 1| < \delta \quad \text{on} \quad U' ,$$

$$\hat{R}_p^A > a - \delta \quad \text{on} \quad U' \cap V ,$$

$$\hat{R}_p^A < (b + \delta)p \quad \text{on} \quad U' \cap W .$$

We may assume without loss of generality that $U' = U$. As before let $C = A \cup W$. Since $z \in W \subset b(W) \subset b(C)$ there exists a neighborhood U_δ of z in U such that for every $x \in U_\delta$,

$$\int p \, d\varepsilon_x^C > 1 - \delta \ , \ \int_{U^*} p \, d\varepsilon_x^C < \delta .$$

Let $x \in U_\delta \cap V$ and $\nu = \varepsilon_x^C$. Since W is finely closed in U the measure ν is supported by $C \cup U^*$. Hence by (9.4)

$$a - \delta < \hat{R}_p^A(x) = \int_A p\, d\nu + \int_{\complement A} \hat{R}_p^A\, d\nu \leq \int_A p\, d\nu + \int_{W \setminus A} \hat{R}_p^A\, d\nu + \int_{U^*} \hat{R}_p^A\, d\nu$$

$$\leq \int_A p\, d\nu + (b+\delta) \int_{W \setminus A} p\, d\nu + \delta \leq (1-b) \int_A p\, d\nu + (b+\delta) \int p\, d\nu + \delta$$

$$\leq (1-b) \int_A p\, d\nu + b + 2\delta$$

and therefore

$$\int_A p\, d\nu > \frac{a - b - 3\delta}{1 - b}.$$

Since ν is supported by $C \cup U^*$ and $\int_{U^*} p\, d\nu < \delta$ there exist a compact subset K of A and a compact subset L of $W \setminus A$ such that

$$\int_{K \cup L} p\, d\nu > 1 - \delta.$$

By (10.7), there exists an open subset G of X such that $L \subset G$, $\bar{G} \subset U \setminus (K \cup \{x\})$ and $\nu' := \varepsilon_x^{C \cup G}$ satisfies

$$\int_K p\, d\nu' > \int_K p\, d\nu - \delta.$$

Moreover,

$$\int_{U^*} p\, d\nu' \leq \int_{U^*} p\, d\nu < \delta.$$

By (9.4), $\nu = \nu'|_C + (\nu'|_{\complement C})^C$ where $\nu'(L) = 0$ since $L \subset G$ and $x \notin G$. Hence

$$\int_L p\, d\nu = \int_{\complement C} \varepsilon_y^C(p1_L)\nu'(dy) \leq \int_{\complement C} p\, d\nu' \leq \int_{V \setminus A} p\, d\nu' + \int_{U^*} p\, d\nu' \leq \int_{V \setminus A} p\, d\nu' + \delta.$$

Moreover, $\varepsilon_x^A = \nu'|_A + (\nu'|_{\complement A})^A$ and therefore

$$\hat{R}_p^A(x) = \int_A p\, d\nu' + \int_{\complement A} \hat{R}_p^A\, d\nu' \geq \int_K p\, d\nu' + \int_{V \setminus A} \hat{R}_p^A\, d\nu' \geq \int_K p\, d\nu' + (a-\delta) \int_{V \setminus A} p\, d\nu'$$

$$\geq \int_K p\, d\nu + a \int_L p\, d\nu - 3\delta = a \int_{K \cup L} p\, d\nu + (1-a) \int_K p\, d\nu - 3\delta$$

$$\geq a + (1-a) \int_A p\, d\nu - 4\delta \geq a + \frac{1-a}{a-b}(a-b-3\delta) - 5\delta.$$

This implies that $a \geq a + \frac{1-a}{1-b}(a - b - 3\delta) - 5\delta$, hence $a \geq a + (1-a)\frac{a-b}{1-b}$, i.e. $a = 1$. ┘

<u>11.17. COROLLARY.</u> $\Lambda_{\complement A}^A(z) \subset \{\varepsilon_z, \varepsilon_z^A\}$ or $\Lambda_{\complement A}^A(z) = \{\alpha\varepsilon_z + (1-\alpha)\varepsilon_z^A : 0 \leq \alpha \leq 1\}$.

Proof. (11.16) and (11.11). ┘

11.18. EXERCISES. 1. Suppose that $(X,\mathcal{W})$ has the local truncation property. Let U be an open subset of X and $z \in \overline{U^*} \cap b(\complement U)$. Furthermore, assume there exists a fundamental system of neighborhoods V of z such that $V \cap U$ is connected. Then $\Lambda_U^{\complement U}(z) = \{\alpha\varepsilon_z + (1-\alpha)\varepsilon_z^{\complement U} : 0 \leq \alpha \leq 1\}$.

2. Classical potential theory. If $z \notin b(A)$ then $\Lambda^A(z) = \{\varepsilon_z^A\}$ or $\Lambda^A(z) = \{\alpha\varepsilon_z + (1-\alpha)\varepsilon_z^A : 0 \leq \alpha \leq 1\}$. More precisely,

$$\overline{b(A)} \setminus b(A) = \{z \in \complement b(A) : \Lambda^A(z) = \{\alpha\varepsilon_z + (1-\alpha)\varepsilon_z^A : 0 \leq \alpha \leq 1\}\},$$

$$X \setminus \overline{b(A)} = \{z \in \complement b(A) : \Lambda^A(z) = \{\varepsilon_z^A\}\} .$$

3. For every $n \in \mathbb{N}$ let $x_n = (\frac{1}{n},0)$, $y_n = (-\frac{1}{n},0)$, $z_n = (0,\frac{1}{n})$, let X be the subspace $\{z_n : n \in \mathbb{N}\} \cup \mathbb{R} \times \{0\}$ of $\mathbb{R}^2$, and $z = (0,0)$. Define $H_{\{z_n\}}(z_n,\cdot) = \frac{1}{2}\varepsilon_{x_n} + \frac{1}{2}\varepsilon_{y_n}$ and $H_{B_{\frac{1}{n}}(z)}(x,\cdot) = \varepsilon_{y_n}$ for every $x \in B_{\frac{1}{n}}(z)$. If $a,b \in \mathbb{R}$ such that $a < b$ and $0 \notin]a,b[$ let $U =]a,b[\times \{0\}$ and $H_U(x,\cdot) = \varepsilon_{(a,0)}$ for every $x \in U$. This definition yields a family of harmonic kernels on X such that $\varepsilon_z^{\{z\}} = 0$, $\Lambda^{\{z\}}(z) = \{\varepsilon_z, \frac{1}{2}\varepsilon_z\}$!

12. Extreme Representing Measures

For every $\nu \in M(P)$ let $M_\nu(P)$ denote the set of all measures μ on X such that $\mu \prec \nu$. By (10.1), $M_\nu(P)$ is a compact convex subset of $M_+(X)$. For every $A \in \mathcal{B}(X)$ let $\mathring{\nu}^A$ be the unique measure in $M_+(X)$ such that $\int p\, d\mathring{\nu}^A = \int R_p^A\, d\nu$ for every $p \in P$. We note that $\mathring{\nu}^A = \nu|_A + (\nu|_{\complement A})^A$. Clearly, this notation is consistent with the definition of $\mathring{\varepsilon}_x^A$ given in (II.5), and $M_{\varepsilon_x}(P) = M_x(P)$ for every $x \in X$. $\mathring{\nu}^A$ is called *reduced measure* of ν on A.

In this section we shall see that the extreme points of the compact convex set $M_\nu(P)$ are given by $\mathring{\nu}^A, A \in \mathcal{B}(X)$. Moreover, it will be proved that $\mu \in M_\nu(P)$ if and only if $\mu = \nu V$ for some dilation V.

12.1. PROPOSITION. Let $\mu,\nu \in M(P)$. Then the set $\{\sigma \in M(P) : \mu \prec \sigma + \nu\}$ contains a smallest element $R_{\mu-\nu}$ with respect to $\prec$. For every $p \in P_{\mu+\nu}$,

$$R_{\mu-\nu}(p) = \sup\{\mu(q)-\nu(q) : q \in P, q \leq p\} .$$

Moreover, $R_{\mu-\nu} \leq \mu$.

Proof. For every $p \in P_{\mu+\nu}$ let

$$T(p) = \sup \{\mu(q) - \nu(q) : q \in P, q \leq p\} .$$

Obviously T is a mapping of $P_{\mu+\nu}$ into $\mathbb{R}_+$ which is positively homogeneous and increasing. From (B_3) we deduce easily that T is additive. Hence by (I.1.4) there exists a measure σ on X such that $T(p) = \int p \, d\sigma$ for every $p \in P_{\mu+\nu}$. Obviously $\mu(p) = \mu(p) - \nu(p) + \nu(p) \leq \sigma(p) + \nu(p)$ for every $p \in P_{\mu+\nu}$ and hence $\mu \prec \sigma + \nu$. If $\sigma' \in M(P)$ such that $\mu \prec \sigma' + \nu$, then

$$\begin{aligned} \sigma(p) &= \sup \{\mu(q) - \nu(q) : q \in P, q \leq p\} \\ &\leq \sup \{\sigma'(q) : q \in P, q \leq p\} \leq \sigma'(p) \end{aligned}$$

for every $p \in P_{\mu+\nu}$, hence $\sigma(p) \leq \sigma'(p)$ for every $p \in P$, i.e. $\sigma \prec \sigma'$.

Finally, let $p_1, p_2 \in P_{\mu+\nu}$ such that $p_2 \leq p_1$. Let $q_1 \in P$ such that $q_1 \leq p_1$ and let $q_2 = \inf(q_1, p_2)$. Then $q_2 \in P$, $q_2 \leq p_2$ and $0 \leq q_1 - q_2 \leq p_1 - p_2$. Therefore

$$\mu(q_1 - q_2) \leq \mu(p_1 - p_2) + \nu(q_1 - q_2)$$

and hence

$$\mu(q_1) - \nu(q_1) \leq \mu(p_1 - p_2) + \mu(q_2) - \nu(q_2) \leq \mu(p_1 - p_2) + \sigma(p_2) .$$

This shows that $\sigma(p_1) \leq \mu(p_1 - p_2) + \sigma(p_2)$, i.e. $\sigma(p_1 - p_2) \leq \mu(p_1 - p_2)$. Using (I.1.2) we conclude that $\sigma \leq \mu$. ┘

<u>12.2. LEMMA.</u> Let $\nu \in M(P)$ and let μ be an extreme point of $M_\nu(P)$. Then $R_{2\mu-\nu} = \mu$.

Proof. Let $\sigma = R_{2\mu-\nu}$. By (12.1), $\sigma \leq 2\mu$. Let $\sigma' = 2\mu - \sigma$. Since $2\mu \prec \sigma + \nu$, we know that $\sigma' \prec \nu$. Since $\mu \in M_\nu(P)$, we obtain that $2\mu(q) - \nu(q) \leq \nu(q)$ for every $q \in P_\mu$, hence $\sigma(p) \leq \nu(p)$ for every $p \in P_\mu$, i.e. $\sigma \prec \nu$. Therefore $\sigma, \sigma' \in M_\nu(P)$ and $\mu = \frac{1}{2}(\sigma + \sigma')$. Thus $\sigma = \sigma' = \mu$. ┘

<u>12.3. LEMMA.</u> Let $\nu \in M(P)$, $\mu \in M_\nu(P)$ and $A \in B(X)$ such that $\mu(\complement A) = 0$. Then $\mu \prec \overset{\circ}{\nu}{}^A$.

Proof. Let $p \in P_\nu$. By (1.9) there exists a decreasing sequence (U_n) of open neighborhoods of A such that $\inf_n \int R_p^{U_n}\, d\nu = \int R_p^A\, d\nu = \overset{\circ}{\nu}{}^A(p)$. Since $\mu(p) = \mu(R_p^{U_n}) \leq \nu(R_p^{U_n})$ for every $n \in \mathbb{N}$, we conclude that $\mu(p) \leq \overset{\circ}{\nu}{}^A(p)$. ┘

Let $(M_\nu(P))_e$ be the set of all extreme points of $M_\nu(P)$.

12.4. THEOREM (G. MOKOBODZKI). Let $\nu \in M(P)$. Then $\{\overset{\circ}{\nu}{}^A : A \in \mathcal{B}(X)\} = (M_\nu(P))_e$.

Proof. Let $A \in \mathcal{B}(X)$ and $\mu_1, \mu_2 \in M_\nu(P)$ such that $\overset{\circ}{\nu}{}^A = \frac{1}{2}(\mu_1 + \mu_2)$. Let B be the fine closure of A . Then $B \in \mathcal{B}(X)$, $\overset{\circ}{\nu}{}^B = \overset{\circ}{\nu}{}^A$, $\overset{\circ}{\nu}{}^B(\complement B) = 0$. By the preceding lemma $\mu_1 \prec \overset{\circ}{\nu}{}^B$, $\mu_2 \prec \overset{\circ}{\nu}{}^B$. Thus $\mu_1 = \mu_2 = \overset{\circ}{\nu}{}^A$ showing that $\overset{\circ}{\nu}{}^A \in (M_\nu(P))_e$.

Now fix $\mu \in (M_\nu(P))_e$ and a strict potential $p \in P_\nu$. By (12.1) and (12.2) there exists a sequence (q_n) in P such that $q_n \leq p$ and

$$\mu(p) \leq 2\mu(q_n) - \nu(q_n) + 2^{-n}$$

for every $n \in \mathbb{N}$. Using $\mu(q_n) \leq \nu(q_n)$ and $\mu(q_n) \leq \mu(p)$ we obtain that

$$\mu(p - q_n) \leq 2^{-n} \quad \text{and} \quad 0 \leq \nu(q_n) - \mu(q_n) \leq 2^{-n}$$

for every $n \in \mathbb{N}$. Let

$$t = \liminf_{n\to\infty} q_n \ , \ A = \{t = p\} \ .$$

Then $t \leq p$ and $\mu(t) = \mu(p)$, hence $\mu(\complement A) = 0$. Therefore $\mu \prec \overset{\circ}{\nu}{}^A$ by (12.3). Moreover, for every compact subset K of A , the measure $\overset{\circ}{\nu}{}^K$ is supported by K and hence

$$\overset{\circ}{\nu}{}^K(p) = \overset{\circ}{\nu}{}^K(t) \leq \liminf_{n\to\infty} \overset{\circ}{\nu}{}^K(q_n) \leq \liminf_{n\to\infty} \nu(q_n) = \liminf_{n\to\infty} \mu(q_n) = \mu(p) \ .$$

Thus $\overset{\circ}{\nu}{}^A(p) \leq \mu(p)$ by (1.9). By the definition of a strict potential we finally conclude that $\mu = \overset{\circ}{\nu}{}^A$. ┘

12.5. COROLLARY. For every $x \in X$,

$$(M_x(P))_e = \{\varepsilon_x\} \cup \{\varepsilon_x^A : A \in \mathcal{B}(X) , x \notin A\} \ .$$

Proof. Immediate consequence of (12.4) and the fact that $\overset{\circ}{\varepsilon}{}_x^A = \varepsilon_x$ if $x \in A$ and $\overset{\circ}{\varepsilon}{}_x^A = \varepsilon_x^A$ if $x \notin A$. ┘

12.6. PROPOSITION. Let $\mu, \nu \in M(P)$. Then the following statements are equivalent:

(1) $\mu \prec \nu$.

(2) There exists a dilation V such that $\mu = \nu V$.

Proof. (2) ⇒ (1): If V is a dilation then for every $p \in P$

$$(\nu V)(p) = \nu(Vp) \leq \nu(p) .$$

(1) ⇒ (2): We fix a strict potential $p \in P$ such that $\nu(p) < \infty$. Given $\rho, \sigma \in M(P)$ we shall write $\rho \lll \sigma$ if there exists a dilation V such that $\rho = \sigma V$. Obviously, $\rho \lll \sigma$ implies that $\rho \prec \sigma$. Moreover, it is easily verified that "$\lll$" is an order relation on $M(P)$. Let

$$N = \{\sigma \in M(P) : \mu \prec \sigma \lll \nu\}$$

and let N' be a non-empty subset of N which is totally ordered with respect to $\lll$. We choose a sequence (σ_n) in N' such that the sequence $(\sigma_n(p))$ is decreasing to $\alpha := \inf \{\sigma(p) : \sigma \in N'\}$. Then $\sigma_{n+1} \lll \sigma_n$ for every $n \in \mathbb{N}$. Indeed, if $n \in \mathbb{N}$ and $\sigma_n \lll \sigma_{n+1}$ then $\sigma_n \prec \sigma_{n+1}$ and $\sigma_n(p) \geq \sigma_{n+1}(p)$, hence $\sigma_{n+1} = \sigma_n$. By (10.1), there exists a measure $\sigma \in M(P)$ such that

$$\sigma(q) = \lim_{n\to\infty} \sigma_n(q)$$

for every $q \in P$. Clearly, $\mu \prec \sigma$. We claim that $\sigma \lll \rho$ for every $\rho \in N'$. Indeed, there exists a sequence (V_n) of dilations such that $\sigma_{n+1} = \sigma_n V_n$ for every $n \in \mathbb{N}$. By (II.1.2) for each $m \in \mathbb{N}$, there exists a unique kernel W_m on X such that

$$W_m q = \lim_{n\to\infty} V_m V_{m+1} \dots V_n q$$

for every $q \in P$. Of course, every W_m is a dilation. For every $m \in \mathbb{N}$ and $q \in P$,

$$\sigma(q) = \lim_{n\to\infty} \sigma_n(q) = \lim_{n\to\infty} (\sigma_m V_m \dots V_n)(q) = (\sigma_m W_m)(q) ,$$

hence $\sigma \lll \sigma_m$. Therefore $\sigma \lll \rho$ if $\rho \in N'$ such that $\sigma_m \lll \rho$ for some $m \in \mathbb{N}$. Finally, let $\rho \in N'$ such that $\rho \lll \sigma_m$ for every $m \in \mathbb{N}$. Then $\rho \prec \sigma$ and $\sigma(p) = \lim_{n\to\infty} \sigma_n(p) = \alpha \leq \rho(p)$, hence $\rho = \sigma$.

Using Zorn's lemma we thus obtain an element $\sigma \in N$ which is minimal with respect to $\lll$. We have to show that $\mu = \sigma$.

Let K be the potential kernel associated with p. Since $\mu \prec \sigma$, we have $\mu K \leq \sigma K$. So by the Radon-Nikodym theorem there exists a function $h \in B^+(X)$ such that $h \leq 1$ and $h(\sigma K) = \mu K$. Fix $0 < \eta < 1$, let $A = \{h > 1 - \eta\}$ and $B = \delta(p_A)$. Defining a dilation V by

$$V(x,\cdot) = (1 - \eta)\varepsilon_x + \eta\, \varepsilon_x^B \qquad (x \in X)$$

we have

$$\sigma V = (1 - \eta)\sigma + \eta\, \sigma^B .$$

We claim that $\mu \prec \sigma V$. Indeed, if $f \in B^+(X)$ such that $f \leq 1_A$ then $\sigma^B(Kf) = \sigma(Kf)$ since $Kf \prec p_A$ and $\delta(Kf) \subset \delta(p_A) = B$, and therefore $(\sigma V)(Kf) = \sigma(Kf) \geq \mu(Kf)$. If $f \in B^+(X)$ such that $f \leq 1_{\complement A}$ then $(\sigma V)(Kf) \geq (1 - \eta)\sigma(Kf) = (\sigma K)((1 - \eta)f) \geq (\sigma K)(hf) = (\mu K)(f) = \mu(Kf)$. So $(\sigma V)(Kf) \geq \mu(Kf)$ for every $f \in B^+(X)$. Using (II.7.12) we conclude that $\mu \prec \sigma V$. Since $\sigma V \lll \sigma$ the minimality of σ yields that $\sigma V = \sigma$, i.e. $\sigma^B = \sigma$.

Now fix $q \in P$. Since $(p_A)_A = p_A$, we know by (II.7.12) that there exists a sequence (f_n) in $B_b^+(X)$ such that the supports $S(f_n)$, $n \in \mathbb{N}$, are contained in A and the sequence (Kf_n) is increasing to R_q^B. Then $(1 - \eta)(\sigma K)(f_n) \leq (\sigma K)(hf_n) = (\mu K)(f_n)$ for every $n \in \mathbb{N}$ and therefore

$$(1 - \eta)\sigma(q) = (1 - \eta)\sigma^B(q) = \lim_{n\to\infty} (1 - \eta)\sigma(Kf_n) \leq \lim_{n\to\infty} \mu(Kf_n) = \mu^B(q) \leq \mu(q) .$$

Thus $(1 - \eta)\sigma \prec \mu$. Letting η tend to zero we finally conclude that $\sigma \prec \mu$ and hence $\sigma = \mu$. ┘

12.7. EXAMPLE. Given $\nu \in M(P)$ and $A \subset X$, we have of course $\nu^A = \nu V$ choosing $V(x,\cdot) = \varepsilon_x^A$ for every $x \in X$.

12.8. EXERCISES. 1. Let $\mu, \nu_1, \nu_2 \in M(P)$ such that $\mu \prec \nu_1 + \nu_2$. Then there exist $\mu_1, \mu_2 \in M(P)$ such that $\mu = \mu_1 + \mu_2$, $\mu_1 \prec \nu_1$, $\mu_2 \prec \nu_2$.

2. Let $\mu, \nu \in M(P)$ such that $\mu \prec \nu$. Then the set $\{\sigma \in M(P) : \mu \prec \sigma \prec \nu\}$ is a face of $M_\nu(P)$ if and only if $\mu \in (M_\nu(P))_e$.

VII. Dirichlet Problem

As an application of balayage theory we shall study various types of the Dirichlet problem in balayage spaces. The properties of Perron sets (section 1) permit to develop the generalized solution of the Dirichlet problem for open sets U by the method of Perron, Wiener, and Brelot (section 2). It turns out that this solution recovers the harmonic kernels H_U. The boundary behavior of the generalized solution is investigated in sections 3 and 4 leading among others to well known classical theorems.

In section 5 we start a different approach motivated by Choquet theory. We show by purely potential-theoretic methods that function cones which are typical for potential theory are simplicial cones. This includes a solution of the weak Dirichlet problem. Furthermore, its relation to the generalized solution is completely settled (section 6). The distinguished rôle of the generalized solution is worked out in section 7. A treatment of the Dirichlet problem for finely open sets yields an extension and improvement of the previous results and leads to a complete characterization of representing measures (section 8). As an application we obtain important approximation theorems (section 9). Results on removable singularities close this chapter (section 10).

1. Perron Sets

Let $(X,\mathcal{W})$ be a balayage space, let $\mathcal{U}$ be the set of all relatively compact open subsets of X, and let U be an open subset of X. By (III.2) the set ${}^*\mathcal{H}(U)$ of all *hyperharmonic* functions on U is defined by

$$^*\mathcal{H}(U) = \{u \in \mathcal{B}(X) : u|_U \text{ l.s.c.}, -\infty < H_V u(x) \le u(x) \quad \forall\, x \in V \in \mathcal{U}(U)\}$$

and the set $\mathcal{H}(U)$ of all *harmonic* functions on U by

$$\mathcal{H}(U) = {}^*\mathcal{H}(U) \cap (-{}^*\mathcal{H}(U)) = \{h \in \mathcal{B}(X) : h|_U \in \mathcal{C}(U), H_V h = h \;\forall\, V \in \mathcal{U}(U)\}.$$

In view of the definition of $\mathcal{S}^+(U)$ given in (III.1) we now define the set $\mathcal{S}(U)$ of all *superharmonic* functions on U by

$$\mathcal{S}(U) = \{s \in {}^*\mathcal{H}(U) : H_V s|_V \in \mathcal{C}(V) \quad \forall\, V \in \mathcal{U}(U)\}$$
$$= \{s \in {}^*\mathcal{H}(U) : H_V s \in \mathcal{H}(V) \quad \forall\, V \in \mathcal{U}(U)\}.$$

If $t \in -\mathcal{S}(U)$ then t is called *subharmonic* on U.

<u>1.1. LEMMA.</u> Let $s \in \mathcal{S}(U)$ and $t \in -\mathcal{S}(U)$ such that $t \le s$. Let (V_n) be an increasing sequence in $\mathcal{U}(U)$ covering U. Then $(H_{V_n} s)$ is a decreasing sequence and $h := \lim_{n\to\infty} H_{V_n} s \in \mathcal{H}(U)$, $h = s$ on $\complement U$. Moreover, $t \le h \le s$, i.e. h is the greatest minorant of s which is subharmonic on U.

Proof. For every $n \in \mathbb{N}$, $t \le H_{V_n} t \le H_{V_n} s \le s$, $H_{V_{n+1}} s = H_{V_n} H_{V_{n+1}} s \le H_{V_n} s$, and similarly $H_{V_n} t \le H_{V_{n+1}} t$. Hence $(H_{V_n} s)$ is decreasing to a function $h \in \mathcal{B}(X)$ such that $t \le H_{V_m} t \le h \le s$ for every $m \in \mathbb{N}$. Let us fix $m \in \mathbb{N}$. Then $(H_{V_n} s - H_{V_m} t)_{n \ge m}$ is a decreasing sequence in $\mathcal{H}^+(V_m)$. Therefore $h - H_{V_m} t \in \mathcal{H}^+(V_m)$ by (III.3.1) and hence $h = (h - H_{V_m} t) + H_{V_m} t \in \mathcal{H}(V_m)$. This shows that $h \in \mathcal{H}(V)$. ┘

Given $s \in \mathcal{S}(U)$ and a covering $\mathcal{V} \subset \mathcal{U}(U)$ of U we define

$$\mathcal{S}_{\mathcal{V}}(s) = \{\hat{H}_{V_1} \cdots \hat{H}_{V_n} s : V_1,\dots,V_n \in \mathcal{V},\; n \in \mathbb{N}\}$$

(where $\hat{H}_V f(z) = \liminf_{x\to z} H_V f(x)$ for every $z \in V^*$ and $\hat{H}_V f = H_V f$ on $\complement V^*$) and

$$s_{\mathcal{V}} = \inf \mathcal{S}_{\mathcal{V}}(s).$$

<u>1.2. PROPOSITION.</u> Let $s \in \mathcal{S}(U)$ and $t \in -\mathcal{S}(U)$ such that $t \le s$ and let $\mathcal{V} \subset \mathcal{U}(U)$ be a covering of U. Then $\mathcal{S}_{\mathcal{V}}(s) \subset \mathcal{S}(U)$, $s_{\mathcal{V}} \in \mathcal{H}(U)$, and $s_{\mathcal{V}}$ is the greatest minorant of s which is subharmonic on U.

Proof. By (1.1) there exists a greatest minorant h of s which is subharmonic on U . Moreover, $h \in H(U)$ and $h = s$ on $\complement U$. Then $\tilde{s} := s - h \in S^+(U)$, $\tilde{s} = 0$ on $\complement U$ and 0 is the greatest minorant of $\tilde{s}$ which is subharmonic on U . An application of (III.5.4) and (III.1.2) to the balayage space $(U, W|_U)$ yields that $S_V(\tilde{s}) \subset S^+(U)$. Moreover, using a countable subcovering $V_0 \subset V$ of U we conclude by (III.6.1) that $\tilde{s}_V = 0$. We finally note that $S_V(s) = S_V(\tilde{s}) + h \subset S(U)$ and $s_V = \tilde{s}_V + h = h$ finishing the proof. ┘

Let $V \subset U(U)$ be an open covering of U . A non-empty set $G \subset S(U)$ is called a *Perron set* with respect to V and U if

(a) G is lower directed;

(b) $\hat{H}_V s \in G$ for every $s \in G$, $V \in V$;

(c) There exists $t \in -S(U)$ such that $t \leq s$ for every $s \in G$.

1.3. PROPOSITION. Let G be a Perron set with respect to V and U and let $g = \inf G$. Then there exists an $h \in H(U)$ such that $g \leq h$ and $g = h$ on U , in particular g is continuous on U .

If moreover $g_1 = g_2$ on $\complement U$ for all $g_1, g_2 \in G$ then $g \in H(U)$.

Proof. By (1.1) there exists a smallest majorant $\tilde{t}$ of t which is superharmonic on U and this function $\tilde{t}$ is harmonic on U. Then $\tilde{G} := G - \tilde{t} \subset S^+(U)$, $\tilde{G}$ is a Perron set with respect to V, and $\tilde{g} := \inf \tilde{G} = g - \tilde{t}$. For every $\tilde{s} \in \tilde{G}$, $\tilde{g} \leq \tilde{s}_V \leq \tilde{s}$, hence $\tilde{F} = \{\tilde{s}_V : s \in \tilde{G}\}$ is by (1.2) a lower directed subset of $H^+(U)$ such that $\tilde{g} = \inf \tilde{F}$. Therefore the assertion follows from (III.3.2). ┘

1.4. EXERCISE. Let U be an open subset of X. Then $H_r^+(U) - H_r^+(U)$ is a vector lattice.

2. Generalized Dirichlet Problem

In this section we study the Dirichlet problem by the PWB-*method*, a method developed by Perron, Wiener and Brelot.

Let $U \subset X$ be an open set. For every $f : X \to \overline{\mathbb{R}}$ we define the set

$$\mathcal{U}_f^U := \{u \in {}^*\mathcal{H}(U) : u \text{ l.s.c. and lower } \mathcal{P}\text{-bounded on } X\,,\ u \geq f \text{ on } \complement U\}$$

of all *upper functions* of f; the set of all *lower functions* of f is defined by $\mathcal{L}_f^U := -\mathcal{U}_{-f}^U$. Then

$$\overline{H}_f^U = \inf \mathcal{U}_f^U \quad (\underline{H}_f^U = \sup \mathcal{L}_f^U \text{ resp.})$$

is called the *upper* (*lower* resp.) *solution of the generalized Dirichlet problem* of f on U.

2.1. REMARKS. 1. $\overline{H}_f^U$ and $\underline{H}_f^U$ depend only on $f_{|\complement U}$, $\overline{H}_f^U = \underline{H}_f^U = f$ on $\complement U$. Indeed, the first statement is obvious and the second follows from the observation that for every $x \in \complement U$ the function u defined by $u = \infty$ on $\complement\{x\}$ and $u(x) = f(x)$ is contained in $\mathcal{U}_f^U$.

2. If

$$'\mathcal{U}_f^U := \{u \in {}^*\mathcal{H}(U) : u \text{ l.s.c. on } X\,,\ u \geq f \text{ on } \complement U,\ \{u < 0\} \text{ rel. compact}\}$$

and

$$^p\mathcal{U}_f^U := \{u \in {}^*\mathcal{H}(U) : u \text{ l.s.c. on } X,\ u \geq f \text{ on } \complement U\,,\ u \geq -\lambda p(\lambda \in \mathbb{R}_+)\}$$

where $p \in \mathcal{P}$ is strictly positive then evidently $'\mathcal{U}_f^U \subset {}^p\mathcal{U}_f^U \subset \mathcal{U}_f^U$. But more is true:

$$\overline{H}_f^U = \inf \mathcal{U}_f^U = \inf {}'\mathcal{U}_f^U = \inf {}^p\mathcal{U}_f^U\,.$$

Indeed, let $u \in \mathcal{U}_f^U$. It suffices to show that $\inf {}'\mathcal{U}_f^U \leq u$. By the very definition of $\mathcal{U}_f^U$ there exists $p \in \mathcal{P}$ such that $u + p \geq 0$. Let $K \subset X$ be compact and define $v := u + R_p^{\complement K}$. Then $v \in {}^*\mathcal{H}(U)$, l.s.c. on X such that $v \geq f$ on $\complement U$. Since $v = u + p \geq 0$ on $\complement K$ we have $v \in {}'\mathcal{U}_f^U$ and therefore

$$u + R_p^{\complement K} = v \geq \inf {}'\mathcal{U}_f^U\,.$$

The assertion now follows from (III.2.8).

2.2. PROPOSITION. Let $f, g, f_n : X \to \overline{\mathbb{R}}$ and $\lambda \in \mathbb{R}_+$. Then:

(1) $\underline{H}_f^U = -\overline{H}_{-f}^U$.

(2) $\underline{H}_f^U \leq \underline{H}_g^U$ and $\overline{H}_f^U \leq \overline{H}_g^U$ if $f \leq g$.

(3) $\overline{H}_{\lambda f}^U = \lambda \overline{H}_f^U$.

(4) $\overline{H}^U_{f+g} \leq \overline{H}^U_f + \overline{H}^U_g$ if $f + g$ and $\overline{H}^U_f + \overline{H}^U_g$ are defined .

(5) $\underline{H}^U_f \leq \overline{H}^U_f$.

(6) If (f_n) is increasing such that $\overline{H}^U_{f_n} \in H(U)$ for every $n \in \mathbb{N}$ then $\overline{H}^U_{\sup f_n} = \sup \overline{H}^U_{f_n}$.

Proof. (1) - (4) are obvious, and (5) follows from the minimum principle (III.6.6).

(6) Let $f := \sup f_n$. Then $\sup \overline{H}^U_{f_n} \leq \overline{H}^U_f$ by (2), and $\sup \overline{H}^U_{f_n} = f = \overline{H}^U_f$ on $\complement U$. So let $x \in U$, assume that $\sup \overline{H}^U_{f_n}(x) < \infty$, and fix $\varepsilon > 0$. Then for every $n \in \mathbb{N}$ there exists a function $u_n \in \mathcal{U}^U_{f_n}$ such that

$$u_n(x) - \overline{H}^U_{f_n}(x) < \frac{\varepsilon}{2^n} .$$

Defining $h_n, v_n : X \to \mathbb{R} \cup \{\infty\}$ by $h_n(y) = \overline{H}^U_{f_n}(y)$ and $v_n(y) = u_n(y) - h_n(y)$ if $|\overline{H}^U_{f_n}(y)| < \infty$ and $h_n(y) = v_n(y) = \infty$ if $|\overline{H}^U_{f_n}(y)| = \infty$ we have $h_n \in H(U)$ and $v_n \in {}^*H(U)$. Let

$$w := \sup h_n + \sum_{n=1}^{\infty} v_n .$$

Then $w \in {}^*H(U)$. Furthermore w is lower P-bounded on X and satisfies $w \geq h_n + v_n \geq u_n$, hence $\hat{w} \geq \hat{u}_n = u_n$ for every $n \in \mathbb{N}$. Therefore $\hat{w} \in \mathcal{U}^U_{f_n}$ and thus $\hat{w} \in \mathcal{U}^U_f$. This implies $\overline{H}^U_f \leq \hat{w} \leq w$, in particular $\overline{H}^U_f(x) \leq w(x) \leq \sup \overline{H}^U_{f_n}(x) + \varepsilon$. Thus $\sup \overline{H}^U_{f_n}(x) = \overline{H}^U_{f_n}(x)$. ┘

<u>2.3. PROPOSITION.</u> Let $u \in \mathcal{W}$. Then $\overline{H}^U_u = R^{\complement U}_u$.

Proof. Let $v \in \mathcal{U}^U_u$. Then $v \geq 0$ by (III.4.3), hence $w = \inf(u,v) \in \mathcal{W}$ by (III.4.6). Since $w = u$ on $\complement U$ we obtain $R^{\complement U}_u \leq w \leq v$, i.e. $R^{\complement U}_u \leq \overline{H}^U_u$. Conversely, let $v \in \mathcal{W}$ such that $v \geq u$ on $\complement U$. Then $v \in \mathcal{U}^U_u$ and therefore $\overline{H}^U_u \leq v$ which implies that $\overline{H}^U_u \leq R^{\complement U}_u$. ┘

The functions in

$$R_n(U) := \{f : X \to \overline{\mathbb{R}} : \underline{H}^U_f = \overline{H}^U_f =: H^U_f , |H^U_f| < +\infty \text{ on } U\}$$

are called *nearly resolutive*, and the functions in

$$R(U) := \{f \in R_n(U) : H^U_f \in H(U)\}$$

are called *resolutive*. Furthermore, U is called *resolutive* if $K(X) \subset R(U)$. By (2.2), the real functions in $R_n(U)$ and $R(U)$ respectively, form a vector space.

2.4. PROPOSITION. Let $\mathcal{U}_f^U \cap S(U) \neq \emptyset$ and $L_f^U \cap (-S(U)) \neq \emptyset$. Then there exist $\overline{h}$, $\underline{h} \in H(U)$ such that $\overline{h} \geq \overline{H}_f^U$, $\underline{H}_f^U \geq \underline{h}$ and $\overline{h} = \overline{H}_f^U$ on U, $\underline{h} = \underline{H}_f^U$ on U. If moreover $f \in R_n(U) \cap B(X)$ then $f \in R(U)$.

Proof. Let $G = \mathcal{U}_f^U \cap S(U)$. Then G is a Perron set with respect $\mathcal{U}(U)$ and U. Indeed, G certainly satisfies the conditions (b) and (c) of the definition of a Perron set. To prove (a), let $s_1, s_2 \in G$. Take $t \in L_f^U \cap (-S(U))$. Then $s_1 \geq t$, $s_2 \geq t$. If $\tilde{t}$ is defined as in the proof of (1.3), then $s_1 - \tilde{t}$, $s_2 - \tilde{t} \in S^+(U)$, hence $\inf(s_1,s_2) - \tilde{t} = \inf(s_1-\tilde{t}, s_2-\tilde{t}) \in S^+(U)$ and therefore $\inf(s_1,s_2) \in G$. The existence of $\overline{h}$, $\underline{h} \in H(U)$ with the required properties follows from (1.3). If $f \in R_n(U) \cap B(X)$ then for every $V \in \mathcal{U}(U)$ and $x \in V$

$$H_V\underline{h}(x) = \underline{h}(x) = H_f^U(x) = \overline{h}(x) = H_V\overline{h}(x)$$

hence $H_V H_f^U(x) = H_f^U(x)$ since $\underline{h} \leq H_f^U \leq \overline{h}$. Therefore f is resolutive. ┘

2.5. PROPOSITION. Every $p \in P$ is resolutive and $H_p^U = R_p^{\complement U} = H_U p$.

Proof. If $p \in P$ then $R_p^{\complement U} \in L_p^U$ by (VI.2.6), hence by (2.2) and (2.3)

$$R_p^{\complement U} \leq \underline{H}_p^U \leq \overline{H}_p^U = R_p^{\complement U}.$$ ┘

2.6. COROLLARY. U is resolutive and for every $f \in K(X)$, $H_f^U = H_U f$.

Proof. We may assume $f \geq 0$. If $q_0 \in P$, $q_0 > 0$ and $\varepsilon > 0$ then by (I.1.2) there exist $p,q \in P$ such that $0 \leq p - q \leq f \leq p - q + \varepsilon q_0$, hence by (2.5) and (2.2)

$$0 \leq H_{p-q}^U \leq \underline{H}_f^U \leq \overline{H}_f^U \leq H_{p-q}^U + \varepsilon H_{q_0}^U$$

whence the assertion using (2.4). ┘

2.7. LEMMA. If $f : X \to \overline{\mathbb{R}}_+$ is l.s.c. on X then $\underline{H}_f^U = \overline{H}_f^U = H_U f$.

Proof. Let $(f_n) \subset K^+(X)$ be an increasing sequence such that $f = \sup f_n$.

By (2.6) $f_n \in R(U)$ and $H^U_{f_n} \leq \underline{H}^U_f$. Therefore by (2.2) $\overline{H}^U_f = \sup H^U_{f_n} \leq \underline{H}^U_f$ and $\overline{H}^U_f = \underline{H}^U_f = \sup H^U_{f_n} = \sup H_U f_n = H_U f$. ┘

<u>2.8. COROLLARY.</u> Let $f : X \to \overline{\mathbb{R}}_+$ be l.s.c. on X and $s \in S^+(U)$ be l.s.c. on X such that $f \leq s$ on $\complement U$. Then $f \in R(U)$ and $H^U_f = H_U f$.

Proof. (2.4) , (2.7) . ┘

<u>2.9. COROLLARY.</u> Let $f \in C(X)$ such that $|f| \leq s$ on $\complement U$ for some $s \in S^+(U)$, l.s.c. on X . Then $f \in R(U)$ and $H^U_f = H_U f$.

<u>2.10. LEMMA.</u> Let $f : X \to \overline{\mathbb{R}}$ be l.s.c. on X . If $s \in S^+(U)$ is l.s.c. and real valued such that $f \geq -s$ on $\complement U$ then $\overline{H}^U_f = \underline{H}^U_f = H_U f > -\infty$.

Proof. By (2.8), $s \in R(U)$ such that $H^U_s = H_U s < \infty$. By (2.7), $\overline{H}^U_{f+s} = \underline{H}^U_{f+s} = H_U(f+s)$, hence by (2.2)

$$H_U f = \underline{H}^U_{f+s} + \underline{H}^U_{-s} \leq \underline{H}^U_f \leq \overline{H}^U_f \leq \overline{H}^U_{f+s} + \overline{H}^U_{-s} = H_U \mathbf{f} .$$ ┘

<u>2.11. PROPOSITION.</u> If $f : X \to \overline{\mathbb{R}}$ then $\overline{H}^U_f = H^*_U f$, $\underline{H}^U_f = (H_U)_* f$.

Proof. Let $x \in X$ and $p \in P$, $p > 0$. Then there exists a decreasing sequence $(u_n) \subset {}^p\mathcal{U}^U_f \subset \mathcal{U}^U_f$ by (2.1.2) such that $\inf u_n(x) = \overline{H}^U_f(x)$. (2.10) implies that $\overline{H}^U_{u_n}(x) = H_U u_n(x)$ for all $n \in \mathbb{N}$. Since $u_n \geq \overline{H}^U_{u_n}$ we obtain that

$$\overline{H}^U_f(x) \geq \inf \overline{H}^U_{u_n}(x) = \inf H_U u_n(x) \geq H^*_U f(x) .$$

There exists a decreasing sequence (f_n) of l.s.c. functions $f_n : X \to \overline{\mathbb{R}}$ such that $f_n \geq f$ and $\lim H_U f_n(x) = H^*_U f(x)$. Let $g_n := \sup(f_n, -np)$, $n \in \mathbb{N}$. Then (g_n) is decreasing and $\lim H_U g_n(x) = H^*_U f(x)$. By (2.10) $H_U g_n(x) = \overline{H}^U_{g_n}(x)$ for every $n \in \mathbb{N}$, hence

$$\overline{H}^U_f(x) \leq \inf \overline{H}^U_{g_n}(x) = \inf H_U g_n(x) = H^*_U f(x) .$$ ┘

Now the *resolutivity theorem* is a simple consequence.

<u>2.12. COROLLARY.</u> Let $f : X \to \overline{\mathbb{R}}$. Then f is nearly resolutive if and only if f is $H_U(x,.)$-integrable for all $x \in U$. In this case $H^U_f = H_U f$.

The PWB-method allows to prove the following minimum principle.

2.13. COROLLARY. Let $u \in {}^*\mathcal{H}(U)$ be l.s.c. on X and $p \in \mathcal{P}$ such that $u \geq -p$. If $A \subset X$ such that $\varepsilon_x^{\complement U}(A) = 0$ for every $x \in U$ and $u \geq 0$ on $(\complement U) \setminus A$ then $u \geq 0$ on U.

Proof. If $v = \inf(u,0)$ then $u \in \mathcal{U}_v^U$ and $v = 0$ on $(\complement U) \setminus A$, hence for every $x \in U$, $0 = H_U v(x) = H_v^U(x) \leq u(x)$.. ⌟

2.14. REMARK. (2.13) can be applied to a polar set $A \subset \complement U$ by (VI.5.6).

2.15. EXERCISES. Suppose that $(X, \mathcal{W})$ has the local truncation property, let U be a relatively compact open subset of X and $f : U^* \to \overline{\mathbb{R}}$.

1. For every $x \in U$,

$$\overline{H}_f^U(x) = \inf \{u(x) : u \in {}^*\mathcal{H}(U),\ u \text{ lower bounded on } U,\ \liminf_{y \to z} u(y) \geq f(z)\ \forall z \in U^*\}\ .$$

2. Let $V \subset U$ be an open set and define $g : V^* \to \overline{\mathbb{R}}$ by

$$g = \begin{cases} f & \text{on } V^* \cap U^*\ , \\ \overline{H}_f^U & \text{on } V^* \cap U\ . \end{cases}$$

If $\overline{H}_f^U \in \mathcal{H}(U)$ then $\overline{H}_g^V = \overline{H}_f^U$ on V.

3. Regular Points

Let U be an open subset of X. A boundary point $z \in U^*$ is called *regular* (with respect to U) if for every $f \in \mathcal{K}(X)$,

$$\lim_{x \to z,\, x \in U} H_f^U(x) = f(z)\ ,$$

i.e. if every filter on U converging to z is regular. Since $H_f^U = H_U f$ by (2.6), the point z is regular if and only if $\lim_{x \to z} H_f^U(x) = f(z)$ for every $f \in \mathcal{K}(X)$. U is called *regular* if every $z \in U^*$ is a regular point.

3.1. PROPOSITION. For every $z \in U^*$, the following statements are equivalent:

(1) z is regular.

(2) $\lim_{x \to z} \varepsilon_x^{\complement U}(f) = f(z)$ for every $f \in \mathcal{C}_P(X)$.

(3) $\lim_{x \to z} H_p^U(x) = p(z)$ for some strict $p \in P$.

(4) $\varepsilon_z^{\complement U} = \varepsilon_z$.

(5) $\complement U$ is not thin at z .

Proof. (1) $\Leftrightarrow$ (2) $\Leftrightarrow$ (3) : (VI.11.4).

(2) $\Leftrightarrow$ (5): (VI.11.2).

(4) $\Leftrightarrow$ (5): Obvious by definition of thinness. ┘

3.2. REMARKS. Let U be an open subset of X .

1. By (3.1), the set U_r of regular points with respect to U can be characterized by

$$U_r = U^* \cap b(\complement U) = \overline{U} \cap b(\complement U) \quad ;$$

hence U_r is a finely closed G_δ-set by (VI.4.1). In particular, U is regular if and only if $\complement U$ is basic.

2. Let $V \subset U$ be open and $z \in V^* \cap U_r$. Then $z \in V_r$ since $b(\complement U) \subset b(\complement V)$.

3. The notion of regularity of a point $z \in U^*$ is a local notion, i.e. $z \in U_r$ if and only if $z \in (U \cap V)_r$ for every open neighborhood V of z ((VI.4.5)).

4. If U and V are regular subsets of X then $U \cap V$ is regular by (1) since the union of two basic sets is basic.

5. Let $p \in P$ be a strict potential and $p^U = p - R_p^{\complement U}$. Then, by (3.1), $z \in U^*$ is regular is and only if $\lim_{x \to z, x \in U} p^U(x) = 0$. Furthermore, $p^U > 0$ on U .

The last property of a regular point motivates the following definition. Let U be open and $z \in U^*$. Then z has a *barrier* (with respect to U) if there exist an open neighborhood V of z and $u \in {}^*H^+(U \cap V)$ such that $u > 0$ on $U \cap V$ and $\lim_{x \to z, x \in U} u(x) = 0$.

3.3. PROPOSITION. Let U be an open subset of X . Then $z \in U^*$ is regular if and only if z has a barrier.

Proof. By (3.2.5) and (3.2.3) it suffices to show that z is regular if there exists a function $u \in {}^*\mathcal{H}^+(U)$, $u > 0$ on U such that $\lim_{x \to z, x \in U} u(x) = 0$. We may assume that $u = 0$ on $\complement U$.

Let $K \subset U$ be compact and $f \in K^+(X)$ such that $f \le u$. By (VI.11.6), there exists a sequence (x_n) in U such that $\lim_{n\to\infty} x_n = z$ and $\lim_{n\to\infty} \varepsilon_{x_n}^{K \cup \complement U} = \varepsilon_z^{K \cup \complement U}$, hence

$$\varepsilon_z^{K\cup\complement U}(f) = \lim_{n\to\infty} \varepsilon_{x_n}^{K\cup\complement U}(f) = \lim_{n\to\infty} H_f^{U\setminus K}(x_n) \le \lim_{n\to\infty} u(x_n) = 0 .$$

Since $u > 0$ on U, the above relations imply $\varepsilon_z^{K\cup\complement U}(K) = 0$ and thus $\varepsilon_z^{K\cup\complement U} = \varepsilon_z^{\complement U}$ by (VI.9.6). If (K_n) is an exhaustion of U then by (VI.10.3), $\lim_{n\to\infty} \varepsilon_z^{K_n \cup \complement U} = \varepsilon_z$. Thus $\varepsilon_z^{\complement U} = \varepsilon_z$, i.e. z is regular. ┘

3.4. EXAMPLES. 1. If A is an absorbing subset of X, then by (VI.4.7.1) A is basic and therefore $\complement A$ is a regular set.

2. Classical potential theory. Using (VI.4.7.3) we obtain the following *strong cone condition* for regularity: Let U be an open subset of $\mathbb{R}^n$. Then $z \in U^*$ is regular if there exist an open cone C with vertex at z, a hyperplane H containing z, and $r > 0$ such that

$$\emptyset \neq C \cap H \cap B_r(z) \subset \complement U .$$

In particular, U is regular if U is convex.

3. Riesz potentials. Using (VI.4.7.4) we obtain that a boundary point z of an open set U is α-regular if the following *cone condition* is satisfied: There exist an open cone C with vertex at z and $r>0$ such that $\emptyset \neq C \cap B_r(z) \subset \complement U$.

If $\alpha < 1$ then hyperplanes are α-polar by (VI.5.4.4), hence the strong cone condition is not sufficient for regularity.

4. Heat equation. Let $c > 0$ and $u(x,t) = c - (\|x\|^2 + nt)$, $(x,t) \in \mathbb{R}^n \times \mathbb{R}$. By (V.6.11), u is harmonic on $\mathbb{R}^n \times \mathbb{R}$. Therefore by (3.3) the set

$$U = \{(x,t) \in \mathbb{R}^n \times \mathbb{R} : \|x\|^2 + nt < c\}$$

is regular. Using example 1) and (V.6.13) we obtain that

$$U_c(0) = \{(x,t) \in U : t > 0\} = U \cap \complement A_0$$

is also a regular set. Hence $\{z + U_c(0) : z \in \mathbb{R}^n \times \mathbb{R}, c > 0\}$ is a base of regular sets.

3.5. PROPOSITION. Let U be a regular set and $f \in C(X)$ with $|f| \leq p$ on $\complement U$ for some potential p on X. Then $H_f^U \in C(X) \cap H(U)$, $H_f^U = f$ on $\complement U$ and $|H_f^U| \leq p$. Moreover, H_f^U is the only function $h \in C(X) \cap H(U)$ such that $h = f$ on $\complement U$ and $|h| \leq q$ for some potential q on X.

Proof. We may assume that $|f| \leq p$ on X. Then by (III.5.6), $p_0 := R_{|f|} \in P$, $p_0 \leq p$. Hence by (3.1) and (2.9), $H_f^U \in C(X) \cap H(U)$, $|H_f^U| \leq p_0 \leq p$ and obviously $H_f^U = f$ on $\complement U$. The uniqueness follows from the minimum principle (III.6.6). ┘

3.6. REMARK. The condition $|f| \leq p$ on $\complement U$ in (3.5) cannot be omitted. For example, consider classical potential theory on $X =]-1,+1[$ and take $U =]0,1[$. Then $h_1 = 0$ and $h_2(x) = x^+$, $x \in X$, are two continuous extensions of $f = 0$ which are harmonic on U. Certainly, h_2 is not bounded by a potential.

3.7. EXAMPLE. Heat equation on $\mathbb{R}^{n+1}$. Let f be a real continuous function on $\mathbb{R}^n \times \{0\}$. The classical *Cauchy problem* asks for conditions such that f can be uniquely extended to a continuous function h on $\mathbb{R}^n \times \mathbb{R}_+$ which is harmonic on $\mathbb{R}^n \times \mathbb{R}_+^*$, i.e. $\Delta h = 0$ on $\mathbb{R}^n \times \mathbb{R}_+^*$. We shall show that (3.5) implies two well known results about solutions of the Cauchy problem. By (V.6.13), $\mathbb{R}^n \times \mathbb{R}_+^*$ is the complement of an absorbing set, hence a regular set by (3.4.1).

(1) If f is bounded then there exists a unique bounded solution of the Cauchy problem.

Proof. Let $|f| \leq M$ for some $M > 0$. By (V.6.12), the function

$$p(x,t) = M \inf(1, e^t), \quad (x,t) \in \mathbb{R}^n \times \mathbb{R},$$

is a potential on $\mathbb{R}^n \times \mathbb{R}$ such that $|f| \leq p$ on $\mathbb{R}^n \times \{0\}$. Thus (1) follows from (3.5). ┘

(2) Let $M > 0$, $\varepsilon > 0$ such that $|f(x,0)| \leq M \cdot e^{-\frac{\|x\|^2}{2\varepsilon}}$ for every $x \in \mathbb{R}^n$.

Then there exists a unique solution h of the Cauchy problem such that for some $M' > 0$,

$$|h(x,t)| \leq \left(\frac{M'}{t+\varepsilon}\right)^{n/2} \cdot e^{-\frac{\|x\|^2}{2(t+\varepsilon)}} \quad \text{for every } (x,t) \in \mathbb{R}^n \times \mathbb{R}_+^* .$$

Proof. Let $z = (0,-\varepsilon) \in \mathbb{R}^n \times \mathbb{R}$. Then (V.6.17) implies that $p = M(2\pi\varepsilon)^{n/2} w^z$ is a potential on $\mathbb{R}^n \times \mathbb{R}$ such that $|f| \leq p$ on $\mathbb{R}^n \times \{0\}$. Thus by (3.5) there exists a unique solution h of the Cauchy problem such that $|h| \leq p$ on $\mathbb{R}^n \times \mathbb{R}_+^*$, whence for all $(x,t) \in \mathbb{R}^n \times \mathbb{R}_+^*$,

$$|h(x,t)| \leq p(x,t) = M(2\pi\varepsilon)^{n/2} w^z(x,t) = \left(\frac{M'}{t+\varepsilon}\right)^{n/2} e^{-\frac{\|x\|^2}{2(t+\varepsilon)}}$$

where $M' = \varepsilon \cdot M^{2/n}$. ┘

In both cases (1) and (2) the unique solution h of the Cauchy problem is given by

$$h(x,t) = \int w(x,t,y,0)\ f(y,0)\ \lambda^n(dy), \qquad (x,t) \in \mathbb{R}^n \times \mathbb{R}_+^* .$$

This follows from (3.5), (2.9) and (V.6.15). Finally, let us note that (3.5) is stronger than any statement of the type (1) or (2) (see exercise (3.8.5)).

3.8. EXERCISES. 1. Let U be a regular subset of X. Then for every compact subset K of U there exists a regular set V such that $K \subset V \subset \overline{V} \subset U$.

2. Classical potential theory. Let U be an open set. Then there exists an increasing sequence (V_n) of relatively compact regular sets such that $U = \bigcup_{n=1}^{\infty} V_n$.

3. Heat equation. The statement of the preceding exercise does not hold for every U.

4. Classical potential theory. The open half space $U = \mathbb{R}^{n-1} \times \mathbb{R}_+^*$ is a regular set and, for every $x \in U$, the harmonic measure $\varepsilon_x^{\complement U}$ has the density $y \mapsto \Gamma(\frac{n}{2})\pi^{-n/2}\ x_n \|x-(y,0)\|^{-n}$ with respect to Lebesgue measure λ^{n-1} on $U^* = \mathbb{R}^{n-1} \times \{0\}$. (Hint: (V.6.2) may be used!)

5. Heat equation. Let $U = \mathbb{R}^n \times]0,\infty[$ and let (f_k) be a sequence of real functions on $U^* = \mathbb{R}^n \times \{0\}$. Then there exists a potential q on $\mathbb{R}^n \times \mathbb{R}$ such that $q_{|\overline{U}} \in C(\overline{U})$, $q_{|U} \in H(U)$ and $q_{|U^*}$ is not bounded by any function f_k, $k \in \mathbb{N}$. (Hint: Choose $x_k \in \mathbb{R}^n$, $k \in \mathbb{N}$, such that $\lim_{k\to\infty} \|x_k\| = \infty$ and $\sum_{k=1}^{\infty} w^{(x_k,0)}$ is

locally uniformly convergent on $\mathbb{R}^n \times \mathbb{R}$. Take $t_k \in]0,1[$ such that $w^{(x_k,t_k)}(x_k,0) > f_k(x_k,0)$, consider $q = \sum_{k=1}^{\infty} w^{(x_k,t_k)}$, and use (III.1.4).)

4. Irregular Points

Let U be an open subset of X. A non-regular boundary point z of U is called *irregular*. The set U_i of all irregular points $z \in U^*$ satisfies by (3.2.1)

$$U_i = U^* \setminus b(\complement U) = \complement U \setminus b(\complement U),$$

hence U_i is a semi-polar K_σ-set by (VI.5.12).

4.1. PROPOSITION. Let U be an open subset of X. Then the following statements are equivalent:

(1) For every $x \in X$, $\varepsilon_x^{\complement U}(U_i) = 0$.

(2) For every $x \in U$, $\varepsilon_x^{\complement U}(U_i) = 0$.

(3) For every $x \in X$, $\varepsilon_x^{\complement U} = \varepsilon_x^{(\complement U) \setminus U_i}$.

(4) For every $x \in X$, $(\varepsilon_x^{\complement U})^{\complement U} = \varepsilon_x^{\complement U}$.

(5) For every $x \in X$, $\varepsilon_x^{\beta(\complement U)} = \varepsilon_x^{\complement U}$.

Proof. (1) ⇒ (2) : Obvious.

(2) ⇒ (3): Consider $p \in P$. Then for every $x \in U$,

$$\hat{R}_p^{\complement U}(x) = \varepsilon_x^{\complement U}(p) = \varepsilon_x^{(\complement U) \setminus U_i}(p) = \hat{R}_p^{(\complement U) \setminus U_i}(x)$$

by (VI.9.6). Since U_i is contained in the fine closure of U, we obtain for every $x \in U \cup U_i$,

$$\hat{R}_p^{\complement U}(x) = \hat{R}_p^{(\complement U) \setminus U_i}(x) = R_p^{(\complement U) \setminus U_i}(x),$$

whereas for every $x \in \complement(U \cup U_i) = (\complement U) \setminus U_i$,

$$\hat{R}_p^{\complement U}(x) = p(x) = R_p^{(\complement U) \setminus U_i}(x).$$

Hence $\hat{R}_p^{\complement U} = R_p^{(\complement U) \setminus U_i}$, so (3) holds.

(3) ⇒ (5) : If $x \in (\complement U) \setminus U_i$ then $\varepsilon_x^{(\complement U) \setminus U_i} = \varepsilon_x^{\complement U} = \varepsilon_x$. Hence

$(\complement U)\setminus U_i \subset b((\complement U)\setminus U_i) \subset b(\complement U)$ which implies that $(\complement U)\setminus U_i \subset \beta(\complement U)$. Therefore, $b(\complement U) = (\complement U)\setminus U_i \subset \beta(\complement U) \subset b(\complement U)$, whence for every $x \in X$,

$$\varepsilon_x^{\beta(\complement U)} = \varepsilon_x^{(\complement U)\setminus U_i} = \varepsilon_x^{\complement U} .$$

(5) ⇒ (1) : For every $x \in X$, the measures $\varepsilon_x^{\complement U} = \varepsilon_x^{\beta(\complement U)}$ are supported by $\beta(\complement U)$, hence

$$0 = \varepsilon_x^{\complement U}(X\setminus\beta(\complement U)) \geq \varepsilon_x^{\complement U}(U^*\setminus b(\complement U)) = \varepsilon_x^{\complement U}(U_i) .$$

(1) ⟺ (4) : Let $p \in P$ be a strict potential and fix $x \in X$. Since $U_i = \complement U \cap \{\hat{R}_p^{\complement U} < p\}$ and $\varepsilon_x^{\complement U}$ is supported by $\complement U$ we have $\varepsilon_x^{\complement U}(U_i) = 0$ if and only if $\varepsilon_x^{\complement U}(\hat{R}_p^{\complement U}) = \varepsilon_x^{\complement U}(p)$, i.e. $(\varepsilon_x^{\complement U})^{\complement U}(p) = \varepsilon_x^{\complement U}(p)$. Since p is strict this equality is satisfied if and only if $(\varepsilon_x^{\complement U})^{\complement U} = \varepsilon_x^{\complement U}$. ┘

4.2. COROLLARY. The following assertions are equivalent:

(1) For every open subsets U of X, the set U_i is polar.

(2) For every open subset U of X and every $x \in U$, $\varepsilon_x^{\complement U}(U_i) = 0$.

(3) $(X,\mathcal{W})$ satisfies the axiom of polarity.

Proof. (1) ⇒ (2) : (VI.5.6).

(2) ⇒ (3) : Using (4.1), we have for every compact subset K of X and all $x \in X$,

$$(\varepsilon_x^K)^K = \varepsilon_x^K .$$

Therefore, every compact totally thin set K is polar. Indeed, if $p \in P$ is a strict potential then $\hat{R}_p^K < p$, hence the equality

$$\int \hat{R}_p^K \, d\varepsilon_x^K = (\varepsilon_x^K)^K(p) = \varepsilon_x^K(p) = \int p \, d\varepsilon_x^K$$

shows that $\varepsilon_x^K = 0$ for every $x \in X$. So the assertion follows from (VI.5.5) and (VI.5.7).

(3) ⇒ (1) : (VI.5.12). ┘

Using the results of section (VI.11) it is possible to obtain a nice classification of the boundary points of an open subset U of X . For that purpose, let $z \in U^*$ and denote $\Lambda_U(z) := \Lambda_U^{\complement U}(z)$ the set of all measures λ on X such that

$(\varepsilon_{x_n}^{\complement U})$ vaguely converges to λ for some sequence (x_n) in U converging to z. By (VI.11.6) and (VI.11.3) we always have

$$\varepsilon_z^{\complement U} \in \Lambda_U(z) \subset \{\alpha\varepsilon_z + (1-\alpha)\varepsilon_z^{\complement U} : \alpha \in [0,1]\} .$$

Evidently, $\Lambda_U(z) = \{\varepsilon_z\}$ if and only if z is a regular point. If z is an irregular point, i.e. $\varepsilon_z^{\complement U} \neq \varepsilon_z$, then z is called *semiregular (weakly irregular, strongly irregular,* resp.) if $\Lambda_U(z) = \{\varepsilon_z^{\complement U}\}$ ($\Lambda_U(z) = \{\varepsilon_z, \varepsilon_z^{\complement U}\}$, $\Lambda_U(z) \not\subset \{\varepsilon_z, \varepsilon_z^{\complement U}\}$, resp.).

We note that a point $z \in U_i$ is weakly irregular if and only if there are an open neighborhood V of z and disjoint open sets V', V'' such that $U \cap V = V' \cup V''$, and $\Lambda_{V'}^{\complement U}(z) = \{\varepsilon_z\}$, $\Lambda_{V''}^{\complement U}(z) = \{\varepsilon_z^{\complement U}\}$. Indeed, let z be weakly irregular. Fix a strict potential $p \in P$ and $\hat{R}_p^{\complement U}(z) < \alpha < p(z)$. Using (VI.11.4) we obtain an open neighborhood V of x such that $\hat{R}_p^{\complement U} \neq \alpha$ on V. Then $V' := U \cap V \cap \{\hat{R}_p^{\complement U} > \alpha\}$ and $V'' := U \cap V \cap \{\hat{R}_p^{\complement U} < \alpha\} = U \cap V \cap \{R_p^{\complement U} < \alpha\}$ are disjoint open sets such that $U \cap V = V' \cup V''$, $\varepsilon_z^{\complement U} \notin \Lambda_{V'}^{\complement U}(z)$, and $\varepsilon_z \notin \Lambda_{V''}^{\complement U}(z)$ proving the statement.

An open subset U of X is called *semiregular* if for every $z \in U^*$, $\Lambda_U(z) = \{\varepsilon_z^{\complement U}\}$, i.e. z is either regular or semiregular. Hence an open set is semiregular if and only if for every $f \in K(X)$, the restriction of the generalized solution H_f^U to U can be continuously extended on $\overline{U}$ and this equivalent to the fact that the mapping $x \mapsto \varepsilon_x^{\complement U}$ is continuous on $\overline{U}$. Furthermore, every regular set is semiregular.

For the remainder of this section we shall assume that (X,W) *satisfies the local truncation property.*

By (VI.11.16) an irregular boundary point z is strongly irregular if and only if $\Lambda_U(z) = \{\alpha\varepsilon_z + (1-\alpha)\varepsilon_z^{\complement U} : \alpha \in [0,1]\}$. As in section (VI.11), define

$$T(\complement U) = \overline{\bigcup_{x \in U} \operatorname{supp}(\varepsilon_x^{\complement U})} .$$

Note that $T(\complement U)$ is a closed subset of U^*.

<u>4.3. PROPOSITION.</u> For every open subset U of X, $U_r \subset T(\complement U)$ and $U^* \setminus T(\complement U)$ is the set of semiregular points of U.

Proof. (VI.11.14). ⌟

4.4. PROPOSITION. Let U be an open subset of X . Then the following statements are equivalent:

(1) U is semiregular.

(2) $U_r = T(\complement U)$.

(3) U_r is closed, and for every $x \in U$, $\varepsilon_x^{\complement U}(U_i) = 0$.

Proof. (1) ⟺ (2): (4.3).

(2) ⇒ (3): Obvious.

(3) ⇒ (2): Since, for every $x \in U$, supp $\varepsilon_x^{\complement U} \subset U^* \setminus U_i = U_r$ we have $T(\complement U) \subset U_r$, hence $T(\complement U) = U_r$ by (4.3). ⌟

4.5. EXAMPLES. 1. Let A be a closed polar set and V be a regular neighborhood of A . Then by (VI.5.6) and (4.4), $U := V \setminus A$ is a semiregular set since $U_r = V_r$.

2. Classical potential theory. Let $L \subset \mathbb{R}^3$ be Lebesgue's spine (see (VI.4.7.3)). Then 0 is the only irregular boundary point of $\complement L$, it is strongly irregular by (4.3) and (VI.11.16).

3. Translation on $\mathbb{R}$. Let $a,b,c \in \mathbb{R}$ such that $a < c < b$. Then by (4.4), $U =]a,b[$ is a semiregular set. Consider the set $V =]a,b[\setminus \{c\}$. Then a is a regular point, b is a semiregular point, whereas by (4.3), c is a weakly irregular point.

4. Riesz potentials. If U is an open subset of $\mathbb{R}^n$ and $\beta < \alpha$ then every α-irregular point of U is β-irregular.

5. Heat equation in $\mathbb{R}^{n+1}$. Consider an open subset U of $\mathbb{R}^{n+1}$ of the form

$$U =]a_1,b_1[\times\ldots\times]a_n,b_n[\times]\alpha,\beta[$$

where $a_i,b_i,\alpha,\beta \in \mathbb{R}$, $a_i < b_i$, $\alpha < \beta$. Then by (4.4), (V.6.8) and (V.6.9) U is a semiregular set such that $U_r = U^*_\Delta$. If $\alpha < \gamma < \beta$ and

$$V = U \setminus \mathbb{R}^n \times \{\gamma\}$$

then $V^* \cap U$ is the set of weakly irregular points of V . This follows from the above considerations.

4.6. EXERCISES. 1. Let U be an open set and $z \in U^*$. Then z is not semiregular if and only if z has a *weak barrier*, i.e. there exists a neighborhood V of z and $u \in {}^*H(V \cap U)$ such that $u > 0$ on $V \cap U$ and $\liminf_{x \to z, x \in V \cap U} u(x) = 0$.

2. Classical potential theory. Let U be an open set and $z \in U_i \cap \overline{\complement U}^f$. Then z is strongly irregular. In particular, every boundary point of an open set U such that $\overset{\circ}{\overline{U}} = U$ is either regular or strongly irregular.

5. Simplicial Cones

A function cone S on X is called *simplicial* if for every $x \in X$ there exists a unique measure $\mu_x \in M_x(S)$ such that $\mu_x(X \setminus Ch_S X) = 0$. In the following we shall prove that some function cones which are typical for potential theory are simplicial cones. The consequences to the Dirichlet problem will be explained in the next section.

Let $(X,\mathcal{W})$ be again a balayage space. Let $\mathcal{T}$ be a family of $\mathcal{W}$-dilations and define

$$S(\mathcal{T}) = \{s \in \mathcal{C}_P(X) : Ts \leq s \text{ for every } T \in \mathcal{T}\},$$

$$H(\mathcal{T}) = \{h \in \mathcal{C}_P(X) : Th = h \text{ for every } T \in \mathcal{T}\}.$$

Since $P \subset S(\mathcal{T})$ the convex cone $S(\mathcal{T})$ is a min-stable function cone. Moreover

$$H(\mathcal{T}) = S(\mathcal{T}) \cap (-S(\mathcal{T}))$$

and

$$Ch_{S(\mathcal{T})}X \subset \bigcap_{T \in \mathcal{T}} b(T) =: b(\mathcal{T}).$$

If $h \in H(\mathcal{T})$ then $|h| \in -S(\mathcal{T})$, hence the following lemma implies in particular that

$$H(\mathcal{T}) = H(\mathcal{T})^+ - H(\mathcal{T})^+.$$

5.1. LEMMA. Let $p \in P$ and $t \in -S(\mathcal{T})$ such that $t \leq p$. Then there exists $q \in P \cap H(\mathcal{T})$ such that $t \leq q \leq p$.

Proof. If $q := R_t$ then $q \in P$ and $t \leq q \leq p$. Let $T \in \mathcal{T}$. Since $Tq \geq Tt \geq t$ we have $Tq \geq R_t = q$, hence $Tq = q$. Thus $q \in H(\mathcal{T})$. ⌟

5.2. LEMMA. Let $q \in P$. Then: $q \in H(\mathcal{T}) \Longleftrightarrow R_q^{b(\mathcal{T})} = q \Longleftrightarrow R_q^{\beta(b(\mathcal{T}))} = q$.

Proof. First let $q \in H(\mathcal{T})$ and consider $u \in \mathcal{W}$ such that $u \geq q$ on $b(\mathcal{T})$. Since $Ch_{S(\mathcal{T})}X \subset b(\mathcal{T})$ and $u - q$ is the limit of an increasing sequence of functions in $P + H(\mathcal{T}) \subset S(\mathcal{T})$ we obtain from the general minimum principle (I.2.2) that $u - q \geq 0$, i.e., $u \geq q$. Hence $R_q^{b(\mathcal{T})} = q$.

Suppose now conversely that $R_q^{b(\mathcal{T})} = q$ and let $T \in \mathcal{T}$. Since $Tq \in \mathcal{W}$ and $Tq = q$ on $b(\mathcal{T})$ we have $Tq \geq R_q^{b(\mathcal{T})} = q$, i.e., $Tq = q$. Hence $q \in H(\mathcal{T})$.

Finally $b(\mathcal{T})$ being finely closed, the second equivalence is a consequence of (VI.6.4). ⌟

5.3. PROPOSITION. $Ch_{S(\mathcal{T})}X = \beta(b(\mathcal{T}))$ and for every closed subset A of $\beta(b(\mathcal{T}))$ there exists a P-dilation D such that $A = b(D)$, $D(C_P(X)) \subset H(\mathcal{T})$ and D is supported by a closed subset of $\beta(b(\mathcal{T}))$.

Proof. Let A be a closed subset of $\beta(b(\mathcal{T}))$. By (VI.8.5) there exists a P-dilation D such that $b(D) = A$ and D is supported by a closed subset B of $\beta(b(\mathcal{T}))$.

If $p \in P$ then $Dp \in P$ and $\delta(Dp) \subset B$, hence $R_{Dp}^{\beta(b(\mathcal{T}))} = Dp$ by (VI.6.4), i.e. $Dp \in H(\mathcal{T})$ by (5.2). Consider now $f \in C_P(X)$. Then there exists $p_0 \in P$ such that for every $\varepsilon > 0$ there are $p, q \in P$ satisfying $|f - (p - q)| \leq \varepsilon p_0$, hence

$$|Df - (Dp - Dq)| \leq \varepsilon Dp_0 \leq \varepsilon p_0 .$$

This implies that $Df \in H(\mathcal{T})$.

Furthermore, we conclude from the preceding part of the proof, (5.1) and (5.2) that for every $p \in P$

$$R_p^{\beta(b(\mathcal{T}))} = \sup\{R_p^K : K \text{ compact} \subset \beta(b(\mathcal{T}))\}$$
$$= \sup\{q \in P \cap H(\mathcal{T}) : q \leq p\} = \sup\{t \in -S(\mathcal{T}) : t \leq p\}$$

and hence by (I.2.4)

$$\mathrm{Ch}_{S(T)}X = \bigcap_{p \in P} \{R_p^{\beta(b(T))} = p\} = \beta(b(T)) \ . \qquad \square$$

Define a W-dilation D^T by

$$D^T(x,\cdot) = \varepsilon_x^{\mathrm{Ch}_{S(T)}X} = \varepsilon_x^{\beta(b(T))} \qquad (x \in X)$$

and let

$$\hat{S}(T) = \{\sup s_n : (s_n) \subset S(T) \ \text{increasing}\} \ .$$

For every $f \in C_p(X)$ we consider

$$\overline{D}^T f = \inf \{s \in \hat{S}(T) : s \geq f \ \text{on} \ \mathrm{Ch}_{S(T)}X\} \ ,$$

$$\underline{D}^T f = \sup \{t \in -\hat{S}(T) : t \leq f \ \text{on} \ \mathrm{Ch}_{S(T)}X\} \ .$$

Then $\underline{D}^T f = -\overline{D}^T(-f)$ and $\underline{D}^T f \leq \overline{D}^T f$ by the general minimum principle. In fact we have the following result.

5.4. PROPOSITION. For every $f \in C_p(X)$, $\overline{D}^T f = \underline{D}^T f = D^T f$.

Proof. Let $p \in P$. Since $W \subset \hat{S}(T)$ we have

$$\overline{D}^T p \leq R_p^{\mathrm{Ch}_{S(T)}X} = D^T p \ .$$

If A is a closed subset of $\mathrm{Ch}_{S(T)}X$ and D is chosen according to (5.3) then

$$R_p^A \leq Dp \leq \underline{D}^T p \ .$$

Therefore $R_p^{\mathrm{Ch}_{S(T)}X} \leq \underline{D}^T p$ and thus $\overline{D}^T p = \underline{D}^T p = D^T p$. Now the usual approximation yields the assertion. $\square$

5.5. REMARK. The proof of (5.4) shows that in the definitions of $\overline{D}^T$ and $\underline{D}^T$ the set $\hat{S}(T)$ can be replaced by the smaller set $W + H(T)$.

5.6. PROPOSITION. For every $x \in X$, $\mu_x := \varepsilon_x^{\mathrm{Ch}_{S(T)}X}$ is the unique measure in $M_x(S(T))$ such that $\mu_x(X \setminus \mathrm{Ch}_{S(T)}X) = 0$, i.e. $S(T)$ is a simplicial cone. In fact, μ_x is the only measure in $M_x(H(T))$ such that $\mu_x(X \setminus \mathrm{Ch}_{S(T)}X) = 0$.

Proof. Using (5.4) we obtain for every $s \in S(T)$

$$\mu_x(s) = D^T s(x) = \overline{D}^T s(x) \leq s(x) \ .$$

Hence $\mu_x \in M_x(S(T))$. Since $Ch_{S(T)}X$ is basic, the measure μ_x is supported by $Ch_{S(T)}X$, i.e. $\mu_x(X \setminus Ch_{S(T)}X) = 0$. Let $\nu \in M_x(H(T))$ such that $\nu(X \setminus Ch_{S(T)}X) = 0$ and let $p \in P$. We recall that the set $\{q \in P \cap H(T) : q \leq p\}$ is increasingly filtered. Therefore

$$\mu_x(p) = D^T p(x) = \sup\{q(x) : q \in P \cap H(T),\ q \leq p\}$$
$$= \sup\{\nu(q) : q \in P \cap H(T), q \leq p\} = \nu(D^T p) = \nu(p)$$

since $D^T p = p$ on $Ch_{S(T)}X$. Thus $\mu_x = \nu$.

5.7. COROLLARY. $H(T) = H(D^T)$.

Proof. If $h \in H(T) = S(T) \cap (-S(T))$ then $\overline{D}^T h \leq h$ and $h \leq \underline{D}^T h$, hence $h = D^T h$, $h \in H(D^T)$. Conversely, let $h \in H(D^T)$ and $T \in T$. If $s \in \hat{S}(T)$ such that $s \geq h$ on $Ch_{S(T)}X$, then $s \geq \overline{D}^T h = h$ and hence $s \geq Ts \geq Th$. Therefore $h = \overline{D}^T h \geq Th$. Analogously $Th \leq h$. Thus $h \in H(T)$.

5.8. COROLLARY. If $H(T)$ is linearly separating then for any $x \in Ch_{S(T)}X$ and any strictly positive $p \in P$ there exists a function $h \in H(T)$ such that $h \leq p$, $h(x) = 0$, and $h(y) > 0$ for every $y \in X \setminus \{x\}$.

Proof. Let $y \in Ch_{S(T)}X$, $y \neq x$. By (5.3) there exists $h_y \in H(T)^+$ such that $h_y \leq p$, $h_y(x) = 0$ and $h_y(y) > 0$. Since X has a countable base there exists a sequence $(y_n) \subset Ch_{S(T)}X \setminus \{x\}$ such that

$$Ch_{S(T)}X \setminus \{x\} \subset \bigcup_{n \in \mathbf{N}} \{h_{y_n} > 0\} .$$

Then

$$h := \sum_{n=1}^{\infty} \frac{1}{2^n} h_{y_n} \in H(T) ,$$

$0 \leq h \leq p$, $h(x) = 0$, $h > 0$ on $Ch_{S(T)}X \setminus \{x\}$. If $y \in X$ such that $h(y) = 0$ then $0 = h(y) = D^T h(y)$, hence $D^T(y,\cdot) = \lambda\varepsilon_x$ for some $\lambda \geq 0$ since $D^T(y,\cdot)$ is supported by $Ch_{S(T)}X$. Hence for every $g \in H(T)$, $g(y) = D^T g(y) = \lambda g(x)$ and therefore $y = x$.

Let $\hat{S}(D^T)$ denote the set of all lower P-bounded l.s.c. numerical functions s on X such that $D^T s \leq s$. Then $S(T) \subset \hat{S}(T) \subset \hat{S}(D^T)$ by (5.4). Moreover, we may conclude from (I.2.7) that every function in $\hat{S}(D^T)$ is the limit of an increasing sequence in $S(D^T)$ (and conversely). However, we will not use this fact.

5.9. PROPOSITION. Let $s \in \hat{S}(D^T)$ and $t \in -\hat{S}(D^T)$ such that $t \leq s$. Then there exists $h \in H(T)$ such that $t \leq h \leq s$.

Proof. There exists $p_0 \in P$ such that $s \geq -p_0$ and $t \leq p_0$. Let $q_0 \in P$ such that $q_0 > 0$, $p_0 \in o(q_0)$, and define

$$s_0 = \inf(s, p_0), \quad t_0 = \sup(t, -p_0) .$$

Then $s_0 \in \hat{S}(D^T)$, $t_0 \in -\hat{S}(D^T)$ and $-p_0 \leq t_0 \leq s_0 \leq p_0$. We claim that there exist sequences $(s_n) \subset \hat{S}(D^T)$ and $(t_n) \subset -\hat{S}(D^T)$ such that

$$t_{n-1} \leq t_n \leq s_n \leq s_{n-1} \quad \text{and} \quad s_n - t_n \leq \frac{2}{n} q_0$$

for every $n \geq 1$.

Indeed, suppose that $n \in \mathbb{N}$, $n \geq 1$, and $s_{n-1} \in \hat{S}(D^T)$, $t_{n-1} \in -\hat{S}(D^T)$ such that

$$-p_0 \leq t_{n-1} \leq s_{n-1} \leq p_0 .$$

Since t_{n-1} is u.s.c. and s_{n-1} is l.s.c., there exists $f \in C(X)$ such that $t_{n-1} \leq f \leq s_{n-1}$. By (VI.8.5) there exists a sequence (D_m) of P-dilations such that $D_m(C_P(X)) \subset H(T)$ and $\lim_{m\to\infty} D_m f = D^T f$. Then

$$t_{n-1} \leq D^T t_{n-1} \leq D^T f = \lim_{m\to\infty} D_m f \leq D^T s_{n-1} \leq s_{n-1}$$

and $|D_m f| \leq D_m |f| \leq D_m p_0 \leq p_0$ for every $m \in \mathbb{N}$. Let K be a compact subset of X such that $p_0 \leq \frac{1}{2n} q_0$ on $\complement K$. Applying (I.1.11) to the convex hull of the set $\{D_m f|_K : m \in \mathbb{N}\}$ we therefore obtain a function $h \in H(D^T)$ such that $|h| \leq p_0$ and

$$t_{n-1} - \frac{1}{n} q_0 < h < s_{n-1} + \frac{1}{n} q_0 \quad \text{on } K .$$

Moreover,

$$t_{n-1} - 2p_0 \leq -p_0 \leq h \leq p_0 \leq s_{n-1} + 2p_0 \quad \text{on } \complement K .$$

Since $2p_0 \leq \frac{1}{n} q_0$ on $\complement K$, we therefore have

$$t_{n-1} - \frac{1}{n} q_0 \leq h \leq s_{n-1} + \frac{1}{n} q_0 \quad \text{on } X .$$

Defining

$$s_n = \inf\,(s_{n-1}, h + \tfrac{1}{n}q_0), \quad t_n = \sup\,(t_{n-1}, h - \tfrac{1}{n}q_0)$$

we have $s_n \in \hat{S}(D^T)$, $t_n \in -\hat{S}(D^T)$,

$$t_{n-1} \le t_n \le s_n \le s_{n-1} \quad \text{and} \quad s_n - t_n \le \tfrac{2}{n}q_0 \quad .$$

This shows the existence of the desired sequences. Obviously, the sequences (s_n) and (t_n) converge to a continuous function h such that $t \le h \le s$, $|h| \le p_0$ and $D^T h = h$, i.e. $h \in H(D^T)$.

5.10. COROLLARY. Let K be a compact subset of $Ch_{S(T)}X$, $f \in C(K)$ and $s \in S(T)$, $t \in -S(T)$ such that $t \le s$ and $t|_K \le f \le s|_K$. Then there exists a function $h \in H(T)$ such that $h|_K = f$ and $t \le h \le s$.

Proof. Defining $s', t' : X \to \mathbb{R}$ by

$$s' = \begin{cases} f & \text{on } K \\ s & \text{on } \complement K \end{cases} , \qquad t' = \begin{cases} f & \text{on } K \\ t & \text{on } \complement K \end{cases}$$

we have $s' \in \hat{S}(D^T)$, $t' \in -\hat{S}(D^T)$, and $t' \le s'$. By (5.9), there exists $h \in H(T)$ such that $t' \le h \le s'$. Then obviously $t \le h \le s$, $h|_K = f$.

5.11. PROPOSITION. The following statements are equivalent:

(1) $Ch_{S(T)}X$ is closed.

(2) $H(T)$ is a lattice, and $H(T)$ linearly separates the points of $\overline{Ch_{S(T)}X}$.

If the points of $Ch_{S(T)}X$ are polar then we have the following equivalent condition:

(2') $H(T)$ is a lattice and for every $z \in \overline{Ch_{S(T)}X}$, there exists a function $h \in H(T)$ such that $h(z) \neq 0$.

Proof. (1) $\Rightarrow$ (2), (2'): (5.3).

(2), (2') $\Rightarrow$ (1): Suppose that $Ch_{S(T)}$ is not closed and that $H(T)$ is a lattice. Let $z \in \overline{Ch_{S(T)}X} \setminus Ch_{S(T)}X$. We shall show that the support T of $\mu_z = \varepsilon_z^{\beta(b(T))}$ does not contain two different points of $Ch_{S(T)}X$. Assume that there are $y_1, y_2 \in T \cap Ch_{S(T)}X$, $y_1 \neq y_2$. By (5.3) there exist $h_1, h_2 \in H(T)^+$ such that $h_i(y_i) > 0$ and $h_i(y_j) = 0$ for $i \neq j$. Since $h_i(z) = \int h_i d\mu_z > 0$, we may suppose that $h_i(z) = 1$. Since $H(T)$ is a lattice, there exists $h \in H(T)$ such that $h \leq \inf(h_1, h_2) =: s$ and $g \leq h$ for every $g \in H(T)$ satisfying $g \leq s$. Since $s \in S(T)$, we know by (5.10) that for every compact subset K of $Ch_{S(T)}X$ there exists a function $g \in H(T)$ satisfying $g \leq s$ and $g = s$ on K . Therefore $h = s$ on $Ch_{S(T)}X$, hence on $\overline{Ch_{S(T)}X}$. In particular, $h(z) = 1 = h_1(z)$. On the other hand , $h \leq h_1$ and $h(y_1) \leq h_2(y_1) = 0 < h_1(y_1)$ which implies the contradiction $h(z) = \int h\, d\mu_z < \int h_1\, d\mu_z = h_1(z)$.

Therefore, there exist a point $y \in Ch_{S(T)}X$ and an $\alpha \geq 0$ such that $\mu_z = \alpha\varepsilon_y$. So we have for every $h \in H(T)$, $h(z) = \int h\, d\mu_z = \alpha h(y)$, i.e. $H(T)$ does not linearly separate z and y . Furthermore, $\mu_z(\complement\{y\}) = 0$, hence $\mu_z = \varepsilon_z^{\{y\}}$ by (VI.9.6). If $\{y\}$ is polar, we conclude that $\mu_z = 0$, so for every $h \in H(T)$, $h(z) = \int h\, d\mu_z = 0$. Thus (2) and (2') do not hold. ┘

5.12. EXERCISES. 1. Suppose that there exists a point $z \in \overline{Ch_{S(T)}X} \setminus Ch_{S(T)}X$ such that $\mathrm{supp}(\mu_z) = \overline{Ch_{S(T)}X}$. Then $H(T)$ is an antilattice, i.e. if $h_1, h_2 \in H(T)$ such that there exists a greatest minorant of $\inf(h_1, h_2)$ in $H(T)$ then $h_1 \leq h_2$ or $h_2 \leq h_1$.

2. Let $q \in P$. Then $P + \mathbb{R}q$ is a min-stable simplicial function cone such that for every $x \in X$, $\mu_x = \varepsilon_x^{\delta(q)}$ and $M_x(P + \mathbb{R}q) = \{\mu \in M_x(P) : \mu^{\delta(q)} = \varepsilon_x^{\delta(q)}\}$. (Hint: Use (II.7.13.2)).

6. Weak Dirichlet Problem

Let $(X, \mathcal{W})$ be a balayage space and $U \subset X$ be an open set. Given a continuous real-valued P-bounded function f on $\complement U$, the "classical Dirichlet problem" asks for a continuous extension of f to a function in $C_P(X)$ which is harmonic in U . Therefore, one is interested in the linear space

$$H(U) = C_P(X) \cap \mathcal{H}(U) .$$

For a non-regular set U , one may ask if at least for some subsets B of $\complement U$ every continuous function on B admits a continuous extension to a function in $H(U)$. Because of the general minimum-principle (I.2.2), a natural candidate for such a set B would be $Ch_{S(U)}X$ where

$$S(U) = C_P(X) \cap S(U) \quad .$$

However, $Ch_{S(U)}X$ may not be closed. Thus the following *weak Dirichlet problem* arises: Given a closed subset A of $Ch_{S(U)}X$ and a continuous P-bounded function f on A , is there a continuous extension of f to a function h in H(U) ? The results of the preceding section give a positive solution to this problem, moreover it will be clarified how this extension h can be assigned to f .

To this end consider the set

$$T := \{(\varepsilon_x^{\complement V})_{x \in X} : V \in \mathcal{U}(U)\}$$

of dilations. Then

$$S(T) = S(U) \ , \ H(T) = H(U) \ , \ b(T) = \complement U \quad .$$

By (5.3) and (3.2.1)

$$Ch_{S(U)}X = \beta(\complement U) \quad \text{and} \quad U_r = U^* \cap b(\complement U) \quad ,$$

hence $U^* \cap Ch_{S(U)}X \subset U_r$. Note that by (VI.6.5) $U^* \cap Ch_{S(U)}X = U_r = U^*$ if U is regular.

(5.3) and (5.6) yield at once the solution to the weak Dirichlet problem:

6.1. PROPOSITION. For every open subset U of X and every closed subset A of $Ch_{S(U)}X$ there exists a P-dilation D such that $D(C_P(X)) \subset H(U)$ and $D(x,\cdot) = \varepsilon_x$ for every $x \in A$.

Furthermore, S(U) is a simplicial cone, and for every $x \in X$, $D^U(x,\cdot) = \varepsilon_x^{\beta(\complement U)}$ is the only measure satisfying $D^U(x,\cdot) \in M_x(S(U))$ and $D^U(x, X \setminus Ch_{S(U)}X) = 0$.

These results will be extended and improved in section 8.

As an immediate consequence of (4.1) and (4.2) the following two propositions clarify the connections between $D^U(x,\cdot)$ and $\varepsilon_x^{\complement U}$.

6.2. PROPOSITION. Let U be an open subset of X and let U_i be the set of irregular boundary points of U . Then the following statements are equivalent:

(1) For every $x \in X$, $D^U(x,\cdot) = \varepsilon_x^{\complement U}$.

(2) For every $x \in U$, $\varepsilon_x^{\complement U}(U_i) = 0$.

6.3. COROLLARY. The following assertions are equivalent:

(1) For every open subset U of X and every $x \in X$, $D^U(x,\cdot) = \varepsilon_x^{\complement U}$.

(2) $(X,\mathcal{W})$ satisfies the axiom of polarity.

In the following we shall characterize the identity between the Choquet boundary $Ch_{S(U)}X$ and the set of regular points U_r.

6.4. LEMMA. Let $S \in \mathcal{B}(X)$ be semipolar. For every $x \in b(S)$ there exists a subset C of S such that $C \cup \{x\}$ is compact and $b(C) = \{x\}$.

Proof. We may assume that $S = \bigcup_{n=1}^{\infty} T_n$ where (T_n) is an increasing sequence of totally thin sets in $\mathcal{B}(X)$. Let $p \in P$ be strict and let $(V_n) \subset \mathcal{U}$ such that $\overline{V}_{n+1} \subset V_n$ and $\bigcap_{n=1}^{\infty} V_n = \{x\}$. Let $n \in \mathbb{N}$. Then $x \in b(S \cap V_n)$, hence

$$p(x) = \hat{R}_p^{S \cap V_n}(x) = \sup_m \hat{R}_p^{T_m \cap V_n}(x) .$$

Thus there exists an integer $m_n \in \mathbb{N}$ such that

$$\hat{R}_p^{T_{m_n} \cap V_n}(x) > p(x) - \frac{1}{n} .$$

By (VI.1.9) there exists a compact subset K_n of $T_{m_n} \cap V_n$ such that $\hat{R}_p^{K_n}(x) > p(x) - \frac{1}{n}$. If $C := \bigcup_{n=1}^{\infty} K_n$, then

$$\hat{R}_p^{C}(x) \geq \hat{R}_p^{K_n}(x) > p(x) - \frac{1}{n} ,$$

for every $n \in \mathbb{N}$. Hence $\hat{R}_p^{C}(x) = p(x)$, i.e. $x \in b(C)$. Since $C \subset V_1$, the set C is relatively compact. Moreover, for every $m \in \mathbb{N}$,

$$C \subset V_m \cup \bigcup_{n<m} K_n ,$$

hence

$$b(C) \subset b(V_m) \cup \bigcup_{n<m} b(K_n) = b(V_m) \subset \overline{V}_m$$

and

$$\overline{C} \subset \overline{V}_m \cup \bigcup_{n<m} K_n \subset \overline{V}_m \cup C .$$

Therefore $b(C) \subset \{x\}$ and $\overline{C} \subset \{x\} \cup C$ which shows that $C \cup \{x\}$ is closed and $b(C) = \{x\}$. ┘

6.5. PROPOSITION. Suppose that the points of X are totally thin. Then the following statements are equivalent:

(1) For every open set U in X , $\beta(\complement U) = b(\complement U)$, i.e. $U^* \cap Ch_{S(U)}X = U_r$.

(2) $(X,\mathcal{W})$ satisfies the axiom of thinness.

Proof. (1) ⇒ (2): Assume that there is a semipolar set S which is not totally thin. We may assume that S is a Borel set. By (6.4) there exists a point $x \in X$ and a subset C of S such that $K := C \cup \{x\}$ is compact and $b(K) = \{x\}$. If $V \subset X$ is open such that $K \subset V$ then $U := V \setminus K$ satisfies $\complement U = K \cup \complement V$, hence $b(\complement U) = b(K) \cup b(\complement V) = \{x\} \cup b(\complement V)$ and $\beta(\complement U) = \beta(\complement V)$. In particular, $x \in b(\complement U) \setminus \beta(\complement U)$.

(2) ⇒ (1): (VI.6.3). ┘

6.6. REMARK. (6.5) and (5.8) yield the following result due to KELDYCH in classical potential theory: If $(X,\mathcal{W})$ satisfies the axiom of thinness then for every open subset U of X such that $H(U)$ is linearly separating, for every regular boundary point z of U there is a function $h \in H(U)$ which has a unique minimum at z .

6.7. EXAMPLE. Heat equation. Let $U = \mathbb{R}^n \times (]0,1[\setminus \{1 - \frac{1}{m} : m \in \mathbb{N}\})$. Then $U^* \cap Ch_{S(U)}\mathbb{R}^{n+1} = \mathbb{R}^n \times \{0\}$, $U_r = \mathbb{R}^n \times \{0,1\}$. (Note that $n = 0$ yields an example for translation on $\mathbb{R}$.)

6.8. EXERCISES. 1. Let $h \in H(U)$ such that $h > 0$ on U . Then $h > 0$ on $\overline{U} \setminus Ch_{S(U)}X$. (Hint: Given $z \in \overline{U} \setminus Ch_{S(U)}X$, choose an exhaustion (K_n) of U , show that the sequence $(\varepsilon_z^{\beta(\complement U) \cup K_n})$ converges to ε_z , and deduce that $\varepsilon_z^{\beta(\complement U) \cup K_n}(K_n) > 0$ for some $n \in \mathbb{N}$. Finally use (VI.4.18.3).)

2. Suppose that $(X,\mathcal{W})$ has the local truncation property and let U be an open subset of X . Then $\mathfrak{S}(U) = S(U)|_{\overline{U}}$ is a simplicial function cone, $Ch_{\mathfrak{S}(U)}\overline{U} = U^* \cap \beta(\complement U)$, and $\mu_x = \varepsilon_x^{\beta(\complement U)}$ for every $x \in \overline{U}$. If $Ch_{\mathfrak{S}(U)}\overline{U}$ is closed then

$\mathfrak{H}(U) := \mathfrak{S}(U) \cap (-\mathfrak{S}(U))$ is a lattice. If there exists $z \in \overline{Ch_{\mathfrak{S}(U)}\overline{U}} \setminus Ch_{\mathfrak{S}(U)}\overline{U}$ such that $\operatorname{supp}(\mu_z) = Ch_{\mathfrak{S}(U)}\overline{U}$ then $\mathfrak{H}(U)$ is an antilattice.

3. Classical potential theory. Let U be an open connected set. Then $\mathfrak{H}(U)$ is a lattice (if U_r is closed) or an antilattice (if U_r is not closed). (Hint: Use (6.8.1) in order to show that $\operatorname{supp}(\mu_z) = \overline{Ch_{\mathfrak{S}(U)}\overline{U}}$ for every $z \in \overline{Ch_{\mathfrak{S}(U)}\overline{U}} \setminus Ch_{\mathfrak{S}(U)}\overline{U}$.)

7. Characterization of the Generalized Solution

Let U be an open subset of X and let $f \in C_P(X)$. The solutions H_f^U and $D^U f$ of the generalized Dirichlet problem and of the weak Dirichlet problem, respectively, define harmonic functions on U. Evidently, $H_f^U = D_f^U = f$ if $f \in H(U)$. For an arbitrary function $f \in C_P(X)$ it is by no means clear which is the "right solution" of the Dirichlet problem. The aim of this section is to show the distinguished rôle of the generalized solution among other ones.

A dilation L on X is called a *Keldych operator* if $L(C_P(X)) \subset H(U)$ such that $Lh = h$ for any $h \in H(U)$. The set U is called a *Keldych set* if for every Keldych operator L on U and every $f \in C_P(X)$ we have $Lf = H_f^U$ on U.

7.1. LEMMA. Let L be a Keldych operator on an open subset U of X. Then for every $p \in P$,

$$D^U p \leq L p \leq H_p^U \quad .$$

Proof. Let $p \in P$. By the proof of (5.3)

$$\begin{aligned} D^U p &= \sup \{q \in P \cap H(U) : q \leq p\} \\ &= \sup \{L q : q \in P \cap H(U),\ q \leq p\} \leq L p \quad . \end{aligned}$$

Let w be an arbitrary Evans function on U and define a hyperharmonic function u on U by $u = H_p^U - Lp + w$. Since $u \geq H_p^U - p + w$, we have $\liminf_{x \to z} u(x) \geq 0$ for every $z \in U^*$, $u \geq 0$ on $\complement U$ and $u \geq -p$ on X, hence by the minimum principle (III.6.6) $u \geq 0$ on X and therefore $Lp \leq H_p^U$ because w was arbitrarily chosen. ∎

7.2. COROLLARY. An open subset U of X is a Keldych set if and only if $D^U(x,\cdot) = \varepsilon_x^{\complement U}$ for every $x \in U$.

The next proposition shows that there exist "many" Keldych sets.

7.3. PROPOSITION. Let U be an open subset of X. If $A \subset U$ is an arbitrary closed set then there is an open set V such that $A \subset V \subset \overline{V} \subset U$ and $\varepsilon_x^{\complement V}(V_i) = 0$ for every $x \in X$. In particular, V is a Keldych set.

Proof. Let $p \in P$ be strict, $x \in U$, and choose $f \in C(X)$ such that $0 \leq f \leq 1$, $f = 1$ on A and the support of f is contained in U. For any $a \in]0,1[$ define $U(a) = \{f > a\}$ and $g(a) = R^{\hat{R}_p^{\complement U(a)}}_{\hat{R}_p^{\complement U(a)}}(x)$. Then $U(a)$ is an open subset of X such that for $0 < a < b < 1$

$$A \subset U(b) \subset \overline{U(b)} \subset U(a) \subset U \quad .$$

Futhermore, $g :]0,1[\to \mathbb{R}$ is an increasing function, hence right continuous at all points of $]0,1[$ except for a countable set $A_x \subset]0,1[$. If $0 < a < b < 1$, $a \in]0,1[\setminus A_x$ then $p = \hat{R}_p^{\complement U(b)}$ on $\overbrace{\complement U(b)}^{\circ} = \complement \overline{U(b)} \supset \complement U(a)$, hence

$$g(a) = \int \hat{R}_p^{\complement U(a)} d\, \varepsilon_x^{\complement U}(a) \leq \int R_p^{\complement U(a)} d\, \varepsilon_x^{\complement U(a)}$$

$$= \int p\, d\, \varepsilon_x^{\complement U(a)} = \int \hat{R}_p^{\complement U(b)} d\, \varepsilon_x^{\complement U(a)} \leq g(b) \quad .$$

Therefore $g(a) \leq \int R_p^{\complement U(a)} d\, \varepsilon_x^{\complement U(a)} \leq \lim\limits_{b \downarrow a} g(b) = g(a)$, thus

$$\varepsilon_x^{\complement U(a)}(R_p^{\complement U(a)} - \hat{R}_p^{\complement U(a)}) = 0 \, ,$$

which implies $\varepsilon_x^{\complement U(a)}(U(a)_i) = 0$.

Let $(x_n) \subset U$ be a dense sequence and let $a \in]0,1[\setminus \bigcup_{n \in \mathbb{N}} A_{x_n}$. Then $V = U(a)$ satisfies $A \subset V \subset \overline{V} \subset U$ and $\varepsilon_{x_n}^{\complement V}(V_i) = 0$ for all $n \in \mathbb{N}$. The assertion now follows from (VI.2.10), (6.2) and (7.2). ┘

7.4. PROPOSITION. Let L be a Keldych operator on an open subset U of X. Then for every $f \in C_P(X)$, $Lf = H_f^U$ on U if and only if L satisfies the following *interior stability condition* : If $f \in C_P(X)$, (U_n) is an exhaustion of U and for every $n \in \mathbb{N}$, L_n is a Keldych operator on U_n, then $\lim\limits_{n \to \infty} L_n f = Lf$ on U.

Proof. Choose for every $n \in \mathbb{N}$ by (7.3) a Keldych set V_n such that $U_n \subset \overline{U}_n \subset V_n \subset \overline{V}_n \subset U_{n+1}$. Obviously, (V_n) is an exhaustion of U. Let $p \in P$ and $n \in \mathbb{N}$. Then for every $x \in U$ by (7.2) and (7.1)

$$\varepsilon_x^{\complement V_n}(p) = D^{V_n}p(x) = \varepsilon_x^{\beta(\complement V_n)}(p) \leq \varepsilon_x^{\beta(\complement U_n)}(p) = D^{U_n}p(x) \leq L_n p(x) \leq \varepsilon_x^{\complement U_n}(p) \quad .$$

The assertion follows since by (VI. 10.5)

$$\lim_{n\to\infty} \varepsilon_x^{\complement V_n}(p) = \lim_{n\to\infty} \varepsilon_x^{\complement U_n}(p) = \varepsilon_x^{\complement U}(p) \quad .$$

□

7.5. EXAMPLE. 1. If $(X,\mathcal{W})$ satisfies the axiom of polarity (e.g. classical potential theory, Riesz potentials) then every open subset of X is a Keldych set by (7.2) and (6.3).

2. Every semiregular set is a Keldych set by (7.2), (4.4) and (6.2).

3. The sets V of (4.5.3) (translation on $\mathbb{R}$) and (4.5.5) (heat equation) are not Keldych sets.

7.6. EXERCISES. 1. Let U be an open subset of X. There exists a unique Keldych operator L^U such that $L^U p = \sup \{u \in \mathcal{W} : u \leq p , \hat{R}_u^{\complement U} = u\}$ for every $p \in P$.

2. Translation on $\mathbb{R}$. If $A = \{1\} \cup \{1 + \frac{1}{n} : n \in \mathbb{N}\}$ and $U = \complement A$ then $D^U \neq L^U \neq H^U$. (Hint: Consider $p(x) = \sup (x,0)$.)

8. Fine Dirichlet Problem

As before let $(X,\mathcal{W})$ be a balayage space. The results of section 5 will now be applied to cones of "finely superharmonic" functions. This will extend and improve our results of section 6.

For that purpose let G be a finely open subset of X and let $\mathcal{V}$ be a local covering of G by finely open sets, i.e. let $\mathcal{V}$ be a family of finely open sets V satisfying $\overline{V} \subset G$ such that for every open subset U of X the family $\mathcal{V}(U) = \{V \in \mathcal{V} : \overline{V} \subset U\}$ is a covering of $G \cap U$. Define

$$S(G,\mathcal{V}) = \{s \in C_P(X) : \varepsilon_x^{\complement V}(s) \leq s(x) \ \forall\, x \in V \in \mathcal{V}\} , \ H(G,\mathcal{V}) = S(G,\mathcal{V}) \cap (-S(G,\mathcal{V})).$$

If $T = \{(\varepsilon_x^{\complement V})_{x \in X} : V \in \mathcal{V}\}$ then

$$S(G,\mathcal{V}) = S(T),\; H(G,\mathcal{V}) = H(T), \text{ and } b(T) = \complement G \;.$$

hence by (5.3)

$$Ch_{S(G,\mathcal{V})}X = \beta(\complement G) \;.$$

Furthermore, $S(G,\mathcal{V})$ is a simplicial cone, i.e. for every $x \in X$ there exists a unique measure $\mu_x \in M_x(S(G,\mathcal{V}))$ which is supported by $Ch_{S(G,\mathcal{V})}X$ and μ_x is given by

$$\mu_x = \varepsilon_x^{\beta(\complement G)} \;.$$

In fact, μ_x is the only measure in $M_x(H(G,\mathcal{V}))$ supported by $\beta(\complement G)$.

Our first aim is to show that $\varepsilon_x^{A \cup \beta(\complement G)} \in M_x(S(G,\mathcal{V}))$ for every subset A of X. To that end we shall define a Dirichlet problem having solutions

$$Df : x \mapsto \overset{o}{\varepsilon}{}_x^{A \cup \beta(\complement G)}(f) \;, \quad f \in C_p(X) \;,$$

such that trivially $Ds \leq s$ for every $s \in S(G,\mathcal{V})$. In detail we shall proceed as follows.

Let $A \in \mathcal{B}(X)$ such that $\beta(\complement G) \subset A$ and let

$$F := S(S(G,\mathcal{V})) - \{\sum_{i=1}^{n} R_{p_i}^{K_i} : n \in \mathbb{N},\; p_i \in P,\; K_i \text{ compact} \subset A\} \;.$$

<u>8.1. LEMMA.</u> Let $u \in F$ such that $u \geq 0$ on A. Then $u \geq 0$.

Proof. There exists a compact subset K of A such that u is contained in the convex cone

$$F_K = S(S(G,\mathcal{V})) - \{\sum_{i=1}^{n} R_{p_i}^{K_i} : n \in \mathbb{N},\; p_i \in P,\; K_i \text{ compact} \subset K\} \;.$$

Obviously $P \subset S(G,\mathcal{V}) \subset F_K$. By (I.2.2) it suffices to show that $Ch_{F_K} X \subset A$.

Let $x \in \complement A$. Then there exists an open neighborhood U of x such that $\overline{U} \cap K = \emptyset$. Furthermore, $x \notin \beta(\complement G)$. If $W = G \cap U$ then $x \notin \beta(\complement W)$, hence $\varepsilon_x^{\beta(\complement W)} \neq \varepsilon_x$ and by (5.6)

$$\varepsilon_x^{\beta(\complement W)} \in M_x(S(W,\mathcal{V}(W))) \subset M_x(S(G,\mathcal{V})) \;.$$

Since K is contained in the interior of $\complement W$ and $x \in \complement K$, we have by (VI.9.1) for every compact subset L of K and every $p \in P$

$$\int R_p^L \, d\,\varepsilon_x^{\beta(\complement W)} = R_p^L(x) \quad .$$

Therefore $\varepsilon_x^{\beta(\complement W)} \in M_x(F_K)$ and thus $x \notin Ch_{F_K} X$, i.e. $Ch_{F_K} X \subset A$. ┘

For every $f \in C_P(X)$ we define

$$\overline{D}^A f = \inf \{u \in F : u \geq f \text{ on } A\} \quad ,$$

$$\underline{D}^A f = \sup \{v \in -F : v \leq f \text{ on } A\} = -\overline{D}^A(-f) \quad .$$

Obviously, the mapping $f \mapsto \overline{D}^A f$ is sublinear and $\underline{D}^A f \leq \overline{D}^A f$ by (8.1).

<u>8.2. PROPOSITION.</u> For every $f \in C_P(X)$ and every $x \in X$,

$$\underline{D}^A f(x) = \overline{D}^A f(x) = \overset{o}{\varepsilon}{}_x^A(f) \quad .$$

Proof. Let $p \in P$. Since $W = S(P) \subset F$ we have $\overline{D}^A p \leq R_p^A$. If K is a compact subset of A then $R_p^K \in -F$ and $R_p^K \leq p$, hence $R_p^K \leq \underline{D}^A p$. Applying (VI.1.9) we conclude that $R_p^A \leq \underline{D}_p^A$, thus $\underline{D}^A p = \overline{D}_p^A = R_p^A$. The proof is finished by the usual approximation procedure based on (I.1.3). ┘

<u>8.3. COROLLARY.</u> For every $x \in X$, $\{\varepsilon_x^A : \beta(\complement G) \subset A \subset X\} \subset M_x(S(G,V))$.

Proof. Let $x \in X$ and $\beta(\complement G) \subset A \subset X$. By (VI.2.2) we may assume that $A \in \mathcal{B}(X)$. If $x \notin A$ then for every $s \in S(G,V)$

$$\varepsilon_x^A(s) = \overset{o}{\varepsilon}{}_x^A(s) = \overline{D}^A s(x) \leq s(x) \quad ,$$

hence $\varepsilon_x^A \in M_x(S(G,V))$. If $x \in \beta(\complement G)$ then $\varepsilon_x^A = \varepsilon_x \in M_x(S(G,V))$. Finally, if $x \in A \setminus \beta(\complement G)$ then $\varepsilon_x^{A \setminus \{x\}} \in M_x(S(G,V))$ by the preceding considerations. Therefore (VI.9.2) implies that

$$\varepsilon_x^A = \varepsilon_x^A(\{x\})\,\varepsilon_x + (1 - \varepsilon_x^A(\{x\}))\,\varepsilon_x^{A \setminus \{x\}} \in M_x(S(G,V)) \quad .$$ ┘

Writing $\mathbb{S}(G)$ instead of $S(G,\{V : V \text{ finely open}, \overline{V} \subset G\})$ we have the following result:

<u>8.4. PROPOSITION.</u> For every local covering V of G by finely open sets, $S(G,V) = \mathbb{S}(G)$.

Proof. Obviously, $S(G) \subset S(G,\mathcal{V})$. Conversely, let $s \in S(G,\mathcal{V})$, $x \in X$ and let V be a finely open set such that $x \in V \subset \overline{V} \subset G$. Then $\beta(\complement G) \subset \complement G \subset \complement V$, hence $\varepsilon_x^{\complement V} \in M_x(S(G,\mathcal{V}))$ by (8.3), i.e. $\varepsilon_x^{\complement V}(s) \leq s(x)$. Thus $s \in S(G)$. ┘

8.5. REMARKS. 1. If U is an open subset of X and if $\mathcal{V} = \{V : V \text{ open}, \overline{V} \subset U\}$ then (8.4) implies that the two definitions of $S(U)$ given in sections 6 and 8 coincide, i.e. every function in $S(U) = C_P(X) \cap \mathcal{S}(U)$ is *finely superharmonic* on U .

2. If $\mathcal{V}$ is a local covering of a finely open subset G of X by finely open sets then by (8.4)

$$H(G,\mathcal{V}) = H(G) \quad ,$$

where $H(G,\mathcal{V}) = S(G,\mathcal{V}) \cap (-S(G,\mathcal{V}))$, $H(G) = S(G) \cap (-S(G))$.

3. It is now possible to characterize the Choquet boundary $Ch_{S(G)}X$, where G is a finely open Borel set, in a similar manner as the regular points of an open subset by the existence of *strong barriers*: A point $z \in \overline{G}$ belongs to $\beta(\complement G)$ if and only if there exists a function $s \in S(G)^+$ such that $s > 0$ on G and $s(z) = 0$.

Indeed, let $z \in \beta(\complement G)$. By (VI.4.16) we may choose a strict potential $p \in P$ and $q \in P$ such that $R_p^{\{z\}} \leq q \leq R_p^{\beta(\complement G)}$ and $R_q^{\beta(\complement G)} = q$. Then $s := p - q \in S(G)^+$, $s > 0$ on G , and $s(z) = 0$. For the converse, we proceed as in the proof of (3.3). For that purpose, let $z \in \overline{G} \setminus \beta(\complement G)$. Then $z \in \beta(G)$. By (VI.10.4) there exists an increasing sequence (K_n) of compact subsets of G such that $\lim_{n\to\infty} \varepsilon_z^{K_n} = \varepsilon_z^G = \varepsilon_z$ and hence $\lim_{n\to\infty} \varepsilon_z^{K_n \cup \beta(\complement G)} = \varepsilon_z$. Then for some $n \in \mathbb{N}$, $\varepsilon_z^{K_n \cup \beta(\complement G)}(K_n) > 0$ since otherwise $\varepsilon_z = \varepsilon_z^{\beta(\complement G)}$, $z \in \beta(\complement G)$. If $s \in S(G)^+$ such that $s > 0$ on G we then obtain by (8.3) that $s(z) \geq \varepsilon^{K_n \cup \beta(\complement G)}(s) > 0$.

8.6. THEOREM. Let U and V be finely open subsets of X . Then the following statements are equivalent:

(1) $S(U) \subset S(V)$.

(2) $H(U) \subset H(V)$.

(3) $H(U) \subset S(V)$.

(4) The constant zero is the only potential $p \in \mathcal{P}$ such that $C(p) \subset V \setminus U$.

(5) $V \setminus U$ is semipolar.

(6) $Ch_{S(U)}X \subset Ch_{S(V)}X$.

Proof. (1) $\Rightarrow$ (2) $\Rightarrow$ (3) : Obvious.

(3) $\Rightarrow$ (4) : Assume that there is a potential $p \in \mathcal{P}$, $p \neq 0$ such that $C(p) \subset V \setminus U$. Let $x \in \delta(p)$ and let K be a compact fine neighborhood of x in V. Then $\varepsilon_x^{\complement K} \neq \varepsilon_x$ and therefore $\varepsilon_x^{\complement K}(p) < p(x)$ by (VI.8.2), hence $-p \notin S(V)$, but evidently $-p \in H(U)$ since $C(p) \subset \complement U$.

(4) $\Rightarrow$ (5) : Assume that $V \setminus U$ is not semipolar. By (VI.5.16) there exist finely open sets $U', V' \in \mathcal{B}(X)$ such that $U' \subset U$, $V' \subset V$ and the sets $U \setminus U'$, $V \setminus V'$ are semipolar. Furthermore, there exists a semipolar set $S \in \mathcal{B}(X)$ such that $U \setminus U' \subset S$. Then $B = V' \setminus (U' \cup S) \in \mathcal{B}(X)$, $B \subset V \setminus U$, and B is not semipolar since $V \setminus U \subset (V \setminus V') \cup B \cup S$. By (VI.8.8) there exists a potential $p \in \mathcal{P}$ such that $p \neq 0$ and $C(p) \subset B \subset V \setminus U$.

(5) $\Rightarrow$ (6) : If $V \setminus U$ is semipolar then $\beta(\complement U) \subset \beta(\complement V)$ since $\complement U \subset (V \setminus U) \cup \complement V$, hence (6) follows from

$$Ch_{S(U)}X = \beta(\complement U) \subset \beta(\complement V) = Ch_{S(V)}X .$$

(6) $\Rightarrow$ (1) : Let $s \in S(U)$, $x \in X$ and let W be a finely open set such that $x \in W \subset \overline{W} \subset V$. Then $\varepsilon_x^{\complement W} \in M_x(S(U))$ by (8.3) since $\beta(\complement U) \subset \beta(\complement V)$. Therefore $\varepsilon_x^{\complement W}(s) \leq s(x)$, i.e. $s \in S(V)$. ┘

<u>8.7. COROLLARY.</u> Let U be a finely open subset of X and let $V = \overline{U} \setminus Ch_{S(U)}X$. Then V is finely open, $S(U) = S(V)$, and $Ch_{S(V)}X = \complement V$. If U is open then in particular, $S(U) = S(\overset{\circ}{\overline{U}} \setminus \overline{Ch_{S(U)}X})$.

Proof. We have $U \subset V \subset \overline{U}$ and $\complement V = \complement \overline{U} \cup Ch_{S(U)}X = \complement \overline{U} \cup \beta(\complement U) = \beta(\complement U)$. Hence V is finely open and $Ch_{S(V)}X = \beta(\complement V) = \beta(\complement U) = Ch_{S(U)}X$. Then $S(U) = S(V)$ by (8.6) and $Ch_{S(V)}X = \complement V$. If U is open then $U \subset \overset{\circ}{\overline{U}} \setminus \overline{Ch_{S(U)}X} \subset V$, and (8.6) implies the last assertion. ┘

<u>8.8. EXERCISES.</u> 1. For every finely open subset G of X there exists a

potential $q \in P$ such that $S(G) = S(\complement\delta(q))$, hence in particular, $Ch_{S(G)}X = \delta(q)$.

2. Let F be a closed subset of X . Then the following statements are equivalent:

(1) $S(\overset{\circ f}{F}) = S(\overset{\circ}{F})$.

(2) $H(\overset{\circ f}{F}) = H(\overset{\circ}{F})$.

(3) $S(\overset{\circ f}{F}) \supset H(\overset{\circ}{F})$.

(4) $F^* \setminus b(\complement F)$ is semipolar.

(5) The fine interior of F^* is empty and for every $x \in \overset{\circ}{F}$ the measure $\varepsilon_x^{\beta(\complement\overset{\circ}{F})}$ is supported by $b(\complement F)$.

(6) $\beta(F^*) \subset b(\complement F)$.

9. Approximation

In this section we shall see that for every finely open subset G the subset $P + H(G)$ of $S(G)$ is "dense" in $S(G)$. This holds as well for the even smaller subset $P - P \cap H(G)$, and then the result can be viewed as an extension theorem for functions in $S(G)$. Moreover, it will turn out that a set $S(G)$ ($H(G)$ resp.) where G is finely open, but in general not open, arises in a natural way if we are interested in continuous functions which are superharmonic (harmonic resp.) in a neighborhood of a given closed set F : The set of these functions is "dense" in $S(G)$ ($H(G)$ resp.) where G is the fine interior of F .

For a precise statement we need the following definitions: For every subset F of $C_P(X)$ let $\overline{F}$ denote the set of all $g \in C_P(X)$ such that there exists $p_0 \in P$ such that for every $\varepsilon > 0$ there exists a function $f \in F$ satisfying $|f - g| \leq \varepsilon p_0$. For every subset A of X let

$$S_0(A) = \bigcup_{U \text{ open} \supset A} S(U) \quad , \quad H_0(A) = \bigcup_{U \text{ open} \supset A} H(U) \quad .$$

Then we shall obtain the following results: For every finely open subset G of X, $\overline{P - P \cap H(G)} = \overline{P + H(G)} = S(G)$. This has interesting consequences for the representing measures. Moreover, if G is the fine interior of a finely closed Borel subset F of X then $\overline{S_0(F)} = S(G)$ and $\overline{H_0(F)} = H(G)$.

<u>9.1. PROPOSITION.</u> Let G be a finely open subset of X . Then $S(G) = \overline{\mathcal{P} + H(G)}$ $= \overline{\mathcal{P} - \mathcal{P} \cap H(G)}$.

If G is the fine interior of a finely closed Borel subset F of X then $S(G) = \overline{\mathcal{P} + H_0(F)} = \overline{\mathcal{P} - \mathcal{P} \cap H_0(F)}$.

Proof. Since $\mathcal{P} - (\mathcal{P} \cap H(G)) \subset \mathcal{P} + H(G) \subset S(G)$ and $\overline{S(G)} = S(G)$, obviously $\overline{\mathcal{P} - \mathcal{P} \cap H(G)} \subset \overline{\mathcal{P} + H(G)} \subset S(G)$.

Conversely, let $s \in S(G)$. Let $p_0 \in \mathcal{P}$ such that $p_0 > 0$ and $|s| \leq p_0$. Choose $q_0 \in \mathcal{P}$ such that $q_0 > 0$ and $p_0 \in o(q_0)$. Fix $\varepsilon > 0$. Then by (I.1.3) there exist $p', q' \in \mathcal{P}$ such that $|s - (p' - q')| \leq \varepsilon q_0$. Let K be a compact subset of X such that $p_0 \leq \varepsilon q_0$ on $\complement K$, let $a \geq 1$ such that $p' + q' \leq a p_0$ on K , and define

$$p := \inf(p', a p_0) , \quad q := \inf(q', a p_0) .$$

Then $p, q \in \mathcal{P}$ and

$$|s - (p - q)| \leq \varepsilon q_0 .$$

Let D denote the $\mathcal{W}$-dilation given by $D(x,\cdot) = \varepsilon_x^{\beta(\complement G)}$ and define

$$u := \inf(p, s + \varepsilon q_0 + Dq) .$$

Then u is a $\mathcal{P}$-bounded l.s.c. function on X . Let $x \in X$ and let U be an open neighborhood of x . If $x \in \beta(\complement G)$ then $\varepsilon_x^{\complement U}(u) \leq \varepsilon_x^{\complement U}(p) \leq p(x) = u(x)$ since $Dq = q$ on $\beta(\complement G)$. If $x \notin \beta(\complement G)$ then $V = U \setminus \beta(\complement G)$ is a fine neighborhood of x , hence $\overset{\circ}{\varepsilon}{}_x^{\complement V} = \varepsilon_x^{\complement V} \neq \varepsilon_x$, and $\varepsilon_x^{\complement V}(s) \leq s(x)$ by (8.3), hence $\varepsilon_x^{\complement V}(u) \leq u(x)$. Thus $u \in \mathcal{W}$ by (II.5.5). Moreover

$$u - Dq = s + \varepsilon q_0 \qquad \text{on } \{p \geq s + \varepsilon q_0 + Dq\}$$

and

$$p - q \leq p - Dq = u - Dq \leq s + \varepsilon q_0 \qquad \text{on } \{p \leq s + \varepsilon q_0 + Dq\} .$$

Therefore

$$|s - (u - Dq)| \leq \varepsilon q_0 .$$

Let L be a compact subset of X such that $a p_0 \leq \varepsilon q_0$ on $\complement L$. Let (p_n) be an increasing sequence in P such that $u = \lim p_n$. By (VI.8.5) there exists an increasing sequence (q_n) in $P \cap H(G)$ such that $Dq = \lim q_n$. Since $|s - (u - Dq)| < 2\varepsilon q_0$ we conclude from (I.1.11) that there exist $n \in \mathbb{N}$ and $\alpha_1, \dots, \alpha_n \in [0,1]$ with $\sum_{j=1}^{n} \alpha_j = 1$ such that $\tilde{p} := \sum_{j=1}^{n} \alpha_j p_j$ and $\tilde{q} := \sum_{j=1}^{n} \alpha_j q_j$ satisfy

$$|s - (\tilde{p} - \tilde{q})| < 2\varepsilon q_0 \qquad \text{on} \quad L .$$

Since $\tilde{p} \leq u \leq p$ and $\tilde{q} \leq Dq \leq q$ we know that

$$|s - (\tilde{p} - \tilde{q})| \leq |s| + |\tilde{p} - \tilde{q}| \leq p_0 + a p_0 \leq 2\varepsilon q_0 \qquad \text{on} \quad \complement L .$$

Therefore $|s - (\tilde{p} - \tilde{q})| \leq 2\varepsilon q_0$. Clearly $\tilde{p} \in P$ and $\tilde{q} \in P \cap H(G)$. Thus $s \in \overline{P - P \cap H(G)}$.

Suppose finally that G is the fine interior of a finely closed Borel subset F . Then G is the complement of $b(\complement F)$, $b(\complement F) = \beta(\complement F)$, and hence $\beta(\complement G) = \beta(\complement F) \supset \complement F$. Therefore we may choose the sequence (q_n) such that even $C(q_n) \subset \complement F$, i.e. $q_n \in P \cap H_0(F)$ for every $n \in \mathbb{N}$, and hence $\tilde{q} \in P \cap H_0(F)$. Thus the preceding considerations show that $S(G) \subset \overline{P - P \cap H_0(F)}$. Moreover, trivially $\overline{P - P \cap H_0(F)} \subset \overline{P + H_0(F)} \subset S(G)$. ⌟

A stronger version of the preceding result is obtained in (9.6.4).

9.2. COROLLARY. Let F be a finely closed Borel subset of X and let G be the fine interior of F . Then $\overline{S_0(F)} = S(G)$ and $\overline{H_0(F)} = H(G)$.

Proof. Since $P + H_0(F) \subset S_0(F) \subset S(G)$ we obtain immediately from (9.1) that $\overline{S_0(F)} = S(G)$. Moreover, $H(U) \subset H(G)$ for every open neighborhood U of F and hence obviously $\overline{H_0(F)} \subset \overline{H(G)} = H(G)$.

Consider now $g \in H(G) = S(G) \cap (-S(G))$. Then there exists $q_0 \in P$ such that for every $\varepsilon > 0$ there exists an open neighborhood U of F and functions $s \in S(U)$, $t \in -S(U)$ satisfying

$$t \leq g \leq s , \quad s - t \leq \varepsilon q_0 .$$

By (5.9), there exists a function $h \in H(U)$ such that $t \leq h \leq s$. Then $h \in H_0(F)$ and $|g - h| \leq s - t \leq \varepsilon q_0$. Thus $g \in \overline{H_0(F)}$. ⌟

9.3. COROLLARY. Let F be a finely closed Borel subset of X. Then the following statements are equivalent:

(1) $\overline{S_0(F)} = C_P(X)$.

(2) $\overline{H_0(F)} = C_P(X)$.

(3) F has no finely interior points.

In classical potential theory the following proposition (for compact sets F) is known as *approximation theorem of* KELDYCH-BRELOT.

9.4. PROPOSITION. Let F be a closed subset of X. Then the following statements are equivalent:

(1) $F^* \subset b(\complement F)$.

(2) $\overline{H_0(F)}|_{\complement \overset{\circ}{F}} = C_P(X)|_{\complement \overset{\circ}{F}}$.

Proof. (1) ⇒ (2): Since $F^* \subset b(\complement F) = \overline{\complement F}^f = \complement \overset{\circ}{F}^f$ we obtain that $\overset{\circ}{F}^f = \overset{\circ}{F}$ and hence $\overline{H_0(F)} = H(\overset{\circ}{F})$ by (9.2). Furthermore, $F^* \subset b(\complement F) \subset b(\complement \overset{\circ}{F})$, hence $\overset{\circ}{F}$ is regular and therefore $H(\overset{\circ}{F})|_{\complement \overset{\circ}{F}} = C_P(X)|_{\complement \overset{\circ}{F}}$.

(2) ⇒ (1): Since

$$C_P(X)|_{\complement \overset{\circ}{F}} = \overline{H_0(F)}|_{\complement \overset{\circ}{F}} = H(\overset{\circ}{F}{}^f)|_{\complement \overset{\circ}{F}} \subset H(\overset{\circ}{F})|_{\complement \overset{\circ}{F}} \subset C_P(X)|_{\complement \overset{\circ}{F}}$$

we conclude first that $\overset{\circ}{F}$ is regular and hence $\complement \overset{\circ}{F} = b(\complement \overset{\circ}{F}) = \beta(\complement \overset{\circ}{F})$. Moreover, given $h \in H(\overset{\circ}{F})$, there exists a function $g \in H(\overset{\circ}{F}{}^f)$ such that $g = h$ on $\complement \overset{\circ}{F}$ and hence $g = h$ on X since $g \in H(\overset{\circ}{F})$. So $H(\overset{\circ}{F}{}^f) = H(\overset{\circ}{F})$ and therefore $\beta(\complement \overset{\circ}{F}{}^f) = \beta(\complement \overset{\circ}{F})$ by (8.6). Thus we finally obtain that

$$F^* \subset \complement \overset{\circ}{F} = \beta(\complement \overset{\circ}{F}) = \beta(\complement \overset{\circ}{F}{}^f) \subset \complement \overset{\circ}{F}{}^f = \overline{\complement F}^f = b(\complement F) .$$ ┘

9.5. THEOREM. Let G be a finely open subset of X. Then for every $x \in X$

$$\begin{aligned} M_x(S(G)) &= M_x(P) \cap M_x(H(G)) \\ &= \{\mu \in M_x(P) : \mu^{\beta(\complement G)} = \varepsilon_x^{\beta(\complement G)}\} = \{\mu \in M_+(X) : \varepsilon_x^{\beta(\complement G)} \prec \mu \prec \varepsilon_x\} . \end{aligned}$$

$M_x(S(G))$ is a closed face of the compact convex set $M_x(P)$, and

$$(M_x(S(G)))_e = \{\varepsilon_x\} \cup \{\varepsilon_x^A : A \in \mathcal{B}(X) , x \notin A , \beta(\complement G) \subset A\} .$$

In particular, for every finely closed Borel subset F of X,

$$(M_x(S_o(F)))_e = \{\varepsilon_x\} \cup \{\varepsilon_x^{\complement B} : B \in \mathcal{B}(X) \ , \ x \in B \subset F\} \ .$$

Proof. Since $S(G) = \overline{P + H(G)}$ we obtain that for every $x \in X$

$$M_x(S(G)) = M_x(P + H(G)) = M_x(P) \cap M_x(H(G)) \ .$$

Fix $x \in X$ and $\mu \in M_x(P)$. Let us note first that for every $s \in S(G)$

$$(*) \quad \mu^{\beta(\complement G)}(s) = \int \varepsilon_y^{\beta(\complement G)}(s)\,\mu(dy) \leq \int s(y)\,\mu(dy) = \mu(s) \ .$$

If $\mu \in M_x(S(G))$ we therefore have $\mu^{\beta(\complement G)} \in M_x(S(G))$ and hence $\mu^{\beta(\complement G)} = \varepsilon_x^{\beta(\complement G)}$ since $\mu^{\beta(\complement G)}$ is supported by $\beta(\complement G)$. If $\mu^{\beta(\complement G)} = \varepsilon_x^{\beta(\complement G)}$ then clearly $\varepsilon_x^{\beta(\complement G)} \prec \mu$. Finally, let $\varepsilon_x^{\beta(\complement G)} \prec \mu$. Then

$$\varepsilon_x^{\beta(\complement G)} = \left(\varepsilon_x^{\beta(\complement G)}\right)^{\beta(\complement G)} \prec \mu^{\beta(\complement G)} \prec \varepsilon_x^{\beta(\complement G)} \ ,$$

hence $\mu^{\beta(\complement G)} = \varepsilon_x^{\beta(\complement G)}$ and by $(*)$ for every $h \in H(G)$,

$$\mu(h) = \mu^{\beta(\complement G)}(h) = \varepsilon_x^{\beta(\complement G)}(h) = h(x) \ .$$

Thus $\mu \in M_x(H(G)) \cap M_x(P) = M_x(S(G))$.

By (VI.10.1), $M_x(P)$ is a compact convex set and $M_x(S(G))$ is a closed convex subset of $M_x(P)$. Let $\mu_1, \mu_2 \in M_x(P)$ and $0 < \lambda < 1$ such that

$$\mu := \lambda\mu_1 + (1-\lambda)\mu_2 \in M_x(S(G)) \ .$$

Then $\mu^{\beta(\complement G)} = \varepsilon_x^{\beta(\complement G)}$ and $\mu_i^{\beta(\complement G)} \prec \varepsilon_x^{\beta(\complement G)}$, $i = 1,2$. Therefore $\mu_i^{\beta(\complement G)} = \varepsilon_x^{\beta(\complement G)}$, $i = 1,2$, i.e. $\mu_1, \mu_2 \in M_x(S(G))$. Thus $M_x(S(G))$ is a face of $M_x(P)$.

In particular, the extreme points of $M_x(S(G))$ are the extreme points of $M_x(P)$ which are contained in $M_x(S(G))$. Hence we conclude from (8.3) and (VI.12.5) that $\varepsilon_x \in (M_x(S(G)))_e$ and $\varepsilon_x^A \in (M_x(S(G)))_e$ for every set $A \in \mathcal{B}(X)$ such that $\beta(\complement G) \subset A \subset X \setminus \{x\}$. Conversely, let $\mu \in (M_x(S(G)))_e$, $\mu \neq \varepsilon_x$. Then there exists a set $A \in \mathcal{B}(X)$ such that $x \notin A$ and $\mu = \varepsilon_x^A$. Moreover, $\varepsilon_x^{\beta(\complement G)} \prec \mu$, hence $\varepsilon_x^{\beta(\complement G) \cup A} = \varepsilon_x^A$ by (VI.9.9). Of course, $x \notin \beta(\complement G)$ since $\varepsilon_x^A \neq \varepsilon_x$. Thus $A' = \beta(\complement G) \cup A \in \mathcal{B}(X)$, $x \notin A'$, $\varepsilon_x^{A'} = \mu$ finishing the proof. ┘

<u>9.6. EXERCISES.</u> 1. Let $F \subset C_p(X)$ and $f \in C_p(X)$ such that $f = \inf f_n$ for some decreasing sequence (f_n) in F. Then $f \in \overline{F}$.

2. Let $F \subset C_p(X)$ be a convex cone containing P. Then $\overline{W(F)}$ is the set of

all F-concave functions in $C_P(X)$. (Hint: Use (I.2.3) and the preceding exercise.)

In the following exercises let G be a finely open subset of X .

3. Use (VI.12.7.2), (VI.12.5), (VI.9.9) and (8.3) in order to give another proof for the fact that $M_x(S(G)) = \{\mu \in M_x(P) : \mu^{\beta(\complement G)} = \varepsilon_x^{\beta(\complement G)}\} = \{\mu \in M_+(X) : \varepsilon_x^{\beta(\complement G)} \prec \mu \prec \varepsilon_x\}$.

4. Let $q \in P$ such that $\delta(q) = \beta(\complement G)$ (VI.8.6). Then $\overline{P + \mathbb{R}q} = S(G)$. (Hint: Use (5.12.2) and exercise 2.)

5. Let $x \in X$ and $A \subset X$. Then: $\varepsilon_x^A \in M_x(S(G)) \Leftrightarrow \varepsilon_x^A = \varepsilon_x^{A \cup \beta(\complement G)} \Leftrightarrow \varepsilon_x^A = \varepsilon_x^B$ for some $\beta(\complement G) \subset B \subset X$.

6. Let $x \in X$. Then $M_x(H(G)) = \{\mu \in M_+(X) : \mu^{\beta(\complement G)} = \varepsilon_x^{\beta(\complement G)}\}$ and the mapping $\nu \mapsto \nu + \varepsilon_x^{\beta(\complement G)} - \nu^{\beta(\complement G)}$ is an affine bijection of $N_x := \{\nu \in M_+(X) : \nu(\beta(\complement G)) = 0$ and $\nu^{\beta(\complement G)} \le \varepsilon_x^{\beta(\complement G)}\}$ onto $M_x(H(G))$.

10. Removable Singularities

Let us note first that finely closed semipolar sets are removable singularities for continuous superharmonic functions. More precisely, by (8.6) we have the following result.

10.1. PROPOSITION. Let U be a finely open subset of X and let F be a finely closed subset of X . Then the following statements are equivalent:

(1) $F \cap U$ is semipolar.

(2) $S(U \setminus F) = S(U)$.

(3) $H(U \setminus F) = H(U)$.

(4) $H(U \setminus F) \subset S(U)$.

10.2. COROLLARY. Let U be an open subset of X , let F be a closed semipolar subset of X , and let $s \in S^+(U \setminus F) \cap C(X)$ such that $s \le s_0$ for some $s_0 \in S^+(U)$. Then $s \in S^+(U)$.

Proof. Let $p \in P$ be strictly positive and define $s_n = \inf(s, np)$. Then

$s_n \in S(U\setminus F)^+ = S(U)^+ \subset S^+(U)$ for every $n \in \mathbb{N}$, hence $s = \sup s_n \in {}^*H^+(U)$. So $s \in S^+(U)$ by (III.1.3.2). ┘

In order to see that a semipolar set may not be a removable singularity for arbitrary positive hyperharmonic functions let us consider the harmonic space given by the translation in $\mathbb{R}$. There the set $\{0\}$ is semipolar, the l.s.c. positive function $1_{]-\infty,0[}$ is hyperharmonic on $\complement\{0\}$, but not hyperharmonic on $\mathbb{R}$. However, the following propositions will show that closed polar sets are removable singularities for arbitrary positive hyperharmonic functions.

<u>10.3. PROPOSITION.</u> Let U be an open subset of X, let F be a closed polar subset of X and $v \in {}^*H^+(U\setminus F)$. Then there exists a unique function $\tilde{v} \in {}^*H^+(U)$ such that $\tilde{v} = v$ on $\complement(U \cap F)$. For every $x \in F$, $\tilde{v}(x) = \liminf\limits_{y\to x, y\in U\setminus F} v(y)$.

Proof. Let $\mathcal{F}$ be the set of all functions $u \in {}^*H^+(U)$ such that $u = 0$ on $\complement U$ and $u = \infty$ on $U \cap F$. If $x \in U\setminus F$ then there exists $w \in \mathcal{W}$ such that $w = \infty$ on F and $w(x) < \infty$, and we have $w1_U \in \mathcal{F}$. Therefore $\inf \mathcal{F} = 0$ on $U\setminus F$. Fix $u \in \mathcal{F}$. We claim that $v + u \in {}^*H^+(U)$. Indeed, since v is l.s.c. on $U\setminus F$ and $\lim\limits_{y\to x} u(y) = \infty$ for every $x \in U \cap F$, we obtain that $v + u$ is l.s.c. on U. If $x \in U\setminus F$ and V is a relatively compact open neighborhood of x such that $\overline{V} \subset U\setminus F$ then $\varepsilon_x^{\complement V}(v) \le v(x)$ and hence $\varepsilon_x^{\complement V}(v+u) \le (v+u)(x)$. If $x \in U \cap F$ then $\varepsilon_x^{\complement V}(v+u) \le \infty = (v+u)(x)$ for every subset V of X. Therefore $v + u \in {}^*H^+(U)$ by (III.4.4).

Let $v_0 = \inf\{v+u : u \in \mathcal{F}\}$. Then $v_0 = v$ on $\complement(U\cap F)$ and $v_0 = \infty$ on $U \cap F$. Define $\tilde{v} : X \to \overline{\mathbb{R}}_+$ by

$$\tilde{v}(x) = \begin{cases} v(x) & , \quad x \in \complement(U\cap F) \\ \liminf\limits_{y\to x, y\in U\setminus F} v(y) & , \quad x \in U \cap F \end{cases} .$$

Then $\tilde{v}(x) = \liminf\limits_{y\to x} v_0(y)$ for every $x \in U$. If V is a relatively compact open set such that $\overline{V} \subset U$ then $H_V v_0 \le v_0$ since $v + \mathcal{F} \subset {}^*H^+(U)$ and hence for every $x \in V$ by (III.3.4)

$$H_V\tilde{v}(x) \le H_V v_0(x) = \liminf_{y\to x} H_V v_0(y) \le \liminf_{y\to x} v_0(y) = \tilde{v}(x) .$$

Thus $\tilde{v} \in {}^*\mathcal{H}^+(U)$. The uniqueness follows from the fact that $U \setminus F$ is finely dense in U and functions in ${}^*\mathcal{H}^+(U)$ are finely continuous on U by (V.1.1). ⌟

<u>10.4. COROLLARY.</u> Let U be an open subset of X , let F be a closed polar subset of X , and let $h \in \mathcal{H}(U \setminus F)$ such that $|h| \leq h_0$ for some real function $h_0 \in \mathcal{H}^+(U)$. Then there exists a unique function $\tilde{h} \in \mathcal{H}(U)$ such that $\tilde{h} = h$ on $\complement(U \cap F)$. For every $x \in F$, $\tilde{h}(x) = \lim_{y \to x, y \in U \setminus F} h(y)$.

Proof. Let us note first that $h_0 \pm h \in \mathcal{H}^+(U \setminus F)$. Hence by (10.3) there exist functions $\tilde{v}_\pm \in {}^*\mathcal{H}(U)$ such that $\tilde{v}_\pm = h_0 \pm h$ on $\complement(U \cap F)$. Then $\tilde{v}_+ + \tilde{v}_- = 2h_0$ on $\complement(U \cap F)$, hence on X . Therefore $\tilde{v}_+ = 2h_0 - v_- \in -{}^*\mathcal{H}(U)$, hence $\tilde{v}_\pm \in \mathcal{H}(U)$. Defining $\tilde{h} = \frac{1}{2}(\tilde{v}_+ - \tilde{v}_-)$ we thus have $\tilde{h} \in \mathcal{H}(U)$ and $\tilde{h} = h$ on $\complement(U \cap F)$. The proof is finished by observing that F has no interior point. ⌟

<u>10.5. COROLLARY.</u> Let F be a closed subset of X and assume that there exists a strictly positive function $h_0 \in \mathcal{H}^+(X)$. Then the following statements are equivalent:

(1) F is polar.

(2) For every $v \in {}^*\mathcal{H}^+(\complement F)$ there exists a unique function $\tilde{v} \in \mathcal{W}$ such that $\tilde{v} = v$ on $\complement F$.

(3) For every $v \in \mathcal{H}^+(\complement F)$ there exists a unique function $\tilde{v} \in \mathcal{W}$ such that $\tilde{v} = v$ on $\complement F$.

Proof. (1) ⇒ (2): (10.3).

(2) ⇒ (3): Trivial.

(3) ⇒ (1): Choose a strict potential $p \in \mathcal{P}$ such that $p \leq h_0$ and let $v_1 = R_p^F$, $v_2 = h_0 - R_p^F$. Then $v_1, v_2 \in \mathcal{H}^+(\complement F)$, $v_1 + v_2 = h_0$. Therefore $\tilde{v}_1 + \tilde{v}_2 = \tilde{h}_0 = h_0$ and hence $\tilde{v}_1, \tilde{v}_2 \in \mathcal{H}^+(X)$. Since $\tilde{v}_1 = \hat{R}_p^F \leq p$, we conclude that $\tilde{v}_1 = 0$, i.e. F is polar. ⌟

<u>10.6. EXERCISES.</u> Let $(X, \mathcal{W})$ be a harmonic space and let F be a closed subset of X .

1. If F is polar then for every function $h \in \mathcal{H}(\complement F)$ satisfying

$\limsup_{y \to x, y \in \complement F} |h(y)| < \infty$ for all $x \in F$ there exists a unique function $\tilde{h} \in \mathcal{H}(X)$ such that $\tilde{h} = h$ on $\complement F$.

2. F is polar if $\overset{\circ}{F} = \emptyset$ and if for every P-bounded function $h \in \mathcal{H}^+(\complement F)$ there exists $\tilde{h} \in \mathcal{H}(X)$ such that $\tilde{h} = h$ on $\complement F$.

VIII. Partial Differential Equations

The theory of harmonic spaces was mainly established with the aim to generalize and unify results and methods of classical potential theory for application to an extended class of elliptic and parabolic differential equations of second order. Originally, the theory started with a sheaf of vector spaces of real continuous functions on a locally compact space, playing the rôle of the sheaf of solutions of a partial differential equation. A convergence property, the boundary minimum principle, the local solvability of the Dirichlet problem are supposed to hold. The most important type of a harmonic space is a Bauer space which is introduced in section 1. In our terminology a Bauer space is locally a harmonic space having a base of regular sets. Semi-elliptic differential operators are treated in section 2. In section 3 we present J. M. BONY's result that a Bauer space whose harmonic functions are smooth is generated by such a differential operator. Section 4 prepares the material (Sobolev spaces, weak solutions, etc.) needed in section 5 to show that elliptic-parabolic differential operators generate Bauer spaces. Besides the deep result of L. HÖRMANDER on the hypoellipticity of such operators the theory is completely selfcontained. For sake of simplicity we mostly assume that the constant function 1 is harmonic.

1. Bauer Spaces

Let X be a locally compact space with a countable base. A *harmonic sheaf* on X is a map H defined on the set of open subsets of X such that:

(a) for any open subset U of X, $H(U)$ is a linear subspace of $C(U)$,

(b) for any two open subsets U, V of X such that $U \subset V$, $H(V)|_U \subset H(U)$,

(c) for any family $(U_i)_{i\in I}$ of open subsets and any numerical function h on $U = \bigcup_{i\in I} U_i$, $h \in H(U)$ if $h|_{U_i} \in H(U_i)$ for every $i \in I$.

The elements of $H(U)$ are called *harmonic functions* on U.

A relatively compact open subset V of X is called *regular* if any $f \in C(V^*)$ possesses a unique continuous extension H_f^V on $\overline{V}$ such that $H^V_{f|V} \in H(V)$ and $H_f^V \geq 0$ provided $f \geq 0$. For any $x \in V$ the map $f \mapsto H_f^V(x)$, $f \in C(V^*)$, is a positive linear functional, hence a measure μ_x^V on V^* called the *harmonic measure* of x with respect to V. If

$$H_V(x,\cdot) = \begin{cases} \mu_x^V, & x \in V \\ \varepsilon_x, & x \notin V \end{cases}$$

then H_V is a sweeping kernel.

The pair (X,H) is called a *Bauer space* if the following axioms hold:

(BS_1) For every $x \in X$ there exists a harmonic function h defined in a neighborhood of x such that $h(x) \neq 0$.

(BS_2) There exists a base V of X of regular sets such that $U \cap V \in V$ for any $U, V \in V$.

(BS_3) ("Convergence property of Bauer"). For any increasing sequence (h_n) of harmonic functions on an open set U, $h = \sup h_n \in H(U)$ if h is locally bounded.

(X,H) is called *normal* if $1 \in H(X)$.

For any open subset U of X let $^*H(U)$ be the convex cone of *hyperharmonic functions* on U, i.e.

$$^*H(U) = \{u : U \to]-\infty, +\infty] \; ; \; u \text{ l.s.c.}, \; H_V u \leq u \; \forall \, V \text{ regular}, \overline{V} \subset U\}$$

<u>1.1. PROPOSITION.</u> Let (X,H) be a normal Bauer space. Then every point $x \in X$ has an open neighborhood U such that $(U, {}^*H^+(U))$ is a harmonic space and $(H_V)_{V \text{ regular}, \overline{V} \subset U}$ is an associated family of harmonic kernels.

Proof. Let U denote the family of all regular subsets of X and let us adopt the notations of chapter III. Clearly, $\lim_{V, U_x} H_V(x,\cdot) = \varepsilon_x$ for every $x \in X$

since $H_V(x,V^*) = 1$ and $H_V(x,\complement V^*) = 0$ if $x \in V \in \mathcal{U}$. Fix $V \in \mathcal{U}$. If $f \in C_b(X)$ and W is regular such that $W \subset V$ then $H_V f \in C_b(X)$ and $H_V f$ is harmonic on W, hence of course $H_W(H_V f) = H_V f$. Given a l.s.c. function $g \in \mathcal{B}_b^+(X)$ we choose a sequence (f_n) in $C_b^+(X)$ which is increasing to g and obtain by (BS_3) that $H_V g = \sup H_V f_n$ is harmonic on V. If $f \in \mathcal{B}_b^+(X)$ then the family $G := \{g \in \mathcal{B}_b^+(X):$ g l.s.c., $g \geq f\}$ is decreasingly filtered and $H_V f = \inf \{H_V g : g \in G\}$. By (I.1.7), there exists a decreasing sequence (g_n) in G such that $H_V f = \inf H_V g_n$ on V. Applying (BS_3) to the sequence $(H_V(g_1 - g_n))|_V$ and using $H_V g_n = H_V g_1 - H_V(g_1 - g_n)$ we conclude that $H_V f$ is harmonic on V. Moreover, the constant 0 is an Evans function since every filter F on V converging to a point $z \in V^*$ is regular. Thus the family $(H_V)_{V \in \mathcal{U}}$ satisfies the axioms (H_1) - (H_4).

Fix a base $\mathcal{V} \subset \mathcal{U}$ such that $V \cap V' \in \mathcal{V}$ for any $V, V' \in \mathcal{V}$ and let W be an open subset of X. If $u \in {}^*\mathcal{H}_{\mathcal{V}}^+(W) \cap C(W)$ and $V \in \mathcal{V}(W)$ then $H_V u \in {}^*\mathcal{H}_{\mathcal{V}}^+(W) \cap C(W)$. Indeed, $H_V u \in C(W)$ and for every $V' \in \mathcal{V}(W)$ we have $H_V u = u \geq H_{V'} u \geq H_{V'} H_V u$ on $V' \setminus V$. The function f on $V' \cap V$ which is equal to $H_V u - H_{V'} H_V u$ on $\overline{V} \cap V'$ and 0 elsewhere is continuous on $\overline{V' \cap V}$, positive on $(V' \cap V)^*$ and harmonic on $V' \cap V$. Since $V' \cap V$ is regular, $f \geq 0$ and hence $H_V u \in {}^*\mathcal{H}_{\mathcal{V}}^+(W)$.

Now fix $x \in X$ and $W \in \mathcal{V}_x$. Since $H_W 1(x) = 1$, the support of $H_W(x,\cdot)$ contains a point x_0. Let $V \in \mathcal{V}$ such that $x_0 \in V$, $x \notin V$ and choose $f \in C^+(X)$ such that $f(x_0) = 1$ and $f = 0$ on $\complement V$. Define $g \in C((W \cap V)^*)$ and $p \in C^+(W)$ by $g = H_W f$ on $W \cap V^*$ and $g = 0$ elsewhere, $p = H_{W \cap V} g$ on $W \cap V$ and $p = H_W f$ on $W \setminus V$. Then $H_{W \cap V} g \leq H_W f$ on $W \cap V$. Furthermore, $p \leq H_W f$, $p(x) > 0$ and $\lim_{y \to z} p(y) = 0$ for every $z \in W^*$. As above we obtain that $p \in {}^*\mathcal{H}_{\mathcal{V}}^+(W)$.

Set $U = \{p > 0\}$. Then $x \in U$, and ${}^*\mathcal{H}_{\mathcal{V}}^+(U)$ separates the points of U. Indeed, let $y, z \in U$, $y \neq z$. Let $F = \{H_{V_1} \cdots\cdots H_{V_n} p : n \in \mathbb{N}, V_i \in \mathcal{V}(W), \{y,z\} \not\subset V_i\}$. Then F is a decreasingly filtered subset of ${}^*\mathcal{H}_{\mathcal{V}}^+(W) \cap C(W)$. Moreover, $\inf F$ is a harmonic function on W tending to zero at every boundary point. Hence $\inf F = 0$ and there exist $f \in F$ and $V \in \mathcal{V}(W)$ such that $\{y,z\} \not\subset V$ and either $H_V f(y) < f(y)$, $f(z) > 0$ or $H_V f(z) < f(z)$, $f(y) > 0$. Then $H_V f(y) \neq H_V f(z)$ or $f(y) \neq f(z)$, and it suffices to note that ${}^*\mathcal{H}_{\mathcal{V}}^+(W)|_U \subset {}^*\mathcal{H}_{\mathcal{V}}^+(U)$.

In order to finish the proof it remains to show that ${}^*\mathcal{H}_{\mathcal{V}}^+(U) = {}^*\mathcal{H}^+(U)$. So fix

$u \in {}^*\mathcal{H}_V^+(U)$, $V \in \mathcal{U}(U)$, and let $f \in C^+(U)$ such that $f \leq u$. Then $v := (u - H_V f)|_V \in {}^*\mathcal{H}_V(V)$ and $\liminf_{x \to z} v(x) = u(z) - f(z) \geq 0$ for every $z \in V^*$. Applying (III.4.1) to the family $(H_V)_{V \in \mathcal{V}(U)}$ of harmonic kernels on U we obtain that $v \geq 0$. Thus indeed $H_V u \leq u$. ┘

1.2. COROLLARY. Let $(X,\mathcal{H})$ be a normal Bauer space. Then for every open subset U of X such that ${}^*\mathcal{H}^+(U)$ separates the points of U , $(U, {}^*\mathcal{H}^+(U))$ is a harmonic space and $(H_V)_{V \text{ regular}, \overline{V} \subset U}$ is an associated family of harmonic kernels.

1.3. EXAMPLE. Every harmonic space $(X,\mathcal{W})$ having a base of regular sets is a Bauer space by (VIII.3.2). Especially, the harmonic sheaves of the solutions of the Laplace equation on $\mathbb{R}^n$ and of the heat equation on $\mathbb{R}^{n+1}$, $n \geq 1$, resp. generate Bauer spaces.

1.4. EXERCISES. 1. For every Bauer space $(X,\mathcal{H})$ the following properties are equivalent:

(1) For every regular region V and every $x \in V$, $\operatorname{supp}(H_V(x,\cdot)) = V^*$.

(2) For every region U and every $h \in \mathcal{H}^+(U)$, either $h > 0$ or $h = 0$.

2. Let $n \in \mathbb{N}$ and

$$X = [0,n+1] \times \{0\} \cup \bigcup_{i=1}^{n} \{i\} \times [-1,+1] \quad .$$

Let $M = \{1, \ldots, n\} \times \{0\}$, $E = \{(0,0), (n+1,0)\}$ and $A = \{1, \ldots, n\} \times \{-1,+1\}$. For every open U in X let $\mathcal{H}(U)$ be the set of all functions $h \in C(U)$ having the following properties: h is locally affine on $U \setminus M$; if $x_0 \in U \cap M$ $\overline{B}_\varepsilon(x_0) \subset U$, $0 < \varepsilon < 1$, and $S_\varepsilon(x_0) \cap U = \{x_1,x_2,x_3,x_4\}$ then $h(x_0) = \frac{1}{4}(h(x_1) + h(x_2) + h(x_3) + h(x_4))$; if $x_0 \in U \cap E$ then h is constant on a neighborhood of x_0 ; if $x_0 \in U \cap A$ then $\frac{df}{ds} = f(x_0)$ on a neighborhood of x_0 .

Then $(X,\mathcal{H})$ is a Bauer space having the properties of excercise 1. Moreover, $\dim \mathcal{H}(X) = n$, but $\mathcal{H}^+(X) = {}^*\mathcal{H}^+(X) = \{0\}$. If U is an open subset of X such that the diameter of U is strictly less than 1 then U is regular and $(U, {}^*\mathcal{H}^+(U))$ is a harmonic space.

2. Semi-Elliptic Differential Operators

In this section let X be an open subset of $\mathbb{R}^n$, $n \geq 1$. We shall consider linear partial differential operators L of second order of the form

$$L = \sum_{i,j=1}^{n} a_{ij} \frac{\partial^2}{\partial x_i \, \partial x_j} + \sum_{i=1}^{n} a_i \frac{\partial}{\partial x_i} + a$$

where the coefficients a_{ij}, a_i, a are real functions on X such that $a_{ij} = a_{ji}$ for all $i,j \in \{1, \ldots, n\}$ and $a \leq 0$. Let us denote by $C^k(U)$ the linear space of k-times continuously differentiable real functions on the open set $U \subset X$, $k = 1,2, \ldots, \infty$. Defining

$$H_L(U) = \{h \in C^2(U) : Lu = 0 \text{ on } U\}$$

we obtain a harmonic sheaf H_L on X. In the following sections we shall show that (X, H_L) is a Bauer space for operators L in a large class of differential operators.

L is called

a) *continuous*, if $a_{ij}, a_i, a \in C(X)$ for all $i,j = 1, \ldots, n$;

b) *smooth* , if $a_{ij}, a_i, a \in C^\infty(X)$ for all $i,j = 1, \ldots, n$;

c) *semi-elliptic*, if for every $x \in X$, the matrix $(a_{ij}(x))$ is positive semidefinite and non-trivial, i.e. $\sum_{i,j=1}^{n} a_{ij}(x) y_i y_j \geq 0$ for all $y \in \mathbb{R}^n$ and $a_{ij}(x) \neq 0$ for some $i,j \in \{1,\ldots,n\}$.

2.1. EXAMPLES. 1. The Laplace operator and the heat operator are smooth semi-elliptic differential operators on $\mathbb{R}^n$, $\mathbb{R}^{n+1}$ respectively.

2. If L is a semi-elliptic differential operator on X then $L_\varepsilon := \varepsilon\Delta + L$ is a semi-elliptic differential operator on X for every $\varepsilon > 0$.

We start our investigations on semi-elliptic differential operators L on X by proving some types of minimum principles. For that purpose let U be a relatively compact open subset of X.

2.2. PROPOSITION. Let $u \in C^2(U)$ and $x \in U$ such that $u(x) = \inf_{y \in U} u(y) \leq 0$. Then $Lu(x) \geq 0$.

Proof. By assumption, $\frac{\partial u}{\partial x_i}(x) = 0$ for every $i = 1, \ldots, n$. Furthermore the matrix $\left(\frac{\partial^2 u}{\partial x_i\, \partial x_j}(x)\right)$ is positive-semidefinite, hence

$$Lu(x) = \sum_{i,j=1}^{n} a_{ij}(x)\, \frac{\partial^2 u}{\partial x_i\, \partial x_j}(x) + a(x)u(x) \geq a(x)u(x) \geq 0 .$$

The relatively compact open set U is called L-*small* if there exists a function $w \in C^2_+(U)$ such that $Lw < 0$ and $\Delta w \leq 0$ on U . We shall say that U is L-*regular* if for every $z \in U^*$ there exists an L-*barrier* at z , i.e. a function $w \in C^\infty(\mathbb{R}^n)$ such that $w(z) = 0$, $w > 0$ on $\overline{U} \setminus \{z\}$, $Lw < 0$ on $\overline{U}$ and $\Delta w \leq 0$ on U .

2.3. REMARKS. 1. If U is L-regular then U is L-small. Every open subset of an L-small set is L-small.

2. If U is L-small (L-regular resp.) then U is L_ε-small (L_ε-regular resp.) for every $\varepsilon > 0$.

3. If U and V are L-regular then $U \cap V$ is L-regular.

The following minimum principle will be used again and again.

2.4. PROPOSITION. Let L be a continuous semi-elliptic differential operator on X , let U be an L-small open subset of X , and $\alpha \in \mathbb{R}_+$. Then for every $u \in C(\overline{U})$ such that $u_{|U} \in C^2(U)$, $Lu \leq 0$ on U and $u \geq -\alpha$ on U^* we have $u \geq -\alpha$ on U . In particular, $u = 0$ if $Lu = 0$ on U and $u = 0$ on U^* .

Proof. Let $x_0 \in U$ and suppose that $u(x_0) < -\alpha$. Choose a function $w \in C^+(\overline{U})$ such that $w_{|U} \in C^2(U)$ and $Lw < 0$ on U . There exists $\varepsilon > 0$ such that the function $v := u + \varepsilon w$ satisfies $v(x_0) < -\alpha$. Since $v \geq u \geq -\alpha$ on U^* , there is a point $x \in U$ such that $v(x) = \inf_{y \in U} v(y)$ and hence $Lv(x) \geq 0$ by (2.2). However, $Lv \leq \varepsilon L w < 0$ on U . This contradiction shows that $u(x_0) \geq -\alpha$.

2.5. PROPOSITION. Let L be a continuous semi-elliptic differential operator on X . Then the L-regular subsets form a base of X .

Proof. Let $y \in X$ and let V be a neighborhood of y . Since L is semi-elliptic there exists $z_0 \in \mathbb{R}^n$ such that $\|z_0\| = 1$ and $\sum_{i,j=1}^{n} a_{ij}(y) z_{0i} z_{0j} > 0$. Since L is continuous there exists a compact neighborhood K of y and $0 < \varepsilon < 1$ such that $K \subset V$ and $\sum_{i,j=1}^{n} a_{ij}(x) z_i z_j > 0$ for every $x \in K$ and every $z \in \overline{B}_\varepsilon(z_0)$. Choose $\delta > 0$ such that the lenticular open neighborhood

$$U := B_{1+\delta}(y+z_0) \cap B_{1+\delta}(y-z_0)$$

of y is contained in $K \cap B_{\frac{\varepsilon}{4}}(y)$. We claim that U is L-regular.

Indeed, let $z \in U^* \cap S_{1+\delta}(y \pm z_0)$ and let x_0 be the point on the line joining $y \pm z_0$ and z such that $\overline{B}_1(x_0) \cap \overline{U} = \{z\}$. Then $\overline{U} - x_0 \subset \overline{B}_\varepsilon(\pm z_0) = \pm \overline{B}_\varepsilon(z_0)$, hence

$$\sum_{i,j=1}^{n} a_{ij}(x)(x_i - x_{0i})(x_j - x_{0j}) > 0$$

for every $x \in \overline{U}$. Take $k \in \mathbb{N}$ and define $w : \mathbb{R}^n \to \mathbb{R}$ by

$$w(x) = e^{-k} - e^{-k\| x-x_0\|^2} .$$

Then $w \in C^\infty(\mathbb{R}^n)$, $w(z) = 0$, and $w > 0$ on $\overline{U} \setminus \{z\} \subset \complement \overline{B}_1(x_0)$. Moreover, for every $x \in X$

$$Lw(x)$$
$$= 2ke^{-\| x-x_0\|^2}\left[-2k \sum_{i,j=1}^{n} a_{ij}(x)(x_i - x_{0i})(x_j - x_{0j}) + \sum_{i=1}^{n} (a_{ii}(x) + a_i(x)(x_i - x_{0i}))\right] + a(x)w(x)$$

and

$$\Delta w(x) = 2ke^{-\| x-x_0\|^2}\left[-2k\| x-x_0\|^2 + n\right]$$

where $a(x)w(x) \leq 0$ if $x \in \overline{U}$. Hence w is an L-barrier at z if k is sufficiently large. ∎

We shall say that a Bauer space (X,H) is *associated* with a semi-elliptic differential operator L on X if $H \subset H_L$, i.e. if $H(V) \subset H_L(V)$ for every open subset V of X. The minimum principle (2.4) allows to prove the following proposition.

2.6. PROPOSITION. Let (X,H) be a Bauer space associated with a continuous semi-elliptic differential operator L on X. Then, for every open subset U of X,

$$H(U) = H_L(U) \quad \text{and} \quad {}^*H(U) \cap C^2(U) = \{u \in C^2(U) : Lu \leq 0 \text{ on } U\} .$$

Moreover, every L-regular subset of X is regular.

Proof. Let $h \in H_L(U)$ and let V be a regular L-small set such that $\overline{V} \subset U$. Then $v = H_V h \in H(V) \subset H_L(V)$, hence $L(v-h) = 0$ on V and $v - h = 0$ on V^*. By (2.4), $v = h$ on V and therefore $h \in H(U)$ since H is a sheaf.

Let $u \in C^2(U)$ such that $Lu \leq 0$ on U. Let V be a regular L-small set such that $\overline{V} \subset U$ and define $v = H_V u$. Then $L(u-v) \leq 0$ on V and $u = v$ on V^*. By (2.4), we conclude that $u - v \geq 0$ on V, i.e. $u \geq v = H_V u$, and thus $u \in {}^*H(U)$.

Conversely, let $u \in {}^*H(U) \cap C^2(U)$. Assume $Lu(x) > 0$ for some $x \in U$, hence $Lu > 0$ on some neighborhood V of x. Therefore by the above result, the function $-u$ is superharmonic on V. Thus $u \in H(V)$, $Lu(x) = 0$ contradicting our assumption.

Finally, every L-regular subset U of X is regular by (VII.3.3) since every L-barrier at $z \in U^*$ is a barrier. ┘

In the next section we shall investigate whether a Bauer space is associated with a semi-elliptic differential operator L on X. However, it is possible by application of (2.6) to prove uniqueness of L up to a multiplicative factor.

2.7. PROPOSITION. Let (X,H) be a normal Bauer space associated with continuous semi-elliptic differential operators L and M. Then there exists $f \in C^+(X)$ such that $L = fM$.

Proof. Let $x \in X$ and define for $\alpha_{ij}, \alpha_i \in \mathbb{R}$, $i,j = 1, \ldots, n$

$$u(y) = \sum_{i,j=1}^{n} \alpha_{ij}(y_i - x_i)(y_j - x_j) + \sum_{i=1}^{n} \alpha_i(y_i - x_i), \quad y \in \mathbb{R}^n .$$

If

$$2 \sum_{1 \leq i \leq j \leq n} a_{ij}(x)\alpha_{ij} + \sum_{i=1}^{n} a_i(x)\alpha_i < 0$$

then $Lu(x) < 0$, hence $Lu < 0$ in a neighborhood V of x, hence $u \in {}^*H(V) \cap C^2(V)$ and therefore $Mu \leq 0$ on V by (2.6). This implies that

$$2 \sum_{1 \leq i \leq j \leq n} b_{ij}(x)\alpha_{ij} + \sum_{i=1}^{n} b_i(x)\alpha_i \leq 0$$

where b_{ij}, b_i are the coefficients of M. We therefore have $f(x) \in \mathbb{R}_+$ such that for all $i,j = 1, \ldots, n$, $b_{ij}(x) = f(x)a_{ij}(x)$ and $b_i(x) = f(x)a_i(x)$.

We have $f(x) > 0$ since L is semi-elliptic. Finally, the continuity of L implies the continuity of f. ┘

2.8. EXERCISES. Let L be a continuous semi-elliptic differential operator on X such that $L1 = 0$ and let U be a relatively compact open subset of X.

1. If $h \in H_L(U)$ then $Lh^2 \geq 0$.

2. Let $w \in C(\overline{U})$ such that $w > 0$, $w|_U \in C^2(U)$, and $Lw \leq -c < 0$ on U for some $c \in \mathbb{R}$. Then $\|u\| \leq \frac{\|w\|}{c} \|Lu\|$ for every $u \in C(\overline{U})$ such that $u|_U \in C^2(U)$ and $u = 0$ on U^*.

3. Smooth Bauer Spaces

Let $X \subset \mathbb{R}^n$, $n \geq 1$, be an open connected set. A Bauer space (X,H) is called *smooth* if $H(U) \subset C^\infty(U)$ for every non-empty open subset U of X. In the sequel we shall show that every smooth Bauer space is associated with a semi-elliptic differential operator. For sake of simplicity we assume that the constant function 1 is harmonic on X.

Let Q be the vector space of all quadratic forms on $\mathbb{R}^n$. This space will be identified with the space of all symmetric $n \times n$ - matrices $A = (a_{ij})$ and equipped with the scalar product

$$\langle A,B\rangle = \sum_{i,j=1}^{n} a_{ij} b_{ij} = \text{trace } AB .$$

We denote by Q^+ the convex cone of all positive semidefinite quadratic forms. Then $A \in Q$ is contained in Q^+ if and only if $\langle A,B\rangle \geq 0$ for every $B \in Q^+$. The interior $\overset{\circ}{Q}{}^+$ of Q^+ consists of all positive definite matrices. For every function $u \in C^2(U)$, where U is an open neighborhood of a point $x \in X$, let

$$Du(x) = \left(\frac{\partial u}{\partial u_1}(x),\dots,\frac{\partial u}{\partial x_n}(x)\right) \text{ and } D^2u(x) = \left(\frac{\partial^2 u}{\partial x_i \partial x_j}(x)\right) \in Q .$$

3.1. PROPOSITION. Every smooth normal Bauer space (X,H) is associated with a semi-elliptic differential operator L on X.

Proof. Fix $x \in X$ and denote by H_x the linear space of all functions h

which are harmonic in a neighborhood of x. Furthermore, let $H_x^0 = \{h \in H_x : Dh(x) = 0\}$ and

$$F_x = \{D^2h(x) : h \in H_x^0\} .$$

Then F_x is a linear subspace of Q such that $F_x \cap \overset{\circ}{Q}{}^+ = \emptyset$ since every $h \in H_x^0$ such that $D^2h(x) \in \overset{\circ}{Q}{}^+$ would have a strict local minimum at x which is impossible. By the theorem of Hahn-Banach there exists a hyperplane H in Q such that $F_x \subset H$ and $H \cap \overset{\circ}{Q}{}^+ = \emptyset$. Let $A = (a_{ij}(x)) \in Q \setminus \{0\}$ such that A is orthogonal to H and on the same side of H as Q^+. Then for every $h \in H_x^0$

$$\sum_{i,j=1}^{n} a_{ij}(x) \frac{\partial^2 h}{\partial x_i \partial x_j}(x) = < A, D^2h(x) > = 0$$

and $< A,B > \geq 0$ for every $B \in Q^+$, hence A is positive semidefinite and non-trivial.

Let p be the dimension of the linear space $\{Dh(x) : h \in H_x\}$ and let $h_1, \ldots, h_p \in H_x$ such that $Dh_1(x), \ldots, Dh_p(x)$ are linearly independent. Then there exist $\alpha_{ij} \in \mathbb{R}$, $1 \leq i \leq n$, $1 \leq j \leq p$, such that

$$Dh(x) = \sum_{j=1}^{p} \left(\sum_{i=1}^{n} \alpha_{ij} \frac{\partial h}{\partial x_i}(x) \right) Dh_j(x)$$

for every $h \in H_x$. Fix $h \in H_x$ and let $v = h - \sum_{j=1}^{p} \left(\sum_{i=1}^{n} \alpha_{ij} \frac{\partial h}{\partial x_i}(x) \right) Dh_j$. Then $v \in H_x^0$, hence $< A, D^2v(x) > = 0$, i.e.

$$\sum_{i,j=1}^{n} a_{ij}(x) \frac{\partial^2 h}{\partial x_i \partial x_j}(x) + \sum_{i=1}^{n} a_i(x) \frac{\partial h}{\partial x_i}(x) = 0$$

where

$$a_i(x) = - \sum_{j=1}^{p} \alpha_{ij} < A, D^2h_j(x) > .$$

Thus

$$L := \sum_{i,j=1}^{n} a_{ij} \frac{\partial^2}{\partial x_i \partial x_j} + \sum_{i=1}^{n} a_i \frac{\partial}{\partial x_i}$$

is a semi-elliptic differential operator on X such that (X,H) is associated with L. ┘

<u>3.2. EXERCISE.</u> Let H be the sheaf on $\mathbb{R}$ of real continuous functions which are locally of the form $x \mapsto ax^3 + b$, $a,b \in \mathbb{R}$. Then $(\mathbb{R}, H)$ is a Bauer space. Find an associated semi-elliptic differential operator L on $\mathbb{R}$. Can L be chosen in such a way that L is continuous or even smooth?

4. Weak Solutions

Let L be a smooth semi-elliptic differential operator on a region X in $\mathbb{R}^n$, $n \geq 1$, and let U be an open relatively compact set such that $\overline{U} \subset X$. First of all we shall extend the notion of a solution of the differential equation

$$Lu = -f \quad \text{on } U \ , \ f \in C(\overline{U}) \ .$$

To this end we introduce the Hilbert space $L^2(U)$ of all $(\lambda^n\text{-})$ square integrable functions on U with scalar product and norm

$$(u,v)_2 = \int_U u\, v\, d\lambda^n \quad \text{and} \quad \|u\|_2^2 = (u,u)_2 \ .$$

Let $\mathcal{D}(U) = C^\infty(U) \cap K(U)$ be the space of *test functions* on U. If $u \in C^2(U)$ and $f \in C(U)$ then $Lu = -f$ on U if and only if

$$(\varphi, Lu)_2 = -(\varphi, f)_2$$

for every $\varphi \in \mathcal{D}(U)$. Partial integration yields that for every $u \in C^2(U)$ and $\varphi \in \mathcal{D}(U)$,

$$(\varphi, Lu)_2 = -\, a_L(\varphi, u) = (L^*\varphi, u)_2$$

where

$$a_L(\varphi,u) := \int_U \Big[\sum_{i,j=1}^n \frac{\partial(a_{ij}\varphi)}{\partial x_i} \cdot \frac{\partial u}{\partial x_j} - \sum_{i=1}^n a_i \varphi \frac{\partial u}{\partial x_i} - a\varphi u \Big]\, d\lambda^n$$

denotes the *Dirichlet form* associated with L and

$$L^*\varphi = \sum_{i,j=1}^n \frac{\partial^2(a_{ij}\varphi)}{\partial x_i\, \partial x_j} - \sum_{i=1}^n \frac{\partial(a_i\varphi)}{\partial x_i} + a\varphi$$

denotes the *adjoint differential operator* of L. Therefore, given $f \in L^2(U)$, a function $u \in L^2(U)$ is called a *weak solution* of the *Poisson equation* $Lu = -f$ if

$$(L^*\varphi, u)_2 = -(\varphi, f)_2$$

for every $\varphi \in \mathcal{D}(U)$.

The Dirichlet form a_L will play an important rôle in the study of the Dirichlet problem. But first we have to extend the notion of a derivative.

Given $u,v \in L^2(U)$ and $1 \leq i \leq n$, we shall write $v = D_i u$ and call v

the *weak partial derivative* of u with respect to x_i if $(v,\varphi)_2 = -(u, \frac{\partial\varphi}{\partial x_i})_2$ for all $\varphi \in \mathcal{D}(U)$. Partial integration shows that $D_i u = \frac{\partial u}{\partial x_i}$ if $u \in C^1(U)$ and $u, \frac{\partial u}{\partial x_i} \in L^2(U)$.

Let $H^1(U)$ denote the linear space of all functions $u \in L^2(U)$ such that $D_i u \in L^2(U)$ exists for all $1 \leq i \leq n$ and define a norm $\|\cdot\|_1$ on $H^1(U)$ by

$$\|u\|_1 = \|u\|_2 + \sum_{i=1}^{n} \|D_i u\|_2 .$$

Finally, let $H_0^1(U)$ be the closure of $\mathcal{D}(U)$ in $H^1(U)$ (with respect to $\|\cdot\|_1$). In the following we shall derive some important facts on the *Sobolev spaces* $H^1(U)$ and $H_0^1(U)$.

4.1. PROPOSITION. Let (u_n) be a sequence in $H^1(U)$ and let $u, v_1, \ldots, v_n \in L^2(U)$ such that $\lim_{n\to\infty} \|u_n - u\|_2 = 0$ and $\lim_{n\to\infty} \|D_i u_n - v_i\|_2 = 0$ for all $1 \leq i \leq n$. Then $u \in H^1(U)$, $D_i u = v_i$ for all $1 \leq i \leq n$, and $\lim_{n\to\infty} \|u_n - u\|_1 = 0$.

In particular, $H^1(U)$ is a Hilbert space with respect to the scalar product

$$(u,v)_1 = (u,v)_2 + \sum_{i=1}^{n} (D_i u, D_i v)_2 .$$

Proof. If $1 \leq i \leq n$ then for every $\varphi \in \mathcal{D}(U)$

$$(v_i,\varphi)_2 = \lim_{n\to\infty} (D_i u_n,\varphi)_2 = -\lim_{n\to\infty} (u_n, \frac{\partial\varphi}{\partial x_i})_2 = -(u, \frac{\partial\varphi}{\partial x_i})_2 ,$$

i.e. $D_i u = v_i$. This shows that $u \in H^1(U)$ and $\lim_{n\to\infty} \|u_n - u\|_1 = 0$. Using the fact that $(L^2(U), \|\cdot\|_2)$ is complete we now conclude easily that $(H^1(U), \|\cdot\|_1)$ is complete. ┘

In order to show that $H^1(U) \cap C^\infty(U)$ is dense in $H^1(U)$ we shall smooth functions in $L^2(U)$. To that end let $j \in C_+^\infty(\mathbb{R}^n)$ such that $j = 0$ outside $B_1(0)$ and $\int j d\lambda^n = 1$, and define functions $j_\varepsilon \in C_+^\infty(\mathbb{R}^n)$, $\varepsilon > 0$, by $j_\varepsilon(x) = \varepsilon^{-n} j(\frac{x}{\varepsilon})$. Note that $j_\varepsilon = 0$ outside $B_\varepsilon(0)$ and $\int j_\varepsilon d\lambda^n = 1$. For every $u \in L^2(U)$ and $\varepsilon > 0$ let

$$J_\varepsilon u(x) = \int_U u(y) j_\varepsilon(x-y)\lambda^n(dy) \qquad (x \in U) .$$

Then $J_\varepsilon u \in C^\infty(U)$ and $\|J_\varepsilon u\|_2 \leq \|u\|_2$ since by Hoelder's inequality

$$\left(\int_U u(y) j_\varepsilon(x-y)\lambda^n(dy)\right)^2 \le \int_U u^2(y) j_\varepsilon(x-y)\lambda^n(dy) \cdot \int_U j_\varepsilon(x-y)\lambda^n(dy)$$

and hence

$$\int_U (J_\varepsilon u(x))^2 \lambda^n(dx) \le \int_U \int_U u^2(y) j_\varepsilon(x-y)\lambda^n(dy)\lambda^n(dx)$$

$$= \int_U u^2(y)\left(\int_U j_\varepsilon(x-y)\lambda^n(dx)\right)\lambda^n(dy) \le \int_U u^2(y)\lambda^n(dy) \quad .$$

If $\varphi \in K(U)$ then $J_\varepsilon \varphi$ tends to φ uniformly on U as ε tends to zero, hence the density of $K(U)$ in $L^2(U)$ yields that $\lim_{\varepsilon\to 0} \| J_\varepsilon u - u\|_2 = 0$ for every $u \in L^2(U)$.

4.2. PROPOSITION. 1. If $u \in H^1(U)$ such that $u = 0$ in a neighborhood of U^* then $u \in H_0^1(U)$. More precisely, if $\varepsilon > 0$ is sufficiently small then $J_\varepsilon u \in \mathcal{D}(U)$ and $\lim_{\varepsilon\to 0} \| J_\varepsilon u - u\|_1 = 0$.

2. If $u \in H^1(U)$ and $\varphi \in \mathcal{D}(U)$ then $\varphi u \in H_0^1(U)$.

3. $H^1(U) \cap C^\infty(U)$ is a dense subspace of $H^1(U)$.

Proof. 1. For every $\varepsilon > 0$ let U_ε denote the set of all $x \in U$ such that $B_\varepsilon(x) \cap \complement U \neq \emptyset$. Let $u \in H^1(U)$ and $d > 0$ such that $u = 0$ on U_d and hence $D_i u = 0$ on U_d for all $1 \le i \le n$. Consider $0 < \varepsilon < \frac{d}{2}$ and $1 \le i \le n$. If $x \in U \setminus U_\varepsilon$ and $\varphi \in \mathcal{D}(U)$ is defined by $\varphi(y) = j_\varepsilon(x-y)$ then

$$D_i J_\varepsilon(x) = \int u(y) D_i j_\varepsilon(x-y)\lambda^n(dy) = -(u, \frac{\partial\varphi}{\partial y_i})_2$$

$$= (D_i u, \varphi)_2 = \int D_i u(y) j_\varepsilon(x-y)\lambda^n(dy) = J_\varepsilon D_i u(x) \quad .$$

If $x \in U_\varepsilon$ then $B_\varepsilon(x) \cap U \subset U_d$, hence $J_\varepsilon u(x) = 0$ and $J_\varepsilon D_i u(x) = 0$. Therefore $J_\varepsilon u \in \mathcal{D}(U)$ and $D_i J_\varepsilon u = J_\varepsilon D_i u$. Moreover, $\lim_{\varepsilon\to 0} \| J_\varepsilon u - u\|_2 = 0$ and $\lim_{\varepsilon\to 0} \| D_i J_\varepsilon u - D_i u\|_2 = \lim_{\varepsilon\to 0} \| J_\varepsilon D_i u - D_i u\|_2 = 0$. Thus $\lim_{\varepsilon\to 0} \| J_\varepsilon u - u\|_1 = 0$.

2. If $u \in H^1(U)$ and $\varphi \in \mathcal{D}(U)$ then obviously $\varphi u \in L^2(U)$ and $D_i(\varphi u) = \frac{\partial\varphi}{\partial x_i} u + \varphi D_i u \in L^2(U)$ for all $1 \le i \le n$.

3. Fix $u \in H^1(U)$ and $\varepsilon > 0$. We take $\varphi_0 = 0$ and choose a sequence (φ_m) in $\mathcal{D}(U)$ such that $\varphi_m = 1$ on $U \setminus U_{\frac{1}{m}}$ for every $m \in \mathbb{N}$. Let $m \in \mathbb{N}$ and define

$$u_m = (\varphi_m - \varphi_{m-1})u \quad .$$

Then $u_m \in H^1(U)$ by (2) and by (1) there exists $0 < \delta_m < \frac{1}{m}$ such that

$\psi_m := J_{\delta_m} u_m \in \mathcal{D}(U)$ and

$$\| \psi_m - u_m \|_1 < \frac{\varepsilon}{2^m} .$$

If $m \geq 2$ then $u_m = 0$ on $U \setminus U_{\frac{1}{m-1}}$ and $\psi_m = 0$ on $U \setminus U_{\frac{2}{m-1}}$. Therefore $u = \sum_{m=1}^{\infty} u_m$, $D_i u = \sum_{m=1}^{\infty} D_i u_m$ for all $1 \leq i \leq n$ and

$$\psi := \sum_{m=1}^{\infty} \psi_m \in C^{\infty}(U) .$$

Moreover, $\psi - u \in H^1(U)$ and $\| \psi - u \|_1 < \varepsilon$ since $\sum_{m=1}^{\infty} \| \psi_m - u_m \|_1 < \varepsilon$. Hence $\psi = (\psi - u) + u \in H^1(U)$ finishing the proof. ┘

4.3. LEMMA. Let $f \in C^1(\mathbb{R})$ such that $f(0) = 0$ and f' is bounded, and let $u \in H^1(U)$. Then $f \circ u \in H^1(U)$ and $D_i(f \circ u) = f' \circ u \, D_i u$ for all $1 \leq i \leq n$.

Proof. Let $\alpha \in \mathbb{R}_+$ such that $|f'| \leq \alpha$. Then $|f \circ u| = |f \circ u - f(0)| \leq \alpha |u|$ and hence $f \circ u \in L^2(U)$. Moreover, $f' \circ u \, D_i u \in L^2(U)$ for every $1 \leq i \leq n$ since $|f' \circ u| \leq \alpha$ and $D_i u \in L^2(U)$.

By (4.2) there exists a sequence (u_m) in $H^1(U) \cap C^{\infty}(U)$ such that $\lim_{m \to \infty} \| u_m - u \|_1 = 0$. We may assume that (u_m) converges λ^n-a.e. to u. Fix $1 \leq i \leq n$. Then $\| f \circ u_m - f \circ u \|_2 \leq \alpha \| u_m - u \|_2$ and

$$\| D_i(f \circ u_m) - f' \circ u \, D_i u \|_2 = \| f' \circ u_m D_i u_m - f' \circ u \, D_i u \|_2$$

$$\leq \| f' \circ u_m (D_i u_m - D_i u) \|_2 + \| (f' \circ u_m - f' \circ u) D_i u \|_2$$

$$\leq \alpha \| D_i u_m - D_i u \|_2 + \| (f' \circ u_m - f' \circ u) D_i u \|_2$$

where $\lim_{m \to \infty} \| (f' \circ u_m - f' \circ u) D_i u \|_2 = 0$ by the convergence theorem of Lebesgue. Hence $f \circ u \in H^1(U)$ and $D_i(f \circ u) = f' \circ u \, D_i u$ by (4.1). ┘

4.4. COROLLARY. 1. Let $u \in H^1(U)$. Then u^+, u^-, $|u| \in H_1(U)$ and, for all $1 \leq i \leq n$, $D_i u^+ = 1_{\{u > 0\}} D_i u$, $D_i u^- = -1_{\{u < 0\}} D_i u$, $D_i |u| = (1_{\{u > 0\}} - 1_{\{u < 0\}}) D_i u$.

2. The mapping $(u,v) \mapsto \inf(u,v)$ from $H^1(U) \times H^1(U)$ into $H^1(U)$ is continuous.

Proof. 1. Obviously $u^+ \in L^2(U)$ and $1_{\{u > 0\}} D_i u \in L^2(U)$ for all $1 \leq i \leq n$. Define functions $f_s \in C^1(\mathbb{R})$, $s > 0$, by

$$f_s(t) = \begin{cases} (t^2 + s^2)^{1/2} - s & \text{if } t > 0 , \\ 0 & \text{if } t \leq 0 , \end{cases}$$

and let $1 \leq i \leq n$. Then $0 \leq f_s(t) \leq t^+$ and $0 \leq f_s'(t) \leq 1$, hence by (4.3) $f_s \circ u \in H^1(U)$ and $D_i(f_s \circ u) = f_s' \circ u D_i u$. Since $\lim\limits_{s\to 0} f_s(t) = t^+$ and $\lim\limits_{s\to 0} f_s' = 1_{\{t>o\}}$, we have $\lim\limits_{s\to 0} \| f_s \circ u - u^+\|_2 = 0$ and $\lim\limits_{s\to 0} \|f_s' \circ u D_i u - 1_{\{u>o\}} D_i u\|_2 = 0$ by the convergence theorem of Lebesgue. Using (4.1) we conclude that $u^+ \in H^1(U)$ and $D_i u^+ = 1_{\{u>o\}} D_i u$. The proof of (1) is finished using $u^- = (-u)^+$ and $|u| = u^+ - u^-$.

2. Let (u_m) be a sequence in $H^1(U)$ such that $\lim\limits_{m\to\infty} \| u_m - u\|_1 = 0$ and $\lim\limits_{m\to\infty} u_m = u$ λ^n-a.e. Fix $m \in \mathbb{N}$. Using estimates as in the proof of (4.3) we obtain that

$$\| f_s \circ u_m - f_s \circ u\|_1 \leq \| u_m - u\|_1 + \sum_{i=1}^{n} \| (f_s' \circ u_m - f_s' \circ u) D_i u\|_2$$

for every $s > 0$ and hence

$$\| u_m^+ - u^+\|_1 \leq \| u_m - u\|_1 + \sum_{i=1}^{n} \|(1_{\{u_m > o\}} - 1_{\{u>o\}}) D_i u\|_2$$

by the convergence theorem of Lebesgue. Since $\lim\limits_{m\to\infty} 1_{\{u_m>o\}} = 1_{\{u>o\}}$ λ^n-a.e. we conclude that $\lim\limits_{m\to\infty} \| u_m^+ - u^+\|_1 = 0$. This shows that the mapping $u \mapsto u^+$ is continuous on $H^1(U)$. The continuity of the mapping $(u,v) \to \inf (u,v)$ is now an immediate consequence of the equality $\inf (u,v) = \frac{1}{2}(u+v - (u-v)^+ - (v-u)^+)$. ⌋

4.5. LEMMA. Let $w \in H^1(U)$, $w \geq 0$, and suppose that $w \leq |u|$ for some $u \in H_0^1(U)$. Let V be an open subset of U such that $\overline{\{w > \varepsilon\}} \cap U \subset V$ for every $\varepsilon > 0$. Then there exists a sequence (φ_m) in $\mathcal{D}^+(U)$ such that $\lim\limits_{m\to\infty} \| \varphi_m - w\|_1 = 0$ and $S(\varphi_m) \subset V$ for every $m \in \mathbb{N}$.

Proof. Let $(u_m) \subset \mathcal{D}(U)$ such that $\lim\limits_{m\to\infty} u_m = u$ in $H^1(U)$. For every $m \in \mathbb{N}$ define

$$w_m = \inf (|u_m| , (w - \tfrac{1}{m})^+) .$$

Since the sequence $(w - \frac{1}{m})$ converges to w , we conclude by (4.4) that the sequence (w_m) converges to $\inf (|u| , w^+) = w$ in $H^1(U)$. Fix $m \in \mathbb{N}$. Then $\overline{\{w_m > 0\}}$ is a compact subset of V . Hence by (4.2) there exists $\varepsilon_m > 0$ such that $\varphi_m = J_{\varepsilon_m} w_m \in \mathcal{D}(U)$, $S(\varphi_m) \subset V$, and $\| \varphi_m - w_m \|_1 < \frac{1}{m}$. Obviously, $\lim_{m\to\infty} \|\varphi_m - w\|_1 = 0$. ┘

4.6. COROLLARY. 1. For every $u \in H_0^1(U)$, u^+ , u^-, $|u| \in H_0^1(U)$.

2. If $w \in H^1(U)$ such that $|w| \le |u|$ in a neighborhood of U^* for some $u \in H_0^1(U)$ then $w \in H_0^1(U)$.

Proof. 1. Immediate consequence of (4.4.1) and (4.5) choosing $V = U$.

2. Let $w \in H^1(U)$ and $u \in H_0^1(U)$ such that $|w| \le |u|$ in a neighborhood of U^* . Then $\inf(w^+,|u|) \in H^1(U)$ by (4.4), hence $\inf(w^+,|u|) \in H_0^1(U)$ by (4.5). Moreover $w^+ - \inf(w^+,|u|) \in H_0^1(U)$ by (4.2.1). Therefore $w^+ \in H_0^1(U)$. Similarly, $w^- \in H_0^1(U)$ and hence $w = w^+ - w^- \in H_0^1(U)$. ┘

The following lemma will show that $\sum_{i=1}^n \| D_i u\|_2$ defines an equivalent norm on $H_0^1(U)$.

4.7. LEMMA (POINCARÉ'S inequality). Let δ be the diameter of U . Then for all $u \in H_0^1(U)$,

$$\| u\|_2 \le \frac{2\delta}{n} \sum_{i=1}^n \|D_i u\|_2 \quad \text{and} \quad \| u \|_2^2 \le \frac{4\delta^2}{n} \sum_{i=1}^n \|D_i u\|_2^2$$

Proof. We may assume $u \in \mathcal{D}(U)$ and $U \subset B_\delta(0)$. Then

$$\begin{aligned} \| u\|_2^2 &= \frac{1}{n} \sum_{i=1}^n \int_U u^2\, d\lambda^n = -\frac{1}{n} \sum_{i=1}^n \int_U D_i u^2(x) x_i \lambda^n(dx) \\ &= -\frac{2}{n} \sum_{i=1}^n \int_U (D_i u(x)) u(x) x_i \lambda^n(dx) \le \frac{2\delta}{n} \| u\|_2 \sum_{i=1}^n \| D_i u\|_2 \end{aligned}$$

which implies the first inequality. The second one follows from the well known inequality $\frac{1}{n}\left(\sum_{i=1}^n |b_i|\right)^2 \le \sum_{i=1}^n b_i^2$ for $b_1, \ldots, b_n \in \mathbb{R}$. ┘

For the remainder of this section assume that L is a smooth semi-elliptic differential operator on X . By Hoelder's inequality the Dirichlet form a_L defined by (**) at the beginning of this section can be extended to a continuous bilinear form on $H^1(U) \times H^1(U)$ by

$$a_L(v,u) = \int_U \left[\sum_{i,j=1}^n D_i(a_{ij} v) D_j u - \sum_{i=1}^n a_i v D_i u - a\, v\, u \right] d\lambda^n \quad .$$

We shall say that L is *coercive* on U if there exists a constant $c > 0$ such that $a_L(u,u) \geq c\|u\|_1^2$ for all $u \in H_0^1(U)$.

4.8. EXAMPLE. Let $\varepsilon > 0$. Then $L_\varepsilon = \varepsilon\Delta + L$ is coercive on U provided U is suffiently small. Indeed, let $C > 0$ such that $|\sum_{i=1}^{n} D_i a_{ij}| \leq C$, $|a_i| \leq C$ and $|a| \leq C$ on U , assume that the diameter δ of U satisfies $(\frac{2C^2n}{\varepsilon} + C)\frac{4\delta^2}{n} \leq \frac{\varepsilon}{4}$, and let $u \in \mathcal{D}(U)$. Since $\sum_{i,j=1}^{n} a_{ij}D_iuD_ju \geq 0$ we have

$$a_{L_\varepsilon}(u,u) \geq \varepsilon \sum_{i=1}^{n} \|D_iu\|_2^2 + R$$

where

$$R = \iint_U\Big[\sum_{i,j=1}^{n}(D_i a_{ij})\, uD_ju - \sum_{i=1}^{n} a_iuD_iu - au^2\Big]d\lambda^n \quad .$$

Since for all real numbers r,s,t such that $t > 0$ the inequality $|rs| \leq \frac{t}{2}r^2 + \frac{1}{2t}s^2$ holds we obtain taking $t = \frac{\varepsilon}{2C}$

$$|R| \leq 2C\sum_{i=1}^{n} \int_U |uD_iu|d\lambda^n + C\|u\|_2^2 \leq \frac{\varepsilon}{2}\sum_{i=1}^{n}\|D_iu\|_2^2 + (\frac{2C^2n}{\varepsilon} + C)\|u\|_2^2$$

$$\leq (\frac{\varepsilon}{2} + (\frac{2C^2n}{\varepsilon} + C)\frac{4\delta^2}{n})\sum_{i=1}^{n}\|D_iu\|_2^2$$

where the last inequality follows from (4.7). Hence $|R| \leq \frac{3\varepsilon}{4}\sum_{i=1}^{n}\|D_iu\|_2^2$ and

$$a_{L_\varepsilon}(u,u) \geq \frac{\varepsilon}{4}\sum_{i=1}^{n}\|D_iu\|_2^2 \geq c\,\|u\|_1^2$$

for some $c > 0$ by (4.7).

Given $\varphi \in \mathcal{D}(U)$, we have $a_L(\varphi,u) = -(L^*\varphi,u)_2$ for every $u \in C^2(U)$ and hence $a_L(\varphi,u) = -(L^*\varphi,u)_2$ for every $u \in H^1(U)$ since the mappings $u \mapsto a_L(\varphi,u)$ and $u \mapsto (L^*\varphi,u)_2$ are continuous on $H^1(U)$ and $H^1(U) \cap C^\infty(U)$ is dense in $H^1(U)$ by (4.2.3). Therefore $u \in H^1(U)$ is a weak solution of $Lu = -f$ if and only if

$$a_L(\varphi,u) = (\varphi,f)_2$$

for every $\varphi \in \mathcal{D}(U)$ or, equivalently, if and only if $a_L(v,u) = (v,f)_2$ for every $v \in H_0^1(U)$.

As an application of the projection theorem of G. STAMPACCHIA (I.5.1) we obtain the following fundamental result.

4.9. PROPOSITION. Let L be coercive on U. Then for every $f \in H^1(U)$ there exists a unique weak solution $u \in H_0^1(U)$ of the Poisson equation $Lu = -f$ on U.

Proof. The mapping $v \mapsto (v,f)_2$ is a continuous linear functional on $H_0^1(U)$. Hence by (I.5.1) there exists a unique $u \in H_0^1(U)$ such that $a_L(v,u) = (v,f)_2$ for every $v \in H_0^1$. ┘

We conclude this section by another useful minimum principle.

4.10 PROPOSITION. Assume that L is coercive on U. Let $w \in H^1(U) \cap C^\infty(U)$ such that $Lw \leq 0$ and $w \geq u$ for some $u \in H_0^1(U)$. Then $w \geq 0$.

Proof. By (4.4), $w^- \in H^1(U)$ and $w^- \leq |u|$. Hence by (4.5) there exists a sequence (φ_m) in $\mathcal{D}^+(U)$ such that $\lim_{m\to\infty} \|\varphi_m - w^-\|_1 = 0$ and $S(\varphi_m) \subset \{w<0\}$ for every $m \in \mathbb{N}$. Since by (4.4)

$$a_L(\varphi_m, w^-) = \int_U [\sum_{i=1}^n \frac{\partial(a_{ij}\varphi_m)}{\partial x_i} D_j w^- - \sum_{i=1}^n a_i\varphi_m D_i w^- - a\varphi_m w^-]d\lambda^n$$

$$= -\int_{\{w<o\}} [\sum_{i=1}^n \frac{\partial(a_{ij}\varphi_m)}{\partial x_i} \frac{\partial w}{\partial x_j} - \sum_{i=1}^n a_i\varphi_m \frac{\partial w}{\partial x_i} - a\,\varphi_m w]d\lambda^n = \int_{\{w<o\}} \varphi_m Lw\, d\lambda^n \leq 0$$

for every $m \in \mathbb{N}$, we obtain that

$$c\|w^-\|_1^2 \leq a_L(w^-,w^-) = \lim_{m\to\infty} a_L(\varphi_m,w^-) \leq 0 .$$

Thus $w^- = 0$, i.e. $w \geq 0$. ┘

4.11. EXERCISES. 1. Let $u,v \in L^2(U)$ and $1 \leq i \leq n$. Then $D_i u = v$ if and only if every $x \in U$ has an open neighborhood V in U such that $D_i(u|_V) = v|_V$.

2. Let $u,v \in H_0^1(U)$. Find conditions for the validity of the product rule $D_i(uv) = (D_i u)v + u(D_i v)$.

3. Determine the Dirichlet form and the adjoint operator associated with the Laplace operator and the heat operator.

5. Elliptic-Parabolic Differential Operators

Let X be an open subset of $\mathbb{R}^n$. A mapping $A = (\alpha_1, \ldots, \alpha_n) : X \to \mathbb{R}^n$ is called a C^∞-*vector field* on X if $\alpha_1, \ldots, \alpha_n \in C^\infty(X)$. A is always identified with the differential operator of first order $Au = \sum_{i=1}^n \alpha_i \frac{\partial u}{\partial x_i}$. For C^∞-vector fields A and B on X we define the *composition* AB and the *Lie-product* $[A,B]$ of A and B by $(AB)u := A(Bu)$ and $[A,B]u := ABu - BAu$ $(u \in C^2(X))$. Then $[A,B]$ is again a C^∞-vector field.

Let $A_1, \ldots, A_r$ be C^∞-vector fields on X. We denote by $L(A_1, \ldots, A_r)$ the Lie algebra generated by $A_1, \ldots, A_r$, i.e. $L(A_1, \ldots, A_r)$ is the smallest $C^\infty(X)$-modul containing $A_1, \ldots, A_r$ which is stable with respect to taking Lie-products. For $x \in X$ we call the dimension of the vector space $\{Z(x) : Z \in L(A_1, \ldots, A_r)\}$ the *rank* of $L(A_1, \ldots, A_r)$ *at* x. We denote it by $\mathrm{rank}_x\ L(A_1, \ldots, A_r)$.

A semi-elliptic differential operator

$$L = \sum_{i,j=1}^n a_{ij} \frac{\partial^2}{\partial x_i \partial x_j} + \sum_{i=1}^n a_i \frac{\partial}{\partial x_i}$$

on X is called *elliptic-parabolic*, if there exist C^∞-vector fields $A_0, A_1, \ldots, A_r$ on X such that

$$L = \sum_{k=1}^r A_k^2 + A_0 ,$$

and for every $x \in X$, $\mathrm{rank}_x\ L(A_1, \ldots, A_r, A_0) = n$.

5.1. EXAMPLES. 1. The Laplace operator on $\mathbb{R}^n$ and the heat operator on $\mathbb{R}^{n+1}$ are elliptic-parabolic differential operators.

2. If L is an elliptic-parabolic differential operator then for every $\varepsilon > 0$, $L_\varepsilon = \varepsilon\Delta + L$ is also an elliptic-parabolic differential operator.

In the sequel we shall use the following non-trivial result whose proof is far beyond the scope of this book.

5.2. THEOREM (L. HÖRMANDER). Let L be an elliptic-parabolic differential operator on X. Then L is *hypoelliptic*, in particular for every open set $U \subset X$ and any weak solution $u \in L^2(U)$ of $Lu = f \in C^\infty(U)$ we have $u \in C^\infty(U)$.

Proof. L. HÖRMANDER [1]. ⌟

In the following let L be an elliptic-parabolic differential operator on X. Our aim is to show that $(X,\mathcal{H}_L)$ is a normal Bauer space.

5.3. PROPOSITION. $(X,\mathcal{H}_L)$ has the convergence property of Bauer.

Proof. Let (h_m) be an increasing sequence of harmonic functions on an open subset U of X and assume that $h = \sup h_m$ is locally bounded. Let V be an open set such that $\overline{V} \subset U$. Then $(h_{m|V})$ is bounded in $L^2(V)$, hence a subsequence converges weakly in $L^2(V)$ to h. Therefore $h \in L^2(V)$ is a weak solution of $Lh = 0$ on V and thus harmonic on V by (5.2). By the sheaf property of $\mathcal{H}_L$ we obtain $h \in \mathcal{H}_L(U)$. ⌟

5.4. PROPOSITION. Let U be an L-regular subset of X such that L is coercive on U. Then U is a regular set.

Proof. 1. Let $f \in C^\infty(X)$. By (4.9) there exists a weak solution $u \in H_0^1(U)$ of the Poisson equation $Lu = -f$ on U. By (5.2), $u \in C^\infty(U)$. Let $z \in U^*$, let $\tilde{w}$ be an L-barrier at z, and $w = \tilde{w}_{|U}$. Then $w \in H^1(U)$. Take $\lambda > 0$ such that $|f| \leq -\lambda Lw$ on $\overline{U}$. Then $L(\lambda w \mp u) = \lambda Lw \pm f \leq 0$ on U, hence $a_L(\varphi,\lambda w \mp u) \geq 0$ for every $\varphi \in \mathcal{D}^+(U)$ and therefore $\lambda w \mp u \geq 0$ on U by (4.10). Thus $0 \leq \limsup_{x \to z} |u|(x) \leq \lambda \lim_{x\to z} w(x) = 0$, i.e. $\lim_{x\to z} u(x) = 0$.

2. Let $f,g \in C^\infty(\mathbb{R}^n)$. Then there exists $u \in C^\infty(U)$ such that $Lu = -f$ on U and $\lim_{x\to z} u(x) = g(z)$ for every $z \in U^*$. Indeed, by (1) there exists $v \in C^\infty(U)$ such that $Lv = -f - Lg$ on U and $\lim_{x\to z} v(x) = 0$ for every $z \in U^*$, and then $u := v + g$ has the desired properties.

3. Let $g \in C(U^*)$. Then there exists a sequence (g_n) in $C^\infty(X)$ such that (g_n) converges uniformly to g on U^*. By (2) there exists a sequence $(u_n) \subset C(\overline{U})$ such that $u_{n|U} \in C^\infty(U)$, $L u_n = 0$ on U and $u_n = g_n$ on U^* for every $n \in \mathbb{N}$. By the minimum principle (2.4), (u_n) converges uniformly on $\overline{U}$ to some function $u \in C(\overline{U})$ which is a weak solution of $Lu = 0$ on $\overline{U}$. Hence $u \in \mathcal{H}_L(U)$ and $u = g$ on U^*. Using (2.4) again we finally conclude that u is uniquely

determined by g and that $u \geq 0$ if $g \geq 0$. Thus U is regular. ┘

5.5. COROLLARY. For every $\varepsilon > 0$, (X, H_{L_ε}) is a Bauer space such that $(U, {}^*H^+_{L_\varepsilon}(U))$ is a harmonic space for every relatively compact open subset U of X .

Proof. Fix $\varepsilon > 0$. By (5.3), (X, H_{L_ε}) has the convergence property of Bauer. By (5.4), (4.8) and (2.6), every sufficiently small L-regular set is regular with respect to H_{L_ε} . Therefore (X, H_{L_ε}) is a normal Bauer space.

Let U be a relatively compact open subset of X . By (1.2) it remains to show that ${}^*H^+_{L_\varepsilon}(U)$ separates the points of U . So let $x, y \in U$ and $1 \leq i \leq n$ such that $x_i \neq y_i$. Let $A > 0$ and $f(z) = e^{Az_i}$, $z \in X$. Then $L_\varepsilon f \geq (\varepsilon A^2 + a_i A) f$, hence $L_\varepsilon f \geq 0$ on U if A is sufficiently large. Let $c > 0$ such that $u = c - f \geq 0$ on U . Then $u \in {}^*H^+_{L_\varepsilon}(U)$ by (2.6) and $u(x) = c - e^{Ax_i} \neq c - e^{Ay_i} = u(y)$. ┘

5.6. THEOREM (J.M. BONY). Let L be an elliptic-parabolic differential operator on X . Then (X, H_L) is a normal Bauer space.

Proof. In view of (5.3) and (2.3.3) it remains to show that every L-regular set is regular. So let $U \subset X$ be L-regular. By (5.5) and (2.6), U is regular with respect to the normal Bauer spaces (X, H_{L_ε}) , $\varepsilon > 0$.

Let $f \in C(U^*)$. Then for each $\varepsilon > 0$ there exists a unique function $u_\varepsilon \in C^\infty(U)$ such that

$$L_\varepsilon u_\varepsilon = 0 \text{ on } U \text{ and } \lim_{x \to z} u_\varepsilon(x) = f(z) \text{ for every } z \in U^* .$$

By the minimum principle (2.5) we have $\| u_\varepsilon \| = \| f \|$ and $u_\varepsilon \geq 0$ if $f \geq 0$. In particular, there exists a sequence (ε_n) in $\mathbb{R}^*_+$ such that $\lim_{n\to\infty} \varepsilon_n = 0$ and (u_{ε_n}) converges in $L^2(U)$ weakly to a function $u \in L^2(U)$. Then u is a weak solution of the differential equation $Lu = 0$, hence $u \in C^\infty(U)$ by (5.2) and therefore $u \in H_L(U)$. Clearly $\| u \| \leq \| f \|$ and $u \geq 0$ if $f \geq 0$.

Now let $z_0 \in U^*$. In order to show that $\lim_{x \to z} u(x) = f(z_0)$ we fix $\delta > 0$ and choose functions $\tilde{f}, \approx{f} \in C(U^*)$ such that

$$f = f(z_0) + \tilde{f} + \overset{\approx}{f} ,$$

$\tilde{f} = 0$ on $U^* \cap V$ for some open neighborhood V of z_0 and $|\tilde{\tilde{f}}| \le \delta$. For each $\varepsilon > 0$ there exist functions $\tilde{u}_\varepsilon, \tilde{\tilde{u}}_\varepsilon \in C^\infty(U)$ such that $L_\varepsilon \tilde{u}_\varepsilon = L_\varepsilon \tilde{\tilde{u}}_\varepsilon = 0$ on U and $\lim\limits_{x\to z} \tilde{u}_\varepsilon(x) = \tilde{f}(z)$, $\lim\limits_{x\to z} \tilde{\tilde{u}}_\varepsilon(x) = \tilde{\tilde{f}}(z)$ for every $z \in U^*$. Obviously, $|\tilde{\tilde{u}}_\varepsilon| \le \delta$ and

$$u_\varepsilon = f(z_0) + \tilde{u}_\varepsilon + \tilde{\tilde{u}}_\varepsilon$$

since $L_\varepsilon(f(z_0) + \tilde{u}_\varepsilon + \tilde{\tilde{u}}_\varepsilon) = 0$ and $\lim\limits_{x\to z} (f(z_0) + \tilde{u}_\varepsilon(x) + \tilde{\tilde{u}}_\varepsilon(x)) = f(z)$ for every $z \in U^*$. There exists a subsequence (ε_n') of (ε_n) such that $(\tilde{u}_{\varepsilon_n'})$ converges in $L^2(U)$ weakly to a function $\tilde{u} \in C^\infty(U)$, and there exists a subsequence (ε_n'') of (ε_n') such that $(\tilde{\tilde{u}}_{\varepsilon_n''})$ converges in $L^2(U)$ weakly to a function $\tilde{\tilde{u}} \in C^\infty(U)$ such that $|\tilde{\tilde{u}}| \le \delta$. Since $(u_{\varepsilon_n''})$ converges weakly to u , we conclude that

$$u = f(z_0) + \tilde{u} + \tilde{\tilde{u}} \ .$$

Let w be an L-barrier at z_0 and $\lambda > 0$ such that $|\tilde{f}| \le \lambda w$ on $U^* \setminus V$. Then

$$L_\varepsilon(\lambda w \pm \tilde{u}_\varepsilon) = \lambda L_\varepsilon w < 0$$

on U and

$$\lim_{x\to z} (\lambda w \pm \tilde{u}_\varepsilon)(x) = \lambda w(z) \pm \tilde{f}(z) \ge 0$$

for every $z \in U^*$, hence $\lambda w \pm \tilde{u}_\varepsilon \ge 0$ on U by (2.4). Therefore $|\tilde{u}| \le \lambda w$ on U . Since $\lim\limits_{x\to z_0} w(x) = w(z_0) = 0$, we obtain that $\lim\limits_{x\to z_0} \tilde{u}(x) = 0$ and hence

$$\limsup_{x\to z_0} |u(x) - f(z_0)| < \|\tilde{\tilde{u}}\| \le \delta \ .$$

Thus $\lim\limits_{x\to z_0} u(x) = f(z_0)$. Finally, the uniqueness of u follows from (2.4). ┘

We close this section with an example which has been studied rather extensively during the past years.

5.7. EXAMPLE. *Laplace-Kohn operator on the Heisenberg group* $\mathbb{H}_n$.

As a topological space the Heisenberg group $\mathbb{H}_n$ of degree n $(n \in \mathbb{N})$ may be identified with $\mathbb{R}^{2n} \times \mathbb{R}$ and the group law is given by

$$(z,t)(z',t') = (z+z' \ , \ t+t'+2\sum_{j=1}^{n}(x_j'y_j - x_jy_j'))$$

if $z = (x_1,y_1, \ldots, x_n,y_n)$, $z' = (x_1', y_1', \ldots, x_n', y_n')$. $\mathbb{H}_n$ is a nilpotent Lie group. Let

$$A_j = \frac{\partial}{\partial x_j} + 2y_j \frac{\partial}{\partial t} \quad , \quad B_j = \frac{\partial}{\partial y_j} - 2x_j \frac{\partial}{\partial t} \qquad (1 \le j \le n) \; .$$

Since $[A_j, B_j] = -4\frac{\partial}{\partial t}$ for every $j = 1, \ldots, n$ we obtain that $\mathrm{rank}_{(z,t)}\, L(A_1, \ldots, A_n, B_1, \ldots, B_n) = 2n+1$ for every $(z,t) \in \mathbb{H}_n$, hence the *Laplace-Kohn operator*

$$\Delta_K = \sum_{j=1}^{n} (A_j^2 + B_j^2) = \sum_{j=1}^{n} \left[\frac{\partial^2}{\partial x_j^2} + \frac{\partial^2}{\partial y_j^2} + 4\, y_j \frac{\partial^2}{\partial x_j \partial t} - 4\, x_j \frac{\partial^2}{\partial y_j \partial t} + 4(x_j^2 + y_j^2) \frac{\partial^2}{\partial t^2} \right]$$

is an elliptic-parabolic differential operator on $\mathbb{H}_n$. Therefore $(\mathbb{H}_n, \mathcal{H}_{\Delta_K})$ is a Bauer space by (5.6).

An elliptic-parabolic differential operator can be rather degenerate as the following exercise shows.

5.8. EXERCISE. Let $X = \mathbb{R}^3$ and $L = \frac{\partial^2}{\partial x_1^2} + x_1 \frac{\partial}{\partial x_2} - \frac{\partial}{\partial x_3}$. Then $(\mathbb{R}^3, \mathcal{H}_L)$ is a normal Bauer space.

Notes

The reader should note that for many statements on balayage spaces, the corresponding results from classical potential theory or from the theory of harmonic spaces are quoted.

INTRODUCTION

Reports on the development of potential theory have been written by M. BRELOT [8],[16], for more modern reviews see the survey articles by H. BAUER [8],[9],[10], [12],[14],[15], J. BLIEDTNER [2], and C. CONSTANTINESCU [1],[4].

CHAPTER 0

Books on classical potential theory are M. BRELOT [14], L.L. HELMS [1], O.D. KELLOGG [1], N.S. LANDKOF [1], J. WERMER [1], whereas the books of K.L. CHUNG [1], M. KAC [2], S.C. PORT - C.J. STONE [1], and M. RAO [1] derive their potential theory as a by-product of probability. The monograph of J.L. DOOB [7] covers both aspects equally. The connection between classical potential theory and complex function theory is extensively studied in W.K. HAYMAN - P.B. KENNEDY [1], M. TSUJI [1].

Section 1. The Poisson integral was introduced by S.D. POISSON [1],[2], a first rigorous proof of (1.2) for continuous boundary functions was given by H.A. SCHWARZ [1]. Superharmonic (hyperharmonic resp.) functions were inaugurated by F. RIESZ [1],[2], for earlier references see T. RADÓ [1]. The properties of (1.7) are due to A. HARNACK [1].

Section 2. The irregular motion of pollen particles in a liquid observed by R. BROWN in 1827 was mathematically treated by L. BACHELIER, A. EINSTEIN, and M. SMOLUCHOWSKI at the beginning of this century. They established that the distribution of the displacement after time t has density g_t .

Section 3. Excessive functions were introduced by G.A. HUNT [1]. More details on the connection between classical potential theory and Brownian motion are discussed by A.W. KNAPP [1].

CHAPTER I

Section 1. The notion of an adapted cone of functions was introduced by G. CHOQUET [2] to solve a certain moment problem. He established the fundamental property (1.4). The importance for potential theory was discovered by G. MOKOBODZKI - D. SIBONY [1] - [4]. The existence of strict elements was first proved in the context of potential theory by R.M. HERVÉ [1]. G. CHOQUET's topological lemma (1.8) can be found in M. BRELOT [14]. (1.9) is from N. BOBOC - A. CORNEA [2].

Section 2. The material of this section is an adaption of the theory of compact convex sets, see e.g. E.M. ALFSEN [1], especially (2.1) is a version of N. BOBOC - A. CORNEA [2] of the original treatment by H. BAUER [1],[2]. (2.2) - (2.7) are due to G. MOKOBODZKI - D. SIBONY [1], see also D. SIBONY [1].

Section 3. The capacitability theorem (3.8) is from G. CHOQUET [1]. The presentation here is essentially due to C. DELLACHERIE [1],[4]. Further properties of capacities can be found in C. DELLACHERIE [2],[3],[5], C. DELLACHERIE - P.A. MEYER [1], and P.A. MEYER [2].

Section 4. S. BERNSTEIN's theorem (4.2) is proved in G. CHOQUET [3]. Its importance for potential theory on locally compact abelian groups is worked out in C. BERG - G. FORST [1].

Section 5. (5.1) is a generalization of the LAX-MILGRAM lemma due to G. STAMPACCHIA [1].

CHAPTER II

Section 1. A detailed study can be found in P.A. MEYER [2]. The statement (1.6) is due to R. CAIROLI [1].

Section 2. The material of this section is from G. MOKOBODZKI [1],[2]. This theory was initiated by G. CHOQUET - J. DENY, see C. DELLACHERIE - P.A. MEYER [3] and P.A. MEYER [2].

Section 3. General references of this section are R.K. GETOOR [1], P.A. MEYER [2], G. MOKOBODZKI [1].

Section 4. In classical potential theory the fine topology with respect to the cone of positive hyperharmonic functions was introduced by M. BRELOT [5] following a suggestion of H. CARTAN. Property (4.2) is due to A. CORNEA (1966, unpublished). For further discussions of the fine topology in a general setting see M. BRELOT [15]. Balayage spaces were introduced by J. BLIEDTNER - W. HANSEN [5]. The notion is based on the theory of potential cones of G. MOKOBODZKI - D. SIBONY [1] and on abstract balayage theory of C. CONSTANTINESCU - A. CORNEA [4]. The rôle of balayage spaces in the more abstract theory of H-cones of N. BOBOC - C. BUCUR - A. CORNEA [1] has been clarified by K. JANSSEN [1]. The main results (4.7) and (4.9) were obtained by J. BLIEDTNER - W. HANSEN [5] where research of G. MOKOBODZKI [1] and J.C. TAYLOR [1] has been used.

Section 5. (5.2) is due to V. DEMBINSKI - K. JANSSEN [1]; the other results are from J. BLIEDTNER - W. HANSEN [5], C. CONSTANTINESCU - A. CORNEA [4].

Section 6. The construction of kernels is due to G. MOKOBODZKI - D. SIBONY [2] (see also D. SIBONY [1]). Another approach can be found in R.-M. HERVÉ [1], an abstract version in C. CONSTANTINESCU - A. CORNEA [4].

Section 7. The relation of potential kernels and their associated resolvents is described in P.A. MEYER [2], D. SIBONY [1], and J.C. TAYLOR [4].

Section 8. (8.2) is an adaption of the HILLE-YOSIDA theorem (see C. CONSTANTINESCU - A. CORNEA [4]) following R.K. GETOOR [1]. Many mathematicians have contributed to the main theorem (8.6): P.A. MEYER [1] for BRELOT's axiomatic theory, N. BOBOC - C. CONSTANTINESCU - A. CORNEA [1], W. HANSEN [1],[2], J.C. TAYLOR [2],[3] for general harmonic spaces, G. MOKOBODZKI - D. SIBONY [3] for adapted potential cones. A corresponding result for harmonic groups is due to J. BLIEDTNER [1]. In this context J. DENY's result on the equivalence of fundamental families associated with a potential kernel and transient convolution semigroups plays an important rôle, see C. BERG - G. FORST [1]. The resolvents in (8.7) are introduced by D.B. RAY [1]. For further properties see R.K. GETOOR [1].

CHAPTER III

Section 1. As a generalization of harmonic measures in harmonic spaces W. HANSEN [3],[7] introduced families of harmonic kernels. The presentation here connects this concept with the use of Evans functions (see C. CONSTANTINESCU - A. CORNEA [4]) avoiding the existence of a base of regular sets.

Section 2. The material of this section is based on J. BLIEDTNER - W. HANSEN [5] and C. CONSTANTINESCU - A. CORNEA [4].

Sections 3 - 6. We refer to H. BAUER [3],[5], C. CONSTANTINESCU - A. CORNEA [2], [4], W. HANSEN [3]. The definition of a potential is due to M. BRELOT [10], the decomposition in (6.2) to F. RIESZ [2].

Section 7. (7.1) is from J. BLIEDTNER - W. HANSEN [5], (7.2) can be found in N. BOBOC - C. BUCUR - A. CORNEA [1].

Section 8. The main theorem (8.5) is in J. BLIEDTNER - W. HANSEN [5].

CHAPTER IV

General references for the material of this chapter are H. BAUER [4],[11], R.M. BLUMENTHAL - R.K. GETOOR [1], C. DELLACHERIE [3], J.L. DOOB [1], E.B. DYNKIN [1],[2], P.A. MEYER [2], in particular for martingales see C. DELLACHERIE - P.A. MEYER [2].

The presentation closely follows R.M. BLUMENTHAL - R.K. GETOOR [1], but note that our definition of a strong Markov process is slightly different. (7.5) and (7.6) are due to W. HANSEN (1980, unpublished). (8.1) is essentially contained in J. BLIEDTNER - W. HANSEN [5], see also J.C. TAYLOR [5].

CHAPTER V

Section 1. Absorbing sets have been introduced by H. BAUER in 1963. (1.2) and

(1.5) collect properties from H. BAUER [5],[7], C. CONSTANTINESCU - A. CORNEA [4].

Section 2. Results on strong Feller semigroups can be found in P.A. MEYER [2], G. MOKOBODZKI [1]. Perturbations of harmonic spaces have been considered by B. WALSH [1]. Results like (2.7) are established by W. HANSEN [4]. For a discussion of the generalized Schrödinger equation by perturbation see A. BOUKRICHA - W. HANSEN - H. HUEBER [1]. (2.10) is due to W. MEIER [1].

Section 3. The concept of subordination by convolution semigroups is due to S. BOCHNER. General references are C. BERG - G. FORST [1], F. HIRSCH [1]. The kernels V_α of (3.4) are connected with M. RIESZ [1], although O. FROSTMAN [1] has studied them before. The statement of (3.6) is due to F. HIRSCH (private communication).

Section 4. The harmonic kernels have been introduced by M. RIESZ [1]. Most results are adaptions of classical potential theory (see L.L. HELMS [1]) and can be found in N.S. LANDKOF [1]. We mention the following properties: (4.11) is due to G.C. EVANS [2], F. VASILESCO [1], (4.13) is contained in O. FROSTMAN [1], A.J. MARIA [1]. The notion of capacity for compact sets was introduced by N. WIENER [3] and extended by O. FROSTMAN [1], C. DE LA VALLÉE-POUSSIN [1]. (4.17) is due to N. WIENER [2] for closed sets. (4.18) is from T. BABA [1].

Section 5. Products of semigroups (harmonic spaces resp.) have been studied by R. CAIROLI [1],(K. GOWRISANKARAN [1], W. MEIER [1], U. SCHIRMEIER [1], L. STOICA [1] resp.). In particular (5.7) and (5.10) are due to U. SCHIRMEIER [1].

Section 6. J.L. DOOB [3]-[5] studied properties of the heat equation connected with the Dirichlet problem. The proof of (6.11) is a modified version of the one given by C. CONSTANTINESCU - A. CORNEA [4]. (6.16) is in H. BAUER [5]. A further characterization of (hyper)harmonic functions is obtained by W. FULKS [1],[2]. The potentialtheoretic interpretation is given by H. BAUER [13].

Section 7. This section is based on the work of C. BERG [2],[3].

Section 8. Images have been studied by C. CONSTANTINESCU - A. CORNEA [1], W. HANSEN [6], and E.B. DYNKIN [2].

Section 9. The construction of (9.1) is a special case of (III.2.11).

CHAPTER VI

The concept of balayage goes back at least as far as C.F. GAUSS. From the two approaches, one by reducing hyperharmonic functions, the other by Hilbert space methods (O. FROSTMAN [1],[2], H. CARTAN [2],[3]), only the first procedure is discussed in this book. H. POINCARÉ's "methode de balayage" was developed further by C. DE LA VALLÉE-POUSSIN [1],[2],[3]. For more details see the historical notes in J.L. DOOB [7] or A.F. MONNA [1].

Section 1. The systematic study of reducing hyperharmonic functions started with M. BRELOT [4]. This section is based on H. BAUER [5], C. CONSTANTINESCU - A. CORNEA [4].

Section 2. The material of this section is essentially taken from C. CONSTANTINESCU - A. CORNEA [4]. Originally, (2.1) and (2.10) are due to M. BRELOT [6], whereas (2.8), (2.9) are from J. BLIEDTNER - W. HANSEN [4].

Section 3. The idea of using G. CHOQUET's capacitability theorem to establish the measurability and approximation of hitting times (3.14) is due to G.A. HUNT [1]. (3.16) is in H. BAUER [7]. The presentation here of these theorems used the capacities connected with reduced functions. We note that the first rigorous construction of Brownian motion is due to N. WIENER [1]. (3.18) is essentially due to P. COURRÈGE - P. PRIOURET [1].

Section 4. Thinness of a set was introduced by M. BRELOT [2]. (4.2) is due to C. CONSTANTINESCU [2]. The condition of non-thinness in (4.7.3) is an improved version of the cone condition of S. ZAREMBA [1] whereas the example of the spine is from H. LEBESGUE [2]. The probabilistic approach to the fine topology and to thinness and the interpretation of superharmonic functions as supermartingales are due to J.L. DOOB [2],[3] for Brownian motion and heat process, see R.M. BLUMENTHAL - R.K. GETOOR [1] for standard processes. (4.8) is contained in H. BAUER [7]. (4.14)-(4.17) are from J. BLIEDTNER - W. HANSEN [1].

Section 5. Polar sets were introduced by M. BRELOT [4] in classical potential theory where they coincide with sets of exterior capacity zero as H. CARTAN [1] has shown. (5.2) is contained in H. BAUER [7]. (5.5) has been proved by H. BAUER in 1965, see H. BAUER [5]. The notion of semipolar sets was introduced by M. BRELOT [13] for proving (5.11). In classical potential theory T. RADO [1] and E. SZPILRAJN [1] have shown that the exceptional set of (5.11) is of Lebesgue measure zero. M. BRELOT [1] showed that all compact subsets are polar, and H. CARTAN [2] succeeded in proving that this set has outer capacity zero. This result was already announced in H. CARTAN [1]. (5.14) is due to J.L. DOOB [6], (5.17) to W. HANSEN [4]. For more equivalent properties in (5.21) see section (VII.4) and C. CONSTANTINESCU - A. CORNEA [4]. (5.23), (5.24) are due to W. HANSEN [5], (5.25.2) to C. BERG [1], (5.25.4) to M. BRELOT [4],[12].

Section 6. The concept of essential base goes back to J. BLIEDTNER - W. HANSEN [1]. A modification of this notion by U. BAUERMANN [1] and contributions of W. HANSEN [9] led to the present notion. The material is essentially taken from J. BLIEDTNER - W. HANSEN [1],[6],[7], W. HANSEN [9].

Section 7. (7.1) is contained in R.M. BLUMENTHAL - R.K. GETOOR [1], (7.3) is from J. BLIEDTNER - W. HANSEN [6]. The final propositions are obtained by W. HANSEN [9], but note that the first equivalence in (7.6) has already been proved by C. DELLACHERIE [1]; extensive study of the penetration times $\tilde{\tau}_A$ has been started by M. KAC in classical potential theory (see Z. CIESIELSKI [1]). For harmonic spaces see R. WITTMANN [1].

Section 8. The study of fine supports has been started by G. MOKOBODZKI - D. SIBONY [2] to construct associated kernels (see section (II.6)). (8.2) is taken from W. HANSEN [8],[9], (8.4)-(8.9) from J. BLIEDTNER - W. HANSEN [7], W. HANSEN [9].

Section 9. (9.2) is due to N. BOBOC - A. CORNEA [1], (9.3) to C. CONSTANTINESCU [2]. The other results have been obtained by J. BLIEDTNER - W. HANSEN [2],[3],[4].

Section 10. The results are from W. HANSEN [10], but note that (10.3), (10.6), (10.8) are essentially contained in C. CONSTANTINESCU - A. CORNEA [4].

Section 11. (11.3) is due to N. BOBOC - A.CORNEA [1], for closed sets A the result has already been proved by O. FROSTMAN [3]. The main results (11.16), (11.17) are due to J. LUKEŠ - J. MALÝ [1] for closed sets and to W. HANSEN [10] for arbitrary sets.

Section 12. (12.2) - (12.5) are due to G. MOKOBODZKI [3]. (12.6) is a potential-theoretic version of P. CARTIER's theorem in convexity (see E.M. ALFSEN [1] for the history of this celebrated theorem). The presented proof is an adaption of the probabilistic one due to H. ROST [1].

CHAPTER VII

Section 1. (1.3) is due to O. PERRON [1].

Section 2. O. PERRON [1] introduced upper and lower functions and the corresponding upper and lower solutions. (2.6) is due to N. WIENER [3] and the resolutivity theorem (2.12) to M. BRELOT [3]. I. PETROWSKY [1], W. STERNBERG [1] applied the PWB-method to the heat equation. The presentation here is essentially taken from W. HANSEN [3].

Section 3. Barriers were first implicitly used by H. POINCARÉ [1]; H. LEBESGUE [2] and finally G. BOULIGAND [1] developed the concept of barriers and proved (3.3). Originally (3.7) was established by A. TYCHONOFF [1]; by H. BAUER [6] and J. KÖHN - M. SIEVEKING [1] it is a consequence of results in harmonic spaces.

Section 4. See C. CONSTANTINESCU - A. CORNEA [4] for (4.2). The classification of irregular boundary points is due to J. LUKEŠ - J. MALÝ [1].

Section 5. The material of this section is taken from J. BLIEDTNER - W. HANSEN [1],[3],[7]. The notion of simplicial function cone is an adaption of convexity theory (see E.M. ALFSEN [1]). An extensive study of these cones is in D. SIBONY [1]. (5.12.1) is originally due to E.G. EFFROS - J.L. KAZDAN [1].

Section 6. H. BAUER [2] proved $U^* \cap Ch_{S(U)}X = U_r$ for classical potential theory. J. KÖHN - M. SIEVEKING [1] found example (6.7). For harmonic spaces satisfying the axiom of domination (see C. CONSTANTINESCU - A. CORNEA [4]), N. BOBOC - A. CORNEA [2] showed that S(U) is simplicial. Using affine dilations, E.G. EFFROS - J.L. KAZDAN [1],[2] and P.D. TAYLOR [1] proved this result for many open sets with respect to the heat equation. The final result (6.1) was obtained by J. BLIEDTNER - W. HANSEN [1]. The presentation here is taken from J. BLIEDTNER - W. HANSEN [7]. A. CORNEA - H. HÖLLEIN [1] have an example showing that the thinness of the points in (6.5) cannot be omitted. For (6.8.3) see E.G. EFFROS - J.L. KAZDAN [1] and

J. BLIEDTNER - W. HANSEN [3].

Section 7. M.V. KELDYCH [1] showed in classical potential theory that every open set is a Keldych set. The generalizations in this section are essentially due to J. LUKEŠ [1],[2],[3], J. LUKEŠ - I. NETUKA [1]. See also I. NETUKA [1] for a survey on this subject.

Section 8. The results are due to J. BLIEDTNER - W. HANSEN [1],[4] and W. HANSEN [11]. Part of (8.8.2) for classical potential theory is in T.W. GAMELIN [1]. B. FUGLEDE [1] studied (not necessarily continuous) finely superharmonic functions in harmonic spaces satisfying the axiom of domination.

Section 9. The material here is an improved version of J. BLIEDTNER - W. HANSEN [4].

Section 10. (10.1) and (10.2) are from J. BLIEDTNER - W. HANSEN [4]. (10.3) is originally due to M. BRELOT [4]. The converse of (10.4) has been obtained by O.D. KELLOGG [1]. The presentation here is an adaption of C. CONSTANTINESCU - A. CORNEA [4]. Further references are J.HYVÖNEN [1], I. NETUKA - J. VESELÝ [1].

CHAPTER VIII

Section 1. See the introduction for a description of the development of the theory of harmonic spaces. The notion of a Bauer space is from C. CONSTANTINESCU - A. CORNEA [4]. The example (1.4.2) is due to W. HANSEN [4].

Section 2. The material is taken from J.-M. BONY [3].

Section 3. (3.1) is due to J.-M. BONY [1],[3] where further results in this direction can be found. An earlier attempt is made by G. TAUTZ [3].

Section 4. The presentation of Sobolev spaces and their properties mostly follows D. GILBARG - N.S. TRUDINGER [1] and R.M. HERVÉ [2]. The consequent use of the coerciveness of the Dirichlet form is due to G. STAMPACCHIA [2].

Section 5. Elliptic-parabolic differential operators are introduced by L. HÖRMANDER [1] where (5.2) is proved and KOLMOGOROV's equation (5.8) is studied. (5.6) is due to J.-M. BONY [2],[3]. A self-contained presentation of the generation of Bauer spaces by elliptic differential operators is given by N. BOBOC - P. MUSTAŢĂ [1]. Many parallels between the potential theory of the Laplace-Kohn operator (5.7) and classical potential theory have been established by G.B. FOLLAND [1], B. GAVEAU [1], P.C. GREINER [1], H. HUEBER [1],[2], A. KORÁNYI [1].

FURTHER ASPECTS

The notions of energy and Dirichlet integral in classical potential theory have been extended to the theory of Dirichlet spaces by A. BEURLING - J. DENY [1], J. DENY [1]. For further investigations see M. FUKUSHIMA [1], F.Y. MAEDA [1].

Ideal boundaries and integral representations of superharmonic functions have been extensively treated in classic potential theory by C. CONSTANTINESCU - A. CORNEA [1] and J.L. DOOB [7]. Attempts in this direction for harmonic spaces are due to C. CONSTANTINESCU - A. CORNEA [3], C. MEGHEA [1]. In this context see also P.A. LOEB [1] where methods of nonstandard analysis are used.

The more abstract standpoint of N. BOBOC - C. BUCUR - A. CORNEA [1] allows not only to treat large parts of balayage theory but also to develop a duality theory. Their theory of H-cones essentially covers the theory of sub-Markov resolvents having a reference measure. A study of the more general class of Ray resolvents and their connection with right processes can be found in R.K. GETOOR [1] and M.J. SHARPE [1].

Bibliography

E.M. ALFSEN:

[1] *Compact convex sets and boundary integrals.* Ergebn. d. Math. 57, Springer-Verlag (1971).

T. BABA:

[1] *Regular points for α-harmonic functions.* Preprint.

H. BAUER:

[1] *Minimalstellen von Funktionen und Extremalpunkte II.* Arch. Math. 11 (1960), 200-205.

[2] *Šilovscher Rand und Dirichletsches Problem.* Ann. Inst. Fourier 11 (1961), 89-136.

[3] *Axiomatische Behandlung des Dirichletschen Problems für elliptische und parabolische Differentialgleichungen.* Math. Ann. 146 (1962), 1-59.

[4] *Markoffsche Prozesse.* Vorlesung an der Universität Hamburg (1963).

[5] *Harmonische Räume und ihre Potentialtheorie.* LN in Math. 22, Springer-Verlag (1966).

[6] *Zum Cauchyschen und Dirichletschen Problem bei elliptischen und parabolischen Differentialgleichungen.* Math. Ann. 164 (1966), 142-153.

[7] *Harmonic spaces and associated Markov processes.* In: "Potential theory" (CIME, 1° Ciclo, Stresa 1969), Ed. Cremonese (1970), 23-67.

[8] *Aspects of modern potential theory.* In: "Proc. Intern. Congress of Math. Vancouver 1974", 1 (1975), 41-51.

[9] *Elliptische harmonische Strukturen.* In: "Elliptische Differentialgleichungen", Tagung in Rostock 1977, Math. Ges. der DDR (1978).

[10] *Halbgruppen und Resolventen in der Potentialtheorie.* L' Enseignement math., 2e série, 25 (1979), 9-22.

[11] *Probability theory and elements of measure theory.* 2nd edition. Academic Press (1981).

[12] *Harmonische Räume.* In: "Jahrbuch Überblicke Mathem.", Bibliograph. Inst. (1981), 9-35.

[13] *Heat balls and Fulks measures.* Ann. Acad. Sc. Fennicae, Sér. A.I. Math. 10 (1985), 67-82.

[14] *Elliptic differential operators and diffusion processes.* Bull. Austral. Math. Soc. 30 (1984), 219-237.

[15] *Zum heutigen Bild der Potentialtheorie.* In: "Zum Werk Leonhard Eulers" (Herausgeg. von E. Knobloch, I.S. Louhivaara, J. Winkler) Birkhäuser-Verlag (1984), 3-20.

U. BAUERMANN:

[1] *Balayage-Operatoren in der Potentialtheorie.* Math. Ann. 231 (1977), 181-186.

C. BERG:

[1] *Quelques propriétés de la topologie fine dans la théorie du potentiel et des processus standard.* Bull. Sci. Math. 95 (1971), 27-31.

[2] *Potential theory on the infinite dimensional torus.* Inventiones math. 32 (1976), 49-100.

[3] *On Brownian and Poissonian convolution semigroups on the infinite dimensional torus.* Inventiones math. 38 (1977), 227-235.

C. BERG, G. FORST:

[1] *Potential theory on locally compact abelian groups.* Ergebn. d. Math. 87, Springer-Verlag (1975).

A. BEURLING, J. DENY:

[1] *Dirichlet spaces.* Proc. Nat. Acad. Sci. USA 45 (1959), 208-215.

J. BLIEDTNER:

[1] *Harmonische Gruppen und Huntsche Faltungskerne.* In: "Seminar über Potentialtheorie", LN in Math. 69, Springer-Verlag (1968), 69-102.

[2] *Axiomatic foundation of potential theory.* In: "Functional Analysis, Surveys and Recent Results II" (editors: K.D. Bierstedt, B. Fuchssteiner). North-Holland Publ. Comp. (1980), 57-67.

J. BLIEDTNER, W. HANSEN:

[1] *Simplicial cones in potential theory.* Inventiones math. 29 (1975), 83-110.

[2] *Cones of hyperharmonic functions.* Math. Z. 151 (1976), 71-87.

[3] *A simplicial characterization of elliptic harmonic spaces.* Math. Ann. 222 (1976), 261-274.

[4] *Simplicial cones in potential theory II (Approximation theorems).* Invent. math. 46 (1978), 255-275.

[5] *Markov processes and harmonic spaces.* Z. Wahrscheinlichkeitstheorie verw. Gebiete 42 (1978), 309-325.

[6] *Bases in standard balayage spaces.* In: "Potential Theory Copenhagen 1979" LN in Math. 787, Springer-Verlag (1980), 55-63.

[7] *The weak Dirichlet problem.* J. Reine Angew. Math. 348 (1984), 34-39.

R.M. BLUMENTHAL, R.K. GETOOR:

[1] *Markov processes and potential theory.* Academic Press (1968).

N. BOBOC, C. BUCUR, A. CORNEA:

[1] *Order and convexity in potential theory: H-cones.* LN in Math. 853, Springer-Verlag (1981).

N. BOBOC, C. CONSTANTINESCU, A. CORNEA:

[1] *Semigroups of transitions on harmonic spaces.* Rev. Roumaine Math. Pures Appl. 12 (1967), 763-805.

N. BOBOC, A. CORNEA:

[1] *Comportement des balayées des mesures ponctuelles. Comportement des solutions du problème de Dirichlet aux points irréguliers.* C.R. Acad. Sci. Paris 264 (1967), 995-997.

[2] *Convex cones of lower semicontinuous functions on compact spaces.* Rev. Roumaine Math. Pures Appl. 12 (1967), 471-525.

N. BOBOC, P. MUSTAŢǍ:

[1] *Espaces harmoniques associés aux opérateurs différentiels linéaires du second ordre de type elliptique.* LN in Math. 68, Springer-Verlag (1968).

J.-M. BONY:

[1] *Détermination des axiomatiques de théorie du potentiel dont les fonctions harmoniques sont différentiables.* Ann. Inst. Fourier 17 (1967), 353-382.

[2] *Principe du maximum, inégalité de Harnack et unicité du problème de Cauchy pour les opérateurs elliptiques dégénérés.* Ann. Inst. Fourier 19 (1969), 277-304.

[3] *Opérateurs elliptiques dégénérés associés aux axiomatiques de la théorie du potentiel.* In: "Potential theory" (CIME, 1° Ciclo, Stresa 1969), Ed. Cremonese (1970), 69-119.

A. BOUKRICHA, W. HANSEN, H. HUEBER:

[1] *Continuous solutions of the generalized Schrödinger equation and perturbation of harmonic spaces.* Bibos Publications 54, Univ. Bielefeld 1985.

G. BOULIGAND:

[1] *Sur le problème de Dirichlet.* Ann. Soc. Polon. Math. 4 (1926), 59-112.

M. BRELOT:

[1] *Sur le potentiel et les suites de fonctions sous-harmoniques.* C.R. Acad. Sci. Paris 207 (1938), 836-838.

[2] *Sur la théorie moderne du potentiel.* C.R. Acad. Sci. Paris 209 (1939), 828-830.

[3] *Familles de Perron et problème de Dirichlet.* Acta scient. Szeged 9 (1939), 133-153.

[4] *Sur la théorie autonome des fonctions sousharmoniques.* Bull. Sci. Math. France 65 (1941), 78-91.

[5] *Sur les ensembles effilés.* Bull. Sci. Math. 68 (1944), 12-36.

[6] *Minorantes sousharmoniques, extrémales et capacités.* Journ. Math. Pures Appl. 24 (1945), 1-32.

[7] *Quelques propriétés et applications du balayage.* C. R. Acad. Sci. Paris 227 (1948), 19-21.

[8] *La théorie moderne du potentiel.* Ann. Inst. Fourier 4 (1952), 113-140.

[9] *Extension axiomatique des fonctions sousharmoniques I.* C. R. Acad. Sci. Paris 245 (1957), 1688-1690.

[10] *Extension axiomatique des fonctions sousharmoniques II.* C. R. Acad. Sci. Paris 246 (1958), 2334-2337.

[11] *Une axiomatique générale du problème de Dirichlet dans les espaces localement compacts.* In: Sém. Brelot-Choquet "Théorie du potentiel" 1 (1958), 6.01-6.16.

[12] *Axiomatique des fonctions harmoniques et surharmoniques dans un espace localement compact.* In: Sém. Brelot-Choquet-Deny "Théorie du potentiel" 2 (1959), 1.01-1.40.

[13] *Quelques propriétés et applications nouvelles de l'effilement.* In: Sém. Brelot-Choquet-Deny "Theorie du potentiel" 6 (1962), 1.27-1.40.

[14] *Éléments de la théorie classique du potentiel.* 4e ed. Centre de Documentation Universitaire Paris (1969).

[15] *On topologies and boundaries in potential theory.* LN in Math. 175, Springer-Verlag (1971).

[16] *Les étapes et les aspects multiples de la théorie du potentiel.* L'Enseignement Math. 18 (1972), 1-36.

R. CAIROLI:

[1] *Produits de semi-groupes de transition et produits de processus.* Publ. Inst. Stat. Univ. Paris 15 (1966), 311-384.

H. CARTAN:

[1] *Capacité extérieure et suites convergentes de potentiels.* C. R. Acad. Sci. Paris 214 (1942), 944-946.

[2] *Théorie du potentiel Newtonien: énergie, capacité, suites de potentiels.* Bull. Soc. Math. France 73 (1945), 74-106.

[3] *Théorie générale du balayage en potentiel newtonien.* Ann. Inst. Fourier 22 (1946), 221-280.

G. CHOQUET:

[1] *Theory of capacities.* Ann. Inst. Fourier 5 (1953-1954), 131-295.

[2] *Le problème des moments.* In: Sém. Choquet "Initiation à l'analyse" 1 (1962), 4.01-4.10.

[3] *Lectures on Analysis I - III.* Benjamin (1969).

K.L. CHUNG:

[1] *Lectures from Markov processes to Brownian motion.* Grundl. d. math. Wiss. 249, Springer-Verlag (1981).

Z. CIESIELSKI:

[1] *Lectures on Brownian motion, heat conduction and potential theory.* Aarhus Univ. Lect. Notes 3 (1965).

C. CONSTANTINESCU:

[1] *Die heutige Lage der Theorie der harmonischen Räume.* Rev. Roumaine Math. Pures Appl. 11 (1966), 1041-1056.

[2] *Some properties of the balayage of measures on a harmonic space.* Ann. Inst. Fourier 17 (1967), 273-293.

[3] *Markov processes on harmonic spaces.* Rev. Roumaine Math. Pures Appl. 13 (1968), 627-654.

[4] *Harmonic spaces and their connections with the semi-elliptic differential equations and with the Markov processes.* In: "Elliptische Differentialgleichungen I", Kolloquium Berlin 1969. Akademie-Verlag (1970), 19-30.

C. CONSTANTINESCU, A. CORNEA:

[1] *Ideale Ränder Riemannscher Flächen.* Ergebn. d. Math. 32, Springer-Verlag (1963).

[2] *On the axiomatic theory of harmonic functions I, II.* Ann. Inst. Fourier 13 (1963), 373-388, 389-394.

[3] *Compactifications of harmonic spaces.* Nagoya Math. J. 25 (1965), 1-57.

[4] *Potential theory on harmonic spaces.* Grundl. d. math. Wiss. 158, Springer-Verlag (1972).

A. CORNEA, H. HÖLLEIN:

[1] *Bases and essential bases in H-cones.* In: "Functional Analysis, Surveys and Recent Results II" (editors: K.D. Bierstedt, B. Fuchssteiner). North-Holland Publ. Comp. (1980), 69-86.

A. CORNEA, G. LICEA:

[1] *Order and potential resolvent families of kernels.* LN in Math. 494, Springer-Verlag (1975).

P. COURRÈGE, P. PRIOURET:

[1] *Axiomatique du problème de Dirichlet et processus de Markov.* In: Sém. Brelot-Choquet-Deny "Théorie du potentiel" 8 (1963/64), 8.01-8.48.

C. DELLACHERIE:

[1] *Ensembles aléatoires I, II.* In: "Séminaire de probabilités III", LN in Math. 88, Springer-Verlag (1969), 97-114, 115-136.

[2] *Ensembles analytiques, capacités, mesures de Hausdorff.* LN in Math. 295, Springer-Verlag (1972).

[3] *Capacités et processus stochastiques.* Ergebn. d. Math. 67, Springer-Verlag (1972).

[4] *Capacités, rabotages et ensembles analytiques.* In: Sém. Choquet "Initiation à l'analyse" 19 (1980), 5.01-5.21.

[5] *Mesurabilité des débuts et théorème de section: le lot à la portée de toutes les bourses.* In: "Séminaire de Probabilités XV", LN in Math. 850, Springer-Verlag (1981).

C. DELLACHERIE, P.-A. MEYER:

[1] *Probabilités et potentiel.* Chapitres I - IV. Acta Sci. Ind. 1372, Hermann (1975).

[2] *Probabilités et potentiel.* Chapitres V - VIII. Acta Sci. Ind. 1385, Hermann (1980).

[3] *Probabilités et potentiel.* Chapitres IX - XI. Acta Sci. Ind. 1410, Hermann (1983).

V. DEMBINSKI, K. JANSSEN:

[1] *Standard balayage spaces and standard Markov processes.* In: "Potential Theory Copenhagen 1979", LN in Math. 787, Springer-Verlag (1980), 84-105.

J. DENY:

[1] *Méthodes Hilbertiennes en théorie du potentiel.* In: "Potential theory" (CIME, 1° Ciclo, Stresa 1969), Ed. Cremonese (1970), 121-201.

J. L. DOOB:

[1] *Stochastic processes.* Wiley 1953.

[2] *Semimartingales and subharmonic functions.* Trans. Amer. Math. Soc. 77 (1954), 86-121.

[3] *A probability approach to the heat equation.* Trans. Amer. Math. Soc. 80 (1955), 216-280.

[4] *Probability methods applied to the first boundary value problem.* In: 3rd Berkely Symp. Math. Statist. and Prob. (1956), 49-80.

[5] *Probability theory and the first boundary value problem.* Ill. J. Math. 2 (1958), 19-36.

[6] *Applications to analysis of a topological definition of smallness of a set.* Bull. Amer. Math. Soc. 72 (1966), 579-600.

[7] *Classical potential theory and its probabilistic counterpart.* Grundl. d. math. Wiss. 262, Springer-Verlag (1983).

E. B. DYNKIN:

[1] *Die Grundlagen der Theorie der Markoffschen Prozesse.* Grundl. d. math. Wiss. 108, Springer-Verlag (1961).

[2] *Markov processes I, II.* Grundl. d. math. Wiss. 121,122, Springer-Verlag (1965).

E.G. EFFROS, J.L. KAZDAN:

[1] *Applications of Choquet simplexes to elliptic and parabolic boundary value problems.* J. Differential Equations 8 (1970), 95-134.

[2] *On the Dirichlet problem for the heat equation.* Indiana Univ. Math. J. 20 (1971), 683-693.

G.C. EVANS:

[1] *Applications of Poincaré's sweeping-out process.* Proc. Nat. Acad. Sci. USA 19 (1933), 457-461.

[2] *On potentials of positive mass.* Trans. Amer. Math. Soc. 37 (1935), 226-253.

[3] *Potentials and positively infinite singularities of harmonic functions.* Monatsh. Math. 43 (1936), 419-424.

G.B. FOLLAND:

[1] *A fundamental solution for a subelliptic operator.* Bull. Amer. Math. Soc. 79 (1973), 373-376.

O. FROSTMAN:

[1] *Potentiel d'équilibre et capacité des ensembles avec quelques applications à la théorie des fonctions.* Medd. Lunds Univ. Mat. Sem. 3 (1935), 1-118.

[2] *Sur le balayage des masses.* Acta Sci. (Szeged) 9 (1938), 43-51.

[3] *Les points irréguliers dans la théorie du potentiel et le critère de Wiener.* Medd. Lunds Univ. Mat. Sem. 4 (1939), 1-10.

B. FUGLEDE:

[1] *Finely harmonic functions.* LN in Math. 289, Springer-Verlag (1972).

M. FUKUSHIMA:

[1] *Dirichlet forms and Markov processes.* North Holland Publ. Comp. (1980).

W. FULKS:

[1] *Regular regions for the heat equation.* Pacific J. Math. 7 (1957), 867-877.

[2] *A mean-value theorem for the heat equation.* Proc. Amer. Math. Soc. 17 (1966), 6-11.

T.W. GAMELIN:

[1] *Criteria for approximation by harmonic functions.* Ill. J. Math. 26 (1982), 353-357.

B. GAVEAU:

[1] *Principe de moindre action, propagation de la chaleur et estimées sous elliptiques sur certains groupes nilpotents.* Acta Math. 139 (1977), 95-153.

R.K. GETOOR:

[1] *Markov processes: Ray processes and right processes.* LN in Math. 440, Springer-Verlag (1975).

D. GILBARG, N.S. TRUDINGER:

[1] *Elliptic partial differential equations of second order.* Grundl. d. math. Wiss. 224, Springer-Verlag (1977).

K. GOWRISANKARAN:

[1] *Multiply harmonic functions.* Nagoya Math. J. 28 (1966), 27-48.

P.C. GREINER:

[1] *Spherical harmonics on the Heisenberg group.* Canad. Math. Bull. 23 (1980), 383-396.

W. HANSEN:

[1] *Konstruktion von Halbgruppen und Markoffschen Prozessen.* Inventiones math. 3 (1967), 179-214.

[2] *Charakterisierung von Familien exzessiver Funktionen.* Inventiones math. 5 (1968), 335-348.

[3] *Potentialtheorie harmonischer Kerne.* In: "Seminar über Potentialtheorie", LN in Math. 69, Springer-Verlag (1968), 103-159.

[4] *Cohomology in harmonic spaces.* In: "Seminar on potential theory, II", LN in Math. 226, Springer-Verlag (1971), 63-101.

[5] *Fegen und Dünnheit mit Anwendungen auf die Laplace- und Wärmeleitungsgleichung.* Ann. Inst. Fourier 21 (1971), 79-121.

[6] *Abbildungen harmonischer Räume mit Anwendung auf die Laplace- und Wärmeleitungsgleichung.* Ann. Inst. Fourier 21 (1971), 203-216.

[7] *Potentialtheorie semi-regulärer harmonischer Kerne.* Z. Wahrscheinlichkeitstheorie u. Verw. Gebiete 18 (1971), 298-304.

[8] *Some remarks on strict potentials.* Math. Z. 147 (1976), 279-285.

[9] *Semi-polar sets and quasi-balayage.* Math. Ann. 257 (1981), 495-517.

[10] *Convergence of balayage measures.* Math. Ann. 264 (1983), 437-446.

[11] *Harmonic and superharmonic functions on compact sets.* Ill. J. Math. 29 (1985), 103-107.

[12] *On the identity of Keldych solutions.* To appear in Czech. Math. J.

A. HARNACK:

[1] *Existenzbeweise zur Theorie des Potentials in der Ebene und im Raume.* Ber. Verhandl. Königl. Sächs. Ges. Wiss. Leipzig 1886 (1886), 144-169.

W.K. HAYMAN, P.B. KENNEDY:

[1] *Subharmonic functions 1.* Academic Press (1976).

L.L. HELMS:

[1] *Introduction to potential theory.* Wiley and Sons (1969).

R.-M. HERVÉ:

[1] *Recherches axiomatiques sur la théorie des fonctions surharmoniques et du potentiel.* Ann. Inst. Fourier 12 (1962), 415-571.

[2] *Un principe du maximum pour les sous-solutions locales d'une équation uniformément elliptique de la forme* $Lu = -\sum_i \frac{\partial}{\partial x_i}\left(\sum_j a_{ij}\frac{\partial u}{\partial x_j}\right) = 0$. Ann. Inst. Fourier 14 (1964), 493-508.

F. HIRSCH:

[1] *Familles d'opérateurs potentiels.* Ann. Inst. Fourier 25 (1975), 263-288.

L. HÖRMANDER:

[1] *Hypoelliptic second order differential equations.* Acta Math. 119 (1967), 147-171.

H. HUEBER:

[1] *The Poisson space of the Korányi ball.* Math. Ann. 268 (1984), 223-232.

[2] *Wiener's criterion in potential theory with applications on nilpotent Lie-groups.* To appear in Math. Z.

G.A. HUNT:

[1] *Markoff processes and potentials I.* Ill. J. Math. 1 (1957), 44-93.

[2] *Markoff processes and potentials II.* Ill. J. Math. 1 (1957), 316-369.

[3] *Markoff processes and potentials III.* Ill. J. Math. 2 (1958), 151-213.

J. HYVÖNEN:

[1] *On the harmonic continuation of bounded harmonic functions.* Math. Ann. 245 (1979), 151-157.

K. JANSSEN:

[1] *Standard H-cones and balayage spaces.* In: "Complex Analysis", 5th Romanian-Finnish Seminar. LN in Math. 1014, Springer-Verlag (1983), 197-203.

M. KAC:

[1] *On some connections between probability theory and differential and integral equations.* In: "Proc. 2nd Berkely Symp. Math. Statist. Prob. (1951), 189-215.

[2] *Aspects probabilistes de la théorie du potentiel.* Les Presses de l'Université de Montréal (1970).

S. KAKUTANI:

[1] *On Brownian motion in n-space.* Proc. Imp. Acad. Tokyo 20 (1944), 648-652.

[2] *Two-dimensional Brownian motion and harmonic functions.* Proc. Imp. Acad. Tokyo 20 (1944), 706-714.

[3] *Markoff process and the Dirichlet problem.* Proc. Japan Acad. 21 (1945), 227-233.

M.V. KELDYCH:

[1] *Sur la résolubilité et la stabilité du problème de Dirichlet.* Dokl. Akad. Nauk SSSR 18 (1938), 315-318.

O.D. KELLOGG:

[1] *Foundations of potential theory.* Grundl. d. math. Wiss. 31 (Nachdruck von 1929) Springer-Verlag (1967).

A.W. KNAPP:

[1] *Connection between Brownian motion and potential theory.* Math. Anal. Appl. 12,2 (1965), 328-349.

J. KÖHN, M. SIEVEKING:

[1] *Reguläre und extremale Randpunkte in der Potentialtheorie.* Rev. Roumaine Math. Pures Appl. 12 (1967), 1489-1502.

[2] *Zum Cauchyschen und Dirichletschen Problem.* Math. Ann. 177 (1968), 133-142.

A. KORÁNYI:

[1] *Kelvin transforms and harmonic polynomials on the Heisenberg group.* J. Funct. Analysis 49 (1982), 177-185.

N.S. LANDKOF:

[1] *Foundations of modern potential theory.* Grundl. d. math. Wiss. 180, Springer-Verlag (1972).

H. LEBESGUE:

[1] *Sur le problème de Dirichlet.* C. R. Acad. Sci. Paris 154 (1912), 335-337.

[2] *Sur des cas d'impossibilité du problème de Dirichlet.* C.R. Séances Soc. Math. France 17 (1913).

[3] *Conditions de régularité, conditions d'irrégularité, conditions d'impossibilité dans le problème de Dirichlet.* C. R. Acad. Sci. Paris 178 (1924), 349-354.

P.A. LOEB:

[1] *Applications of nonstandard analysis to ideal boundaries in potential theory.* Israel J. Math. 25 (1976), 154-187.

J. LUKEŠ:

[1] *Principal solution of the Dirchlet problem in potential theory.* Comm. Math. Univ. Carolinae 14 (1973), 773-778.

[2] *Théorème de Keldych dans la théorie axiomatique de Bauer des fonctions harmoniques.* Czech. Math. J. 24 (1974), 114-125.

[3] *Functional approach to the Brelot-Keldych theorem.* Czech. Math. J. 27 (1977), 609-616.

J. LUKEŠ, J. MALÝ:

[1] *On the boundary behaviour of the Perron generalized solution.* Math. Ann. 257 (1981), 355-366.

J. LUKEŠ, I. NETUKA:

[1] *The Wiener type solution of the Dirichlet problem in potential theory.* Math. Ann. 224 (1976), 173-178.

F.-Y. MAEDA:

[1] *Dirchlet integrals on harmonic spaces.* LN in Math. 803, Springer-Verlag (1980).

A.J. MARIA:

[1] *The potential of a positive mass and the weight function of Wiener.* Proc. Nat. Acad. Sci. U.S.A. 20 (1934), 485-489.

C. MEGHEA:

[1] *Compactification des espaces harmoniques.* LN in Math. 222, Springer-Verlag (1971).

W. MEIER:

[1] *Charakterisierung von Produkten harmonischer Räume.* Math. Ann. 257 (1981), 199-218.

P.-A. MEYER:

[1] *Brelot's axiomatic theory of the Dirichlet problem and Hunt's theory.* Ann.Inst. Fourier 13 (1963), 357-372.

[2] *Probability and potentials.* Blaisdell Publ. Comp. (1966).

[3] *Processus de Markov.* LN in Math. 26, Springer-Verlag (1967).

[4] *Processus de Markov: la frontière de Martin.* LN in Math. 77, Springer-Verlag (1968).

G. MOKOBODZKI:

[1] *Cônes de potentiels et noyaux subordonnés.* In: "Potential theory" (CIME, 1° Ciclo, Stresa 1969), Ed. Cremonese (1970), 207-248.

[2] *Densité relative de deux potentiels comparables.* In: "Séminaire de Probabilités IV", LN in Math. 124, Springer-Verlag (1970), 170-194.

[3] *Eléments extrémaux pour le balayage.* In: Sém. Brelot-Choquet-Deny "Théorie du potentiel" 13 (1970), 5.01-5.14.

G. MOKOBODZKI, D. SIBONY:

[1] *Cônes adaptés de fonctions continues et théorie du potentiel.* In: Sém. Choquet "Initiation à l'analyse" 6 (1968), 5.01-5.35.

[2] *Cônes de fonctions et théorie du potentiel I: Les noyaux associés à un cône de fonctions.* In: Sém. Brelot-Choquet-Deny "Théorie du potentiel" 11 (1968), 8.01-8.35.

[3] *Cônes de fonctions et théorie du potentiel, II: Résolvantes et semi-groupes subordonnés à un cône de fonctions harmoniques.* In: Sém. Brelot-Choquet-Deny "Théorie du potentiel" 11 (1968), 9.01-9.29.

[4] *Familles additives de cônes convexes et noyaux subordonnés.* Ann. Inst. Fourier 18 (1969), 205-220.

A.F. MONNA:

[1] *Dirichlet's principle. A mathematical comedy of errors and its influence on the development of analysis.* Oosthoek, Scheltema & Holkema (1975).

I. NETUKA:

[1] *The classical Dirichlet problem and its generalizations.* In: "Potential Theory Copenhagen 1979" LN in Math. 787, Springer-Verlag (1980), 235-266.

I. NETUKA, J. VESELÝ:

[1] *Harmonic continuation and removable singularities in the axiomatic potential theory.* Math. Ann. 234 (1978), 117-123.

O. PERRON:

[1] *Eine neue Behandlung der Randwertaufgabe für* $\Delta u = o$. Math. Z. 18 (1923), 42-54.

I. PETROWSKY:

[1] *Zur ersten Randwertaufgabe der Wärmeleitungsgleichung.* Comp. Math. 1 (1935), 383-419.

H. POINCARÉ:

[1] *Sur les équations aux dérivées partielles de la physique mathématique.* Amer. J. Math. 12 (1890), 211-294.

S.D. POISSON:

[1] *Addition au mémoire précédent et au mémoire sur la manière d'exprimer les fonctions par les séries de quantités périodiques.* J. éc. royal polyt. 12 (1823), 145-162.

[2] *Suite du mémoire sur les intégrales définies et sur la sommation des séries.* J. éc. royal polyt. 12 (1823), 404-509.

S.C. PORT, C.J. STONE:

[1] *Brownian motion and classical potential theory.* Academic Press (1978).

T. RADÓ:

[1] *Subharmonic functions.* Ergebn. d. Math. 5, Springer-Verlag (1937).

M. RAO:

[1] *Brownian motion and classical potential theory.* Aarhus Univ. Lect. Notes 47 (1977).

D.B. RAY:

[1] *Resolvents, transition functions, and strongly Markovian processes.* Ann. Math. 70 (1959), 43-72.

F. RIESZ:

[1] *Sur les fonctions subharmoniques et leur rapport à la théorie du potentiel. I.* Acta Math. 48 (1926), 329-343.

[2] *Sur les fonctions subharmoniques et leur rapport à la théorie du potentiel. II.* Acta Math. 54 (1930), 321-360.

M. RIESZ:

[1] *Intégrales de Riemann-Liouville et potentiels.* Acta Sci. Szeged 9 (1938), 1-42.

H. ROST:

[1] *Die Stoppverteilungen eines Markoff-Prozesses mit lokalendlichem Potential.* Manuscripta math. 3 (1970), 321-329.

U. SCHIRMEIER:

[1] *Produkte harmonischer Räume.* Sitz.ber. math. nat. Kl. Bayer. Akad. Wiss. 1978 (1979), 5-22.

B.-W. SCHULZE, G. WILDENHAIN:

[1] *Methoden der Potentialtheorie für elliptische Differentialgleichungen beliebiger Ordnung.* Birkhäuser-Verlag (1977).

H.A. SCHWARZ:

[1] *Zur Integration der partiellen Differentialgleichung* $\frac{\partial^2 u}{\partial x^2} + \frac{\partial^2 u}{\partial y^2} = 0$. J. Reine Angew. Math. 74 (1872), 218-253.

M.J. SHARPE:

[1] *General theory of Markov processes.* Forthcoming book.

D. SIBONY:

[1] *Cônes de fonctions et potentiels.* Lecture Notes McGill Univ. Montreal (1968).

G. STAMPACCHIA:

[1] *Formes bilinéaires coercitives sur les ensembles convexes.* C.R. Acad. Sc. Paris 258 (1964), 4413-4416.

[2] *Le problème de Dirichlet pour les équations elliptiques du second ordre à coefficients discontinus.* Ann. Inst. Fourier 15 (1965), 189-258.

W. STERNBERG:

[1] *Über die Gleichung der Wärmeleitung.* Math. Ann. 101 (1929), 394-398.

L. STOICA:

[1] *Local operators and Markov processes.* LN in Math. 816, Springer-Verlag (1980).

E. SZPILRAJN:

[1] *Remarques sur les fonctions sousharmoniques.* Ann. Math. 34 (1933), 588-594.

G. TAUTZ:

[1] *Zur Theorie der elliptischen Differentialgleichungen II.* Math. Ann. 118 (1941-43), 733-770.

[2] *Zur Theorie der ersten Randwertaufgabe.* Math. Nachr. 2 (1949), 279-303.

[3] *Zum Umkehrproblem bei elliptischen Differentialgleichungen I, II, Bemerkungen.* Arch. Math. 3 (1952), 232-250, 361-365.

J.C. TAYLOR:

[1] *Balayage de fonctions excessives.* In: Sém. Brelot-Choquet-Deny "Théorie du potentiel" 14 (1971), 2.01-2.11.

[2] *Strict potentials and Hunt processes.* Inventiones math. 16 (1972), 249-259.

[3] *Potential kernels of Hunt processes.* Indiana Math. J. 22 (1973), 1091-1102.

[4] *A characterization of the kernel* $\lim_{\lambda\downarrow 0} V_\lambda$ *for submarkovian resolvents* (V_λ). Ann. Prob. 3 (1975), 355-357.

[5] *The harmonic space associated with a "reasonable" standard process.* Math. Ann. 233 (1978), 89-96.

P.D. TAYLOR:

[1] *Some open sets for which the heat equation is simplicial.* Canad. J. Math. 26 (1974), 455-472.

M. TSUJI:

[1] *Potential theory in modern function theory.* Maruzen (1959).

A. TYCHONOFF:

[1] *Théorème d'unicité pour l'équation de la chaleur.* Mat. Sbornik 42 (1935), 199-216.

C. DE LA VALLÉE-POUSSIN:

[1] *Extension de la méthode du balayage de Poincaré et problème de Dirichlet.* Ann. Inst. H. Poincaré 2 (1932), 169-232.

[2] *Potentiel et problème généralisé de Dirichlet.* Math. Gazette 22 (1938), 17-36.

[3] *Points irréguliers. Détermination des masses par les potentiels.* Acad. Belg. Bull. Cl. Sci. (5) 24 (1938), 368-384.

F. VASILESCO:

[1] *Sur la continuité du potentiel à travers les masses et la démonstration d'une lemme de Kellog.* C. R. Acad. Sci. Paris 200 (1935), 1173-1174.

[2] *La notion de point irrégulier dans le problème de Dirichlet.* Actual. Sci. Ind. 660, Masson (1938).

B. WALSH:

[1] *Perturbation of harmonic structures and an index-zero theorem.* Ann. Inst. Fourier 20 (1970), 317-359.

J.B. WALSH:

[1] *Some topologies connected with Lebesgue measure.* In: "Sém. de probabilités V", LN in Math. 191, Springer-Verlag (1971), 290-310.

J. WERMER:

[1] *Potential theory.* LN in Math. 408, Springer-Verlag (1974).

D.V. WIDDER:

[1] *Positive temperatures on an infinite rod.* Trans. Amer. Math. Soc. 55 (1944), 85-95.

N. WIENER:

[1] *Differential space.* J. Math. Phys. M.I.T. 2 (1923), 131-174.

[2] *Certain notions in potential theory.* J. Math. Phys. M.I.T. 3 (1924), 24-51.

[3] *The Dirichlet problem.* J. Math. Phys. M.I.T. 3 (1924), 127-146.

[4] *Note on a paper of O. Perron.* J. Math. Phys. M.I.T. 4 (1925), 21-32.

R. WITTMANN:

[1] *Kacsche Potentialtheorie für Resolventen, Markoffsche Prozesse und harmonische Räume.* Sitz.ber. math. nat. Kl. Bayer. Akad. Wiss. 161 (1982), 3-164.

S. ZAREMBA:

[1] *Sur le principe de Dirichlet.* Acta Math. 34 (1911), 293-316.

Index of Symbols

In addition to the basic notations the following symbols are used:

Subject Index

Guide to Standard Examples

Property \ Example	Classical Potential Theory	Translation	Discrete Potential Theory	Riesz Potentials	Heat Equation
balayage space	64, 130, 386	64, 130	64, 130	186	211, 386
potentials	65, 198 - 203, 206, 207	18, 64, 65	104	198 - 203, 206, 207	224, 226
hyperharmonic, excessive functions	5, 11, 12	44, 96	45, 64		211, 214, 223, 226, 227
semigroup	7 - 10, 41, 207, 225	44, 176, 209 - 211, 213	44	186	211
potential kernel	48	48	48	187	212
harmonic kernel	3, 39, 95, 98, 336	95	96	192 - 196	216 - 223
Hunt process	266, 271, 272, 310	138, 150, 155, 161	267		266
fine topology	64, 203, 206	64		188, 203, 206	
thinness, base	275, 294, 296, 336	275		276	277
reduced measure		69, 190, 191	69		225, 227
absorbing sets	174	174	174	197	224, 225
polar sets	283, 294	284	284	284	284, 321
capacity	205, 206, 291, 296, 297			205, 206, 291, 296, 297	
semipolar sets	289, 290	285, 380	286	290	286, 289
essential base	298	298		298	
Dirichlet problem	351, 352				351, 352
regular points or sets	3, 350, 367	356, 366		350	350, 366
irregular points	356, 357	356		356	356
Choquet boundary	26	366			366
Keldych sets	369	369		369	369
perturbation	177	184			
miscellaneous	110, 304, 236 - 238	53, 124, 316	240 - 241	237, 238, 304	237, 238

Further examples can be found on the following pages:

92, 114, 125, 130, 144 (161, 168), 191, 232, 236 - 238, 239 - 241, 305, 336, 404, 405.

Erratum

Universitext

J. Bliedtner, W. Hansen

Potential Theory

ISBN 3-540-16396-4
ISBN 0-387-16396-4

In the front matter the text of the pages X and XI has been exchanged.

Springer-Verlag
Berlin Heidelberg New York Tokyo 1986